FUNDAMENTALS OF ELECTRIC CIRCUITS

FUNDAMENTALS OF ELECTRIC CIRCUITS

DAVID A. BELL

Lambton College of
Applied Arts and Technology
Sarnia, Ontario, Canada

RESTON PUBLISHING COMPANY, INC.
Reston, Virginia
A Prentice-Hall Company

Library of Congress Cataloging in Publication Data

Bell, David A
Fundamentals of electric circuits.

Includes index.
1. Electric circuits. 2. Electric networks.
I. Title.
TK454.B3987 621.319′2 77-17625
ISBN 0-87909-318-8

Reston Publishing Company, Inc.
A Prentice-Hall Company
Reston, Virginia 22090

10 9 8 7 6 5 4 3 2

Printed in the United States of America.

CONTENTS

PREFACE

This text is intended for use in first-year electricity courses in two-year and three-year community college programs. Assuming that the reader has had no previous electrical studies, the first chapter offers explanations for basic electrical phenomena, thus laying the foundation upon which an understanding of electricity can be constructed. Throughout the book, an attempt is made to give clear, concise explanations of the various concepts covered. Illustrations and step-by-step worked examples are included at every stage. A Glossary of formulas is provided at the end of appropriate chapters, and Review Questions and Problems are also included.

Since all technical personnel must be able to use the International System of Units, Chapter 2 introduces the most important thermal and mechanical SI units, and gives examples of applications. Chapter 3 treats the SI electrical units, defining each one and explaining its origin. Ohm's Law is also introduced at this stage, and examples are given for calculating voltage, current, resistance, and power.

The first laboratory experiments performed by students in electricity involve the use of voltmeters, ammeters, and ohmmeters. Therefore, it is very important for the student to acquire an early knowledge of how to use these instruments. Chapter 4 covers how to correctly connect each instrument, precautions to be observed, reading the pointer position on the scale, and so on. Digital instruments are also treated. Further explanation of how the instruments work is left to later chapters in the book.

Conductors, insulators, and resistors are covered in Chapter 5. The differences between the various materials is explained, and insulator breakdown, conductor resistivity, thermal effects, etc., are all discussed.

Voltage cells and batteries are the topic of Chapter 6. Following this are three chapters covering resistive circuits as: series circuits, parallel circuits, and series-parallel combinations. This leads into network analysis techniques and network theorems, treated in two separate chapters.

In introducing magnetism (Chapters 12 and 13), an attempt is again made to explain the basic phenomena in a simple understandable way. As always, the various formulas are derived, and illustrations and worked examples are employed at every stage. Once an understanding of magnetism has been achieved, the operation of dc measuring instruments is explained in Chapter 14.

Inductance and capacitance are introduced in Chapters 15 and 16 respectively, and the dc performance of inductive and capacitive components is treated in the succeeding chapter.

The generation of an alternating voltage is explained in terms of a basic two-conductor generator. The sine wave is discussed, and its frequency, phase-angle, rms value, and so on are all explained.

At this point, the student is at a stage where (to study ac) it is necessary to become familiar with a new laboratory instrument—the cathode ray oscilloscope. To facilitate this, the front panel of a typical double-beam oscilloscope is presented, its controls are discussed, and its basic application to the study of waveforms is explained. No attempt is made to explain how the oscilloscope works.

The study of ac circuits is further pursued in chapters on: phasors and complex numbers, inductance and capacitance in ac circuits, power in ac circuits, and ac network analysis. Resonance is treated in a separate chapter, and a single chapter is also devoted to transformers. Finally, the operation of ac measuring instruments is covered.

Every effort has been made to present the various topics in this book with the greatest possible clarity and the fewest possible words.

David A. Bell

1

THE NATURE OF ELECTRICITY

INTRODUCTION

Man's knowledge of electricity began with the study of the phenomenon known to the early experimenters as electrification by friction. The early concept of an electric fluid being transferred from one body to another is now understood as a motion of electrons that have become detached from their atoms. The behavior of an electron in relation to an atom is best described in terms of the planetary atom. A flow of electrons constitutes an electric current, and the pressure that produces the electron flow is the result of a potential difference between two bodies, or between two terminals. An excess of electrons on one body, and/or a deficiency of electrons on another body, constitutes a potential difference between the two. In the simplest source of electricity, the excess and deficiency of electrons are both the result of chemical action. When an electric current flows, heat is generated by the electron motion. In certain circumstances, light may also be produced—hence the electric lamp.

The various components and interconnections of an electrical system comprise what is described as an electric circuit, and a circuit diagram is a graphic representation of an electric circuit.

Because electricity can be dangerous, it is important to use it carefully and to take immediate action when someone suffers an electric shock.

1-1 ELECTRIFICATION BY FRICTION

The study of electricity began with investigations of what became known as *electrification by friction*. The ancient Greeks knew that when amber was rubbed with wool, it could attract lightweight particles of other material, such as feathers or lint. The Greek word *elektron*, which means amber, is the origin of the word *electricity*.

Early in the seventeenth century it was discovered that amber was not the only material that had this property. Glass rubbed with silk, and ebonite rubbed with fur, were both found capable of attracting small particles of other materials. It was also discovered that when a silk-rubbed glass rod was suspended on thread and a fur-rubbed ebonite rod was brought near it, the glass was attracted to the ebonite. When another silk-rubbed glass rod was brought near the suspended glass rod, it was found that the two were repelled from each other (see Figure 1-1). Similarly, a suspended ebonite rod that had first been rubbed with fur was found to be repelled from another fur-rubbed ebonite rod and attracted to silk-rubbed glass. The effect is most easily demonstrated by

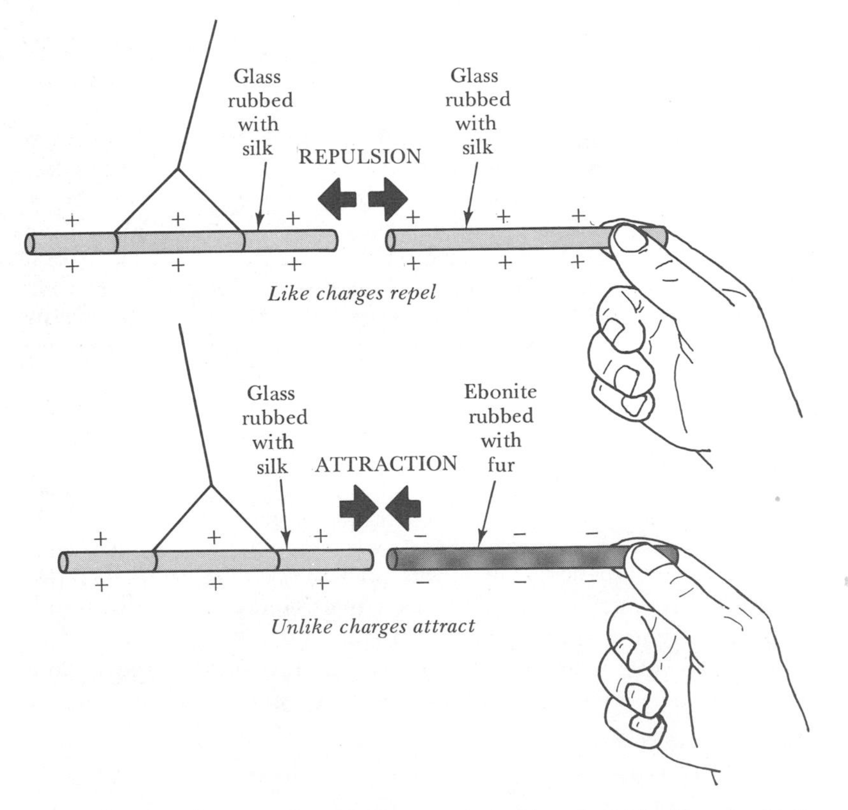

FIG. 1-1. Repulsion and attraction produced by electrification by friction

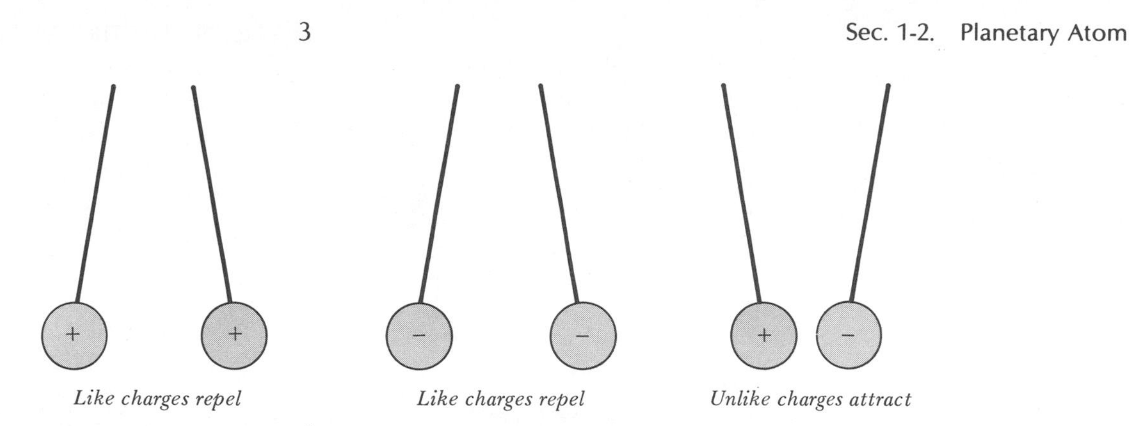

FIG. 1-2. Thread-suspended lightweight pith balls demonstrating that *like charges repel* and *unlike charges attract*

charging lightweight pith balls from the silk-rubbed and fur-rubbed rods, as illustrated in Figure 1-2.

From these results it was concluded that there were two types of electricity, and that those materials which were *charged* with the same type of electricity repelled each other, while those which were charged with different types of electricity attracted each other. This conclusion produced a fundamental law:

Like charges repel; unlike charges attract.

Around the middle of the eighteenth century, Benjamin Franklin suggested that when glass was rubbed with silk, some kind of *electric fluid* passed from the silk to the glass, and that this gave the glass an increased amount of electric fluid, or *positive charge*. Conversely, when the ebonite was rubbed with fur, the electric fluid passed from the ebonite to the fur, he argued. Thus, the ebonite acquired a reduced amount of electricity, or *negative charge*.

1-2 PLANETARY ATOM

The modern explanation of electrification by friction utilizes our present understanding of the atom. The atom is believed to consist of a central *nucleus* surrounded by orbiting *electrons*.* The plane diagram in Figure 1-3(a) illustrates the concept of the *planetary atom*. The electrons have a negative charge, and relative to the nucleus they are extremely small particles. The nucleus [Figure 1-3(b)] is largely a cluster of two types of particles, *protons* and *neutrons*, each of which has a mass approximately

*This was first proposed in 1913 by the Danish physicist Neils Bohr.

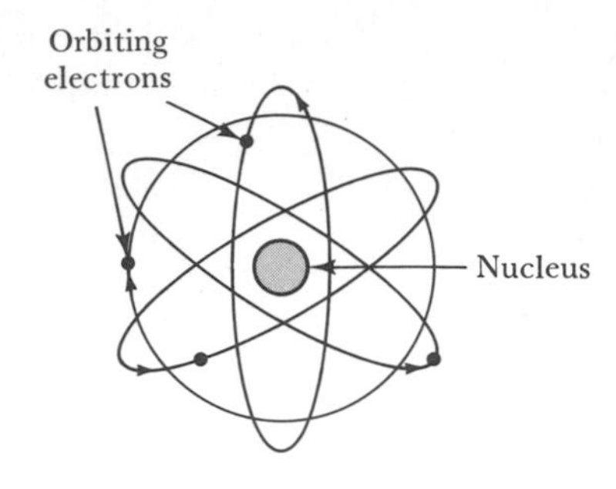

(a) Nucleus with orbiting electrons

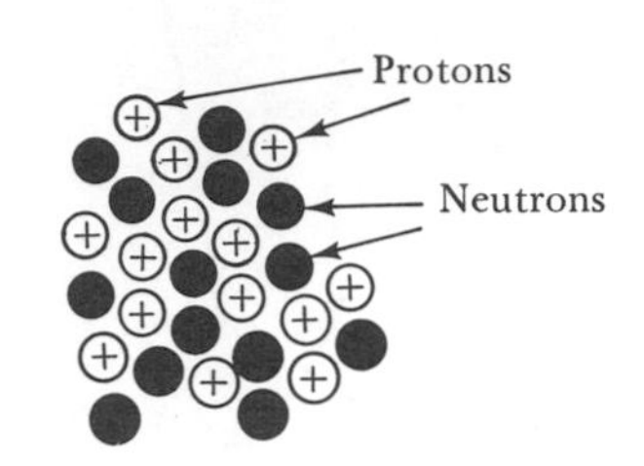

(b) The nucleus is largely a cluster of protons and neutrons

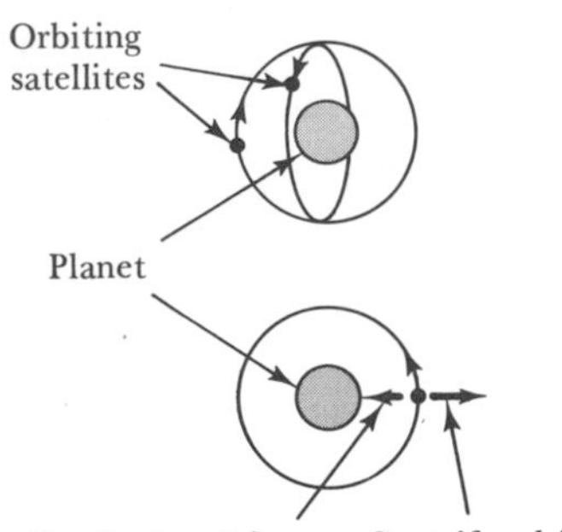

(c) Forces on satellite orbiting a planet

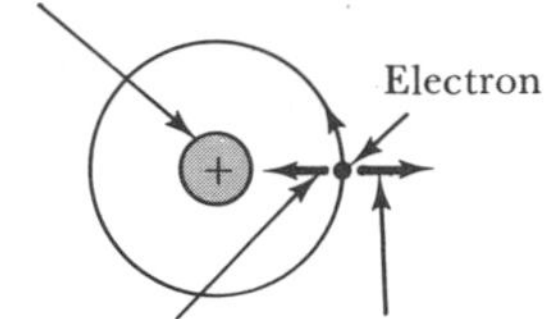

(d) Forces on electron orbiting a nucleus

FIG. 1-3. Planetary atom

1800 times the mass of the electron. Neutrons have no charge at all, and protons have a positive charge equal in magnitude to the negative charge on an electron. Thus, the nucleus is positively charged. Since opposite charges attract, the negatively charged electrons are attracted to the positively charged nucleus.

The atom may be compared to a planet with orbiting satellites. The attractive electric force between the electron and the nucleus holds the electron in orbit, in the same way as the attractive force of the planet's gravity keeps the satellite from drifting off into space. As in the case of the satellite, the force of attraction is balanced by the centrifugal force that results from the motion of the electron around the nucleus [Figure 1-3(c) and (d)]. The three basic particles—*proton*, *neutron*, and *electron*—are similar from one atom to another. Differences between atoms are due to dissimilar numbers and arrangements of the three particles. Different materials are made up of different types of atoms, or combinations of several types of atoms.

It has been found that electrons can occupy only certain orbital rings or *shells* at fixed distances from the nucleus, and that each shell can contain only a certain number of electrons. The outer shell is termed the *valence shell*, and the electrons that occupy it are referred to as *valence electrons*. These outer-shell electrons largely determine the electrical (and chemical) characteristics of an atom.

Since the protons and orbital electrons of an atom are equal in number and equal and opposite in charge, they neutralize each other electrically. Consequently, each atom is normally electrically neutral; that is, it exhibits neither a positive nor a negative charge. If an atom loses an electron, thus losing some negative charge, its charge is tipped to the positive side and it is now a *positive ion*. Similarly, an atom that gains an additional electron becomes negatively charged and is then termed a *negative ion*.

Benjamin Franklin's *electric fluid* can now be looked upon as being composed of electrons passing from one material to another. That material which loses electrons becomes positively charged as the result of the loss of negative charges. The material that gains electrons becomes negatively charged. Thus, glass rubbed with silk acquires its positive charge by losing some electrons to the silk, and the negative charge on the ebonite is acquired by its gaining electrons from the fur with which it was rubbed.

1-3 ELECTRIC CURRENT AND POTENTIAL DIFFERENCE

When two oppositely charged bodies make contact, electrons flow from the negatively charged body to the positively charged body. Thus, the electrons flow from a location with an excess of electrons to one with a deficiency of electrons. Electrons will also flow from a negatively charged body to an uncharged body, or to one with a lower negative charge. This movement of *charge carriers* (i.e., electrons) constitutes a flow of *electric current*. The flow of charge carriers also occurs if the two charged bodies are connected by a piece of metallic material (Figure 1-4). The flow does not occur when nonmetallic material is employed to connect the bodies together. Those materials which readily permit electrons to pass through them are termed *conductors*, and those which do not permit electron flow are known as *insulators*. Since some materials are better conductors than others, it can be said that they offer less *resistance* to the flow of electric current. Thus, conductors obviously have much less resistance than insulators.

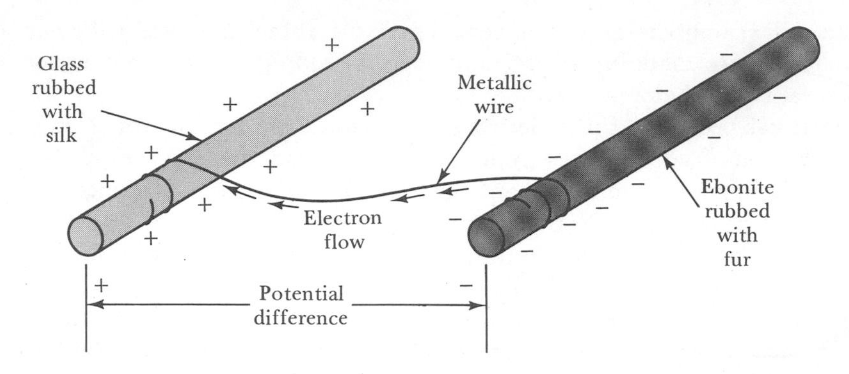

FIG. 1-4. Potential difference and flow of electrons from a negatively charged body to a positively charged body

The ability of two oppositely charged bodies to cause a flow of electricity between them may be thought of as a *potential* for the production of electric current. Thus, a positively charged body is said to have a *positive potential*, and a plus sign is used to identify this potential on all diagrams. Similarly, a negatively charged body is described as having a *negative potential*, and a minus sign is used in its identification. Two oppositely charged bodies are said to have a *potential difference* between them (see Figure 1-4). It is also possible for a potential difference to exist between two similarly charged bodies if one is charged to a higher potential than the other. Potential difference can best be thought of as an electrical pressure tending to cause a current flow. When a suitable conducting path is provided, current will flow as a result of the electrical pressure.

1-4 SOURCE OF ELECTRICITY

The first really useful source of electricity was invented by the Italian physicist Alessandro Volta. He discovered that the action of saltwater on certain metals could produce either an excess or a deficiency of electrons on the metals. This led to his construction of the *voltaic pile*, which, as illustrated in Figure 1-5, consists of a stack of pairs of alternate silver and zinc discs with saltwater-soaked cloths between them. The acid effect of the saltwater on the metals creates a negative charge on the zinc and a positive charge on the silver. Each pair of metal discs in the voltaic pile constitutes an *electric cell*, and the complete device is a *battery of cells*. Therefore, the voltaic pile was the first electric battery.

The top and bottom plates of the pile are the *terminals* of the battery. The potential difference that results at the terminals comprise the sum of the potential differences of the individual cells. Current flow may be maintained from the voltaic pile while the chemical action lasts.

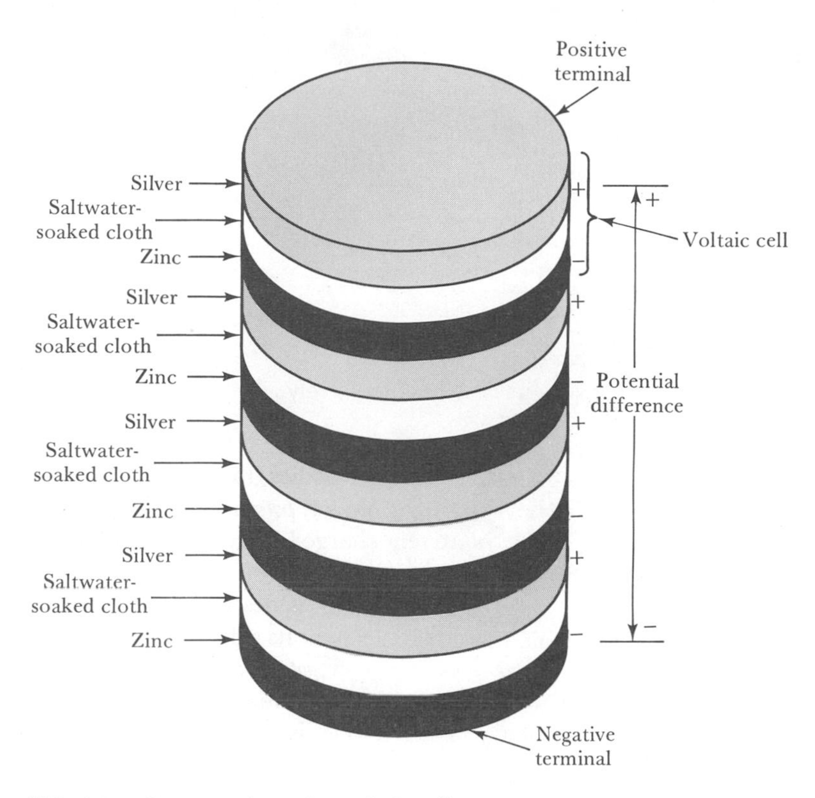

FIG. 1-5. Construction of a voltaic pile

1-5 ELECTRIC LAMP

Figure 1-6(a) illustrates the movement of an electron in a conductor when a potential difference is applied to the ends of the conductor. The electron is repelled from the negative terminal and attracted toward the positive terminal. However, the atoms within the conductor impede the motion of the electron, so that it cannot simply accelerate from the negative end to the positive end. Instead, the electron bounces about from one atom to another and merely drifts toward the positive end. Each time the electron strikes an atom, it dissipates energy in the form of heat. When a great many electrons are involved in a flow of current, the heat dissipation can be considerable.

An electric lamp consists of a thin tungsten wire, or *filament*, contained in a glass bulb from which the air has been evacuated [see Figure 1-6(b)]. When a current is passed through the filament, many electron–atom collisions occur within the filament. The resulting heat makes the filament glow white hot, thus emitting light.

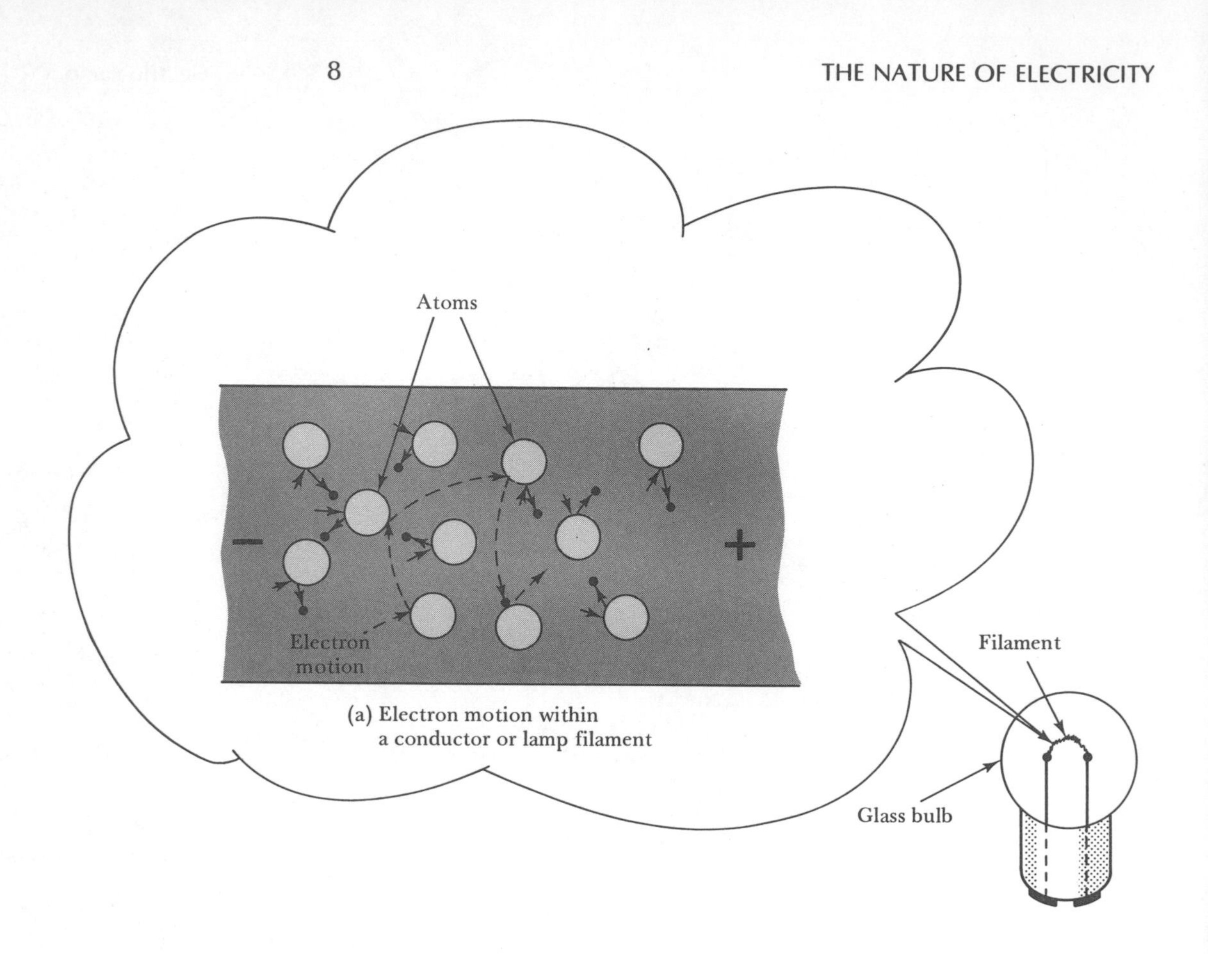

(a) Electron motion within a conductor or lamp filament

(b) Electric lamp

FIG. 1-6. When an electric current flows, electron–atom collisions produce heat

1-6 CIRCUIT DIAGRAMS

An *electric circuit* usually consists of a source of electricity, and other components, such as a switch and a lamp. Conductors to connect the components also form part of the circuit. A *circuit diagram* is a drawing that uses graphic symbols to represent circuit components and lines to represent the conductors.

Figure 1-7 shows the diagram for a very simple electric circuit made up of a battery, a switch, and a lamp. For the battery symbol, the long bar always represents the positive terminal, and the short bar is the negative terminal, as illustrated. When the switch is closed, electrons flow from the negative terminal of the battery, through the lamp filament, to the battery positive terminal. When the switch is open, current cannot

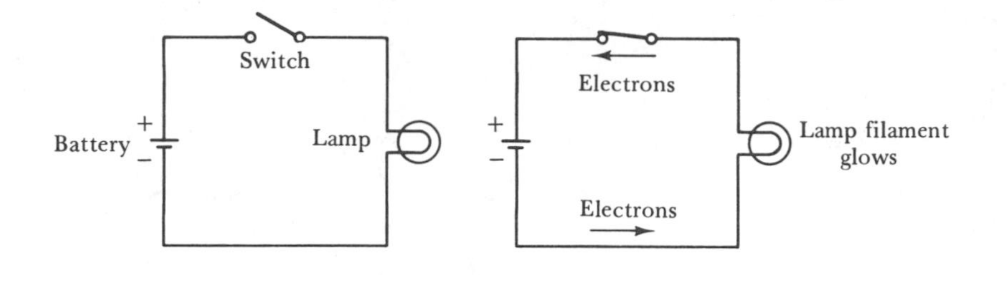

(a) Switch open—no complete path for conduction of electrons

(b) Switch closed—electrons flow continuously through lamp filament

FIG. 1-7. Circuit diagram for a simple electric circuit

flow and the lamp remains unlighted. Other circuit symbols are shown in Appendix 1.

1-7 ELECTRIC SHOCK*

Electricity can kill, and for that reason it is very important to treat electrical supplies and appliances with great care. Perhaps the greatest domestic danger exists in bathrooms and kitchens, because of the water and water taps there. Someone touching a faulty appliance while reaching for a water tap can get a very severe shock [Figure 1-8(a)]. In such a situation it is frequently impossible for the person to open his hands to disconnect himself from the appliance or water tap. Similarly, a person using a portable appliance such as an electric drill can be in danger if, for example, the supply cable is worn or has temporary joins. While holding the appliance in one hand and pulling the cable with the other, his hand could close around a bare wire. He may find it impossible to open either hand as the electricity flows through his body [Figure 1-8(b)].

An even more lethal situation exists when someone taking a bath reaches for an electrical appliance [Figure 1-8(c)]. In this case the person's whole body is grounded via the water. If the appliance is faulty, death is virtually inevitable.

When someone has received or is receiving a shock, the most important thing to do is to get him away from the source of the shock. The supply should immediately be switched off or the plug of the appliance disconnected. Medical help should be summoned immediately, and the victim should be made comfortable and kept calm. If the victim's heart has stopped beating, artificial respiration should be begun, and continued until medical help arrives.

*For further information, see Edward A. Lacy, *Handbook of Electronic Safety Procedures* (Englewood Cliffs, N.J.: Prentice-Hall, Inc., 1976).

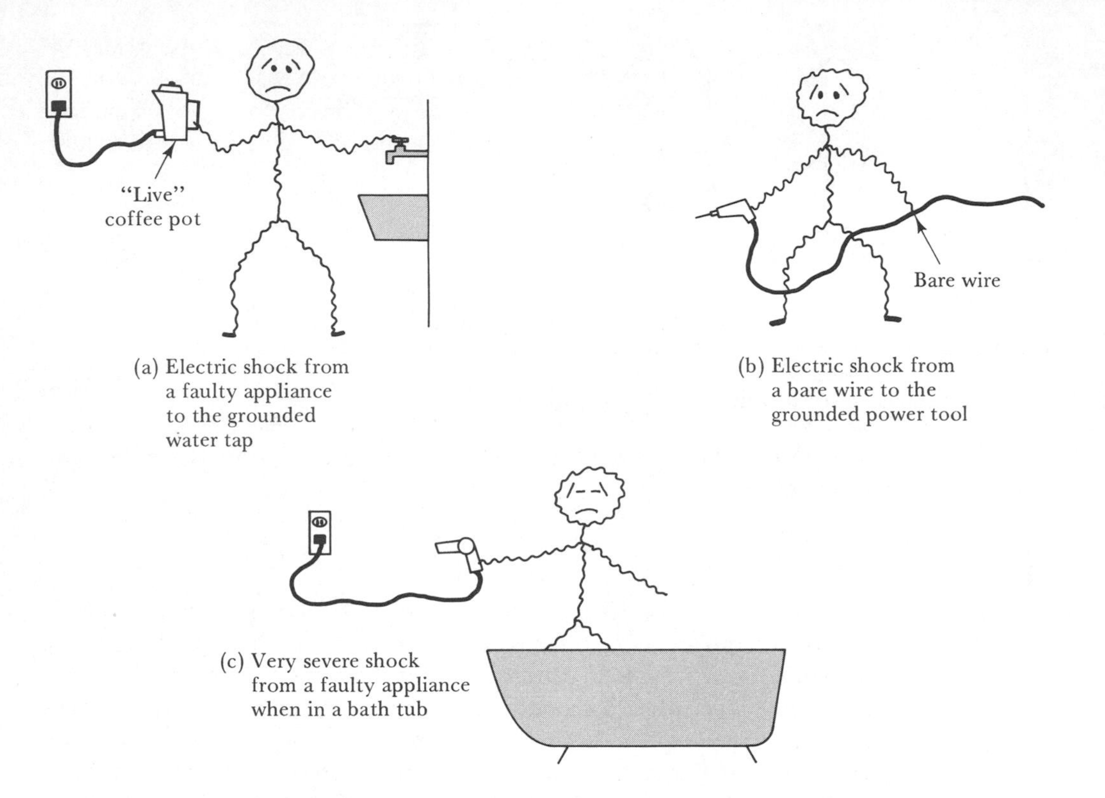

"Live"
coffee pot
(a) Electric shock from
a faulty appliance
to the grounded
water tap
Bare wire
(b) Electric shock from
a bare wire to the
grounded power tool
(c) Very severe shock
from a faulty appliance
when in a bath tub

FIG. 1-8. Some dangerous electrical situations

REVIEW QUESTIONS

1-1 Explain what is meant by electrification by friction. Also define positive and negative electricity as related to electrification by friction.

1-2 State the fundamental law that originated from the study of electrification by friction. Briefly explain the law.

1-3 Draw a plane diagram of the planetary atom. Identify the three basic particles that constitute an atom and define their relative quantities of charge and mass.

1-4 Describe a valence electron and discuss the effect on an atom when it loses an electron and the effect when it gains an electron.

1-5 Describe what occurs when an electric current flows between two charged bodies. Briefly define a conductor, an insulator, and resistance.

1-6 Explain what constitutes a potential difference and how it is related to electric current.

1-7 Sketch the construction of a voltaic pile and identify its components. Explain how a voltaic pile works.

1-8 Explain how electric current can produce light and briefly describe an electric lamp.

1-9 Sketch a diagram for an electric circuit consisting of a battery, a lamp, and a switch. Briefly explain the operation of the circuit.

1-10 Describe some of the most dangerous situations in which a person can suffer an electric shock. List the emergency actions that should be taken to aid a victim of electric shock.

2

THE INTERNATIONAL SYSTEM OF UNITS (SI UNITS)

INTRODUCTION

Before standard systems of measurement were invented, many approximate units were used. A long distance was often measured by the number of *days* it would take to *ride* a horse over the distance; a horse's height was measured in *hands*; liquid was measured by the *bucket* or *barrel*.

With the development of science and engineering, more accurate units had to be devised. The English-speaking peoples adopted the *foot* and the *mile* for measuring distances, the *pound* for mass, and the *gallon* for liquid. Other nations followed the lead of the French in adopting a *metric system*, in which large and small units are very conveniently related by a factor of 10.

With the increase of world trade and the exchange of scientific information between nations, it became necessary to establish a single system of units of measurement that would be acceptable internationally. After several world conferences on the matter, a metric system which uses the *meter*, *kilogram*, and *second* as fundamental units has now been generally adopted around the world. This is known, from the French term "système international," as the *SI* or *international system*.

2-1 FUNDAMENTAL UNITS

In the SI system the three basic units are:

unit of *length*:	the **meter** (m)
unit of *mass*:	the **kilogram** (kg)
unit of *time*:	the **second** (s)

These are known as *fundamental units*. Other units derived from the fundamental units are termed *derived units*. Because of the use of the meter, kilogram, and second, the system is sometimes referred to as an *MKS system*.

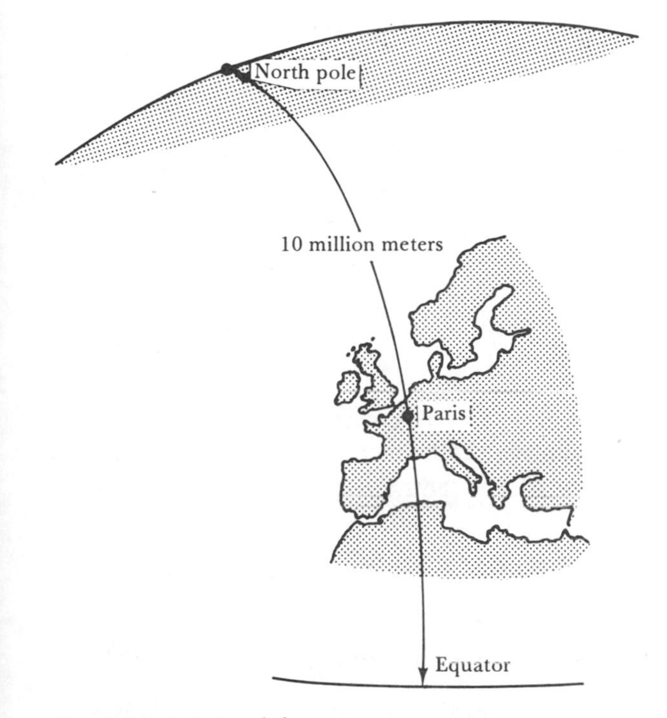

FIG. 2-1. Origin of the meter

The *meter* was originally defined as 1 ten-millionth of a meridian passing through Paris from the North Pole to the equator (see Figure 2-1). The *kilogram* was defined as 1000 times the mass of 1 cubic centimeter of distilled water. The *liter*, which is a measure of volume of liquid, is 1000 times the volume of 1 cubic centimeter of liquid (see Figure 2-2). Consequently, 1 liter of water has a mass of 1 kilogram. Because of the possibility of error in the original definitions, the meter was redefined in terms of atomic radiation. Also, the kilogram is now defined as the mass of a certain platinum–iridium standard bar kept at the International Bureau of Weights and Measures in France. The second is, of course, 1/(86 400) of a mean solar day, but it is also accurately defined by atomic radiation.

Other fundamental SI units include the unit of electrical current (see Chapter 3), the unit of temperature, and the unit of luminous intensity.

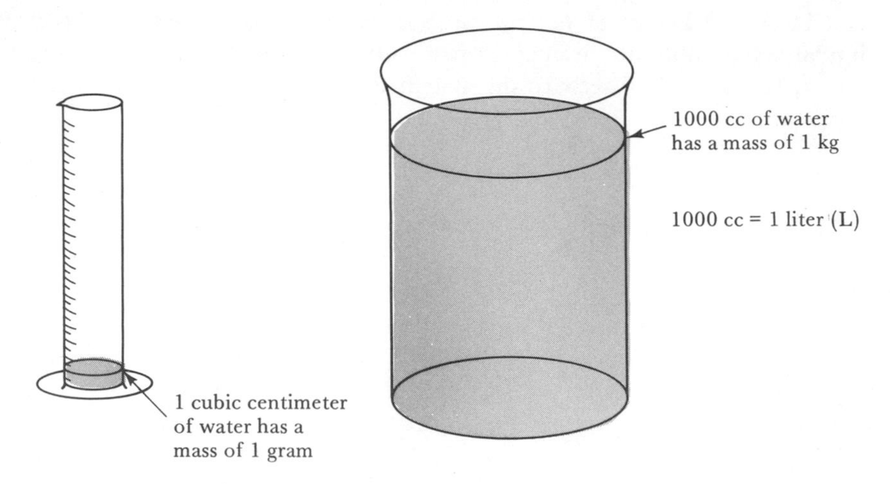
1000 cc of water
has a mass of 1 kg
1000 cc = 1 liter (L)
1 cubic centimeter
of water has a
mass of 1 gram

FIG. 2-2. Origin of the kilogram and liter

2-2 SCIENTIFIC NOTATION, METRIC PREFIXES, AND SIGNIFICANT FIGURES

Very large or very small numbers can be more conveniently written as a number multiplied by 10 raised to a power:

$$100 = 1 \times 10 \times 10$$

$$= 1 \times 10^2$$

$$10\,000 = 1 \times 10 \times 10 \times 10 \times 10$$

$$= 1 \times 10^4$$

$$0.001 = 1/(10 \times 10 \times 10)$$

$$= 1/10^3$$

$$= 1 \times 10^{-3}$$

$$1500 = 1.5 \times 10^3$$

$$0.015 = 1.5 \times 10^{-2}$$

This method of writing numbers is known as *scientific notation*. Prefixes and symbols are also available for the various multiples and submultiples of 10. Those that are most frequently used in electrical calculations are listed in Table 2-1. For example, *1000 meters* can be written as *1 kilometer* or *1 km*. Here *kilo* is the prefix employed to represent *1000 times* and *k* is the symbol for *kilo*.

TABLE 2-1.

VALUE	SCIENTIFIC NOTATION	PREFIX	SYMBOL
1 000 000 000 000	10^{12}	tera	T
1 000 000 000	10^{9}	giga	G
1 000 000	10^{6}	mega	M
1 000	10^{3}	kilo	k
0.001	10^{-3}	milli	m
0.000 001	10^{-6}	micro	μ
0.000 000 001	10^{-9}	nano	n
0.000 000 000 001	10^{-12}	pico	p

When a number is expressed in scientific notation, 10 is known as the *base*. The number to which 10 is raised is termed the *power*, *index*, or *exponent*:

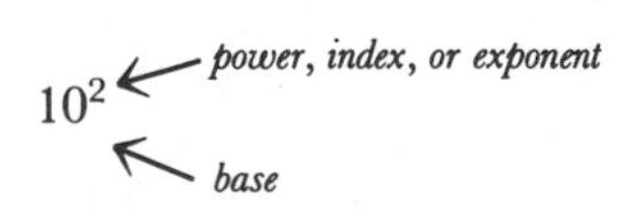

For addition or subtraction of numbers expressed in scientific notation, it is necessary to convert to a common power of 10:

$$(1.5\times10^2)+(2.5\times10^3)=(0.15\times10^3)+(2.5\times10^3)$$
$$=2.65\times10^3$$

and

$$(4.4\times10^6)-(2\times10^5)=(4.4\times10^6)-(0.2\times10^6)$$
$$=4.2\times10^6$$

Multiplication and division with numbers expressed in scientific notation involves addition and subtraction of the exponent:

$$(5\times10^3)\times(1.2\times10^2)=5000\times120$$
$$=600\ 000$$

or

$$(5\times10^3)\times(1.2\times10^2)=5\times1.2\times10^{(3+2)}$$
$$=6\times10^5$$

and

$$(4\times10^6)/(2\times10^3)=\frac{4\ 000\ 000}{2000}$$
$$=2000$$

or

$$(4\times10^6)/(2\times10^3)=\frac{4}{2}\times10^{(6-3)}$$
$$=2\times10^3$$

Consider the number 816, which is said to be accurate to three significant figures. If the number is rewritten as accurate to two significant figures, it becomes 820. This is because 816 is clearly closer to 820 than it is to 810. If the number 814 were rewritten accurate to two significant figures, it would be 810. In the case of 820 and 810, the zero is not a significant figure.

The number 2.03, written accurate to two significant figures, becomes 2.0. In this case the zero is a significant figure, and it obviously implies that the number is greater than 2.0 but less than 2.1.

Now consider the result of using an electronic calculator to resolve 5.4/2.3:

$$\frac{5.4}{2.3} = 2.347826087$$

Clearly, it does not make sense to have an answer containing 10 significant figures when each of the original numbers has only two significant figures. Suppose that the two original numbers are assumed to be accurate to ±1%, which would not be unusual in electrical calculations. Then the accuracy of the answer can be shown to be not better than ±2%. Therefore,

$$\frac{5.4 \pm 1\%}{2.3 \pm 1\%} = 2.347826087 \pm 2\%$$

$$= 2.347826087 \pm 0.0469565$$

The only reasonable way to write this answer is

$$2.35 \pm 0.05$$

Consequently, the problem and answer become

$$\frac{5.4}{2.3} = 2.35$$

The answer now has three significant figures, although the numerator and denominator each have only two significant figures but were assumed accurate to ±1%. Where the accuracy is known to be better than ±1%, the answer may have a larger number of significant figures.

2-3 UNIT OF FORCE

The SI unit of force is the **newton*** *(N), defined as that force which will give a mass of 1 kilogram an acceleration of 1 meter per second per second.*

---

*Named for the great English scientist Sir Isaac Newton (1642–1727).

When a body is to be accelerated, a force must be applied that is proportional to the desired rate of change of velocity, that is, proportional to the acceleration (or deceleration).

$$\boxed{\begin{aligned}&\textbf{force} = \textbf{mass} \times \textbf{acceleration}\\ &\boldsymbol{F = ma}\end{aligned}} \tag{2-1}$$

When the mass is in kg and the acceleration is in m/s², Eq. (2-1) gives the force in newtons.

If the body is to be accelerated vertically from the earth's surface, the *acceleration due to gravity* (g) must be overcome before any vertical motion is possible. In SI units:

$$\boxed{\boldsymbol{g = 9.81\ \mathrm{m/s^2}}}$$

Thus, as illustrated in Figure 2-3, a mass of 1 kg has a gravitational force of 9.81 N.

EXAMPLE 2-1 Calculate the force required to lift a 5000-kg elevator.

SOLUTION

From Eq. (2-1),

$$\begin{aligned}F &= m \times a\\ &= 5000\ \mathrm{kg} \times 9.81\ \mathrm{m/s^2}\\ \boldsymbol{F} &\boldsymbol{= 49.05 \times 10^3\ \mathrm{N}}\end{aligned}$$

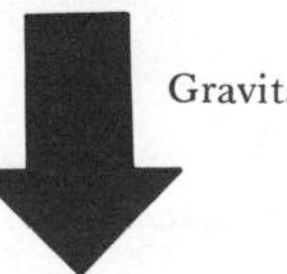

FIG. 2-3. Gravitational force of one kilogram

In Example 2-1, the force calculated is just sufficient to overcome the gravitational force. To move the elevator vertically, a slightly larger force is required. Any force in excess of that calculated will tend to impart a vertical acceleration to the elevator.

2-4 WORK, ENERGY, AND POWER

WORK. When a body is moved, a force is exerted to overcome the body's resistance to motion.

*The **work** done in moving the body is the product of the force and the distance through which the body is moved in the direction of the force.*

$$\text{work} = \text{force} \times \text{distance}$$
$$W = Fd \tag{2-2}$$

*The SI unit of work is the **joule*** (J), defined as the amount of work done when a force of 1 newton acts through a distance of 1 meter.*

Thus, the *joule* may also be termed a *newton-meter. For Eq. (2-2), work is expressed in joules when F is in newtons and d is in meters.*

EXAMPLE 2-2 Calculate the work done in raising a 2-kg mass of metal through a height of 20 m.

SOLUTION

From Eq. (2-1),

$$F = m \times a$$
$$= 2\ \text{kg} \times 9.81\ \text{m/s}^2$$
$$= 19.62\ \text{N}$$

From Eq. (2-2),

$$W = F \times d$$
$$= 19.62\ \text{N} \times 20\ \text{m}$$
$$W = 392.4\ \text{J}$$

ENERGY.

Energy *is defined as the capacity for doing work.*

---

*Named after the English physicist James P. Joule (1818–1899).

Thus, energy is measured in the same units as work. Energy can exist in several forms.

***Potential energy** is the energy due to position.*

Water in a dam has the potential for doing work as it falls to a lower level (Figure 2-4). A body of mass M falling from a height h is accelerated at the rate of g (acceleration due to gravity). Since

$$\text{force} = \text{mass} \times \text{acceleration}$$
$$F = m \times g$$

and since

$$\text{work} = \text{force} \times \text{distance}$$
$$W = F \times h$$
$$= m \times g \times h$$

or

$$\boxed{\textbf{\textit{potential energy, }} E_p = mgh} \qquad (2\text{-}3)$$

In Eq. (2-3), E_p is in joules when m is in kg, g is in m/s^2, and h is in meters.

***Kinetic energy** is the energy possessed by a body in motion.*

The kinetic energy of a moving body can be determined by calculating the amount of work done in bringing the body to its state of motion.

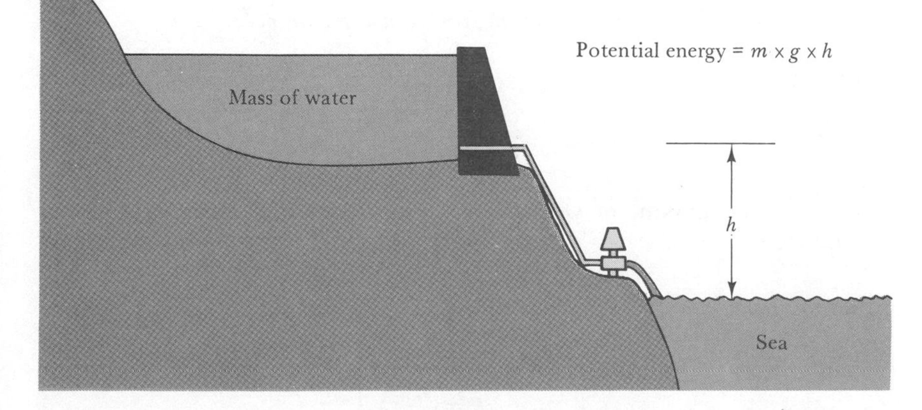

FIG. 2-4. Potential energy possessed by a mass of water above sea level

When the velocity of a body is increased from zero to v in a time t, its average acceleration a is v/t, or

$$a = \frac{v}{t}$$

Also, the distance traveled in time t is (*average velocity*)$\times t$; that is, $(v-0)/2 \times t$, or

$$d = \tfrac{1}{2}vt$$

Since force = mass × acceleration

$$F = m \times \frac{v}{t}$$

and since work = force × distance

$$W = F \times \tfrac{1}{2}vt$$

or

$$W = \frac{mv}{t} \times \tfrac{1}{2}vt$$

$$= mv \times \frac{v}{2}$$

or

$$\textbf{\textit{kinetic energy}}, E_k = \tfrac{1}{2}mv^2 \qquad \textbf{(2-4)}$$

In Eq. (2-4), E_k is in joules when m is in kg and v is in m/s.

POWER.

***Power** is the time rate of doing work.*

If a certain amount of work W is to be done in a time t, the power required is

$$\textit{power} = \frac{\textit{work}}{\textit{time}}$$

$$P = \frac{W}{t} \qquad \textbf{(2-5)}$$

The SI unit of power is the ***watt**** *(W), defined as the power developed when 1 joule of work is done in 1 second.*

For Eq. (2-5), P is in watts when W is in joules and t is in seconds.

EXAMPLE 2-3

A lake that is 10 km long, 1 km wide, and 50 m deep is situated at an average height of 1000 m above sea level. Calculate the potential energy of the water.

SOLUTION

$$\begin{aligned} \textit{total volume of water} &= 10\ \text{km} \times 1\ \text{km} \times 50\ \text{m} \\ &= 50 \times 10^{7}\ \text{cubic meters} \\ &= 50 \times 10^{7} \times 100 \times 100 \times 100\ \text{cubic centimeters (cm}^3\text{)} \\ &= 50 \times 10^{13}\ \text{cm}^3 \end{aligned}$$

$$1\ \text{liter} = 1000\ \text{cm}^3$$

$$\begin{aligned} \textit{total volume} &= \frac{50 \times 10^{13}}{1000}\ \text{liters} \\ &= 50 \times 10^{10}\ \text{liters} \end{aligned}$$

As 1 liter *has a mass of* 1 kg,

$$\textit{total mass of water} = 50 \times 10^{10}\ \text{kg}$$

Since

$$\begin{aligned} \textit{force} &= \textit{mass} \times (\textit{acceleration due to gravity}) \\ F &= m \times g \\ &= 50 \times 10^{10}\ \text{kg} \times 9.81\ \text{m/s}^2 \\ &= 4.905 \times 10^{12}\ \text{N} \end{aligned}$$

and

$$\begin{aligned} \textit{work} &= \textit{force} \times (\textit{height above sea level}) \\ W &= F \times h \\ &= 4.905 \times 10^{12}\ \text{N} \times 1000\ \text{m} \\ &= 4.905 \times 10^{15}\ \text{J} \end{aligned}$$

$$\boldsymbol{\textit{potential energy} = 4.905 \times 10^{15}\ \text{J}}$$

---

*Named after the Scottish engineer James Watt (1736–1819).

or, from Eq. (2-3),

$$E_p = mgh$$

$$= 50 \times 10^{10}\ \text{kg} \times 9.81\ \text{m/s}^2 \times 1000\ \text{m}$$

$$\mathbf{E_p = 4.905 \times 10^{15} J}$$

EXAMPLE 2-4

The lake referred to in Example 2-3 is replenished continuously by streams, so the total mass of water can be allowed to flow to sea level over a period of 8 weeks. Calculate the power developed by the water as it falls to sea level.

SOLUTION

$$t = \textit{time}, \text{ in seconds}$$

$$= (8\ \text{weeks} \times 7 \times 24)\ \text{hours}$$

$$= (1344\ \text{hours} \times 60 \times 60)\ \text{seconds}$$

$$= 4.84 \times 10^6\ \text{s}$$

and

$$\textit{power} = \frac{\textit{work}}{\textit{time}}$$

$$P = \frac{W}{t}$$

$$= \frac{4.905 \times 10^{15}\ \text{J}}{4.84 \times 10^6\ \text{s}}$$

$$= 1 \times 10^9\ \text{W}$$

$$\mathbf{P = 10^6\ kW = 1000\ MW}$$

2-5 TEMPERATURE AND HEAT

There are two SI temperature scales, the *Celsius scale** and the *Kelvin scale*.† These are illustrated in Figure 2-5.

The Celsius scale has 100 equal divisions (or *degrees*) between the freezing temperature and the boiling temperature of water. At normal

---

*Invented by the Swedish astronomer and scientist Anders Celsius (1701–1744).

†Named for the Irish-born scientist and mathematician William Thomson, who became Lord Kelvin (1824–1907).

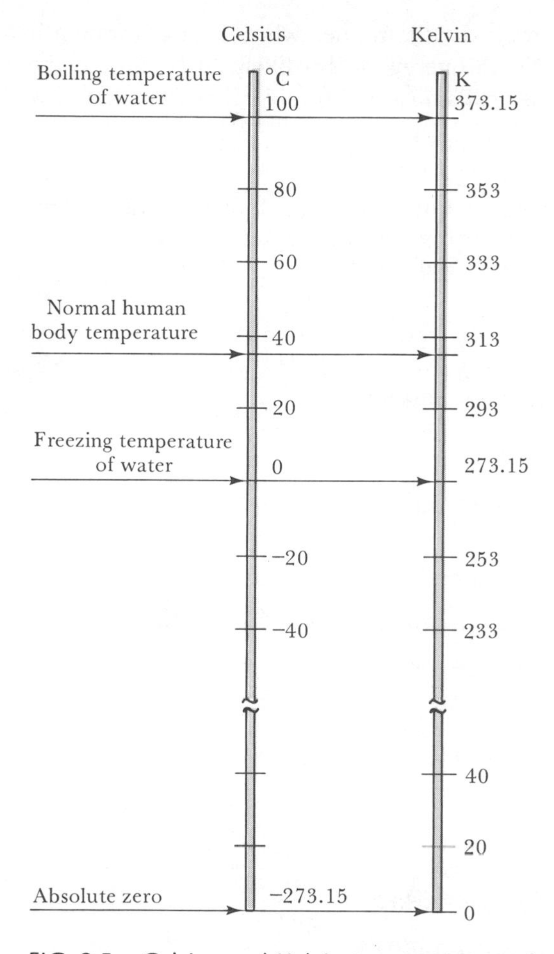

FIG. 2-5. Celsius and Kelvin temperature scales

atmospheric pressure, water freezes at 0°C (*zero degrees Celsius*) and boils at 100°C.

The Kelvin temperature scale, also known as the *absolute scale*, commences at absolute zero of temperature, which corresponds to −273.15°C. Therefore, 0°C is equal to 273.15 K, and 100°C is the same temperature as 373.15 K. A temperature difference of 1 K is the same as a temperature difference of 1°C.

To raise 1 *liter of water through* 1°*C requires an energy input of* 4187 J. Using this figure, the energy required to raise any quantity of water through a given temperature change can be easily calculated.

When water is heated, the container must also be raised to the same temperature as the water. The container is usually defined as having a certain *water equivalent*. This is a quantity of water that would absorb the same amount of energy as the water container when heated through a given temperature change. Some energy is also likely to be lost to the surrounding atmosphere. This is taken care of by specifying the efficiency of the heater.

EXAMPLE 2-5 Calculate the amount of energy required to raise 1 liter of water from 25°C to 100°C. Take the water equivalent of the kettle as 0.1 liter and assume that the water heater is 90% efficient.

SOLUTION

$$\textit{temperature increase} = 100°\text{C} - 25°\text{C}$$

$$\Delta T = 75°\text{C}$$

$$\textit{total water equivalent of water and kettle} = 1 + 0.1 \text{ liters}$$

$$= 1.1 \text{ liters}$$

$$\textit{energy required} = (\textit{total water equivalent}) \times \Delta T \times 4187 \text{ J}$$

$$= 1.1 \text{ liters} \times 75°\text{C} \times 4187 \text{ J}$$

$$= 345\ 427.5 \text{ J}$$

$$\cong 345 \text{ kJ}$$

Since the heater is 90% *efficient,*

$$\textit{total energy input} = 345 \text{ kJ} \times \frac{100\%}{90\%}$$

$$\textbf{\textit{total energy input}} = \mathbf{383\ kJ}$$

EXAMPLE 2-6 If the water referred to in Example 2-5 is to be heated in a period of 5 minutes, determine the power required.

SOLUTION

$$\textit{power} = \frac{\textit{energy}}{\textit{time}}$$

$$= \frac{383 \times 10^3 \text{ J}}{5 \times 60 \text{ s}}$$

$$\boldsymbol{P} \cong \mathbf{1.28\ kW}$$

GLOSSARY OF FORMULAS

Force = mass × acceleration:

$$F = ma$$

Acceleration due to gravity:

$$g = 9.81 \text{ m/s}^2$$

Work = force × distance:

$$W = F \times d$$

Potential energy:

$$E_p = mgh$$

Kinetic energy:

$$E_k = \frac{1}{2}mv^2$$

Power = work / time:

$$P = \frac{W}{t}$$

REVIEW QUESTIONS

2-1 Define SI. List the three basic SI units and briefly explain their origin.

2-2 List the names of the various metric prefixes and the corresponding symbols. Also, list the value represented by each prefix, both in the long form and by scientific notation.

2-3 When an electronic calculator is used to determine 7.2/3.1, the answer displayed is 2.322580645. How would you give the answer if 7.2 and 3.1 were both known to be accurate to ±1%? Explain.

2-4 Define work, energy, power, kinetic energy, and potential energy. Identify the SI units in which each is measured. Define g and give its numerical value in SI units.

2-5 Name the two SI temperature scales. Sketch the two scales, showing the corresponding temperatures on each. Identify the boiling and freezing points of water.

PROBLEMS

2-1 Determine the force that must be exerted by a crane that is required to lift a load having a mass of 20 000 kg.

2-2 An automobile having a mass of 2000 kg is accelerated to a speed of 70 km/h in a time of 20 s. Neglecting all friction effects, calculate the force exerted by the engine.

2-3 A 1000-kg elevator with a load of 1500 kg is raised through a height of 60 m. Calculate the work done.

2-4 If the elevator in Problem 2-3 travels the distance in a time of 1 min, calculate the power involved.

2-5 One thousand liters of water is pumped through a height of 20 m in a period of 30 min. Calculate the work done and the power required.

2-6 A lake situated at a height of 750 m above sea level is completely replenished by rivers over a period of 6 weeks. If the lake is approximately 20 km by 7 km and 100 m deep, determine the potential energy of the water and the power developed by the water flowing to sea level.

2-7 Ten liters of water in a coffee urn is to be raised from 24°C to boiling point in a period of 30 min. The water equivalent of the urn is 1 liter, and the water heater is 90% efficient. Determine the power required and the energy consumed.

3

THE ELECTRICAL UNITS

INTRODUCTION

As well as the mechanical and thermal SI units discussed in Chapter 2, there are SI electrical units. The *unit of current* is closely related to the *unit of charge*. However, the unit of current is the fundamental electrical unit from which other units are derived. *Current direction* can be defined in two ways: *conventional current direction* and direction of *electron flow*. It is important to understand both. It is also important to be able to correctly use an *ammeter* for the measurement of current.

The electrical pressure produced by a source of electricity is termed *emf*, *potential difference*, or *voltage*, and it is measured by means of a *voltmeter*. Current is the result of an applied voltage. But in every electric circuit, there is opposition to current flow. The opposition is termed *resistance*, and its converse is *conductance*. The relationship between voltage, current, and resistance is defined by the simple but extremely important *Ohm's law*.

Electrical power and energy can be converted to mechanical or thermal quantities. The electrical units involved are related to the units of current voltage and resistance.

3-1 UNITS OF CURRENT AND CHARGE

As explained in Chapter 1, electric current (I) is a flow of charge carriers. Therefore, current could be defined in terms of the quantity of electricity (Q) that passes a given point in a conductor during a time of 1 s.

The ***coulomb* (C)*** *is the unit of electrical charge or quantity of electricity.*

*Named after the French physicist Charles Augustin de Coulomb (1736–1806).

The coulomb was originally selected as the fundamental electrical unit from which all other units were derived. However, since it is much easier to measure current accurately than it is to measure charge, the unit of current is now the *fundamental electrical unit* in the SI system. Thus, the coulomb is a *derived unit*, defined in terms of the unit of electric current.

The ***ampere**** *(A) is the unit of electric current.*

The ***ampere*** *is defined as that constant current which, when flowing in each of two infinitely long parallel conductors 1 meter apart, exerts a force of* 2×10^{-7} *newtons per meter of length on each conductor.* (Note that this definition involves a magnetic force. The appropriate calculations are considered in Section 12-6.)

The ***coulomb*** *is defined as that charge which passes a given point in a conductor each second, when a current of 1 ampere flows.*

These definitions show that the coulomb could be termed an *ampere-second*. Conversely, the ampere can be described as a *coulomb per second.*

$$\text{amperes} = \frac{\text{coulombs}}{\text{seconds}}$$

or

$$\boxed{I = \frac{Q}{t}} \qquad \textbf{(3-1)}$$

where I is current in amperes, Q is charge in coulombs, and t is time in seconds.

It has been established experimentally that *1 coulomb is equal to the total charge carried by* 6.24×10^{18} *electrons.* Therefore, the charge carried by one electron is

$$Q = \frac{1}{6.24 \times 10^{18}}$$
$$= 1.602 \times 10^{-19}\ \text{C}$$

The coulomb as a unit of electrical quantity is much too small for many practical purposes. Since the coulomb is an *ampere-second*, it is convenient to measure larger electrical quantities in *ampere-hours*. For example, if a battery supplies 6 A for a period of 5 h, the quantity of electricity supplied is 6 A×5 h, that is, 30 ampere-hours.

---

*Named after the French physicist and mathematician André Marie Ampère (1775–1836).

In the case of the charge contained in some electronic devices, the coulomb is a relatively large unit. Therefore, the *microcoulomb* (μC) is frequently used to express such charges. The ampere is also an inconveniently large unit of current for most electronic circuits. Consequently, in electronics, currents are usually measured in *milliamps* (mA), *microamps* (μA), or *nanoamps* (nA).

EXAMPLE 3-1

Determine the number of electrons that pass through the filament of an electric lamp when a current of 5 A flows for 30 min. Also calculate the quantity of electricity used, in ampere-hours.

SOLUTION

From Eq. (3-1),

$$I = \frac{Q}{t}$$

$$Q = I\,t = 5\ \text{A} \times 30\ \text{min}$$

$$= 5 \times 30 \times 60\ \text{As}$$

$$\mathbf{Q = 9000\ C}$$

$$1\ \text{C} = 6.24 \times 10^{18}\ \text{electrons}$$

$$9000\ \text{C} = 6.24 \times 10^{18} \times 9000\ \text{electrons}$$

$$\mathbf{9000\ C = 5.616 \times 10^{22}\ electrons}$$

$$Q = I\,t = 5\ \text{A} \times 0.5\ \text{h}$$

$$\mathbf{Q = 2.5\ Ah}$$

3-2 CURRENT DIRECTION

CONVENTIONAL CURRENT AND ELECTRON FLOW. In the early days of electrical engineering it was believed that a positive charge represented an increased amount of electricity, while a negative charge was seen as a reduced quantity of electricity. Consequently, it was assumed that current flowed from positive to negative. This is a convention that remains in use today, even though current flow has long been accepted as a movement of electrons from negative to positive.

Current flow from positive to negative is known as ***the conventional direction of current flow****. Electron flow from negative to positive is known as the* ***direction of electron flow****.*

It is important to understand and to be able to think both in terms of conventional current direction and in terms of electron flow. *Every graphic symbol employed to represent an electronic device in a circuit has an arrowhead which indicates conventional current direction.* Figure 3-1 shows the symbol for the simplest electronic device, a *diode*. The diode is a *one-way device*. Current flows through the device only when the polarity of the applied voltage is such that the arrowhead points from positive to negative. When the applied voltage is reversed (arrowhead pointing from negative to positive), current cannot flow. Thus, the arrowhead indicates the direction of conventional current flow.

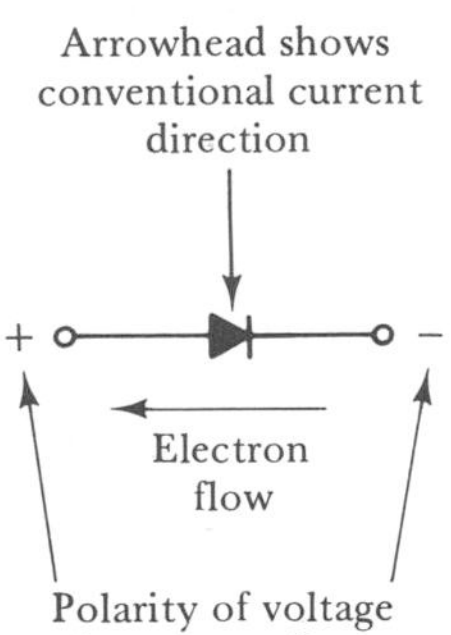

FIG. 3-1. Conventional current direction and electron flow in a diode

DIRECT CURRENT AND ALTERNATING CURRENT.

Current that flows continuously in one direction is referred to as ***direct current****, abbreviated* ***dc****.*

This is the kind of current supplied from a battery. Since the voltage at the terminals of a battery does not reverse its polarity but remains substantially constant, a battery can be termed a *direct voltage source*. Figure 3-2(a) shows the circuit symbol for a direct voltage source and the graph of direct current plotted versus time.

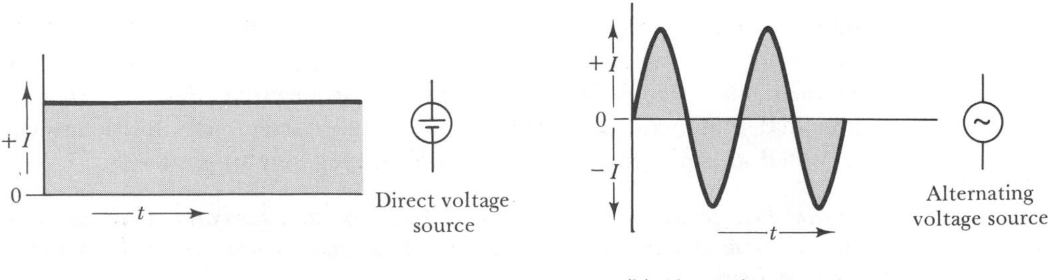

FIG. 3-2. Direct and alternating current graphs, and direct and alternating voltage sources

***Alternating current**, abbreviated **ac**, is current that flows first in one direction for a brief time and then reverses to flow in the opposite direction for a similar time.*

This is illustrated in Figure 3-2(b), where the graph of alternating current is plotted versus time. The circuit symbol for an alternating voltage source is also shown in Figure 3-2(b). The study of alternating voltage and current begins in Chapter 18.

3-3 EMF, POTENTIAL DIFFERENCE, AND VOLTAGE

In Chapter 2 a *potential difference* between two charged bodies is described as *an electrical pressure that tends to cause current flow when a suitable conducting path is provided*. Another more descriptive name for the electrical pressure from a battery or other source is *electromotive force*, abbreviated *emf. Emf* is the name usually applied to the electrical supply connected to a circuit. The differences between levels of emf at various points throughout a circuit is referred to as a *potential difference* (PD).

*The **volt*** (v) is the unit of emf and potential difference.*

Because emf and potential difference are measured in volts, the term *voltage* is frequently substituted in both cases.

Consider a negatively charged particle situated midway between two metal plates 1 cm apart, as illustrated in Figure 3-3. With an emf (E) applied as shown, the particle experiences a force repelling it from the negative plate and attracting it toward the positive plate. If the particle had no charge, there would be no force tending to cause motion. Neither would there be any force if there were no applied emf. If either the applied emf or the charge is doubled, the force acting on the particle is doubled. Thus, it is seen that the force acting on the particle is directly proportional to both the emf and the charge.

Now consider the effect of increasing the distance between the plates to 1 m. The force tending to move the particle is obviously weakened. Therefore, the force on the particle can be said to be inversely proportional to the distance between the plates.

Since the force acting on the charged particle is directly proportional to emf and charge, and inversely proportional to distance between plates, the equation for force on the charged particle is

$$\text{force} = \frac{\text{emf} \times \text{charge}}{\text{distance}}$$

*Named in honor of the Italian physicist Alessandro Volta (1745–1827).

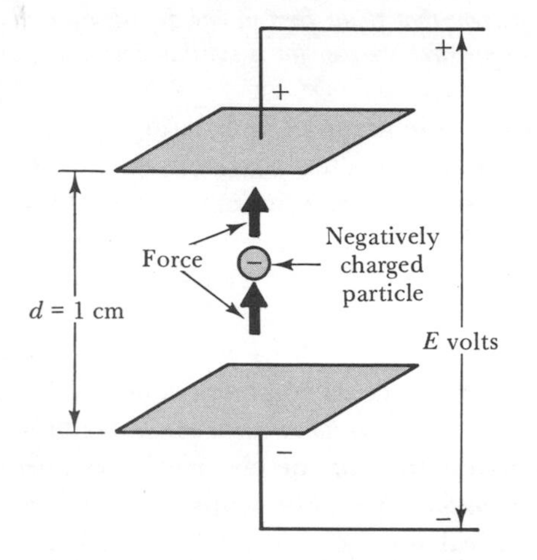

FIG. 3-3. Force on negatively charged particle situated between two metal plates

or

$$\boxed{F = \frac{E}{d} \times Q} \qquad (3\text{-}2)$$

The term E/d is the *electric field strength* and is expressed in *volts per meter*. For a given value of E, the field strength is greatest when d is very small, and least when d is large. Since

$$\text{work} = \text{force} \times \text{distance}$$

$$W = \frac{E}{d} Q d$$

or

$$\boxed{W = EQ} \qquad (3\text{-}3)$$

From the derivation of Eq. (3-3) it can be stated: *When a charge of Q coulombs is moved through an applied emf of E volts, the work done is EQ joules.* Also, from Chapter 1,

$$\text{power} = \frac{\text{work}}{\text{time}}$$

or

$$p = \frac{W}{t}$$

Therefore,

$$\boxed{p=\frac{EQ}{t}} \qquad (3\text{-}4)$$

and since

$$I=\frac{Q}{t} \qquad \text{[Eq. (3-1)]}$$

$$\boxed{P=EI} \qquad (3\text{-}5)$$

Using the SI units of power and current (the *watt* and the *ampere*, respectively), the SI unit of emf is defined in terms of $P=EI$:

The ***volt (v)*** *is defined as the potential difference between two points on a conductor carrying a constant current of 1 ampere when the power dissipated between these points is 1 watt.*

As already noted, the coulomb is the charge carried by 6.24×10^{18} electrons. Therefore, 1 J of work is done when 6.24×10^{18} electrons are moved through a potential difference of 1 V [see Eq. (3-3)]. If only one electron is moved through 1 V, the energy involved is an *electron-volt* (eV).

$$1\ \text{eV}=\frac{1}{6.2\times10^{18}}\ \text{J}$$

The electron-volt is frequently used in the case of the very small energy levels associated with electrons in orbit around the nucleus of an atom.

EXAMPLE 3-2

Two metal plates situated 2 cm apart in a vacuum have a potential difference of 100 V.

a. Calculate the field strength in the space between the plates.

b. Determine the work done in joules when a current of 100 mA flows from one plate to the other for a period of 5 min.

c. Calculate the work done in electron-volts when 100 electrons travel from the negative plate to the positive plate.

SOLUTION

a.

$$\textit{electric field strength}=\frac{E}{d}=\frac{100\ \text{V}}{2\times10^{-2}\ \text{m}}$$

$$\frac{E}{d}=5000\ \text{V/m}$$

b. *From* Eq. (3-3),

$$W = EQ$$

and from Eq. (3-1),

$$Q = I\,t$$

Therefore,

$$W = E\,I\,t$$

$$= 100\text{ V} \times 100\text{ mA} \times (5 \times 60)\text{ s}$$

$$\boldsymbol{W = 3000\text{ J}}$$

c.

$$W = E \times (\text{number of electrons})\text{ eV}$$

$$= 100\text{ V} \times 100\text{ electrons}$$

$$\boldsymbol{W = 10\ 000\text{ eV}}$$

3-4 RESISTANCE AND CONDUCTANCE

When a potential difference is applied to the ends of a metal conductor, the amount of current that can flow depends upon the applied voltage, the type of metal used, the length of the conductor, and the cross-sectional area of the conductor. The type of metal is important, because different metals offer different resistances to current flow. (This is investigated further in Chapter 5.) The cross-sectional area is involved, because, just as a thick water pipe can carry a larger flow of water than a thin pipe, so a thick conductor passes more electric current than a thin one of the same material.

How the conductor length affects current flow is best understood by taking another look at Figure 1-6. It is seen that the resistance of the conductor has a lot to do with the atoms with which the electrons are continually colliding. Obviously, an electron passing through a conductor that is 2 m long has to get past twice as many atoms than would be the case if the conductor were only 1 m long (assuming the same cross-sectional area for each conductor). Since there are likely to be twice as many electron–atom collisions in the longer conductor, it can be said that the 2-m length of conductor has twice the resistance of a 1-m length.

The ***ohm**** *is the unit of resistance, and the symbol used for ohms is* **Ω**, *the Greek capital letter* omega. *The* ***ohm*** *is defined as that resistance which permits a current flow of 1 ampere when a potential difference of 1 volt is applied to the resistance.*

---

*Named after the German physicist George Simon Ohm (1787–1854), whose investigations led to his statement of "Ohm's law of resistance."

As in the case of other units, the ohm is frequently too small or too large for some practical measurements. It is usual to express large resistances as *kilohms* (kΩ) or *megohms* (MΩ). In the case of resistances smaller than 1 Ω the common practice is to express the resistance either using scientific notation or as a decimal quantity; for example, 1/1000 is written 1×10^{-3} Ω or 0.001 Ω.

In some circumstances it is convenient to use the reciprocal of resistance rather than the actual resistance value. The term *conductance (G)* is applied to the reciprocal of resistance. While resistance is a measure of opposition to current flow, conductance is a measure of the ease with which a conductor or resistor will pass current.

The ***siemens**** *(S) is the unit of conductance.*

$$\text{conductance} = \frac{1}{\text{resistance}}$$

$$\boxed{G = \frac{1}{R}} \tag{3-6}$$

where G is in siemens and R is in ohms.

3-5 OHM'S LAW

Consider Figure 3-4(a), in which an emf of E volts is shown producing a current of I amperes through a resistance of R ohms. Note the circuit symbol for a resistor. If two similar resistances are connected one after the other, [Figure 3-4(b)], and each has a potential difference of E volts across it, the current remains I amperes. In this case the total resistance has been doubled, and the total applied emf has also been doubled. If the applied emf is reduced to E volts once again, as shown in Figure 3-4(c), only half the total emf appears across each resistance. Consequently, the current flowing through each resistor is reduced to $I/2$ A. Another way of looking at this is: Because the total resistance is doubled, the current flow is halved.

The discussion above shows that the current flow through a resistance is directly proportional to the applied emf and inversely proportional to the value of the resistance: When the emf is doubled, the current is doubled. When the resistance is doubled, the current is halved. The proportionality of current to emf and resistance is stated in the important fundamental law of electricity known as *Ohm's law.*

---

*Named after Sir William Siemens (1823–1883), a British engineer who was born Karl William von Siemens in Germany.

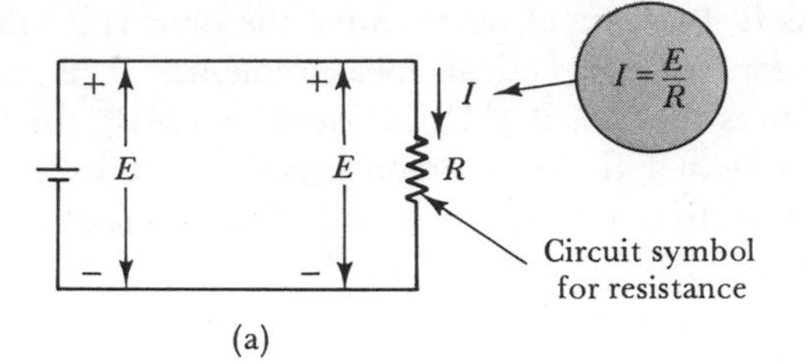

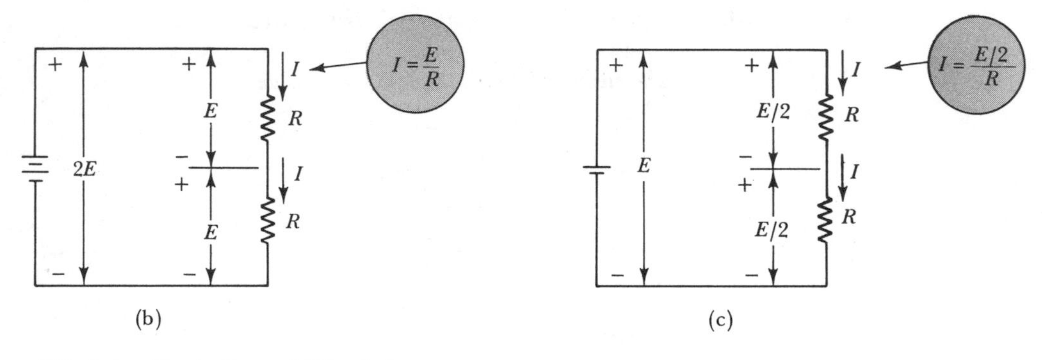

FIG. 3-4. Illustration of Ohm's Law

Ohm's Law:

$$\text{current} = \frac{\text{emf}}{\text{resistance}}$$

or

$$\boxed{I = \frac{E}{R}} \tag{3-7}$$

where *I* is in amperes, *E* is in volts, and *R* is in ohms.

Equation (3-7) can be rewritten

$$\boxed{E = IR} \quad \text{and} \quad \boxed{R = \frac{E}{I}}$$

EXAMPLE 3-3 A resistance of 100 Ω has an emf of 10 V applied across its terminals.

a. Calculate the current that flows through the resistance.

b. If the resistance is changed to 250 Ω, determine the new emf that must be applied to give the same level of current.

c. If the emf is altered to 15 V, calculate the new value of resistance that will give a current of 300 mA.

SOLUTION

a. Eq. (3-7):

$$I=\frac{E}{R}=\frac{10\ \text{V}}{100\ \Omega}$$

$$\boldsymbol{I=0.1\ \text{A}}$$

b.

$$E=IR=0.1\ \text{A}\times 250\ \Omega$$

$$\boldsymbol{E=25\ \text{V}}$$

c.

$$R=\frac{E}{I}=\frac{15\ \text{V}}{300\ \text{mA}}$$

$$\boldsymbol{R=50\ \Omega}$$

EXAMPLE 3-4

Calculate the conductance of a circuit that passes 250 mA when an emf of 50 V is applied to it. Also determine the current that flows if the conductance becomes 0.01 S.

SOLUTION

$$R=\frac{E}{I} \quad \text{and} \quad G=\frac{1}{R}$$

$$G=\frac{I}{E}=\frac{250\ \text{mA}}{50\ \text{V}}$$

$$\boldsymbol{G=5\times 10^{-3}\ \text{S}}$$

$$I=EG=50\ \text{V}\times 0.01\text{S}$$

$$\boldsymbol{I=500\ \text{mA}}$$

3-6 ELECTRICAL POWER AND ENERGY

In Section 3-3 it is shown that

electrical power = potential difference × current

From Eq. (3-5),

$$P=EI$$

From Ohm's law,

$$I = \frac{E}{R} \quad \text{and} \quad E = IR$$

Substituting for I in Eq. (3-5) gives

$$P = E\left(\frac{E}{R}\right)$$

or

$$\boxed{\mathbf{P = \frac{E^2}{R}}} \tag{3-8}$$

and substituting for E in Eq. (3-5) gives

$$P = (IR)I$$

or

$$\boxed{\mathbf{P = I^2R}} \tag{3-9}$$

It is seen that the power dissipated in a resistance can be calculated from a knowledge of any two of I, E, and R.

EXAMPLE 3-5

An electric lamp consumes 100 W of power when a current of 0.9 A flows through its filament. Determine the required emf and calculate the work done in the filament over a period of 15 min.

SOLUTION

From Eq. (3-5),

$$E = \frac{P}{I} = \frac{100\ \text{W}}{0.9\ \text{A}}$$

$$\mathbf{E \cong 111\ V}$$

From Eq. (3-3),

$$W = EQ$$

$$Q = I\,t$$

$$W = E\,I\,t$$

$$= 100\ \text{W} \times (15 \times 60)\ \text{s}$$

$$\mathbf{W = 90\ 000\ J}$$

Power dissipated in a conductor is wasted power, so conductors are usually selected to have the lowest possible resistance. If excessive power is dissipated in a conductor, it may overheat to the extent that the insulation is damaged.

Each electrical device or appliance is designed to dissipate a certain maximum amount of power. Power dissipation in excess of the maximum may damage or destroy the device. To protect equipment and circuits, a *fuse* is usually connected in the circuit. In its simplest form, a fuse is a thin copper or aluminum wire or strip of fixed dimensions. When the circuit current flowing through the fuse exceeds the designed limit, the power dissipation in the fuse causes it to melt and thus interrupt the current flow.

Not all the power supplied to electrical equipment is converted into usable energy. In the case of an electric motor, for example, some of the power is converted into heat, which is wasted power. How good the motor is at converting the electrical power into mechanical power is defined as the *efficiency* of the motor. Thus, if 80% of the electrical power is converted to mechanical power and 20% is wasted in heat, the motor has an efficiency of 80%.

Electrical heaters convert electricity into heat with 100% efficiency. All other electrical appliances (industrial or domestic) have efficiencies that must be taken into consideration when determining their power requirements. The symbol employed for efficiency is η (Greek lowercase letter eta).

$$\text{efficiency} = \frac{\text{power output}}{\text{power input}} \times 100\%$$

or

$$\boxed{\eta = \frac{P_o}{P_i} \times 100\%} \qquad \textbf{(3-10)}$$

When 1 watt of power is delivered to an electrical appliance for a period of 1 hour, the energy consumed is *1 watt-hour* (Wh). The watt-hour is too small for most practical purposes, so the kilowatt-hour (kWh) is normally used as the unit of electrical energy or work. From Eq. (2-5), energy consumed (or work done) is

$$\boxed{W = Pt} \qquad \textbf{(3-11)}$$

EXAMPLE 3-6

An electric heater with a resistance of 6 Ω is supplied from a 115-V source via conductors that have a total resistance of 0.05 Ω. Calculate the power dissipated in the heater and the power wasted in the conductors.

SOLUTION

$$\textit{total resistance} = (\textit{heater resistance}) + (\textit{conductor resistance})$$
$$= 6\ \Omega + 0.05\ \Omega$$
$$\mathbf{R = 6.05\ \Omega}$$

$$I = \frac{E}{R} = \frac{115\ \text{V}}{6.05\ \Omega}$$
$$\mathbf{I = 19\ A}$$

For heater,

$$P = I^2R = (19\ \text{A})^2 \times 6\ \Omega$$
$$\mathbf{P = 2.166\ kW}$$

For conductors,

$$P = (19\ \text{A})^2 \times 0.05\ \Omega$$
$$\mathbf{P = 18.05\ W}$$

EXAMPLE 3-7 A water pump driven by an electric motor moves water through a height of 50 m at the rate of 1000 liters/min. Calculate the electric power required if the motor is 90% efficient.

SOLUTION

1 liter *of water has a mass of* 1 kg.

$$1000\ \text{liters} = 1000\ \text{kg}$$

Eq. (2-1):

$$F = ma = 1000\ \text{kg} \times 9.81\ \text{m/s}^2$$
$$= 9.81 \times 10^3\ \text{N}$$

Eq. (2-2):

$$W = Fd = 9.81 \times 10^3\ \text{N} \times 50\ \text{m}$$
$$= 49.05 \times 10^4\ \text{J}$$

Eq. (2-5):

$$P = \frac{W}{t} = \frac{49.05 \times 10^4\ \text{J}}{60\ \text{s}}$$
$$= 8.175\ \text{kW}$$

Eq. (3-10):

$$\eta = \frac{P_o}{P_i} \times 100\%$$

or

$$P_i = \frac{P_o}{\eta} \times 100\% = \frac{8.175\ \text{kW}}{90\%} \times 100\%$$

$$\mathbf{P_i = 9.08\ kW}$$

EXAMPLE 3-8 If the motor in Example 3-7 runs for 8 h each day, how much energy does it consume in 1 week?

SOLUTION

$$\textit{running time for } 1 \text{ week} = 8\ \text{h} \times 7\ \text{days}$$
$$= 56\ \text{h}$$

$$\textit{energy consumed} = P \times t$$
$$= 9.08\ \text{kW} \times 56\ \text{h}$$
$$\textbf{\textit{energy consumed}} = \mathbf{508.48\ kWh}$$

EXAMPLE 3-9 A 10-*horsepower* motor is to be used to pump water through a height of 50 m, as described in Example 3-7. Determine the rate at which it can pump the water.

SOLUTION

The horsepower *is not an SI unit, but there is a direct relationship between horsepower and watts. From Appendix* 2,

$$1\ \text{horsepower} = 746\ \text{W}$$
$$10\ \text{horsepower} = 10 \times 746\ \text{W}$$
$$= 7460\ \text{W}$$

From Eq. (2-5),

$$\textit{work done/min, } W = Pt = 7460\ \text{W} \times 60\ \text{s}$$
$$= 4.476 \times 10^5\ \text{J}$$

From Eq. (2-2),

$$\textit{force, } F = \frac{W}{d} = \frac{4.476 \times 10^5\ \text{J}}{50\ \text{m}}$$
$$= 8.952 \times 10^3\ \text{N}$$

From Eq. (2-1),

$$\textit{mass, } m=\frac{F}{a}=\frac{8.952\times10^3\ \text{N}}{9.81\ \text{m/s}^2}$$

$$=913\ \text{kg}$$

or

$$\textit{quantity of water}=913\ \text{liters}$$

Therefore, the 10-horsepower *motor can pump the water at a rate of* 913 liters/min.

GLOSSARY OF FORMULAS

Current:

$$\textit{current}=\frac{\text{coulombs}}{\text{time}} \qquad I=\frac{Q}{t}$$

Force:

$$\textit{force}=\frac{\text{emf}\times\text{charge}}{\text{distance}} \qquad F=\frac{E}{d}\times Q$$

Work:

$$\textit{work}=\text{emf}\times\text{charge} \qquad W=EQ$$

Power:

$$\textit{power}=\frac{\text{emf}\times\text{charge}}{\text{time}} \qquad P=\frac{EQ}{t}$$

Power:

$$\textit{power}=\text{emf}\times\text{amperes} \qquad P=EI$$

Conductance:

$$\textit{conductance}=\frac{1}{\text{resistance}} \qquad G=\frac{1}{R}$$

Current:

$$\textit{current}=\frac{\text{emf}}{\text{resistance}} \qquad I=\frac{E}{R}$$

Power:

$$\textit{power}=\frac{(\text{emf})^2}{\text{resistance}} \qquad P=\frac{E^2}{R}$$

Power:

$$\text{power} = (\text{current})^2 \times \text{resistance} \qquad P = I^2R$$

Efficiency:

$$\text{efficiency} = \frac{\text{power output}}{\text{power input}} \times 100\% \qquad \eta = \frac{P_o}{P_i} \times 100\%$$

Energy consumed:

$$\text{energy consumed} = \text{power} \times \text{time} \qquad W = Pt$$

REVIEW QUESTIONS

3-1 State the SI definitions of the ampere and the coulomb and discuss their relationship.

3-2 Explain the difference between conventional current direction and direction of electron flow. Sketch the circuit diagram of a lamp supplied from a battery and identify the directions of conventional current flow and electron flow.

3-3 Define direct current, alternating current, emf, potential difference, and voltage.

3-4 Derive the equation for work in terms of voltage and charge, and the equation for power in terms of voltage and current.

3-5 State the SI definition of the volt and explain the electron-volt.

3-6 Define the SI unit of resistance. State Ohm's law and explain the relationships among current, voltage, and resistance.

PROBLEMS

3-1 Calculate the number of electrons that pass through a resistor in a period of 1.5 h when a current of 500 mA flows. Also determine the amount of electricity used in ampere-hours.

3-2 A glass tube from which the air has been evacuated has a nickel plate at each end. The plates are 10 cm apart and have a potential difference of 2000 V.

a. Calculate the electric field strength in the space between the plates.

b. Calculate the work done in electron-volts when one electron travels from one plate to the other.

c. Determine the work done in joules when a current of 5 mA flows through the tube for a period of 10 min.

3-3 An emf of 100 V is applied across a resistor of 5 kΩ.

a. Calculate the current that flows through the resistor.

b. If the emf is changed to 110 V, determine the new value of resistance required to maintain the original current level.

c. If the resistance is changed to 3.3 kΩ, calculate the emf required to pass 10 mA through the resistor.

3-4 A current of 3 A is to be passed through a 100-Ω resistor.

a. Calculate the required emf.

b. Determine the ampere-hours taken from the supply if the current flows for a period of 7 h.

3-5 Determine the level of current flowing through a 2.2-kΩ resistor when it is connected across a 12-V battery. Also, calculate the quantity of electricity (in coulombs) consumed when the resistor has been connected to the battery for 20 min.

3-6 Determine the work done in joules, when a current of 2 A flows for 45 min through a 12-Ω resistor.

3-7 A field strength of 1000 V/m exists between two metal plates 1 cm apart in a vacuum. Calculate the applied emf and determine the force on an electron passing between the plates.

3-8 Calculate the conductance of a circuit that passes 45 mA of current when 25 V is applied to it. Also, determine the current that flows when the conductance is doubled.

3-9 An electric appliance consumes 1250 W of power when a supply of 115 V is connected to it. Determine the current taken from the supply, and calculate the electric energy consumed by the appliance over a period of 5 h.

3-10 An electric cooking element on the top of a stove takes 5 min to boil a pot containing 1 liter of water. If the element takes 11 A from a 115-V source, calculate the power input and the efficiency.

3-11 If the electric element in Problem 3-10 is used for 4 h/day, calculate the energy consumed in kWh.

4

MEASURING CURRENT, VOLTAGE, AND RESISTANCE

INTRODUCTION

The first experiments performed by students of electricity and electronics involve measurement of current, voltage, and resistance. Consequently, it is necessary to be able to use the basic electrical measuring instruments. This requires knowledge of the correct method of connecting each type of meter and selection of the most suitable range. It is also important to be able to read the meter scale correctly and to be aware of the possible error in the measured quantity as specified by the instrument accuracy.

The use of multifunction instruments involves setting the function switch and the range switch, as well as making correct connections and reading the scale. Digital multimeters are easier to read than deflection instruments. However, correct connection of the meter is still important, and the accuracy of measurement must be carefully assessed.

4-1 CURRENT MEASUREMENT

The instrument used to measure electric current is called an *ammeter* (shortened form of *amp–meter*). Different types of ammeters are usually required for measuring *direct current* (dc) and *alternating current* (ac). However, many *multifunction instruments* can be employed to measure either type of current.

Ignoring its internal components, a *deflection type* ammeter consists basically of a *calibrated scale* over which a pointer is deflected to indicate the measured current, two *terminals* identified as + and −, and a *range*

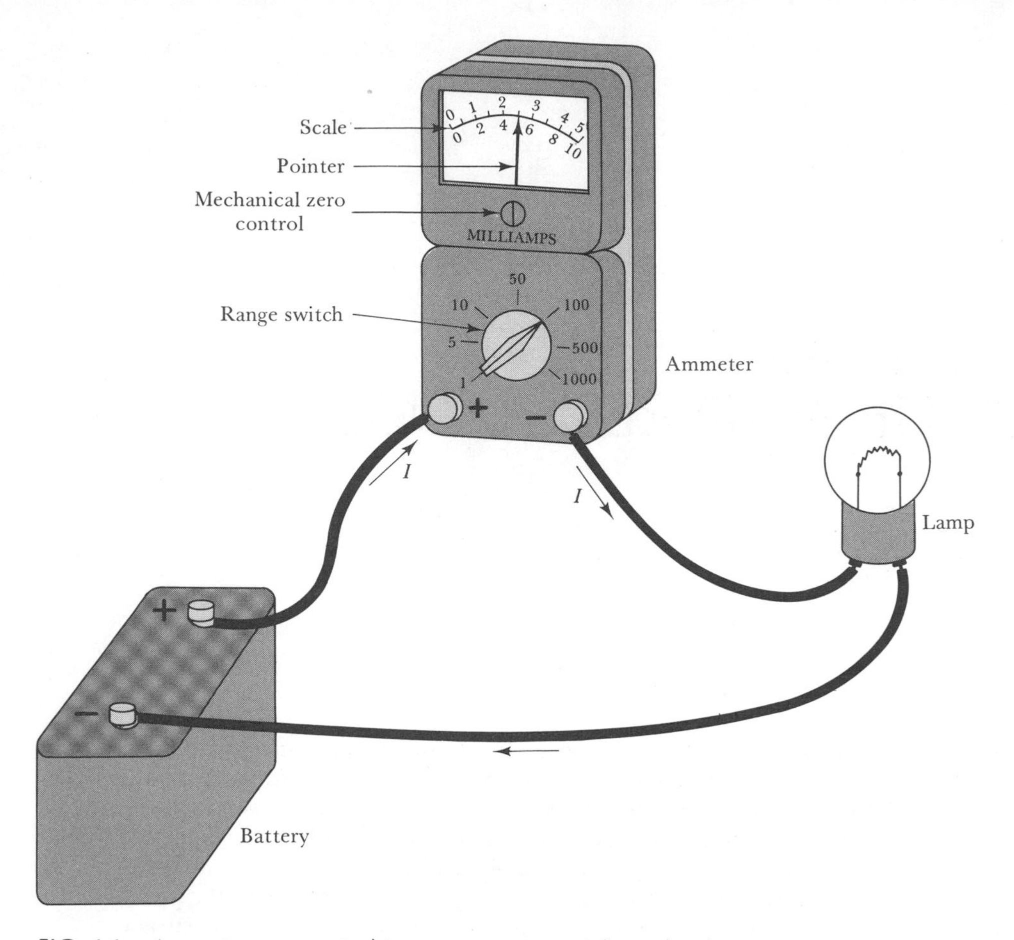

FIG. 4-1. Ammeter connected to measure current in a circuit

switch to select the current range for each particular measurement. Such an instrument is shown in Figure 4-1.

An ammeter must be connected so that the current to be measured flows through the instrument. Consequently, *ammeters are always connected in series with the component in which current is to be measured.* In Figure 4-1, a dc ammeter is shown measuring the current I passing through an electric lamp supplied from a battery. Note that the circuit is arranged so that the current flow (in the conventional direction) is from the battery + terminal, into the + terminal of the ammeter, and out of its − terminal. From the ammeter, the current continues through the lamp back to the battery − terminal.

If the instrument were connected so that the current flow was into the ammeter – terminal, and out of its + terminal, the pointer would deflect to the left of the 0 mark on the scale, instead of moving to the right over the scale. In this case, the ammeter is said to be connected with the wrong *polarity*. For an ac ammeter, the terminal polarity is not important; the pointer normally deflects over the scale no matter what the polarity of the terminals.

Before using any deflection instrument, the pointer position should always be checked to see that it indicates exactly zero when the instrument is still unconnected. If the pointer is not at zero, it should be *zeroed* by adjusting the *mechanical zero control* (see Figure 4-1). When zeroing, the glass cover on the instrument scale should be lightly tapped with a finger to relieve friction within the meter.

The ammeter illustrated in Figure 4-1 has its scale marked in *milliamps* (*mA*), and its range switch set to the *100* position. This means that when the pointer is at the right hand side of the scale (full scale deflection), the current measured is 100 mA. Using the scale calibrations of *0 to 10* (bottom scale), 10 represents 100 mA, and the pointer indication of *5* gives the measured current as 50 mA. When the range switch is set to 50 mA, the *0 to 5* calibration (upper scale) is used, and *5* represents full scale deflection of 50 mA. The *0 to 5* scale is always used on the 5-mA, 50-mA, and 500-mA ranges, while the *0 to 10* scale is used with the 1-mA, 10-mA, 100-mA, and 1000-mA ranges.

When the approximate level of current to be measured is unknown, the instrument should be set to its highest range before being connected into the circuit. (The ammeter illustrated in Figure 4-1 should be set to the 1000-mA range.) The range can then be switched down until the greatest on-scale deflection is obtained.

Figure 4-2 shows the pointer positions that might result when a current of approximately 0.9 mA is measured using five different ranges of the instrument. In Figure 4-2(a) the range is 100 mA, and there is almost no readable movement of the pointer away from the 0 position. On the 50-mA range, Figure 4-2(b), the current can be read only as something less than 2.5 mA. On the 10-mA range illustrated in Figure 4-2(c), the pointer indicates 0.8 mA to 0.9 mA; while on the 5-mA range, Figure 4-2(d), the meter can be read with a little more certainty as approximately 0.9 mA. The 1-mA range indication illustrated in Figure 4-2(e), clearly shows that the current is greater than 0.9 mA. The reading could be reliably stated as 0.93 mA. (The process of estimating the exact position of the pointer between two scale markings is referred to as *interpolation*.) Obviously, *for the greatest measurement accuracy the best range to use is the one that gives the largest deflection not exceeding full scale*.

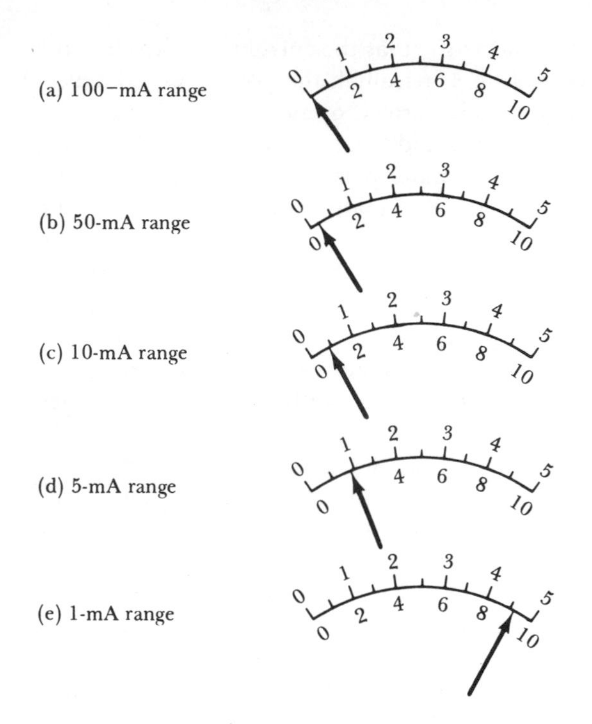

FIG. 4-2. Effect of range selection upon meter reading

EXAMPLE 4-1 Determine the measured current for each of the pointer positions shown in Figure 4-3.

SOLUTION

The range switch is at 500 mA; *therefore use the upper scale and take* 500 mA *as full scale deflection.*

$$\textit{measured current at } A = 125 \text{ mA}$$
$$\textit{measured current at } B = 210 \text{ mA}$$
$$\textit{measured current at } C = 330 \text{ mA}$$
$$\textit{measured current at } D = 450 \text{ mA}$$

The circuit diagram for the battery, ammeter, and lamp shown in Figure 4-1 is drawn in Figure 4-4(a). Note the circuit symbol for the ammeter. In Figure 4-4(b), a circuit diagram is shown for two lamps supplied from a single battery. Two ammeters are included; A_1 measures the current I_1 through L_1, while A_2 indicates the total current I supplied from the battery, where $I = I_1 + I_2$.

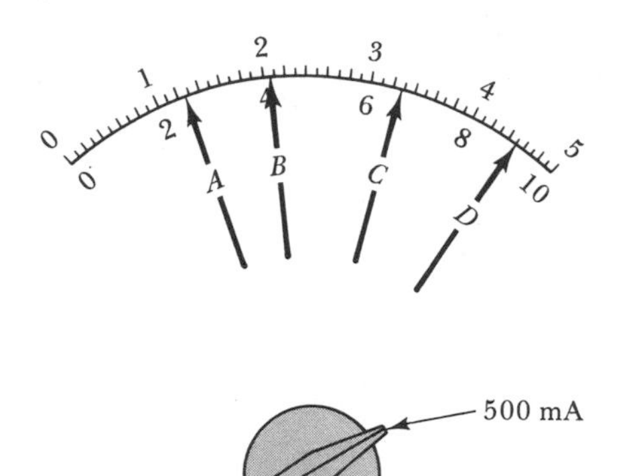

FIG. 4-3. Meter indications for Example 4-1

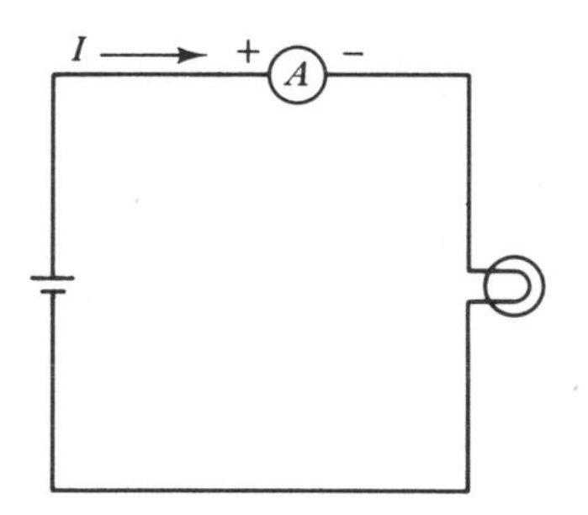

(a) Circuit diagram for Fig. 4-1

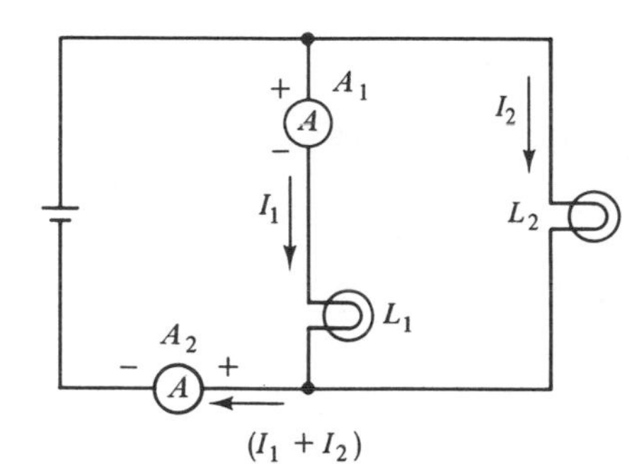

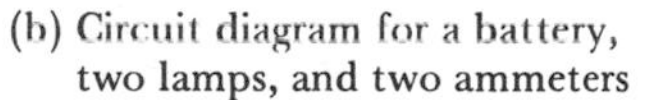

(b) Circuit diagram for a battery, two lamps, and two ammeters

FIG. 4-4. Circuit diagrams involving ammeters

4-2 VOLTAGE MEASUREMENT

The instrument used to measure the potential difference, or electrical pressure, between two points in a circuit is called a *voltmeter*. As in the case of ammeters, different types of instruments are usually employed for measuring *direct voltage* (dc volts) and *alternating voltage* (ac volts). The basic deflection-type dc voltmeter (see Figure 4-5) is very similar in appearance to a dc ammeter. The instrument has two terminals identified as + and −, a range switch, and a pointer which moves over a scale calibrated in volts.

Figure 4-5 shows a dc voltmeter connected to measure the terminal voltage V of a battery that is supplying current to an electric lamp. Note that the voltage being measured appears directly *across*, or in parallel with, the voltmeter terminals. *Voltmeters are always connected in parallel with*

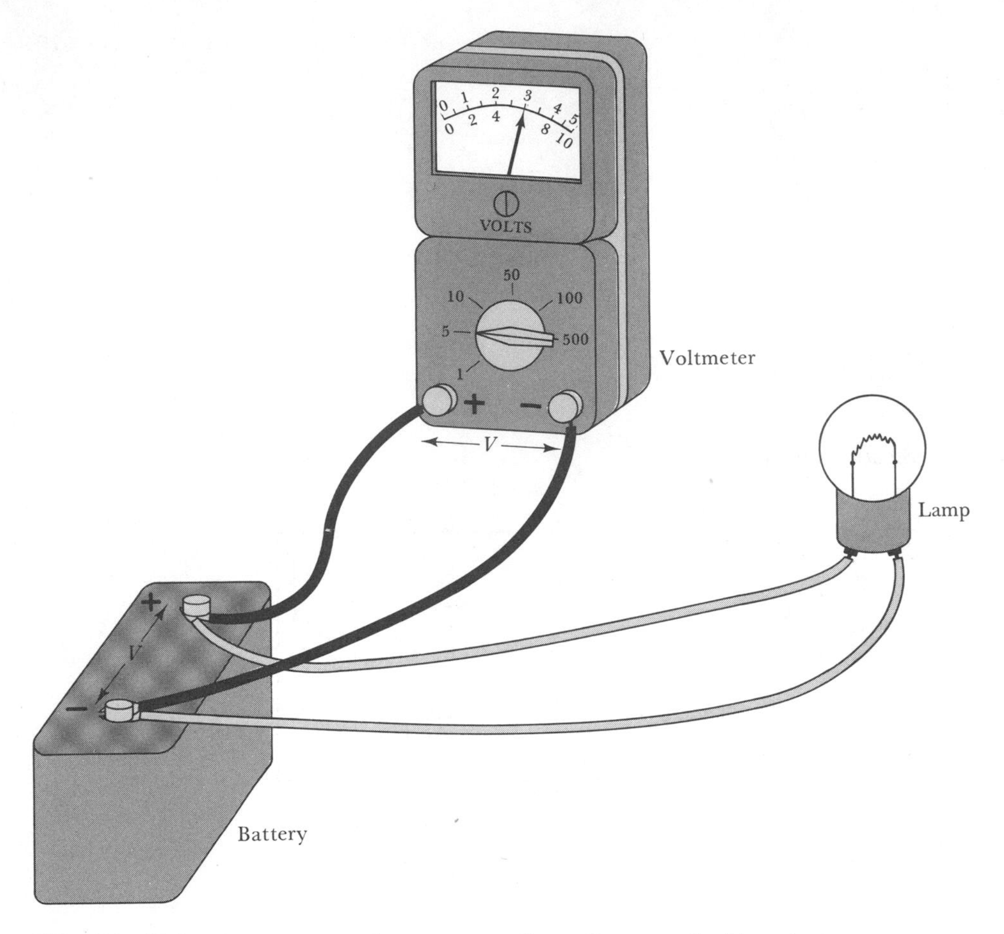

FIG. 4-5. Voltmeter connected to measure the voltage applied to a lamp

the voltage to be measured. Note also that the + terminal of the battery is connected directly to the + terminal of the voltmeter, and that the battery − terminal is connected to the voltmeter − terminal. Thus, as in the case of the dc ammeter, a dc voltmeter must be connected with the correct terminal polarity, in order to obtain a positive (on-scale) deflection. With an ac voltmeter, a positive deflection is obtained no matter what the terminal polarity.

Like all deflection instruments, the voltmeter pointer should be checked before use to ensure that it indicates exactly zero. Where necessary, the mechanical zero control should be adjusted. Also, when the approximate value of the voltage to be measured is not known, the instrument should always be set to its highest range before connecting it into a circuit. The range can then be switched down until the greatest on-scale deflection is obtained.

The instrument shown in Figure 4-5 has six voltage ranges: 1 V, 10 V, and 100 V for use with the *0 to 10* scale (lower scale); and 5 V, 50 V, and 500 V for use with the *0 to 5* scale (upper scale). With the meter on its 5-V range, as illustrated, it is clear that the battery voltage is 3 V. As explained for the ammeter, *the greatest voltmeter accuracy is obtained when the range is selected to give a deflection approaching full scale.*

EXAMPLE 4-2

Figure 4-6 shows two pointer positions on a voltmeter scale which is marked *0 to 3* and *0 to 10*. Determine each indicated voltage:

a. When the range switch is set to 0.1 V.

b. When the range is 30 V.

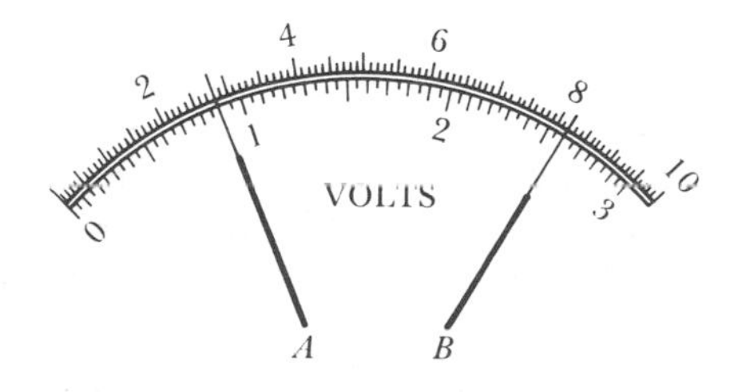

FIG. 4-6. Meter indications for Example 4-2

SOLUTION

a. *0.1 range*

$$\textit{pointer position A indication} = 0.028 \text{ V}$$
$$\textit{pointer position B indication} = 0.082 \text{ V}$$

b. *30 V range*

$$\textit{pointer position A indication} = 9 \text{ V}$$
$$\textit{pointer position B indication} = 26 \text{ V}$$

The circuit diagram for the circuit in Figure 4-5 is shown in Figure 4-7(a). The circuit symbol for the voltmeter is seen to be simply a V with a circle around it. In Figure 4-7(b) a two-lamp circuit is shown, with a battery and two voltmeters. Voltmeter V_1 is seen to be measuring voltage E_1 across lamp L_1, while voltmeter V_2 indicates the total battery voltage $E = E_1 + E_2$.

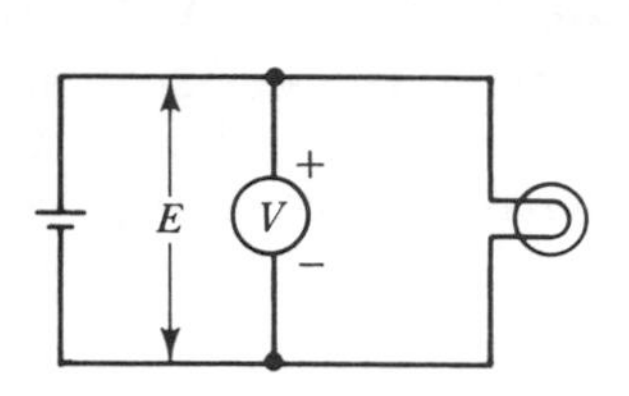

(a) Circuit diagram for Fig. 4-5

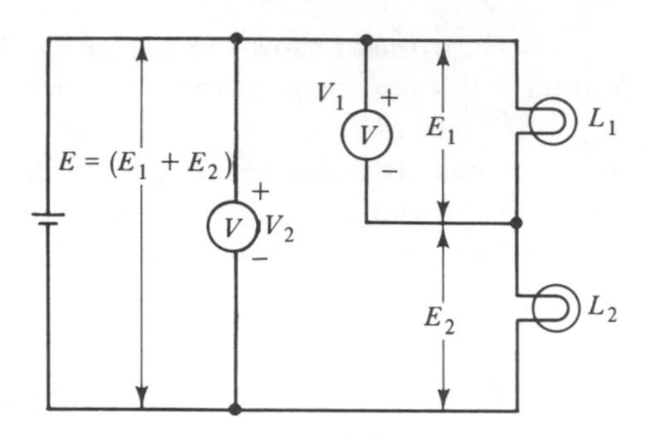

(b) Circuit diagram for a battery, two lamps, and two voltmeters

FIG. 4-7. Circuit diagrams involving voltmeters

4-3 RESISTANCE MEASUREMENT

Resistance is measured by means of an *ohmmeter*, (ohm–meter). The ohmmeter has its own battery or internal power supply for the purpose of passing a current through the resistance to be measured. Thus, *an ohmmeter must never be connected to a circuit in which a supply is switched ON.* Any externally applied voltage may damage the instrument.

The use of an ohmmeter to measure the resistance of a component is illustrated in Figure 4-8. The ohmmeter terminals are connected in parallel with the resistance to be measured [Figure 4-8(a)]. The polarity of the connection is unimportant. In the case of a component that forms part of a circuit, the circuit supply must be OFF, and one terminal of the component must be disconnected from the circuit before the ohmmeter is connected [see Figure 4-8(b)].

Like ammeters and voltmeters, ohmmeters usually have a range switch and a pointer that moves over a calibrated scale. However, as Figure 4-8 shows, the ohmmeter scale is quite different from that on other instruments. The *0 ohms* position is seen to be on the right-hand side of the scale instead of the left. The left-hand end of the scale is marked *INF*, which means *infinity*—that is, an extremely large value. Note that the length of the scale representing a 10 Ω change from 0 Ω to 10 Ω is greater than the scale length from 10 Ω to 20 Ω, which also represents a 10 Ω change in resistance. Moving from right to left, the ohmmeter scale becomes more and more cramped, and this is said to be *nonlinear*. The ammeter and voltmeter scales already discussed are termed *linear scales* because equal lengths on each scale represent equal current or voltage changes. However, some ammeters and voltmeters have nonlinear scales.

The ohmmeter has a *ZERO OHMS* control, as well as a mechanical zero control (see Figure 4-8). As with other deflection instruments, the

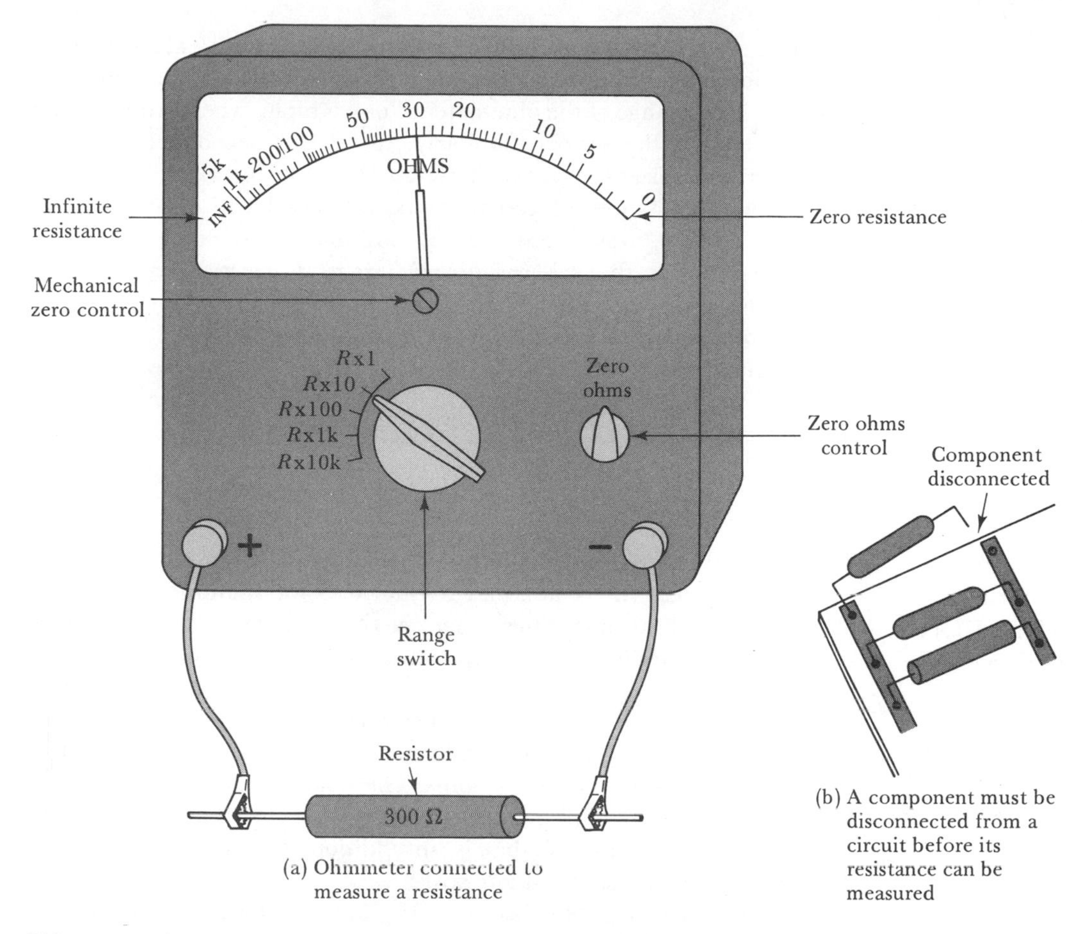

(a) Ohmmeter connected to measure a resistance

(b) A component must be disconnected from a circuit before its resistance can be measured

FIG. 4-8. Ohmmeter

mechanical zero control should be used to set the pointer exactly at the left-hand end of the scale when the instrument is unconnected. Before measuring a resistance, the ohmmeter terminals should be connected together (that is, *short-circuited*), and the *ZERO OHMS* control should be adjusted until the pointer indicates exactly 0 Ω on the right-hand side of the scale. This procedure should also be repeated whenever the range of the ohmmeter is altered. The ohmmeter terminals should never be left in a short-circuited condition because this would discharge the internal battery.

When the ohmmeter range switch is at $R \times 1$, the pointer indicates the measured resistance value directly in ohms. With the range switch at $R \times 10$ and the pointer indicating 30, as illustrated in Figure 4-8(a), the measured resistance is $(30 \times 10)\ \Omega = 300\ \Omega$. On this range, the small length of scale from 200 to INF represents all resistance values above 2 kΩ. Obviously, resistances greater than 2 kΩ cannot be measured accurately on this range of the ohmmeter. (For example, where would 5 MΩ be indicated on the scale?) Similarly, the *0 to 5* section of the scale represents all resistance values below 50 Ω, when the range is $R \times 10$. Again, these resistance values cannot be measured with any degree of accuracy. (Where would the 0.05 Ω position be found?) From this discussion it is seen that *the ohmmeter measures resistances most accurately when its pointer position is around center scale.*

The ohmmeter described above is normally not available as a single instrument. Instead, it is usually found as part of a *multimeter*, which is described in the following section. Ohmmeters are not shown as part of a circuit diagram because they are never left connected permanently in a circuit.

4-4 THE DEFLECTION MULTIMETER

The *multimeter* is a multifunction instrument capable of measuring voltage (dc or ac), current (dc or ac), and resistance. Some multimeters can also measure additional quantities, such as inductance and capacitance. Another name applied to the instrument is *VOM*, which is short for *volt–ohm–milliameter*.

A typical deflection-type multimeter is shown in Figure 4-9. The instrument illustrated has three scales, one of which (the upper scale) is used only for resistance measurements. The other two scales may be used for either voltage or current measurements. A *VOM* that is capable of measuring other quantities (that is, in addition to voltage, current, and resistance) may have several more scales.

Two terminals identified as + and − are used for all measurements. The terminal polarity is important only when measuring dc quantities. Some multimeters have additional terminals for use when unusually high voltages or currents are to be measured.

The main controls on the multimeter shown in Figure 4-9 are a *function switch* and a *range switch*. The function switch sets the instrument to the type of quantity which is to be measured: *dc mA* (direct current in milliamps), *dc V* (direct voltage), *R* (resistance), *ac V* (alternating voltage), or *ac mA* (alternating current in milliamps). The range switch selects the instrument range to be used. Moving clockwise around the range switch from the top, the *0.1* through *1000* positions are current and voltage ranges, both ac and dc. If the function switch were at *dc V* and the range switch at *250*, the meter could measure up to 250 V dc at full-scale deflection. When the function switch is set to *ac mA* and the range switch to *1000*, alternating currents up to a level of 1 A (1000 mA) can be measured.

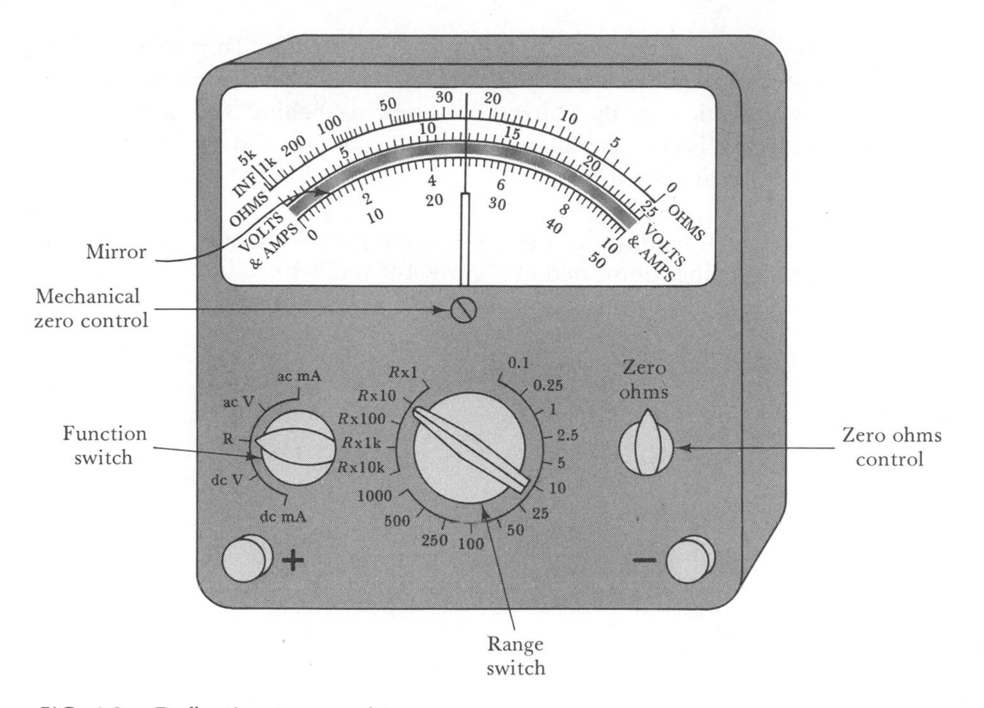

FIG. 4-9. Deflection-type multimeter

Resistance measurements are made using the $R \times 10K$ to $R \times 1$ range positions. With the function switch set at R and the range switch at $R \times 1$, the resistance scale reads directly in ohms. If the function switch is at $R \times 1K$, all the resistance values indicated on the scale must be multiplied by 1 kΩ. For example, a pointer indication of *30* would be read as $(30 \times 1\ k\Omega) = 30\ k\Omega$. The *ZERO OHMS* control is used as described in Section 4-3.

Some multimeters are equipped with an overload protection circuit. When the current passing through the instrument exceeds a maximum safe level, internal connections are automatically open-circuited, and a *cut-out* button pops up. The cut-out device is reset by pushing the button down again.

A *REVERSE* button is also sometimes provided on a multimeter. This can be used when a dc current or voltage is not connected with the correct polarity. Pushing (and holding) the *REVERSE* button causes the meter polarity to be reversed internally, so that the pointer deflects to the right of *0*.

The multimeter illustrated in Figure 4-9 has a fine *knife-edge* pointer for accurate scale readings. A mirror is also provided alongside the scale; the purpose of this is to eliminate the scale-reading error that occurs when the viewer's eye is not directly in front of the pointer. To take a reading, the operator closes one eye and then moves his head until he sees the pointer reflection in the mirror disappearing behind the pointer. The operator's eye is now directly in front of the pointer, and the meter can be read with great accuracy.

EXAMPLE 4-3 Determine the readings of the multimeter for the pointer positions and the switch settings illustrated in Figures 4-9 and 4-10.

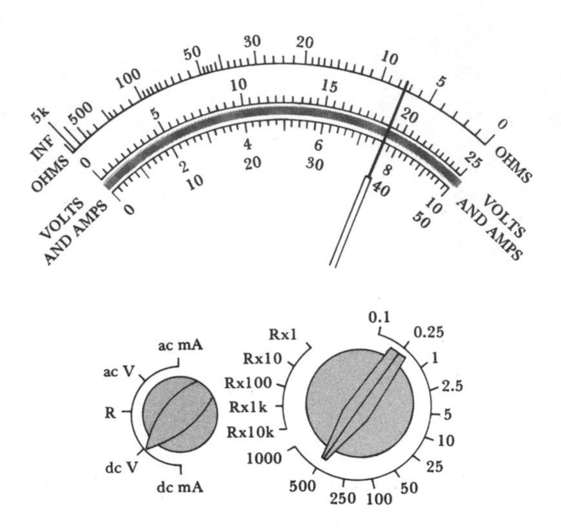

FIG. 4-10. Meter indication for Example 4-3

SOLUTION

In Figure 4-9 the function switch is at R and the range switch is at $R \times 10$. Therefore, read the pointer position on the resistance scale and multiply by 10.

$$\textit{meter reading} = 25\ \Omega \times 10 = 250\ \Omega$$

In Figure 4-10,

$$\textit{full-scale deflection} = 500\ \text{V dc}$$

Using the bottom scale,

$$\textit{the meter reading} = 380\ \text{V dc}$$

4-5 THE DIGITAL MULTIMETER

All the instruments previously discussed have been deflection-type, or *analog*, meters. The multimeter illustrated in Figure 4-11(a) is known as a *digital instrument*, simply because the pointer and scales are replaced by electronic *digital display devices*.

Like the deflection-type multimeter, the digital multimeter shown has a *function switch*, a *range switch*, and two terminals. It also has an *ON/OFF* switch because this instrument either has an external supply of electricity or is battery operated.

The quantity measured is displayed in the form of a four-digit number (sometimes more than four) with a properly placed decimal point. When dc quantities are measured, the polarity is identified by means of a + or − sign displayed to the left of the number.

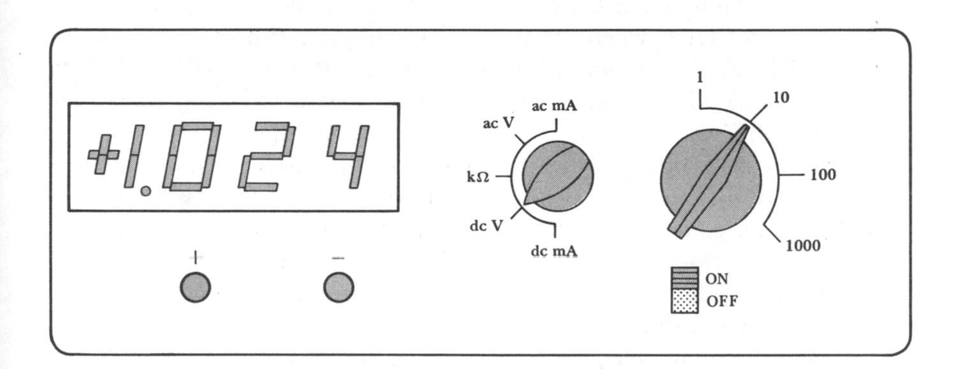

(a) Front panel of a digital instrument

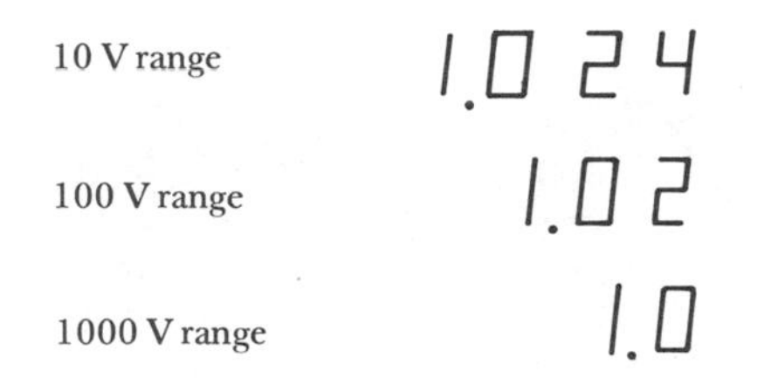

(b) 1.024 V measured on different ranges

FIG. 4-11. Digital multimeter

Figure 4-11(b) shows the displays that would be obtained when measuring 1.024 V on each of several ranges. As in the case of the deflection instrument, the most accurate measurements are achieved when the lowest possible range is used. Some digital instruments do not have a range switch, but employ *automatic ranging circuits* that automatically change the instrument range to give the best display for a given input. Push-button type switches for range and function selection are also employed on many digital instruments.

Digital instruments are obviously easier to read than deflection-type meters. They are also usually more accurate and less likely to be damaged by overload conditions.

4-6 INSTRUMENT ACCURACY

The accuracy of all deflection instruments is specified by the manufacturer as a percentage of full-scale deflection. For example, a voltmeter with an FSD of 100 V might have its accuracy stated as ±2%. In this case, the maximum possible error at all points on the scale is ±2% of 100 V, or ±2 V. Thus, when the pointer is indicating exactly 100 V, as in Figure 4-12(a), the measured voltage is correctly stated as 100 V±2 V, or 98 V to 102 V. The actual measured voltage is seen to be anything from 98 V to 102 V.

When the pointer of this voltmeter indicates 50 V, as illustrated in Figure 4-12(b), the actual measured voltage must be taken as 50 V±2% of full scale, or 50 V±2 V. This means that the measured voltage is somewhere between 48 V and 52 V. The ±2 V error is ±4% of 50 V. Consequently, the measurement error increases as the pointer moves further away from full-scale deflection.

Deflection-type multimeters typically have an accuracy of ±2% for voltage and current measurements, although more accurate instruments are available. As illustrated in Figure 4-12(c), the error becomes considerably greater when the instrument is used for resistance measurements. It

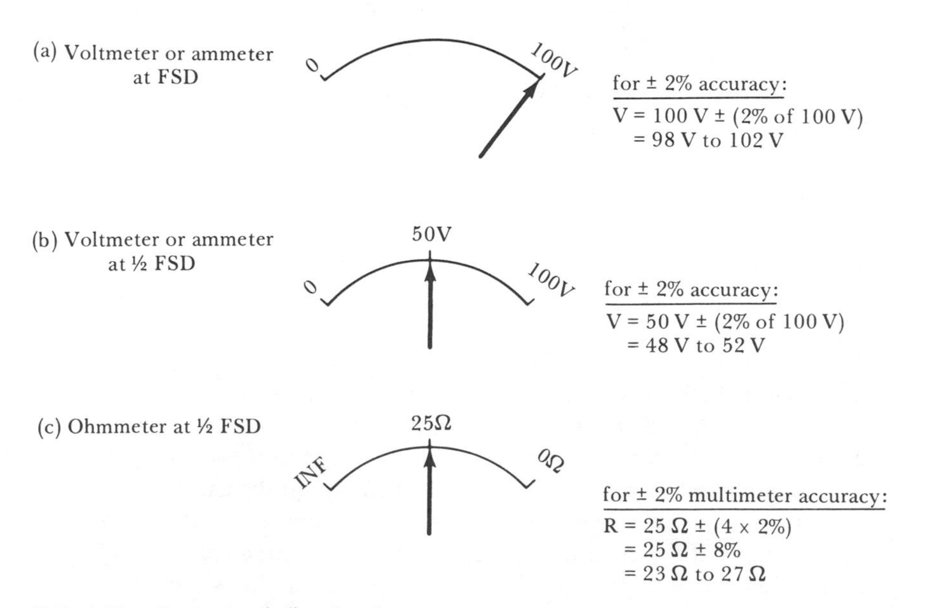

FIG. 4-12. Errors in deflection instrument

has already been explained in Section 4-3 that ohmmeter measurements are most accurate when the pointer is near mid-scale. However, even at mid-scale, the resistance measurement error is typically $\pm(4x\%$ of mid-scale resistance), where x is the multimeter specified accuracy. Thus, for a multimeter with an accuracy of $\pm 2\%$, the mid-scale accuracy of the ohmmeter function is $\pm$(8% of mid-scale resistance). When measurements are made away from mid-scale, the error is increased.

The accuracy of a digital multimeter might be stated as *±0.5% Rdg ±1d*. This means ± *(0.5% of the reading)* ± *1 digit*, where the *1 digit* refers to the right hand numeral of the display. For example, a display of 100.0 V would be accurate to:

$$\pm(0.5\% \text{ of } 100.0 \text{ V}) \pm 000.1 \text{ V}$$
$$= \pm(0.5 \text{ V}) \pm 000.1 \text{ V}$$
$$= \pm 0.6 \text{ V}$$

Therefore, the measured voltage could be stated as:

$$100 \text{ V} \pm 0.6 \text{ V}$$
$$= 99.4 \text{ V to } 100.6 \text{ V}$$

EXAMPLE 4-4

A digital multimeter indicates 5.432 V. If the instrument accuracy is specified as ±0.1% Rdg ± 1d, calculate the upper and lower limits of the measured voltage.

SOLUTION

$$error = \pm(0.1\% \text{ of } 5.432 \text{ V}) \pm 0.001 \text{ V}$$
$$\cong \pm 0.005 \text{ V} \pm 0.001 \text{ V}$$
$$\cong \pm 0.006 \text{ V}$$

The measured voltage becomes:

$$5.432 \text{ V} \pm 0.006 \text{ V}$$
$$= 5.426 \text{ V to } 5.438 \text{ V}$$

REVIEW QUESTIONS

4-1 How should an ammeter be connected in a circuit? Explain briefly. Discuss the terminal polarity for dc and ac ammeters, and the effect of incorrect connection in each case.

4-2 Explain the purpose and use of the mechanical zero control on a deflection instrument.

4-3 An ammeter has the following ranges: 1 mA, 2.5 mA, 50 mA, and 100 mA. Which range should be selected to give the most accurate measurement of:

a. 55 mA,

b. 3.9 mA,

c. 0.9 mA,

d. 19 mA.

What would be the effects of selecting a range which is too high and a range which is too low?

4-4 Sketch a circuit diagram showing an ammeter measuring the current supplied from a battery to three series-connected electric lamps. Identify the polarity of the instrument, and show the (conventional) direction of current flow.

4-5 Briefly explain how a voltmeter should be connected into a circuit. Discuss the importance of terminal polarity for ac and dc voltmeters, and state the effect of incorrect connection in each case.

4-6 A voltmeter has the following ranges: 0.3 V, 1 V, 3 V, 10 V, and 30 V. The meter is to be used to measure the terminal voltage of a battery, which is known to be around 1.5 V. What range should be selected? What would be the effect of selecting each of the other ranges?

4-7 Sketch a circuit diagram showing a voltmeter measuring the terminal voltage of a battery that is supplying current to two electric lamps in parallel. Show the current direction in the circuit and the terminal polarity of the instrument.

4-8 Discuss the precautions necessary when using an ohmmeter to measure the resistance of a component connected in a circuit. How are the mechanical zero control and the *ZERO OHMS* control used?

4-9 What is meant by linear and nonlinear meter scales? At what part of a scale are the most accurate measurements made:

a. On a voltmeter?

b. On an ohmmeter?

c. On an ammeter?

Explain briefly.

4-10 Sketch the scale of an ohmmeter suitable for measuring resistance values around 100 Ω. Show the zero, infinity, and 100-Ω positions.

4-11 List the quantities that can be measured on a typical multimeter. Also list the controls usually available and the purpose of each, and explain the use of a mirror scale.

4-12 Explain how a digital multimeter differs from a deflection-type instrument.

4-13 Explain the accuracy specification of a measuring instrument:

a. For a deflection instrument specified as accurate to ±1%.

b. For a digital instrument with an accuracy of ±0.3%Rdg±1d.

PROBLEMS

4-1 Determine the measured current for each of the pointer positions shown in Figure 4-13 when the range switch is at 1000 mA.

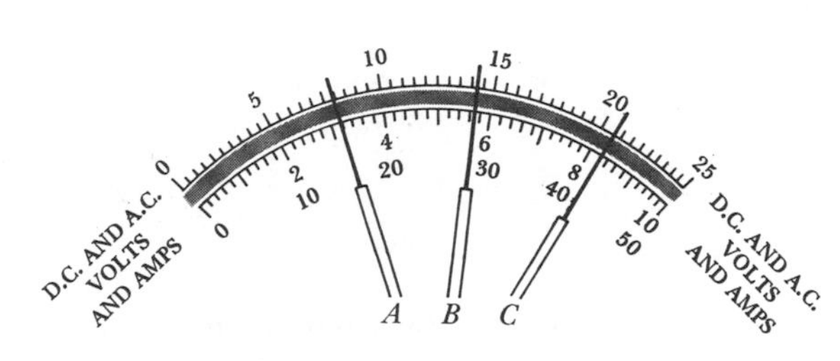

FIG. 4-13.

4-2 If the range of the instrument illustrated in Figure 4-13 were 250 mA, what would be the current indicated by each pointer position shown?

4-3 For each pointer position shown in Figure 4-14, write the measured voltage:

a. When the voltmeter range is 100 V.

b. When the range is 300 mV.

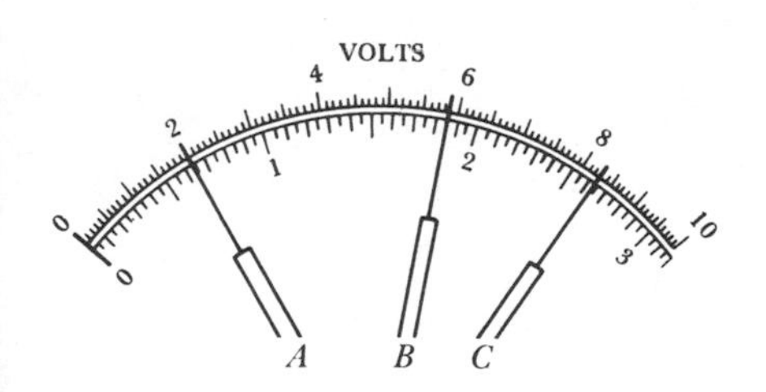

FIG. 4-14.

4-4 Figure 4-15 shows an ohmmeter scale and range switch. Which ranges should be used when measuring resistances that have the following approximate values:

a. 560 Ω?

b. 12 kΩ?

c. 2 MΩ?

d. 10 Ω.

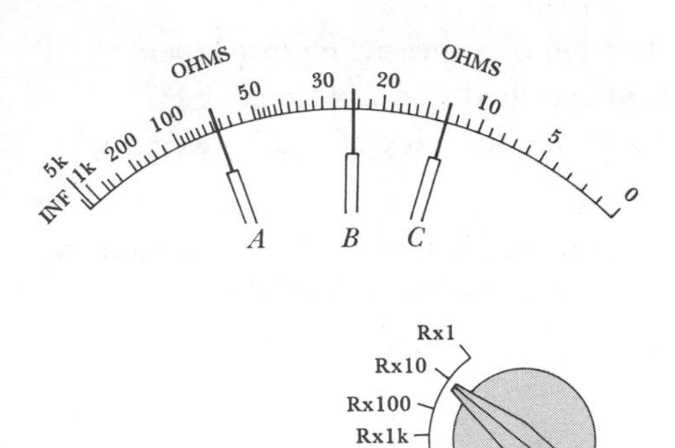

FIG. 4-15.

4-5 What are the resistance values indicated by each of the pointer positions shown in Figure 4-15 when the instrument range is:

a. $R \times 10$.

b. $R \times 10\text{K}$.

4-6 Determine the readings of the multimeter for the pointer positions and switch settings in Figure 4-16.

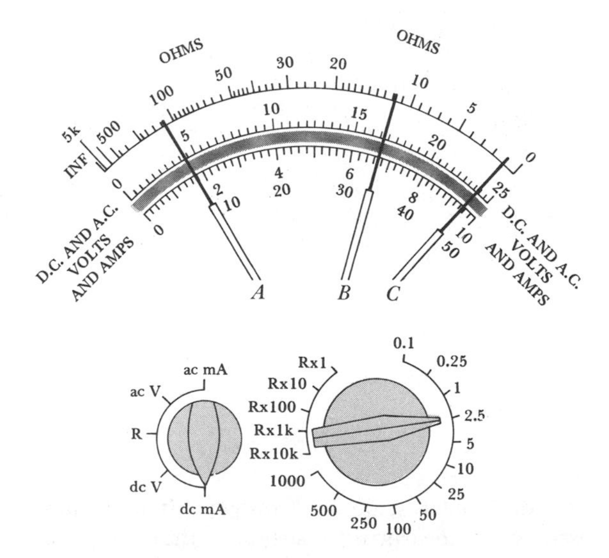

FIG. 4-16.

4-7 Calculate the maximum and minimum values of the quantities indicated in Figure 4-16 if the multimeter accuracy is specified as $\pm 2\%$.

4-8 If the resistance values determined in Problem 4-5 were made on a multimeter with an accuracy of $\pm 3\%$, calculate the maximum and minimum resistances for pointer position *B*.

4-9 A digital instrument has a specified accuracy of $\pm 0.2\%$ Rdg $\pm$ 1d. Calculate the maximum and minimum values of the voltage when the instrument indicates:

a. 9.418 V.

b. 1.290 V.

c. 3.067 V.

4-10 The accuracy of the resistance ranges on a digital multimeter are specified as $\pm 0.6\%$ Rdg $\pm$ 1d. Determine the upper and lower limits of the measured resistance when the instrument indication is:

a. 4.663 kΩ.

b. 560.1 Ω.

c. 27.77 Ω.

5

CONDUCTORS, INSULATORS, AND RESISTORS

INTRODUCTION

Conductors carry electric current. Insulators protect conductors, and protect people from conductors. Whether a material is a conductor or an insulator depends upon its atoms, and upon the relationship of each atom to its surrounding atoms.

Insulators may break down if subjected to excessive voltages. Similarly, conductors may be destroyed if too much current is passed through them. The best conductors have the lowest *resistivity*, that is, resistance per cubic meter. The best insulators have the highest breakdown voltage.

The resistance of any conductor can be calculated from a knowledge of its cross-sectional area, its length, and the resistivity of the material. Once the resistance is known, the voltage drop along a conductor and the power dissipated in it can be determined for any given current.

A device constructed to have a certain value of resistance is known as a *resistor*. Large quantities of resistors are used in electronic circuits for providing voltage drops, current limiting, and the like. These are manufactured in standard resistance values and various ranges of accuracy and power dissipation capability. Small resistors use a *color code* to identify the resistance value and accuracy.

The *power rating* of a resistor is important because it limits the maximum voltage that may be applied and the maximum current that may flow through the component. How a resistor value varies with temperature is defined by its *temperature coefficient*.

5-1 ATOMIC BONDING

Consider the planetary atom again as described in Section 1-2, and recall that all atoms consist of a central nucleus surrounded by orbiting electrons. It has been found that electrons can occupy only certain orbital rings or *shells* at fixed distances from the nucleus, and that each shell can contain a particular number of electrons. The force that holds each electron in orbit around its nucleus depends upon the distance of the electron from the nucleus. Those electrons which are closest to the nucleus are securely held in orbit. Those that are in the outermost orbit are less tightly bound to the atom.

The atoms of two important electrical materials are illustrated by the two-dimensional diagrams in Figure 5-1. Since it is the electrons in the outer shell (or *valence shell*) that determine the electrical characteristics of

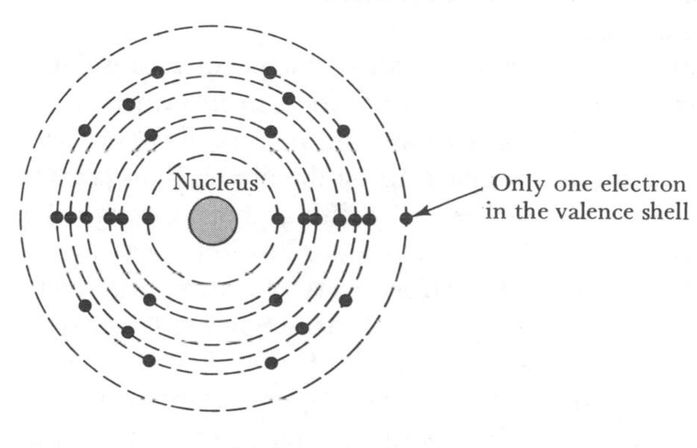

(a) Copper atom

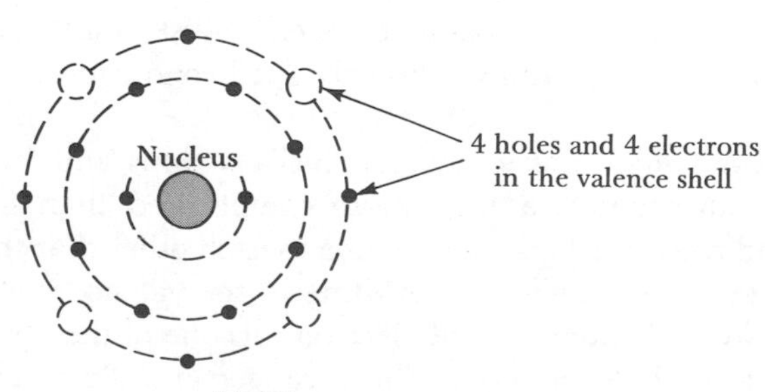

(b) Silicon atom

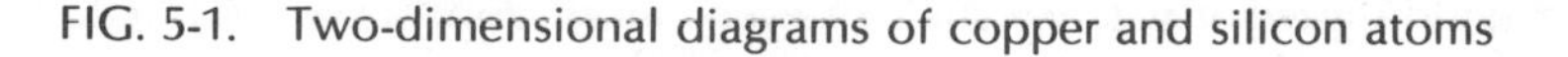
FIG. 5-1. Two-dimensional diagrams of copper and silicon atoms

an atom, only the outer shells need be examined. Referring to Figure 5-1(a), it is seen that the outer shell of the copper atom contains only one electron. This electron is so weakly attached to the nucleus that it easily escapes and drifts off through the spaces between the atoms that make up a piece of copper.

Now look at the silicon atom illustrated in Figure 5-1(b). Observe that the outer shell of the silicon atom contains four electrons but has the facility to contain eight electrons. Thus, the valence shell of silicon is said to have four electrons and four holes. A *hole* is defined simply as the absence of an electron in a shell where one could exist. The outer-shell electrons are more strongly attached to the silicon atom than in the case of copper. However, the electrons can be separated from the atom by the application of small amounts of energy. Thus, the silicon atom can either accept electrons into the holes in the valence shell, or give up electrons to drift about within the material.

Whether a material is a conductor, a semiconductor, or an insulator depends largely upon what happens to the outer-shell electrons when the atoms bond themselves together to form a solid. In the case of copper, the easily detached valence electrons are given up by the atoms. This creates a great mass of free electrons (or *electron gas*) drifting about through the spaces between the copper atoms. Since each atom has lost a (negative) electron, it becomes a *positive ion*. The electron gas is, of course, negatively charged; consequently, an electrostatic force of attraction exists between the positive ions and the electron gas. This is the *bonding force* that holds the material together in a solid. In the case of copper and other metals, the bonding force is termed *metallic bonding*, or sometimes *electron gas bonding*. This type of bonding is illustrated in Figure 5-2(a).

Since the free electrons in the electron gas can be given motion by the application of an electric field, current flow is easily achieved. Thus, it is obvious that the electron gas makes copper an excellent conductor of electricity. In general, all metals are good conductors, but some are better than others. Silver is the very best conductor (i.e., lowest resistance) followed closely by copper. Gold is also a relatively good conductor, as is aluminum.

In the case of silicon, which has four outer-shell electrons and four holes, the bonding arrangement is a little more complicated than for copper. Atoms in a solid piece of silicon are so close to each other that the outer-shell electrons behave as if they were orbiting in the valence shells of two atoms. In this way each valence-shell electron fills one of the holes in the valence shell of a neighboring atom. This arrangement, illustrated in Figure 5-2(b), forms a bonding force known as *covalent bonding*. In covalent bonding every valence shell of every atom appears to be filled, and consequently there are no holes and no free electrons drifting about within the material.

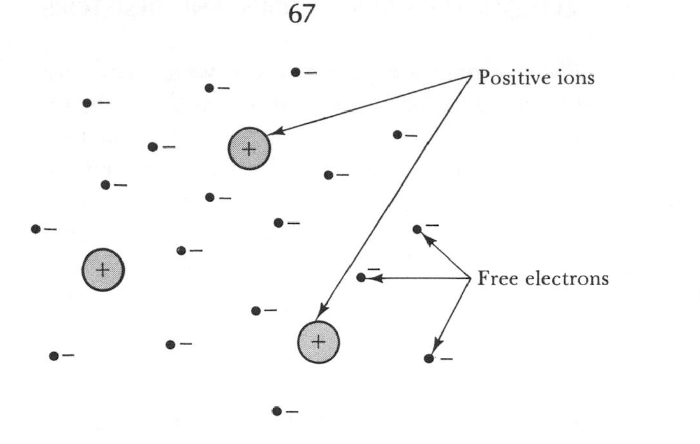

(a) Metallic bonding

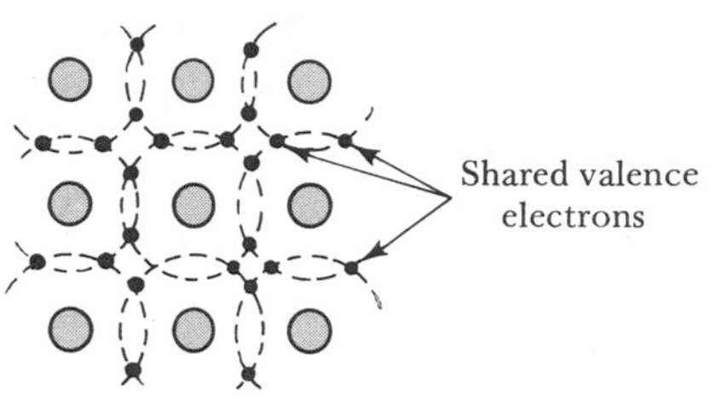

(b) Covalent bonding

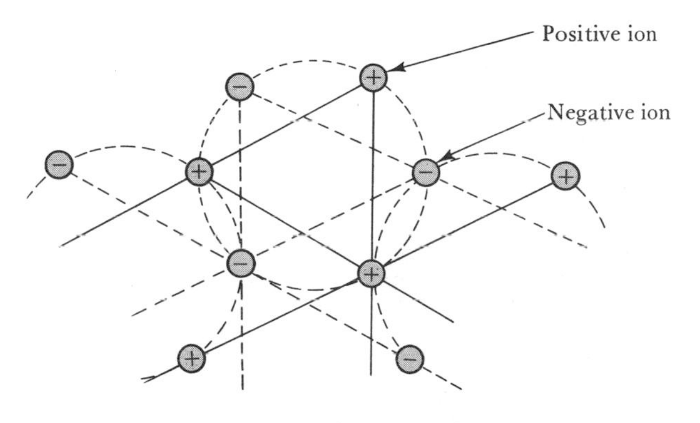

(c) Ionic bonding

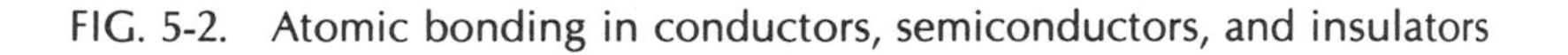

FIG. 5-2. Atomic bonding in conductors, semiconductors, and insulators

Because there are no free electrons to permit conduction of an electric current, silicon should be a very poor conductor of electricity. At very low temperatures, silicon does, in fact, behave like an insulator. However, even at normal room temperature, some of the silicon electrons have been given enough thermal energy to escape from their atoms. Consequently, small electric currents can be made to flow through the silicon, and so it is said to behave as a *semiconductor*. For the manufacture of semiconductor devices, the resistance of silicon is reduced by the addition of quantities of a different kind of atom. This is a process known as *doping*.

In some insulating materials, notably rubber and plastics, the bonding process is also covalent. The valence electrons in these bonds are very strongly attached to their atoms, so the possibility of current flow is virtually zero. In other types of insulating materials, some atoms have parted with outer-shell electrons, but these have been accepted into the orbit of other atoms. Thus, the atoms are *ionized*; those which gave up electrons have become *positive ions*, and those which accepted the electrons become *negative ions*. This creates an electrostatic bonding force between the atoms, termed *ionic bonding*. The situation is illustrated in Figure 5-2(c), which shows how the negative and positive ions are arranged together in groups. Ionic bonding is found in such materials as glass and porcelain. Because there are virtually no free electrons, no current can flow, and the material is an insulator.

5-2 INSULATORS

Figure 5-3 shows some typical arrangements of conductors and *insulators*. Electric cable usually consists of conducting copper wire surrounded by an insulating sheath of rubber or plastic [Figure 5-3(a)]. Sometimes there

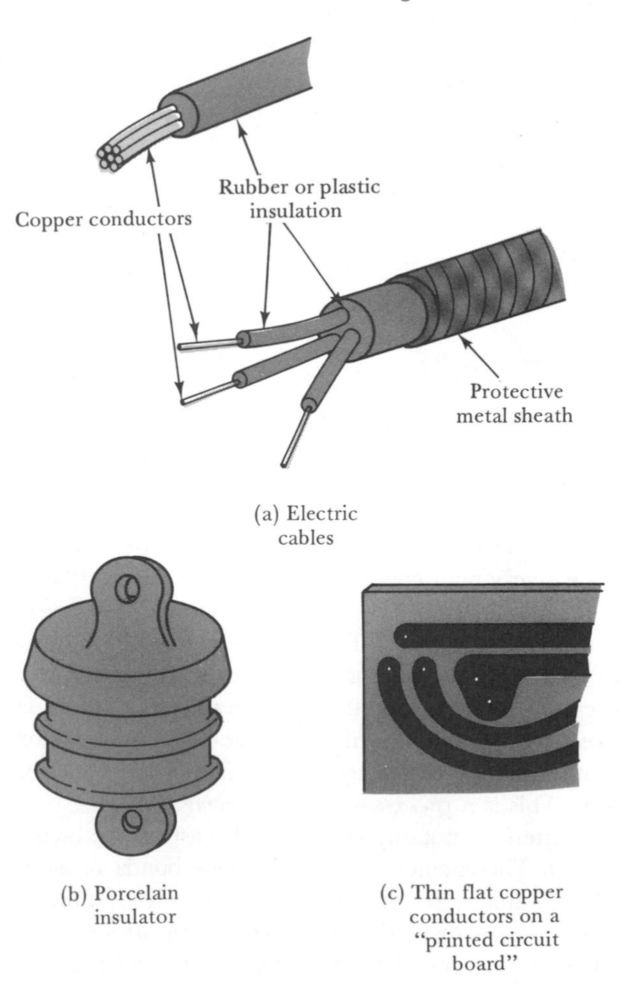

(a) Electric cables

(b) Porcelain insulator

(c) Thin flat copper conductors on a "printed circuit board"

FIG. 5-3. Conductors and insulators

is more than one conductor, and these are, of course, individually insulated. Frequently, a protective metal sheath covers the cable exterior. Electric cables which are suspended from poles must be insulated from the poles, even in the wettest weather conditions. Porcelain insulators, as illustrated in Figure 5-3(b), are usually employed for this purpose.

Even in the best of insulating materials there are some free electrons drifting about between the atoms. Therefore, a very small electric current can flow through an insulator. In normal circumstances this current is so small that it is absolutely negligible. A current can also flow along the surface of an insulator, especially if the insulator is dirty or wet. This is particularly evident with porcelain insulators of high-voltage overhead cables during a rain storm.

If a sufficiently high potential difference is applied across an insulator, electrons can be *pulled out* of the atoms, and substantial current flow may occur. This is known as *insulator breakdown*, and it normally results in the destruction of the insulator. In some circumstances insulator breakdown may produce a fire or other dangers. To help avoid breakdown, all insulating materials (e.g., sheaths of electric cables, etc.) are rated according to the maximum voltage that may be safely applied. Table 5-1 lists typical *breakdown voltages* for various insulating materials. It is seen that air has the lowest breakdown voltage, at 30 kV/cm, while mica has the highest, at 2000 kV/cm. The figures given vary to some extent with the actual thickness of the insulating material.

TABLE 5-1. Typical Breakdown Voltages of Some Insulating Materials

MATERIAL	BREAKDOWN VOLTAGE (kV/cm)
Air	30
Porcelain	70
Rubber	270
Glass	1200
Mica	2000

5-3 CONDUCTORS

The function of a *conductor* is to conduct current from one point to another in an electric circuit. As already discussed, electric cables usually consist of copper conductors sheathed with rubber or plastic insulating material. Cables that have to carry large currents must have relatively thick conductors. Where very small currents are involved, the conductor may be a thin strip of copper or even an aluminum film [see Figure 5-3(c)]. Between these two extremes a wide range of conductors exist for various applications.

Because each conductor has a finite resistance, a current passing through it causes a voltage drop from one end of the conductor to the other. When conductors are long and/or carry large currents, the conductor voltage drop may cause unsatisfactory performance of the equipment supplied. Power (I^2R) is also dissipated in every current-carrying conductor, and this is, of course, wasted power. In extreme cases the power dissipated in the conductors may generate sufficient heat to destroy the insulation or even melt the conductor. Where the resistance per unit length of a conductor is known, the conductor volt drop and power dissipation are easily calculated.

EXAMPLE 5-1

An electric motor is supplied with 1 A of current from a 115-V source (E_1), as shown in Figure 5-4. Each of the cables used is 40 m long, and the resistance of the cables is 0.025 Ω/m. Determine the potential difference (E_2) at the motor terminals. Also calculate the power dissipated in the cables.

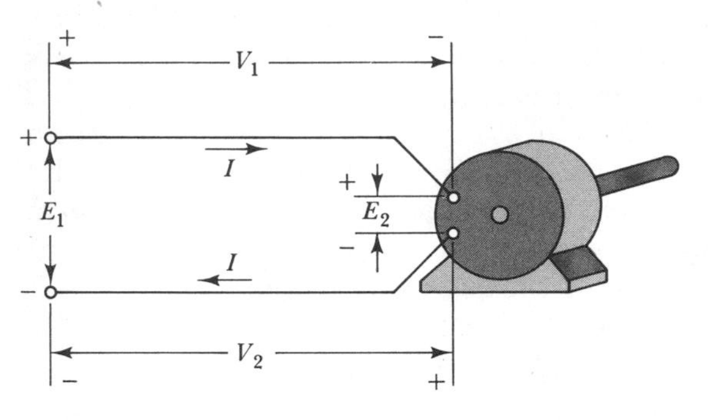

FIG. 5-4. Volt drops along conductors supplying an electric motor

SOLUTION

For 40 m *of cable*:

$$R = 40\ \text{m} \times 0.025\ \Omega/\text{m}$$
$$= 1\ \Omega$$

Voltage drop along each cable:

$$V_1 = V_2 = I \times R$$
$$= 1\ \text{A} \times 1\ \Omega$$
$$= 1\ \text{V}$$

Motor terminal voltage:

$$E_2 = E_1 - V_1 - V_2$$
$$= 115\ \text{V} - 1\ \text{V} - 1\ \text{V}$$
$$\mathbf{E_2 = 113\ V}$$

Power dissipated in the cables:

$$P = E \times I$$

$$= 2\ \text{V} \times 1\ \text{A}$$

$$\mathbf{P = 2W}$$

5-4 CONDUCTOR RESISTIVITY

The resistance per unit length of a given conductor is not always available. Therefore, to allow for all possible combinations of cable length and cross-sectional area, the *specific resistance* of the conducting metal is employed.

*The **specific resistance** (or **resistivity**) of a material is the resistance of 1 cubic meter of the material.*

Obviously, since some metals are better conductors than others, each has its own specific resistance.

*The symbol used for **specific resistance** is* **ρ** (*Greek lowercase letter* rho), *and the units of* ρ *are ohm meters* **(Ω·m)**.

The origin of the units will become apparent shortly.

Consider the cubic meter of copper illustrated in Figure 5-5(a). From Table 5-2, the specific resistance of copper is $\rho = 1.72 \times 10^{-8}\ \Omega \cdot \text{m}$.

If the length of the copper is doubled or tripled by putting cubic meter blocks in series, as shown in Figure 5-5(b), the resistance is increased. Obviously, the resistance of two 1-m^3 blocks is $2\rho\ \Omega$, that of three blocks is $3\rho\ \Omega$, and so on. From this it is seen that the total resistance of any length of copper with a cross-sectional area of 1 m^2 is

$$R = \rho \times (\text{length in meters}) \qquad \text{ohms}$$

Now consider the effect of putting two blocks in parallel, as illustrated in Figure 5-5(c). In this case the cross-sectional area is doubled; consequently, the resistance is halved. If four blocks are placed in parallel [Figure 5-5(d)], the resistance is quartered. In the case of parallel blocks, the resistance can be written as

$$R = \frac{\rho}{\text{cross-sectional area in m}^2} \qquad \text{ohms}$$

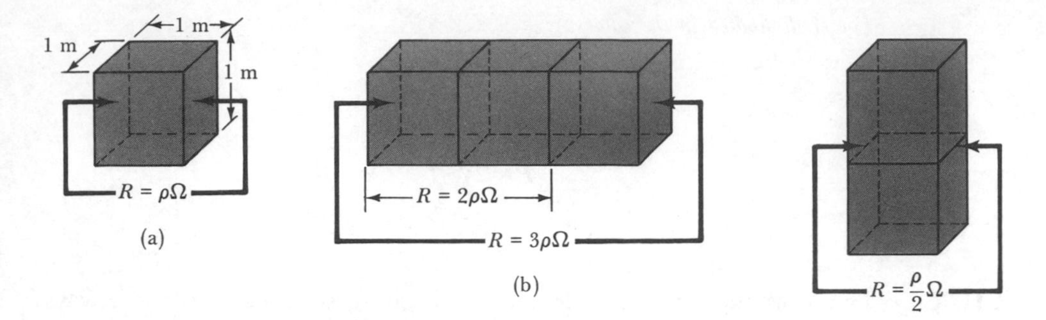

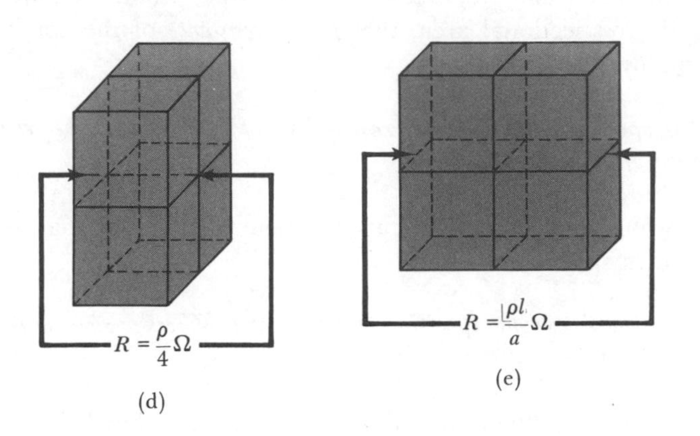

FIG. 5-5. Resistance of copper blocks in series and in parallel

TABLE 5-2. Specific Resistance for Electric Conducting Metals

METAL	SPECIFIC RESISTANCE AT 20°C (Ω·m)
Silver	1.64×10^{-8}
Copper (annealed)	1.72×10^{-8}
Gold	2.45×10^{-8}
Aluminum	2.83×10^{-8}
Tungsten	5.5×10^{-8}
Nickel	7.8×10^{-8}
Constantan	49×10^{-8}

When blocks are put in series and parallel [Figure 5-5(e)], the two equations above are combined to give

$$\boxed{R = \frac{\rho l}{a}} \tag{5-1}$$

where $\boldsymbol{R}$ *is the resistance in ohms,* $\boldsymbol{l}$ *is the length in meters, and* $\boldsymbol{a}$ *is the cross-sectional area in* m^2.

The resistance of any length of cable of any cross-sectional area can be calculated from Eq. (5-1). Where R, l, and a are known, the equation can be rewritten to determine the specific resistance:

$$\rho = \frac{Rl}{a}$$

The units of ρ are

$$\rho = \frac{\Omega \times \text{m}^2}{\text{m}} = \Omega \cdot \text{m} = \text{ohm meters}$$

EXAMPLE 5-2 Determine the resistance of 20 m of annealed copper wire 2 mm in diameter.

SOLUTION

$$\textit{cross-sectional area, } a = \frac{\pi D^2}{4}$$

Therefore,

$$a = \frac{\pi \times (2 \times 10^{-3})^2}{4}$$

$$= \pi \times 10^{-6} \text{ m}^2$$

Eq. (5-1):

$$R = \frac{\rho l}{a}$$

$$= \frac{1.72 \times 10^{-8} \times 20}{\pi \times 10^{-6}} \ \Omega$$

$$\boldsymbol{R = 0.109\ \Omega}$$

EXAMPLE 5-3 Calculate the resistance per meter length of annealed copper wire 5 mm in diameter.

SOLUTION

$$\textit{cross-sectional area, } a = \frac{\pi D^2}{4}$$

$$= \frac{\pi \times (5 \times 10^{-3})^2}{4}$$

$$= 6.25\ \pi \times 10^{-6}$$

$$R = \frac{\rho l}{a}$$

$$= \frac{1.72 \times 10^{-8} \times 1 \text{ m}}{6.25\ \pi \times 10^{-6}}$$

$$\boldsymbol{R = 8.76 \times 10^{-4}\ \Omega/\text{m}}$$

EXAMPLE 5-4

An electric heater takes a current of 15 A from a 115-V source. The cables connecting the heater to the supply are each 43 m long. If the total voltage drop along the cables is not to exceed 12 V, determine the diameter of suitable copper wire.

SOLUTION

The total allowable resistance of the cable is

$$R = \frac{E}{I} = \frac{\text{voltage drop along cable}}{I}$$

$$= \frac{12\ \text{V}}{15\ \text{A}} = 0.8\ \Omega$$

$$\textit{total length of wire} = 2 \times 43\ \text{m}$$

$$= 86\ \text{m}$$

From Eq. (5-1),

$$a = \frac{\rho l}{R} = \frac{1.72 \times 10^{-8} \times 86}{0.8}\ \text{m}^2$$

$$= 1.849 \times 10^{-6}\ \text{m}^2$$

and

$$a = \frac{\pi}{4} D^2$$

$$D = \sqrt{\frac{4a}{\pi}} = \sqrt{\frac{4 \times 1.849 \times 10^{-6}}{\pi}}$$

$$\mathbf{D = 1.53\ mm}$$

5-5 TEMPERATURE EFFECTS ON CONDUCTORS

The values of specific resistance stated in Table 5-2 refer only to conducting materials at a temperature of 20°C. Thus, in all resistance calculations using the specific resistance, the conductor temperature is assumed to remain constant at 20°C. Since the resistance of metals changes with temperature, it is necessary to be able to calculate the new resistances at higher or lower temperature levels.

The resistance of all pure metals tends to increase as the temperature of the metal rises. This may be explained in terms of the atoms actually vibrating at elevated temperatures, and thus becoming greater obstructions in the path of moving electrons. Because the resistance increases with increasing temperatures, metals are said to have a *positive temperature coefficient* (PTC). Some materials, notably semiconductors, exhibit a decrease in resistance as their temperature rises. These have a *negative temperature coefficient* (NTC). Over the normal range of operating temperatures, all metals exhibit a nearly linear relationship between resistance and temperature.

In Figure 5-6 the variation in resistance of copper is plotted versus temperature. Note that the resistance of copper appears to go to zero at a temperature of $-234.5°C$. The actual graph is not completely linear, as shown by the dashed line in Figure 5-6. The straight line passing through $-234.5°C$ is a projection of the linear portion of the graph. If the resistance R_1 at a temperature of 20°C is known, the new value of resistance at any other temperature can be calculated.

Let $R_1 = 1\ \Omega$. If the temperature is reduced to $-234.5°C$, the temperature change is

$$\Delta T = 20°C - (-234.5°C)$$
$$= 254.5°C$$

and the resistance change per degree of temperature change is

$$\frac{\Delta R_1}{\Delta T} = \frac{1\ \Omega}{254.5°C}$$
$$= 0.003\ 93\ \Omega/°C$$

If R_1 were 10 Ω at 20°C, the resistance change with temperature would be

$$\frac{\Delta R_1}{\Delta T} = \frac{10\ \Omega}{254.5°C}$$
$$= 10\ \Omega \times 0.003\ 93\ \Omega/°C$$
$$= 0.0393\ \Omega$$

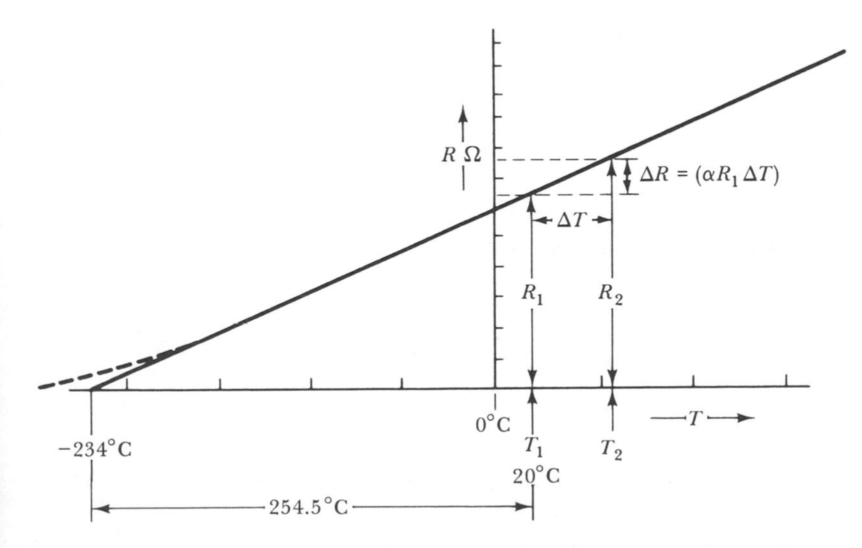

FIG. 5-6. Variation of copper resistance with temperature change

Therefore, for any value of R_1,

$$\frac{\Delta R_1}{\Delta T} = R_1 \times 0.003\ 93\ \Omega/°\text{C}$$

The quantity 0.003 93 Ω/°C is the *temperature coefficient for copper at 20° C.* The symbol α (Greek lowercase letter *alpha*) is used for temperature coefficient. Table 5-3 lists the values of temperature coefficients for various metals at 20°C. Note that the alloy *constantan* has an extremely small temperature coefficient. This characteristic makes constantan useful in applications where resistance is required to remain as constant as possible when the temperature changes. Temperature coefficients can also be determined at temperatures other than 20°C. Column 3 in Table 5-3 lists temperature coefficients at 0°C.

TABLE 5-3. Temperature Coefficient for Metals

MATERIAL	α AT 20°C	α AT 0°C
Silver	0.003 8	0.004 12
Copper	0.003 93	0.004 26
Gold	0.003 4	0.003 65
Aluminum	0.003 9	0.004 24
Tungsten	0.004 5	0.004 95
Nickel	0.006	
Constantan	0.000 008	

For any conducting material having a resistance R_1,

$$\frac{\Delta R_1}{\Delta T} = \alpha R_1$$

or

$$\Delta R_1 = \alpha R_1 \Delta T$$

and resistance R_2 at a new temperature T_2 is

$$R_2 = R_1 + \Delta R_1$$

$$= R_1 + \alpha R_1 \Delta T$$

Therefore

$$\boxed{\mathbf{R_2 = R_1 (1 + \alpha \Delta T)}} \qquad \textbf{(5-2)}$$

where R_1 is the resistance at 20° C, α is the temperature coefficient of the material, *and* ΔT *is the temperature change from* 20° *C*. The temperature coefficient at 0°C may also be used in Eq. (5-2) if the resistance R_1 is measured at 0°C and ΔT is the temperature increase from 0°C.

EXAMPLE 5-5 The resistance of a length of copper cable is 5.8 Ω at 20°C. Determine the new resistance of the cable at 125°C.

SOLUTION

Eq. (5-2):

$$R_2 = R_1(1 + \alpha \Delta T)$$

$$\Delta T = T_2 - T_1$$

$$= 125°\text{C} - 20°\text{C}$$

$$= 105°\text{C}$$

For copper at 20°C, $\alpha = 0.003\ 93$ (*see* Table 5-3). *Therefore,*

$$R_2 = 5.8\ \Omega[1 + (0.00393 \times 105°\text{C})]$$

$$\mathbf{R_2 = 8.2\ \Omega}$$

EXAMPLE 5-6 The resistance of a coil of nickel wire is 1 kΩ at 20°C. After being submerged in a liquid for some time, the resistance falls to 880 Ω. Calculate the temperature of the liquid.

SOLUTION

Eq. (5-2):

$$R_2 = R_1(1 + \alpha \Delta T)$$

Therefore,

$$\Delta T = \frac{1}{\alpha}\left(\frac{R_2}{R_1} - 1\right)$$

For nickel at 20°C, $\alpha = 0.006$ (*see* Table 5-3). *Therefore,*

$$\Delta T = \frac{1}{0.006}\left(\frac{880\ \Omega}{1\ k\Omega} - 1\right)$$

$$= -20°\text{C}$$

and

$$\Delta T = T_2 - T_1$$

Therefore,

$$T_2 = \Delta T + T_1$$

$$= -20°\text{C} + 20°\text{C}$$

$$\mathbf{T_2 = 0°C}$$

EXAMPLE 5-7

A certain length of wire has its resistance measured as 330 Ω at 20°C and 448.8 Ω at 100°C. Calculate the temperature coefficient and identify the material.

SOLUTION

Eq. (5-2):

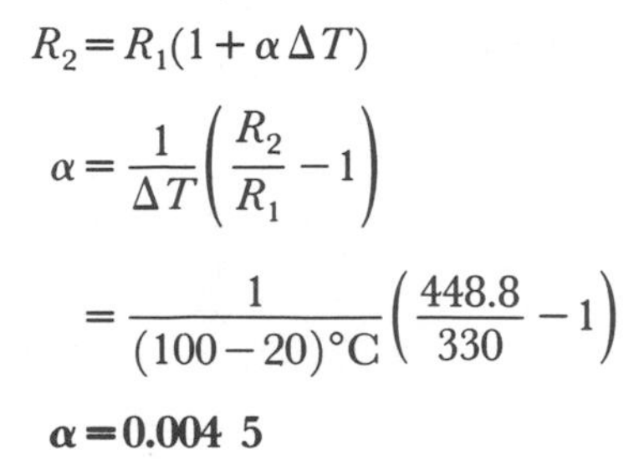

$$R_2 = R_1(1 + \alpha \Delta T)$$

$$\alpha = \frac{1}{\Delta T}\left(\frac{R_2}{R_1} - 1\right)$$

$$= \frac{1}{(100-20)°\text{C}}\left(\frac{448.8}{330} - 1\right)$$

$$\boldsymbol{\alpha = 0.004\ 5}$$

From Table 5-3, $\alpha = 0.004\ 5$ *identifies the material as tungsten.*

5-6 RESISTOR CONSTRUCTION

The two commonest types of electronics resistors are *wire-wound* and *carbon composition* construction. A typical wire-wound resistor consists of a length of nickel wire wound on a ceramic tube and covered with porcelain. Low-resistance connecting terminals are provided, and the resistance value is usually printed on the side of the component. Figure 5-7(a) illustrates the construction of a typical wire-wound resistor. Carbon composition resistors are constructed by molding mixtures of powdered carbon and insulating materials into a cylindrical shape [see Figure

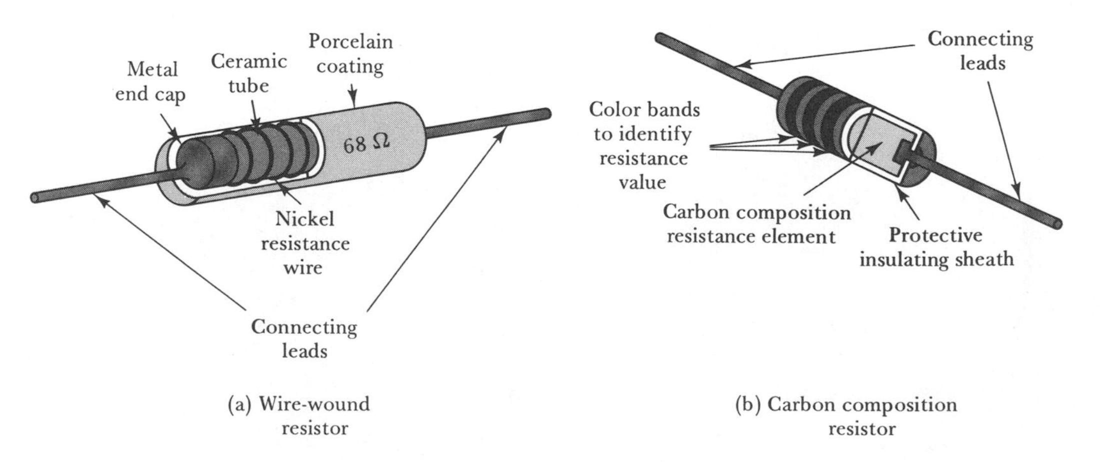

FIG. 5-7. Construction of wire-wound and carbon composition resistors

5-7(b)]. An outer sheath of insulating material affords mechanical and electrical protection, and copper connecting wires are provided at each end. Carbon composition resistors are smaller and less expensive than the wire-wound type. However, the wire-wound type are the more rugged of the two, and are able to survive much larger power dissipations than the carbon composition type.

Most resistors have standard fixed values, and therefore they can be termed *fixed resistors*. *Variable resistors*, or *adjustable resistors*, are also used a great deal in electronics. The construction of a typical variable resistor is illustrated in Figure 5-8(a), together with two frequently employed graphic symbols for a variable resistor in Figure 5-8(b).

The illustration in Figure 5-8(a) shows a coil of closely wound insulated resistance wire formed into a partial circle. The coil has a low-resistance terminal at each end, and a third terminal is connected to a movable contact with a shaft adjustment. The movable contact can be set to any point on a connecting track which extends over one edge of the

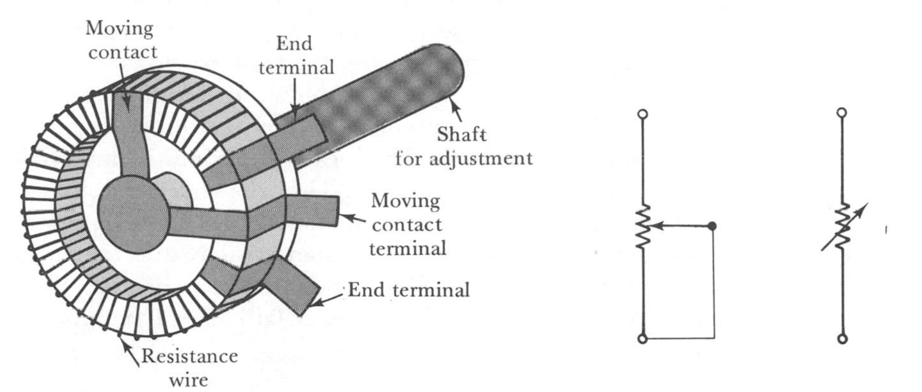

(a) Typical construction of a variable resistor (and potentiometer)

(b) Circuit symbols for a variable resistor

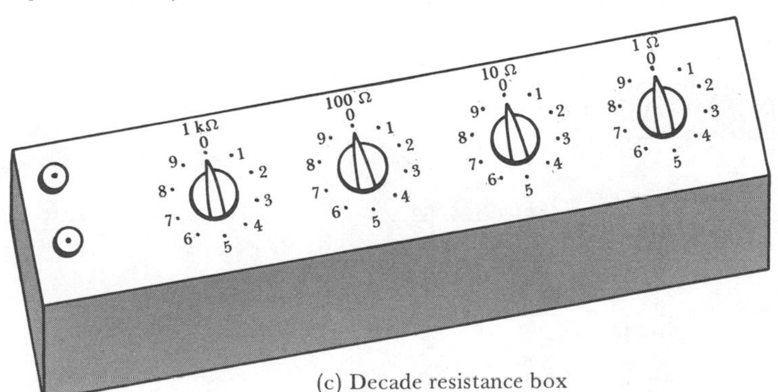

(c) Decade resistance box

FIG. 5-8. Variable resistors and circuit symbols

coil. Using the adjustable contact, the resistance from either end terminal to the center terminal may be adjusted from zero to the maximum coil resistance.

Another type of variable resistor, known as a *decade resistance box*, is shown in Figure 5-8(c). This is a laboratory component which contains precise values of switched series-connected resistors. As illustrated, the first switch (from the left) controls resistance values in 1-Ω steps from 0 Ω to 9 Ω, while the second switches values of 10 Ω, 20 Ω, 30 Ω, and so on. The decade box shown can be set to within ±1 Ω of any value from 0 Ω to 9999 Ω.

5-7 COLOR CODE

Standard (fixed-value) resistor values normally range from 2.7 Ω to 22 MΩ. The resistance *tolerances* on these standard values are typically ±20%, ±10%, ±5%, or ±1%. A tolerance of ±10% on a 100-Ω resistor means that the actual resistance may be as high as 100 Ω+10% (i.e., 110 Ω), or as low as 100 Ω−10% (i.e., 90 Ω). Obviously, the resistors with the smallest tolerance are the most accurate, and the most expensive.

Because carbon composition resistors are physically small (some are less than 1 cm in length), it is not convenient to print the resistance value on the side. Instead, a *color code* in the form of colored bands is employed to identify the resistance value and tolerance. The color code is illustrated in Figure 5-9. Starting from one end of the resistor, the first two bands identify the first and second digits of the resistance value, and the third band indicates the number of zeros. An exception to this is when the third band is either silver or gold, which indicates a 0.01 or 0.1 multiplier,

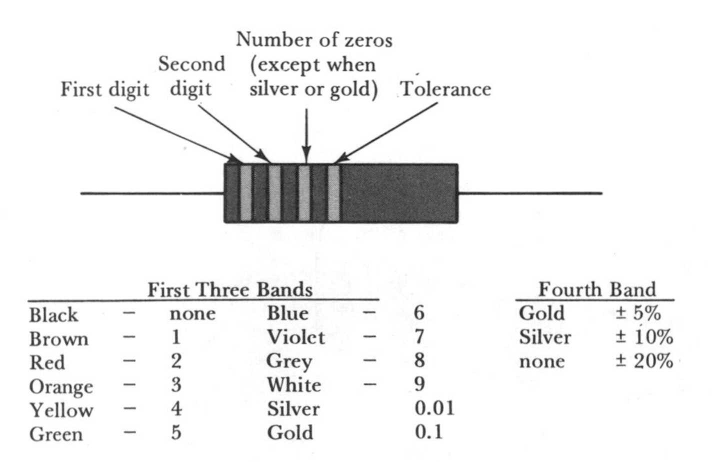

First Three Bands					
Black	–	none	Blue	–	6
Brown	–	1	Violet	–	7
Red	–	2	Grey	–	8
Orange	–	3	White	–	9
Yellow	–	4	Silver		0.01
Green	–	5	Gold		0.1

Fourth Band	
Gold	± 5%
Silver	± 10%
none	± 20%

FIG. 5-9. The resistor color code

respectively. The fourth band is always either silver or gold, and in this position silver indicates a ±10% tolerance and gold indicates ±5% tolerance. Where no fourth band is present, the resistor tolerance is ±20%.

EXAMPLE 5-8

Identify the values of resistors with the following color codes: brown–black–red–silver, red–red–black–gold, and red–violet–gold–silver.

SOLUTION

$$\begin{array}{cccc} \text{brown} & \text{-black-} & \text{red} - & \text{silver} \\ \downarrow & \downarrow & \downarrow & \downarrow \\ 1 & 0 & 00 & \pm 10\% \end{array} = 1000\ \Omega \pm 10\%$$

$$\begin{array}{cccc} \text{red} & \text{-red-} & \text{black} & -\ \text{gold} \\ \downarrow & \downarrow & \downarrow & \downarrow \\ 2 & 2 & \overline{\text{no zeros}} & \pm 5\% \end{array} = 22\ \Omega \pm 5\%$$

$$\begin{array}{cccc} \text{red} & \text{-violet-} & \text{gold} & -\ \text{silver} \\ \downarrow & \downarrow & \downarrow & \downarrow \\ 2 & 7 & \overline{\times 0.1} & \pm 10\% \end{array} = (27\ \Omega \times 0.1) \pm 10\%$$

$$= 2.7\ \Omega \pm 10\%$$

5-8 RESISTOR POWER RATINGS

Typical power ratings for wire-wound resistors start at 1 W and range to 10 W and much larger. Large surface areas are required to dissipate the heat generated; consequently, high power ratings result in physically large resistors. Carbon composition resistors are essentially low-power components; excessive power dissipation destroys them rapidly. The usual range of power ratings for carbon composition resistors are $\frac{1}{8}$ W, $\frac{1}{4}$ W, $\frac{1}{2}$ W, 1 W, and 2 W.

The maximum current that may be permitted to flow through a resistor, and the maximum voltage that may be applied across it, are limited by the specified maximum power dissipation. Using Eq. (3-8), $P = E^2/R$, and Eq. (3-9), $P = I^2R$, the maximum levels of current and voltage are easily calculated.

EXAMPLE 5-9

A 2.2-kΩ resistor has a specified maximum power dissipation of 1 W. Determine the maximum current that may be passed through the resistor and the maximum voltage that may be applied to its terminals.

SOLUTION

Eq. (3-9):

$$P=I^2R$$

Therefore,

$$I=\sqrt{\frac{P}{R}}$$

and

$$I_{\text{max}}=\sqrt{\frac{1\text{ W}}{2.2\text{ k}\Omega}}$$

$$\boldsymbol{I=21.3\text{ mA}}$$

Eq. (3-8):

$$P=\frac{E^2}{R}$$

Therefore,

$$E=\sqrt{PR}$$

and

$$E_{\text{max}}=\sqrt{1\text{ W}\times 2.2\text{ k}\Omega}$$

$$\boldsymbol{E=46.9\text{ V}}$$

5-9 TEMPERATURE COEFFICIENT OF RESISTORS

Wire-wound resistors behave much the same as conductors when their temperature increases or decreases (see Section 5-5). Thus, they have the kind of resistance/temperature characteristic illustrated in Figure 5-6. Carbon composition resistors, on the other hand, tend to have the characteristic illustrated in Figure 5-10. In this case the resistance value tends to increase when the temperature varies *above or below* room temperature.

The *temperature coefficient* (T.C.) of resistance is usually expressed in *parts per million per degree Celsius* (*ppm/°C*). To understand what this means, consider a resistor with a resistance of R Ω and a T.C. of ±100 ppm/°C. For every 1°C change in temperature, the maximum resistance change (ΔR) is

$$\Delta R=\pm\frac{R}{1\ 000\ 000}\times 100$$

i.e., a change of 100 parts in 1 000 000 parts of R.

When the temperature change is ΔT°C, the resistance change is

$$\boxed{\boldsymbol{\Delta R=\pm\frac{R}{1\ 000\ 000}\times\text{T.C.}\times\Delta T}} \qquad (5\text{-}3)$$

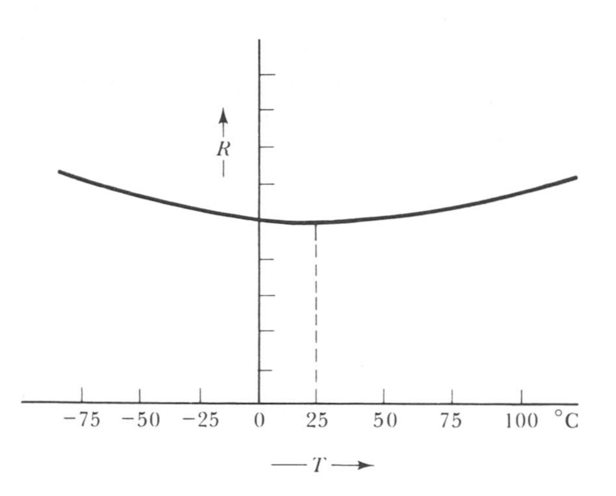

FIG. 5-10. Variation of resistance with temperature for carbon composition resistors

EXAMPLE 5-10 Calculate the maximum change in resistance for a 470-Ω resistor with a T.C. of ±500 ppm/°C, when its temperature increases from 20°C to 75°C.

SOLUTION

$$\Delta T = 75°\text{C} - 20°\text{C}$$
$$= 55°\text{C}$$
$$\Delta R = \pm \frac{470\ \Omega}{1\ 000\ 000} \times 500 \times 55°\text{C}$$
$$\mathbf{\Delta R \simeq \pm 12.9\ \Omega}$$

The resistance change calculated from the temperature coefficient is the maximum change in resistance that should occur over a given temperature range. With some resistors, the resistance change may be only a small fraction of the maximum, while other resistors of the same type may exhibit the maximum possible change. There are circuit applications in which it is important that two (or more) resistors change by approximately the same proportion as temperature increases or decreases. In this case it is said that the resistors are required to *track* each other. When resistors are constructed to fulfill this requirement, the manufacturer specifies the *matching* of the temperature coefficients. Sometimes the term *tracking temperature coefficient* is used. Thus, the T.C. for a certain type of resistors may be specified as:

temperature coefficient = ±20 *ppm/°C matched to* ±2 *ppm/°C*

or

tracking temperature coefficient = ±2 *ppm/°C*

5-10 LINEAR AND NONLINEAR RESISTORS

In Figure 5-11(a), a graph is plotted of V versus I for a 1-kΩ fixed resistor. Several points on the graph can be quickly calculated by use of Ohm's law:

for $V=1$ V:

$$I=\frac{1\text{ V}}{1\text{ k}\Omega}=1\text{ mA}$$

for $V=3$ V:

$$I=\frac{3\text{ V}}{1\text{ k}\Omega}=3\text{ mA}$$

for $V=5$ V:

$$I=\frac{5\text{ V}}{1\text{ k}\Omega}=5\text{ mA}$$

Thus, it is seen that the graph of V and I for a normal fixed resistor is a straight line. Consequently, the resistor may be termed a *linear resistor*. The V/I graph for a resistor is sometimes termed the *V/I characteristic* of the resistor.

Figure 5-11(b) shows a graph of V versus I which is clearly *nonlinear*. The device that this graph refers to can be termed a *nonlinear resistor*. The graph shown is, in fact, the V/I characteristic for a device known as a *thermistor*. (The term is constructed from *thermal* and *resistor*.) It is clear

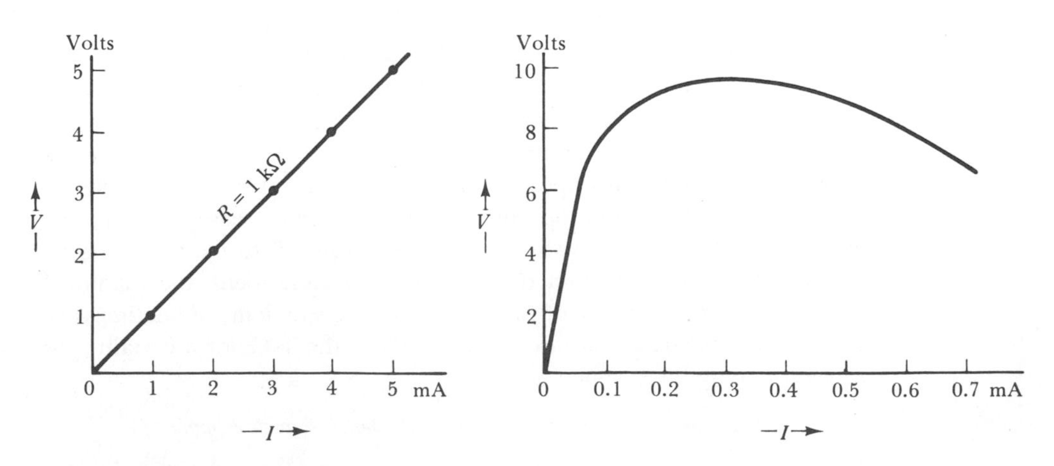

FIG. 5-11. Graphs of V/I for linear and nonlinear resistors

that a thermistor has a resistance which is substantially altered with temperature change.

A thermistor is a semiconductor device constructed of metallic oxide. The resistance of the thermistor changes when the surrounding temperature increases or decreases. When the ambient temperature remains constant, the resistance of the thermistor may be altered by the heating effect of the current flowing through the device. Unlike a conductor, the resistance of a semiconductor tends to fall when its temperature increases. This is because more electrons are released from the semiconductor atoms by the effects of thermal energy, and these increase the current flow. Thus, most thermistors have a negative temperature coefficient.

Referring to the V/I characteristic for the thermistor, it is seen that the first portion of the graph is near-linear. This is because the currents flowing are too small to produce any significant heating of the device. As the current increases further, however, the heating effect causes the resistance of the thermistor to fall. This results in a reduced voltage drop across the device for a given current level. Further increase in current causes progressive reductions in resistance, and consequently further reductions in terminal voltage.

GLOSSARY OF FORMULAS

Resistance:

$$\textit{resistance} = \frac{\text{resistivity} \times \text{length}}{\text{cross-sectional area}} \qquad R = \frac{\rho l}{a}$$

Temperature effect on resistance:

$$R_2 = R_1 (1 + \alpha \Delta T)$$

Resistance change:

$$\Delta R = \pm \frac{R}{1\ 000\ 000} \times \text{T.C.} \times \Delta T$$

REVIEW QUESTIONS

5-1 Sketch the plane diagram of a copper atom and of a silicon atom. Identify the valence shell and the electrons and holes.

5-2 Using illustrations, explain the various kinds of atomic bonding, and identify the kinds of bonding found in conductors, insulators, and semiconductors.

5-3 Explain the function of various types of insulators and discuss insulator breakdown.

5-4 Explain the function of a conductor and discuss the reasons for the various sizes of available conductors.

5-5 Using illustrations, explain how temperature affects the resistance of a metallic conductor. Derive an equation relating conductor resistance at an elevated temperature to resistance at 20°C.

5-6 Describe the two commonest types of electronic resistor construction and discuss their relative advantages and disadvantages.

5-7 Sketch the construction of a variable resistor, identify each part of the device and explain its operation.

5-8 Define linear and nonlinear resistors. Sketch the V/I characteristics for a linear and for a nonlinear resistor and explain the shape of each graph.

5-9 Explain what is meant by T.C. = 100 ppm/°C. Sketch the approximate shape of the resistance/temperature characteristic of a carbon composition resistor. Define tracking temperature coefficient.

PROBLEMS

5-1 An electric heater is supplied with 5 A of current from a 120-V source. The cables used to connect the heater are each 50 m long and have a resistance of 0.01 Ω/m. Calculate the potential difference at the heater terminals and the power dissipated in the cables.

5-2 Calculate the resistance per meter of copper conductors with the following diameters: 1 mm, 2 mm, 3 mm, 4 mm, and 5 mm.

5-3 Determine the resistance of copper and aluminum conductors 6 mm in diameter and 100 m long.

5-4 A piece of electronic equipment takes 2.5 A from a 120-V supply. The cables connecting the supply to the equipment are 10 m long. If the terminal voltage at the equipment is to be not less than 115 V, determine the smallest diameter of suitable copper conductors.

5-5 The copper conductors on a printed circuit board are 0.05 mm thick and 2 mm across. Determine the volts drop per centimeter along the conductors when a current of 300 mA flows.

5-6 An aluminum conductor has a resistance of 10 Ω at 20°C. Determine the resistance of the conductor at 100°C.

5-7 The printed circuit board conductors described in Problem 5-5 are raised to a temperature of 65°C from the normal temperature level of 20°C. Determine the new value of voltage drop per centimeter when the current is 300 mA.

5-8 A coil of nickel wire is used to monitor the temperature of a liquid in which it is submerged. The coil resistance is 500 Ω at 20°C. Prepare a table of coil resistance values for liquid temperatures at 10° intervals ranging from 0°C to 100°C.

5-9 Two lengths of wire each have resistances of 1 kΩ at 20°C. At 75°C the resistance values are measured as $R_a = 1.209$ kΩ and $R_b = 1.187$ kΩ. Calculate the temperature coefficient of each wire and identify the materials.

5-10 Identify the values of resistors with the following color codes: green–blue–orange, gray–red–brown–silver, and orange–white–green–gold.

5-11 Calculate the maximum current and voltage limits for:

a. A $\frac{1}{4}$-W 820-Ω resistor.

b. A $\frac{1}{2}$-W 1-MΩ resistor.

5-12 A 3.3-kΩ precision resistor with a T.C. of 10 ppm/°C is to operate over a temperature range of 20°C to 100°C. Calculate the maximum change in resistance that can occur over the temperature range.

5-13 Calculate the new value of resistance for each of the resistors referred to in Problem 5-10 when its temperature changes from 20°C to 75°C. Take the T.C. of each resistor as +150 ppm/°C.

6

VOLTAGE CELLS AND BATTERIES

INTRODUCTION

A voltage cell consists basically of two different metal plates immersed in an acid solution. The action of the acid causes electrons to be removed from one plate and to be accumulated on the other. In this way a potential difference is produced between the two plates, and a current can be made to flow through an external circuit.

Many types of voltage cells exist for different applications. Some can be recharged and used again many times, others can be used only once and are then discarded. One of the most important of the rechargeable cells is the *lead–acid cell*, which is the basic unit of the automobile battery.

Batteries of voltage cells can be operated in series or in parallel, or in series–parallel combinations. The performance of a voltage cell may be defined in terms of the maximum voltage and current it can supply, its *ampere-hour* rating, and the *cell equivalent circuit.*

6-1 SIMPLE VOLTAGE CELL

OPERATION. The construction of a simple voltage cell is illustrated in Figure 6-1. The complete cell consists of copper and zinc *electrodes* immersed in a chemical solution known as the *electrolyte*. Diluted sulfuric acid is frequently used as the electrolyte, but any one of several other chemical compounds may be employed.

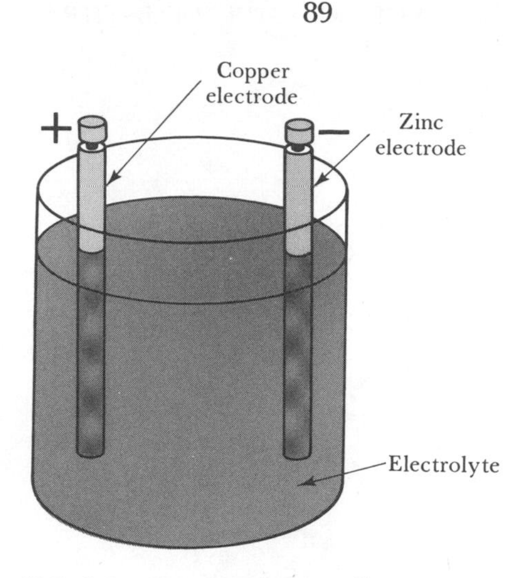

FIG. 6-1. Simple voltage cell

Sulfuric acid is a combination of *hydrogen* and *sulfate* in which each hydrogen atom has given up an electron to the sulfate molecules. The solution remains chemically stable until the electrodes are introduced. The surface of the zinc electrode readily dissolves in the sulfuric acid, and the zinc atoms combine with the sulfate to form *zinc sulfate*. In this process the zinc atoms leave electrons behind them on the zinc electrode [see Figure 6-2(a)]. This is because the sulfate already has an excess of electrons acquired from the hydrogen. As the zinc sulfate forms, hydrogen ions are released, and each hydrogen ion is an atom that is short of one electron (i.e., a positive ion). These ions travel to the copper electrode, from which they acquire electrons, as illustrated.

In giving up electrons, the copper electrode becomes positively charged. Similarly, the zinc electrode becomes negatively charged as it accumulates excess electrons. Thus, a potential difference is created between the copper and zinc electrodes.

If no current is drawn from the cell, the chemical action eventually ceases when the zinc becomes so negative that it repels the (negative) sulfate molecules, and the positive charge on the copper repels the (positive) hydrogen ions. In this condition the potential difference at the cell terminals is typically 1.1 V.

An external circuit connected to the cell terminals provides a path for electrons to flow from the (negative) zinc electrode to the (positive) copper electrode [Figure 6-2(b)]. Thus, the zinc electrode loses negative charges and the copper electrode gains negative charges (i.e., becomes less positive). Once the electrodes begin to lose some of their accumulated charges, the chemical action resumes, and the electrolyte continues to provide electrons to supply the output current from the cell.

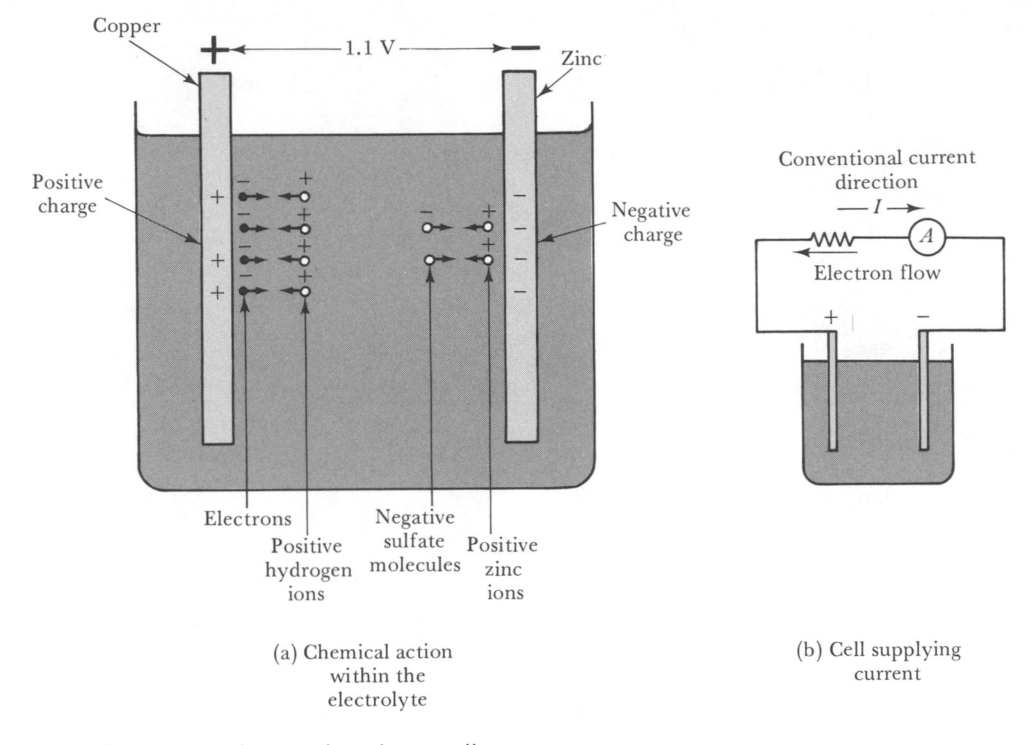

FIG. 6-2. Action of a simple voltage cell

POLARIZATION. The hydrogen released in the simple voltage cell frequently remains clinging to the copper electrode in the form of bubbles. When large areas of the electrode become coated with hydrogen bubbles, the active surface area is reduced and the cell's ability to supply current is diminished. The effect is known as *polarization*. In more complex cells, several methods are employed to combat polarization. Materials used for this purpose are termed *depolarizers*.

OUTPUT VOLTAGE AND CURRENT. The potential difference produced at the terminals of a cell depends upon the electrode materials. For the zinc and copper electrodes of the simple voltage cell, the terminal voltage is approximately 1.1 V. This potential difference is produced no matter what the physical size of the electrodes or the quantity of the electrolyte.

The output current from a cell depends upon the output voltage and the external load resistance. As always,

$$I = \frac{E}{R}$$

where I is the output current, E is the cell terminal voltage, and R is the external *load resistance*. However, there is a limit to the maximum current that a given cell can supply. When this limit is approached, the cell's terminal voltage begins to fall, so that the output current cannot be increased. The maximum current that can be drawn from a cell is directly related to the physical size of the cell's components. Obviously, electrodes with large surface areas can release more atoms and generate more free electons than is possible with small surface areas. Consequently, physically large voltage cells can sustain larger output currents than small cells.

The maximum output current from a voltage cell can be determined in terms of the cell's *internal resistance*, which is discussed in Section 6-2.

AMPERE-HOUR RATING. The amount of energy that can be supplied by any cell (or battery of cells) is defined in terms of its rating in *ampere-hours* (Ah).

If a cell can supply a current of 5 A for a maximum time of 1 h, its rating is

$$\boxed{\text{Ah rating} = I \times t} \qquad \textbf{(6-1)}$$

$$= 5\ \text{A} \times 1\ \text{h}$$

$$= 5\ \text{Ah}$$

When the same cell is supplying a current of only 1 A, it can be expected to sustain this current for a time of

$$t = \frac{\text{Ah rating}}{I}$$

$$= \frac{5\ \text{Ah}}{1\ \text{A}}$$

$$= 5\ \text{h}$$

For a time of 50 h, a cell with a rating of 5 Ah can be expected to supply a continuous current of

$$I = \frac{\text{Ah rating}}{t}$$

$$= \frac{5\ \text{Ah}}{50\ \text{h}}$$

$$= 100\ \text{mA}$$

Usually a cell's Ah rating applies only to a certain range of load currents. For a given cell, the Ah rating when supplying a high current is not as great as that at a low current level. One reason for this is that polarization occurs more rapidly when the load current is high. Also, as already discussed, every cell has a limit to the maximum current it can supply.

EXAMPLE 6-1

A voltage cell supplies a load current of 0.5 A for a period of 20 h until its terminal voltage falls to an unacceptable level. Calculate the Ah rating of the cell and determine how long it could be expected to supply a current of 200 mA.

SOLUTION

Eq. (6-1):

$$\text{Ah } rating = I \times t$$
$$= 0.5 \text{ A} \times 20 \text{ h}$$
$$= 10 \text{ Ah}$$

$$t = \frac{\text{Ah } rating}{I} = \frac{10 \text{ Ah}}{200 \text{ mA}}$$

$$\boldsymbol{t = 50 \text{ h}}$$

PRIMARY AND SECONDARY CELLS. The simple voltage cell is classified as a *primary cell*. Primary cells can supply current until the negative electrode is completely dissolved, the electrolyte is exhausted, or perhaps until polarization renders the cell useless. A *secondary cell*, on the other hand, can be *recharged*. That is, the electrode surface can be re-formed and the electrolyte returned to its original condition. This is done by passing a current through the cell in an opposite direction to the current that normally flows from the cell.

The *lead–acid cell* discussed in Section 6-6 is a secondary cell, because it is rechargeable. Most *dry cells*, such as the *zinc–carbon cell* (see Section 6-3) cannot be recharged; consequently, they are primary cells.

6-2 CELL EQUIVALENT CIRCUIT

As already explained, when no current is being drawn from a voltage cell charges are accumulated on the electrodes, and the chemical action ceases within the electrolyte. At this time, the potential difference between the cell terminals is known as the *open-circuit output voltage* or the *no-load output voltage*.

When current flows from the cell, electrons are transferred externally from the negative electrode to the positive electrode. Thus, the potential difference between the electrodes falls, and the chemical action within the electrolyte commences to replace the lost charges. If the load current is very small, the terminal voltage under load may not be noticeably different from the open-circuit output voltage. When the load current is large, the terminal voltage falls below the open-circuit voltage level. If the load current is made too large, the cell terminal voltage falls to near zero, because the chemical action simply cannot occur fast enough to replace the charges removed from the electrodes. The maximum current that may be taken from a voltage cell without a substantial drop in output voltage is proportional to the surface area of the electrodes and to the composition of the electrolyte.

To account for the terminal voltage drop of a cell when supplying current, each cell is said to have a certain *internal resistance*. The cell can then be represented by an equivalent circuit consisting of its internal resistance together with an *ideal cell* which has a terminal voltage equal to the actual cell's open-circuit terminal voltage. An ideal cell is a theoretical cell that has a terminal voltage which is assumed to remain constant. The equivalent circuit of a voltage cell is illustrated in Figure 6-3.

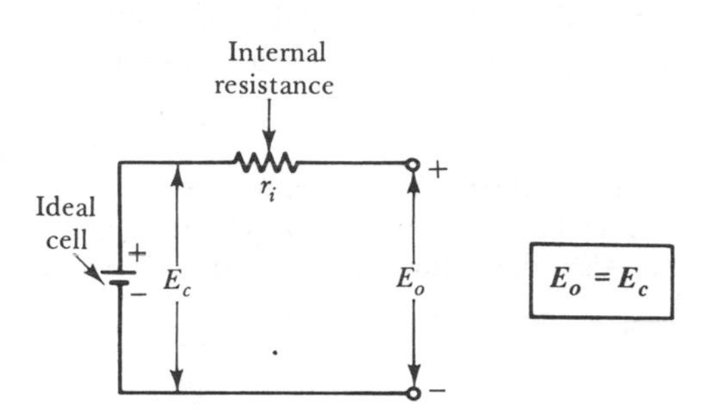

(a) Equivalent circuit of cell ($I_o = 0$)

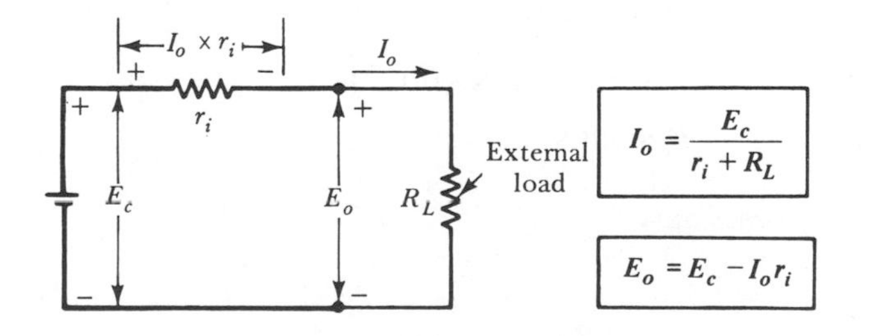

(b) Equivalent circuit of cell supplying a load current

FIG. 6-3. Equivalent circuit of a voltage cell under no-load and load conditions

When no current is flowing from the equivalent circuit, there is no voltage drop across the internal resistance (r_i) and the terminal voltage is equal to the ideal cell voltage. That is, the output voltage is the open circuit voltage E_c of the cell [see Figure 6-3(a)]. When a load resistance (R_L) is connected to the cell terminals [Figure 6-3(b)], the current that flows is

$$I_o = \frac{E_c}{r_i + R_L} \tag{6-2}$$

When supplying current, a voltage drop occurs across r_i, and the output voltage becomes

$$E_o = E_c - I_o r_i \tag{6-3}$$

Obviously, a cell that has a low internal resistance is capable of supplying a larger current than one that has a relatively high internal resistance.

EXAMPLE 6-2

A certain voltage cell has an open-circuit terminal voltage of $E_c = 2$ V and an internal resistance of $r_i = 0.1\ \Omega$. Determine the output voltage of the cell when the load current is 2 A. Also, determine the cell current when the terminals are short-circuited.

SOLUTION

When $I_o = 2$ A:
Eq. (6-3):

$$E_o = E_c - I_o r_i$$
$$= 2\text{ V} - (2\text{ A} \times 0.1\ \Omega)$$
$$\mathbf{E_o = 1.8\ V}$$

When the cell is short-circuited:
Eq. (6-2):

$$I = \frac{E_c}{r_i + R_L} = \frac{2\text{ V}}{0.1\ \Omega + 0}$$
$$\mathbf{I = 20\ A}$$

6-3 DRY CELLS

ZINC–CARBON CELL. The *zinc–carbon* cell is a primary cell commonly used in electric flashlights and other portable equipment. As illustrated in Figure 6-4(a), this cell basically consists of a zinc can containing electrolyte with a carbon rod at the center. The zinc can is the negative electrode of the cell, and the carbon rod is the positive electrode. The electrolyte is in the form of a moist paste, so that the *dry cell* is, in fact, not dry at all. The cell becomes useless when the electrolyte drys out because the necessary chemical reactions cannot occur.

The operation of the zinc–carbon cell is similar to that of the zinc–copper cell described in Section 6-1. Electrons accumulate on the zinc electrode, and electrons are removed from the carbon electrode as

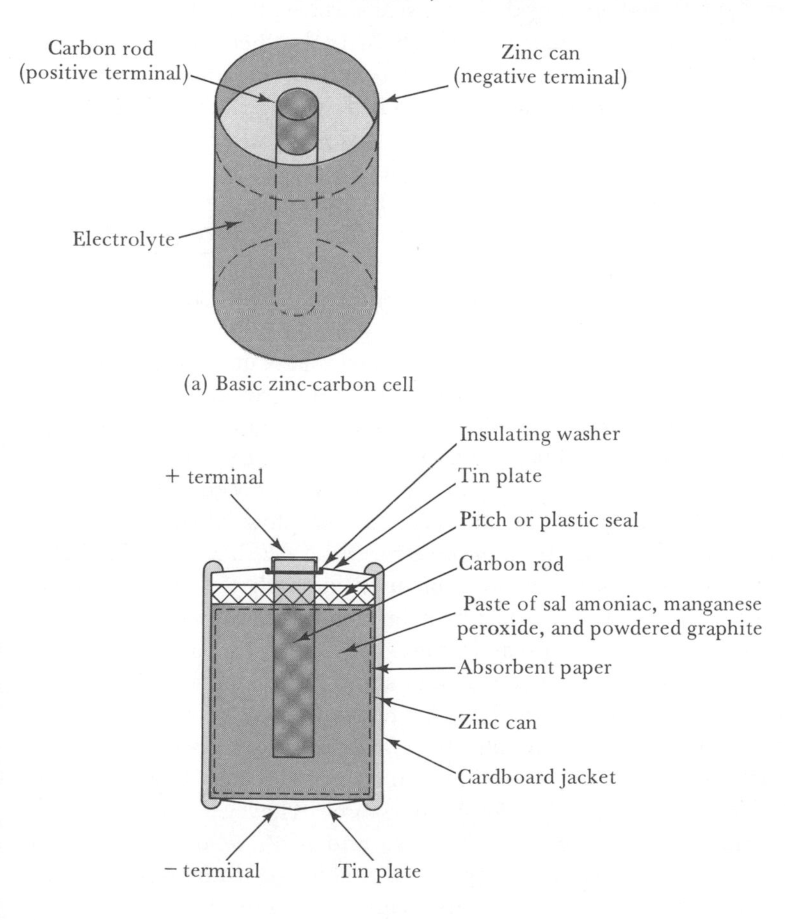

(a) Basic zinc-carbon cell

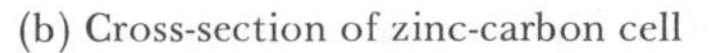

(b) Cross-section of zinc-carbon cell

FIG. 6-4. Construction of a zinc-carbon D cell

zinc atoms combine with the electrolyte. Thus, a potential difference occurs between the electrodes, with the zinc being negative and the carbon positive.

Figure 6-4(b) shows the construction of the zinc–carbon cell in more detail. The electrolyte consists of *sal amoniac* (ammonium chloride) mixed with *manganese peroxide* and powdered graphite. The manganese peroxide acts as a depolarizer by combining with the hydrogen, which appears at the positive electrode. The combination results in *manganous oxide* and water. This eliminates the polarizing effect of the hydrogen and helps to keep the electrolyte moist.

The electrolyte must be hermetically sealed in the can to prevent it from drying out quickly. A layer of pitch or plastic at the top of the can is the usual type of seal employed. The inside surface of the zinc is protected by a layer of absorbent paper [see Figure 6-4(b)] to help prevent the zinc from reacting directly with the carbon and other impurities in the electrolyte. Such reactions tend to set up little voltage cells at the surface of the zinc, and this is referred to as *local action*. Local action reduces the output voltage of the cell and shortens its life by using up the zinc. To combat local action due to impurities in the zinc, the zinc surface is usually coated with mercury, or else mercury is mixed with the zinc during the manufacturing process.

Most dry cells have a surrounding outer jacket of cardboard or tin plate. Also, tin-plate discs are used at the top and bottom, as illustrated. As well as providing protection for the cell, the outer jacket helps to protect battery-powered equipment from the corrosive action that occurs on the outside of the zinc can when dry cells are unused for a long time.

The typical output voltage from a zinc–carbon dry cell is 1.5 V, but it could be anything from 1.4 V to 1.6 V in a good cell. This terminal voltage (1.5 V) is available no matter what the physical size of the cell. The very small *C cell* and the much larger *No. 6 cell* illustrated in Figure 6-5 each have a terminal voltage of approximately 1.5 V. For larger output voltages, several cells must be connected in series (see Section 6-4). In continuous use, the terminal voltage of the zinc–carbon cell tends to fall off rapidly unless the current is very small. Consequently, this type of cell is best applied where it will be used intermittently.

The output current that may be drawn from a zinc–carbon cell is very much dependent upon the physical size of the cell. The No. 6 cell can supply as much as 20 A for a short time. However, at this rate it would tend to discharge very quickly, and 1 A would be a more appropriate maximum load. The *D cell*, which is commonly used in flashlights, cannot supply more than about 5 A. Again, this would be an excessive load, and a maximum of a few hundred milliamps is normal. The *C cells* supply even smaller maximum currents, corresponding to their smaller size.

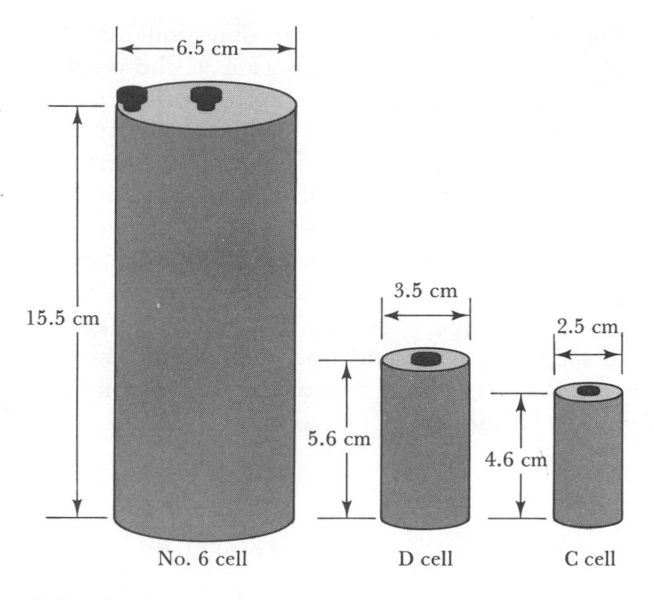

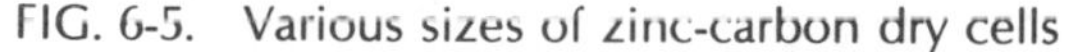
FIG. 6-5. Various sizes of zinc-carbon dry cells

MANGANESE–ALKALINE CELL. This is a rechargeable (or secondary) cell which can sustain a constant terminal voltage for heavy loads over a much larger period of time than the less expensive carbon–zinc cell.

Figure 6-6 shows the construction of a typical *button-type* manganese–alkaline cell as used in hearing aids. The electrolyte is *potassium hydroxide*, and the negative electrode is a *zinc–mercury* amalgam. The positive electrode is a layer of *manganese dioxide*. As illustrated in the figure, the cell is enclosed in a steel container. The open-circuit terminal voltage of this cell is typically 1.5 V.

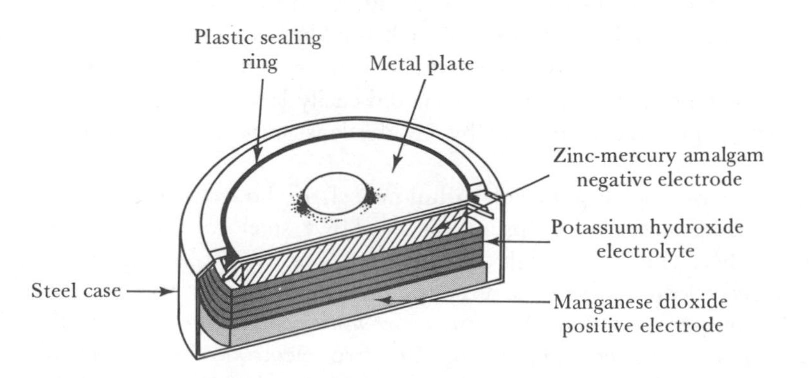

FIG. 6-6. Construction of a button-type manganese–alkaline cell

MERCURY CELL. The mercury cell is a modification of the manganese–alkaline cell, in which *mercuric oxide* is used as the positive electrode (i.e., instead of manganese dioxide). The open-circuit terminal voltage of the mercury cell is usually around 1.4 V, falling to about 1.3 V under load conditions. The major advantage of this type of cell is that its terminal voltage is maintained constant for much longer time periods than either the manganese–alkaline or carbon–zinc cells. Figure 6-7 shows a graph of output voltage plotted versus time for comparable zinc–carbon, manganese–alkaline, and mercury cells. It is obvious from the graph that the terminal voltage of the zinc–carbon cell falls off relatively rapidly when supplying a constant load. The performance of the manganese–alkaline cell is better than that of the zinc–carbon cell, but the mercury cell gives by far the best performance.

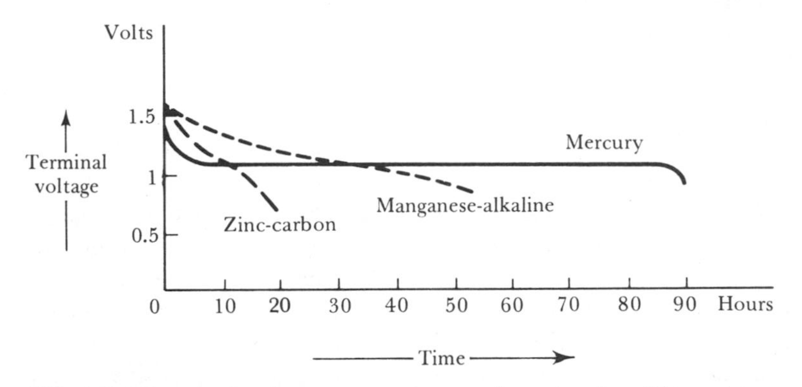

FIG. 6-7. Terminal voltage versus hours of service for different types of D cells

NICKEL–CADMIUM CELL. The nickel–cadmium cell is a rechargeable cell widely used today in electronic calculators and other such portable equipment. The initial cost of a set of nickel–cadmium cells and charger is much greater than the cost of a set of zinc–carbon cells. However, given the fact that the nickel–cadmium cells could easily last 10 years, at about two chargings per week, they are by far the least expensive in the long run.

The construction of a nickel–cadmium cell is shown in Figure 6-8. The two plates are long strips of nickel-plated steel to which nickel powder has been sintered at high temperatures. This results in a mechanically rugged electrode with a very porous surface. For the positive electrode, the porous surface contains *nickel–hydroxide*. The negative plate contains *metallic cadmium*. Separating the two electrodes is a strip of absorbent material impregnated with the *potassium hydroxide* electrolyte. The two plates and the separator are rolled up together and enclosed in a

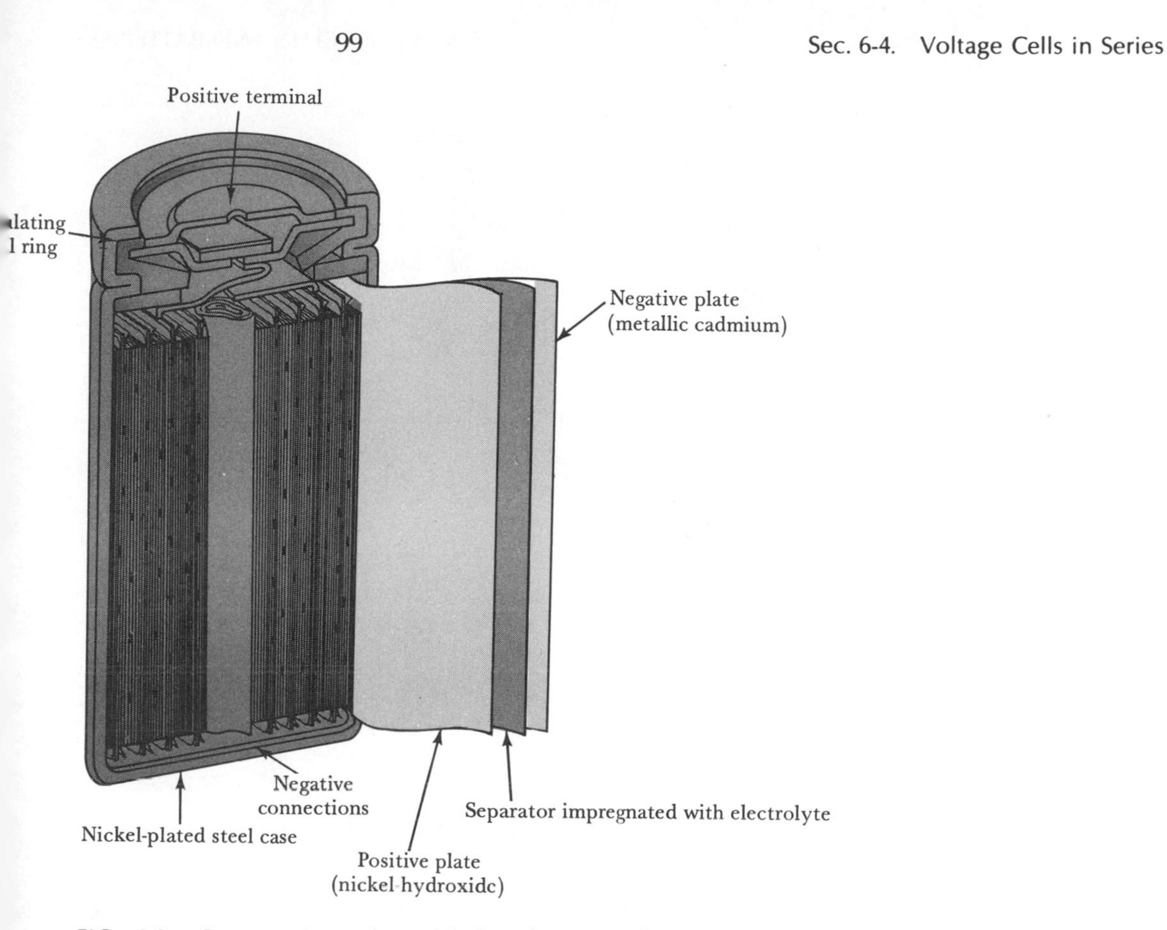

FIG. 6-8. Construction of a nickel–cadmium cell

hermetically sealed steel container. Connections are provided from the two electrodes to the outside terminals.

When discharging (i.e., supplying an output current), electrons flow through the external circuit from the cadmium to the nickel. During this time the electrolyte atoms are combining with cadmium atoms, which leave electrons behind them on the cadmium electrode. Also, electrons are being absorbed from the nickel–hydroxide. The reverse of this occurs during the recharging process (see Section 6-6).

The output voltage from a nickel–cadmium cell is about 1.3 V on open circuit, dropping to around 1.25 V under load.

6-4 VOLTAGE CELLS IN SERIES

When a number of voltage cells are connected together in *series* or *parallel*, or in a *series–parallel* combination, the group is usually referred to as a *battery*.

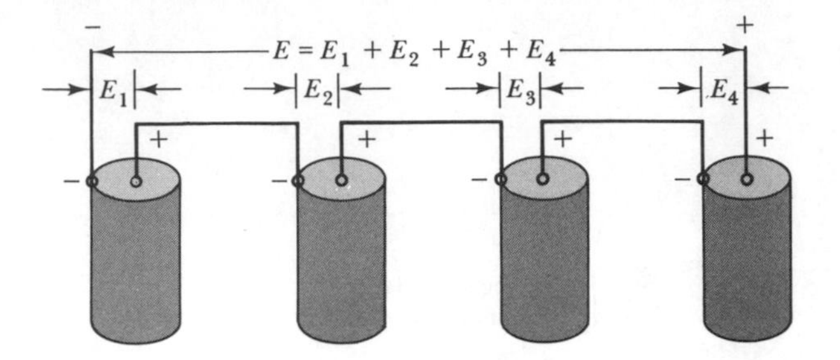

(a) Four cells correctly connected in series

(b) Circuit diagram for series-connected cells

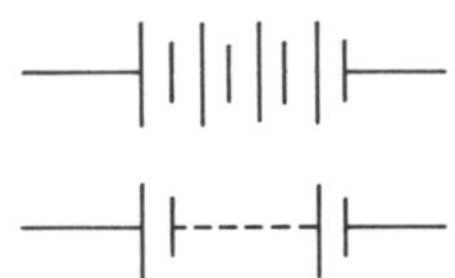

(c) Alternative circuit diagrams for series-connected cells

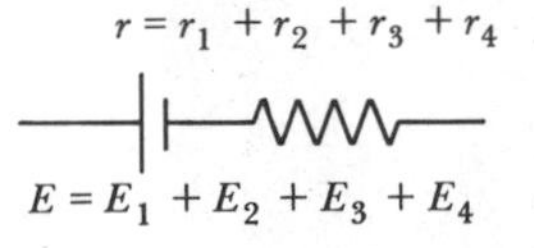

(d) Equivalent circuit for series-connected cells

FIG. 6-9. Series-connected cells, with circuit diagrams and equivalent circuit for series-connected cells

Figure 6-9(a) shows a battery of four cells connected in series. The negative terminal of each cell is connected to the positive terminal of the previous cell. With this arrangement, the cell voltages add together to give a relatively large battery terminal voltage. Thus, if each cell in Figure 6-9(a) has an emf of 1.5 V, the total output voltage is

$$\boxed{E = E_1 + E_2 + E_3 + E_4 + \cdots} \quad \text{(6-4)}$$

$$= 1.5\text{ V} + 1.5\text{ V} + 1.5\text{ V} + 1.5\text{ V}$$

$$= 6\text{ V}$$

Obviously, if the cell emf's were not all 1.5 V, the output voltage from the battery would still be the sum of the individual cell voltages.

Figure 6-9(b) shows the circuit diagram for the four cells in series. Alternative ways of graphically representing a battery of series-connected cells are shown in Figure 6-9(c).

The total internal resistance of the battery of series-connected cells is, of course, the sum of the internal resistances of the individual cells:

$$r_i = r_1 + r_2 + r_3 + \cdots \tag{6-5}$$

Consequently, the equivalent circuit of the four-cell battery consists of cells having an emf of $E = (E_1 + E_2 + E_3 + E_4)$ and an internal resistance of $r_i = (r_1 + r_2 + r_3 + r_4)$ [see Figure 6-9(d)].

When an external load resistance (R_L) is connected to a group of series-connected cells (Figure 6-10), the output current is

$$I = \frac{E}{r + R_L}$$

or

$$I = \frac{E_1 + E_2 + E_3 + E_4}{R_L + r_1 + r_2 + r_3 + r_4} \tag{6-6}$$

Note from the circuit that the output current flows through each individual cell. Therefore, if each cell is capable of supplying a maximum current of 1 A, for example, the output current from the group of cells cannot exceed 1 A. The actual output voltage and current from the battery can be calculated from Eqs. (6-4) and (6-6).

Series-connected cells produce an output voltage equal to the sum of the individual cell voltages and can supply a maximum current equal to the maximum that can be taken from any one cell.

EXAMPLE 6-3

The battery shown in Figure 6-10(a) is made up of four individual cells each of which has an open-circuit terminal voltage of $E_c = 2$ V and an internal resistance of $r = 0.1\ \Omega$. Determine the terminal voltage of the

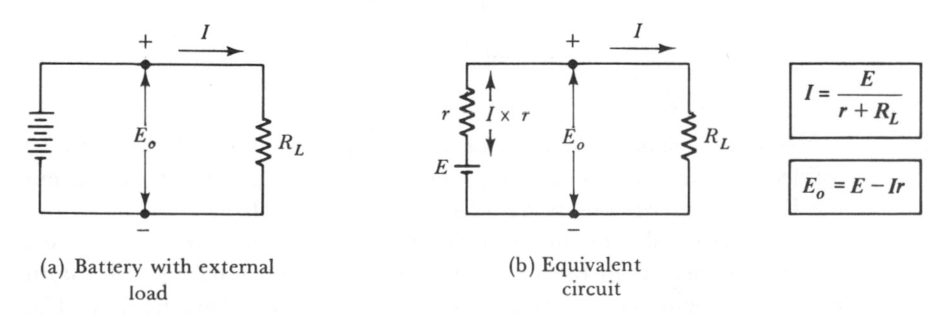

(a) Battery with external load

(b) Equivalent circuit

FIG. 6-10. Battery of cells with an external load, and equivalent circuit

battery for no-load and for $R_L = 9.6\ \Omega$. Also determine the current that flows when the battery terminals are short-circuited.

SOLUTION

$$E = E_1 + E_2 + E_3 + E_4 = 4 \times 2\text{ V}$$
$$= 8\text{ V}$$

$$r = r_1 + r_2 + r_3 + r_4 = 4 \times 0.1\ \Omega$$
$$= 0.4\ \Omega$$

For no-load:

$$I_o = 0$$

and

$$E_o = E - I_o r = 8\text{ V} - 0$$
$$\mathbf{E_o = 8\text{ V}}$$

For $R_L = 9.6\ \Omega$:

$$I_o = \frac{E}{r + R_L} = \frac{8\text{ V}}{0.4 + 9.6\ \Omega}$$
$$\mathbf{I_o = 0.8\text{ A}}$$

$$E_o = E - I_o r = 8\text{ V} - (0.8\text{ A} \times 0.4\ \Omega)$$
$$\mathbf{E_o = 7.68\text{ V}}$$

For short circuit:

$$R_L = 0$$
$$E_o = \frac{E}{r + 0} = \frac{8\text{ V}}{0.4\ \Omega + 0}$$
$$\mathbf{I_o = 20\text{ A}}$$

Note that short circuiting of a battery or cell terminals is to be avoided, because it rapidly discharges the cells. Also, the large current that flows can permanently damage rechargeable cells.

The series cells illustrated in Figure 6-11(a) each have their positive terminal connected to the negative terminal of the adjacent cell. With this arrangement they assist each other to produce a current flow. The connection method is termed *series-aiding*. When cells are series-connected

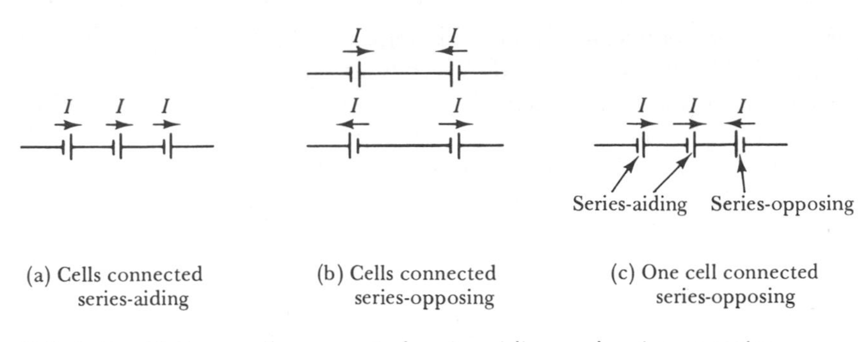

FIG. 6-11. Voltage cells connected series-aiding and series-opposing

to produce currents in opposite directions, as shown in Figure 6-11(b), the connection is referred to as *series-opposing*. Series-opposing connections are normally avoided, and usually occur only as a result of error. The total output voltage is still the sum of the individual cell voltages, but the signs of the voltages must be carefully considered. As illustrated in Figure 6-11(c), when one of three 1.5-V cells is connected series-opposing, the resultant terminal voltage of the battery is

$$\begin{aligned} E &= E_1 + E_2 + E_3 \\ &= 1.5\text{ V} + 1.5\text{ V} + (-1.5\text{ V}) \\ &= 1.5\text{ V} \end{aligned}$$

If a discharged cell is connected in series with fully charged cells, the discharged cell adds little or nothing to the total output voltage. However, it does increase the total internal resistance of the battery; therefore, a discharged cell should *not* be connected in series with good fully charged cells.

In Figure 6-12 two batteries, each having three series-connected cells are shown with their common terminal grounded. If each cell has an emf

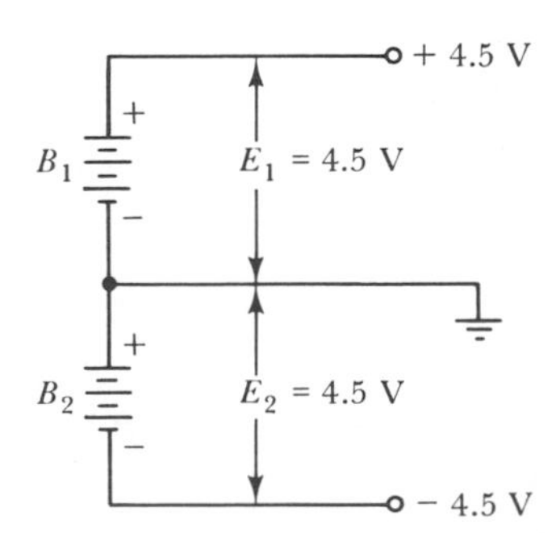

FIG. 6-12. Two batteries of cells connected to provide ±4.5 V

of 1.5 V, each battery has a terminal voltage of 4.5 V. Since the negative terminal of battery B_1 is grounded, its output voltage is $E_1 = 4.5$ V, that is, positive with respect to ground. The positive terminal of B_2 is grounded; therefore, the output voltage of B_2 is $E_2 = -4.5$ V, that is, negative with respect to ground. The combined output of the two batteries is usually designed ± 4.5 *V*. Many electronics circuits use this *plus–minus* type of supply.

6-5 VOLTAGE CELLS IN PARALLEL

Consider two voltage cells connected in parallel as illustrated in Figure 6-13(a). If each cell has a terminal voltage of exactly 1.5 V, the output voltage from the parallel combination is also 1.5 V.

$$\boldsymbol{E = E_1 = E_2 = E_3 = \cdots} \tag{6-7}$$

If each cell is capable of supplying a maximum output current of 1 A, the maximum output current that can be drawn from the two-cell parallel combination is 2 A. Similarly, for three such cells in parallel the maximum output current is three times the maximum current per cell. Of course, the actual output current still depends upon the output voltage and the load resistance. As shown by Figure 6-13(b), neglecting cell

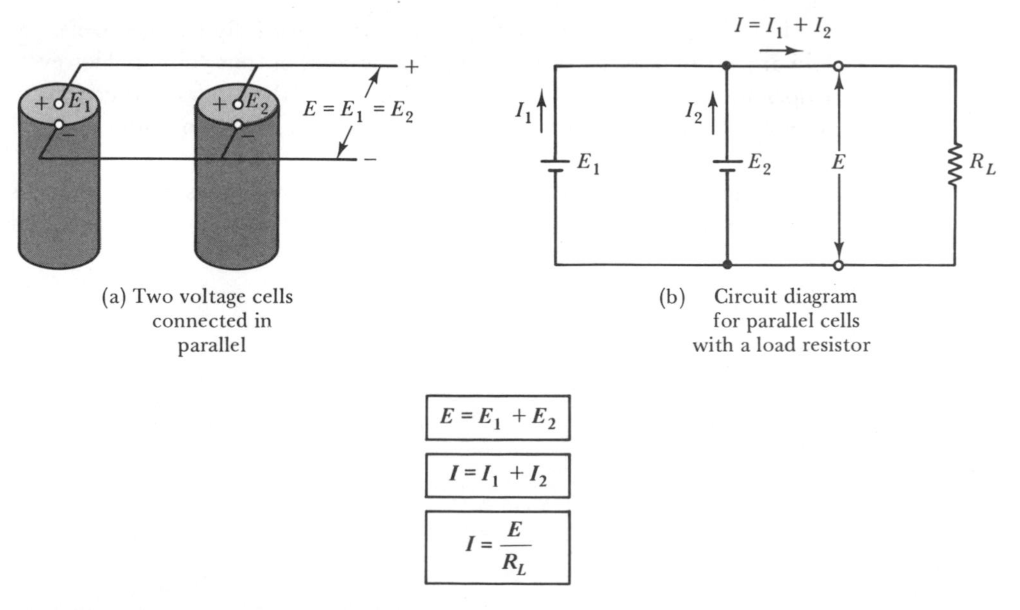

FIG. 6-13. Voltage cells in parallel

internal resistances the output current is

$$I_o = \frac{E}{R_L}$$

But the output current I_o is the sum of the individual currents from each cell. Therefore,

$$\boxed{I_o = I_1 + I_2 + I_3 + \cdots + I_n} \qquad \textbf{(6-8)}$$

For the case of two 1.5-V cells in parallel supplying a load resistor of 5 Ω:

$$I_o = \frac{1.5\text{ V}}{5\ \Omega} = 0.3\text{ A}$$

and $$I_1 = I_2 = 0.15\text{ A}$$

giving $$I_o = 0.15\text{ A} + 0.15\text{ A} = 0.3\text{ A}$$

EXAMPLE 6-4

A 0.2-Ω resistance is to be supplied from a parallel-connected battery of 2-V cells. If each cell can supply a maximum current of 0.5 A, determine the number of cells that should be connected in parallel. Assume the cell internal resistances to be negligible.

SOLUTION

$$I_o = \frac{E}{R_L} = \frac{2\text{ V}}{0.2\ \Omega} = 10\text{ A}$$

$$\textit{number of parallel cells} = \frac{I_o}{\textit{current per cell}} = \frac{10\text{ A}}{0.5\text{ A}}$$

$$\boldsymbol{\textit{number of parallel cells} = 20}$$

Each cell in a parallel combination of cells can be represented by its own equivalent circuit [Figure 6-14(a)]. The equivalent circuit of the battery of cells can then be seen to be a voltage source equal to the terminal voltage of each cell, in series with an internal resistance that

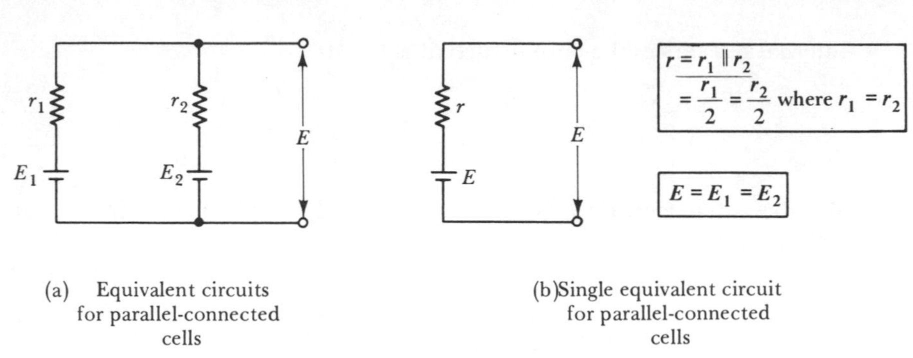

(a) Equivalent circuits for parallel-connected cells

(b) Single equivalent circuit for parallel-connected cells

FIG. 6-14. Equivalent circuits for voltage cells in parallel

equals the parallel combination of the cell internal resistances. This is illustrated in Figure 6-14(b). For the case of two 1.5-V cells, each having an internal resistance of 0.1 Ω, the equivalent circuit has

$$E = E_1 = E_2 = 1.5 \text{ V}$$

and $$r = r_1 \text{ in parallel with } r_2$$

or $$r = r_1 \| r_2 = 0.05\ \Omega$$

(This is explained in Chapter 8, "Parallel Resistive Circuits.")

If there were three identical cells in parallel, the internal resistance of one cell would be divided by 3 to find the internal resistance of the equivalent circuit, or

$$\boxed{r_i = r_1 \| r_2 \| r_3} \tag{6-9}$$

Taking the cell internal resistance into consideration, the output current from a group of parallel cells supplying a load resistance R_L is

$$\boxed{I_o = \frac{E}{R_L + (r_1 \| r_2 \| r_3 \cdots)}} \tag{6-10}$$

Parallel-connected cells produce an output voltage equal to the terminal voltage of one cell and can supply a maximum output current equal to the sum of the maximum currents from each cell.

EXAMPLE 6-5

Four cells connected in parallel each have terminal voltage of $E = 2$ V and internal resistances of $r_i = 0.1\ \Omega$. Each cell can normally supply a maximum load current of 500 mA for a time period of 4 h. Determine the output voltage of the parallel combination and the maximum load current that can be supplied. Also, calculate the Ah rating and the internal resistance for the battery of cells.

SOLUTION

$$\textit{output voltage} = \textit{cell terminal voltage}$$

$$\mathbf{E = 2\ V}$$

Eq. (6-8):

$$I_o = I_1 + I_2 + I_3 + I_4$$

$$= 4 \times 500\ \text{mA}$$

$$\mathbf{I_o = 2\ A}$$

$$\text{Ah } \textit{rating of battery} = 4 \times (\text{Ah } \textit{rating per cell})$$

$$= 4 \times (500\ \text{mA} \times 4\ \text{h})$$

$$\textbf{Ah } \boldsymbol{\textit{rating of battery}} \mathbf{= 8\ Ah}$$

$$r = r_1 \| r_2 \| r_3 \| r_4$$

$$= \frac{0.1\ \Omega}{4}$$

$$\mathbf{r = 0.025\ \Omega}$$

Figure 6-15 shows the circuit diagram of two parallel-connected groups of cells, each group consisting of three 1.5 V-cells in series. The terminal voltage of each group of series cells is obviously

$$E = E_1 + E_2 + E_3$$

$$= 1.5\ \text{V} + 1.5\ \text{V} + 1.5\ \text{V}$$

$$= 4.5\ \text{V}$$

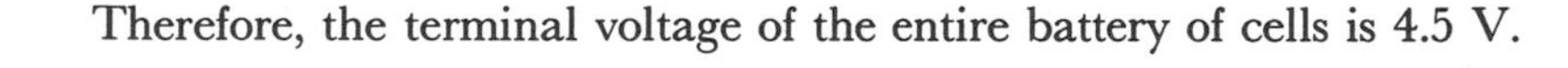

Therefore, the terminal voltage of the entire battery of cells is 4.5 V.

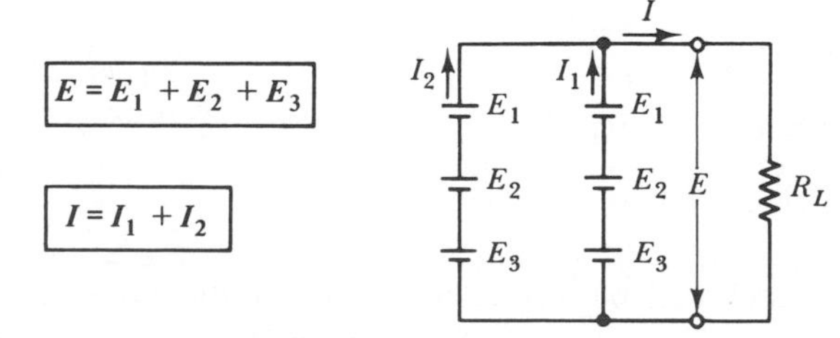

FIG. 6-15. Voltage cells connected in series-parallel

If each cell is capable of supplying a maximum current of 1 A, the maximum output current that can be drawn from each series group is also 1 A. For the entire battery of cells, the maximum output current is

$$I = I_1 + I_2$$
$$= 1\ \text{A} + 1\ \text{A}$$
$$= 2\ \text{A}$$

This kind of *series–parallel* arrangement of cells can be employed to give any desired combination of output voltage and maximum current. Where a greater output voltage is required, more cells are added to each series group. For larger maximum output currents, the number of parallel groups is increased.

In some cases cells connected in parallel may not have voltages which are exactly equal. For example, in Figure 6-13 E_1 may be 1.54 V while E_2 is 1.49 V. When this occurs, the cell with the largest terminal voltage will tend to discharge through the one with the smaller terminal voltage, until both have equal voltages. If secondary (i.e., rechargeable) cells are used, one cell is charged from the other. *Cells that do not have closely equal terminal voltages should not be connected together.*

It is very important to ensure that parallel cells are correctly connected. Correct and incorrect ways of connecting cells in parallel are shown in Figure 6-16. All positive terminals should be connected together, and all negative terminals should be connected together, as shown in Figure 6-16(a). Where positive and negative terminals are connected together as in Figure 6-16(b), *the cells short-circuit each other and discharge rapidly*.

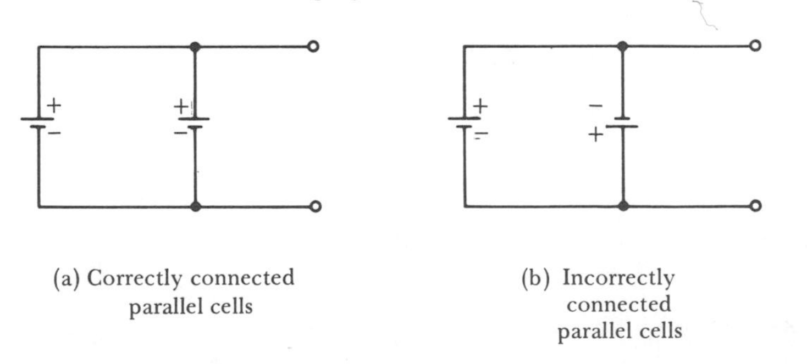

(a) Correctly connected parallel cells

(b) Incorrectly connected parallel cells

FIG. 6-16. Correctly and incorrectly connected parallel voltage cells

6-6 LEAD–ACID BATTERY

CONSTRUCTION. The lead–acid battery is the most commonly used type of storage battery and is well known for its application in automobiles. The battery is made up of several cells, each of which consists of lead plates immersed in an electrolyte of dilute sulfuric acid. The voltage per cell is typically 2 V to 2.2 V. For a 6-V battery, three cells are connected in series, and for a 12-V battery six cells are series-connected.

The construction of a lead–acid automobile-type battery is illustrated in Figure 6-17. The electrodes are lead–antimony alloy plates with a pattern of recesses, so that they are in the form of grids [see Figure 6-17(a)]. Lead oxide is pressed into the recesses of the plates. Each electrode consists of several plates connected in parallel with porous rubber separators in between, as illustrated in Figure 6-17(b). This arrangement, and the shape of the plates, gives the largest possible electrode surface area within the size limitations of the battery.

The complete 12-V battery, illustrated in Figure 6-17(c), has an outer case of hard rubber. The case is divided into six sections for the six

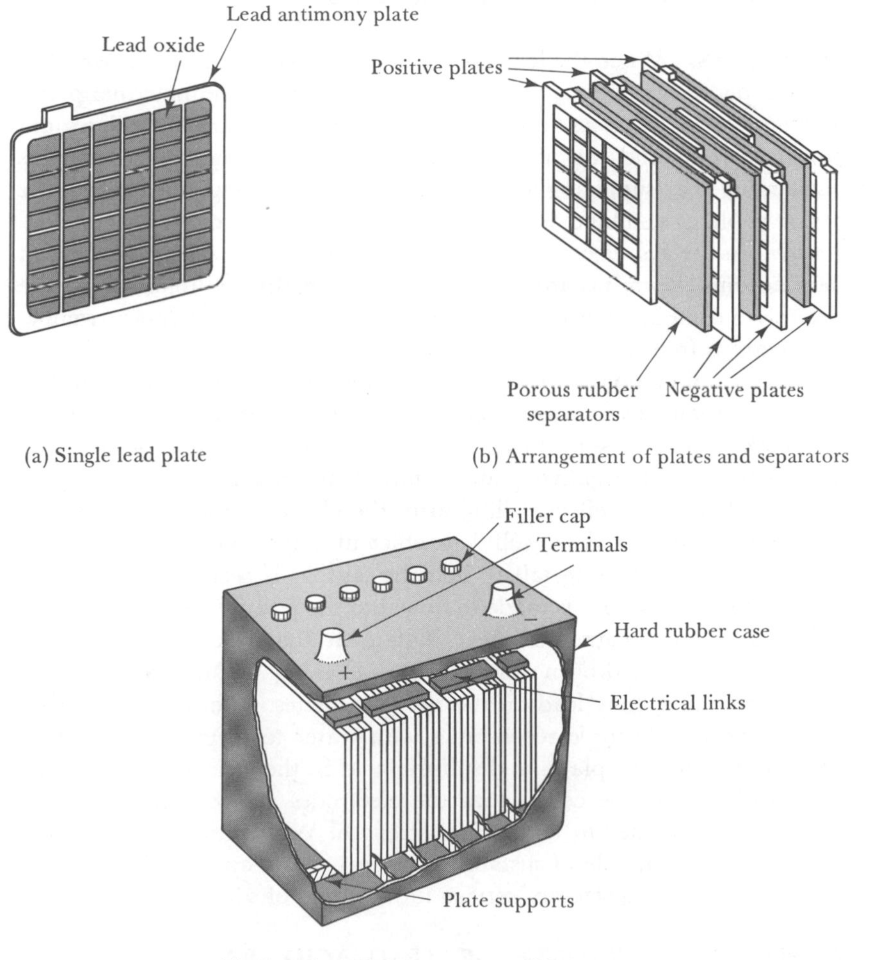

(a) Single lead plate

(b) Arrangement of plates and separators

(c) Complete 12-V battery assembly

FIG. 6-17. Construction of a lead–acid battery

separate cells. Projections are provided on the inside at the bottom of the case to support the plates. These projections ensure that the lower edges of the plates are normally well above the level of any active material that falls to the bottom of a cell. Such material can short out the positive and negative plates, and render a cell useless.

Every cell has a threaded *filler cap* with a small hole in its center. The filler caps provide access for adding electrolyte, and the holes allow gases to be vented to the atmosphere.

Thick electrical links connect the cells in series, and hefty battery terminals are provided. Since the battery may be required to supply a very heavy current, it is important that the resistance of all electrical connections be very low, to minimize voltage drops. A current of 250 A is not unusual for a battery driving an automobile starter.

OPERATION. When the lead–acid cell is charged, the lead oxide on the positive plates changes to *lead peroxide*, and that on the negative plates becomes *spongy* or *porous* lead. In this condition, the positive plates are brown in color and the negative plates are gray. When the battery is discharging (i.e., supplying a current), atoms from the spongy lead on the negative plates combine with sulfate molecules to form *lead sulfate* and *hydrogen*. As always, electrons are left behind on the negative plates so that they accumulate a negative potential. The hydrogen released in the electrolyte combines with the lead peroxide on the positive plate, removing electrons from the plate to give it a positive potential.

The combination of lead peroxide and hydrogen at the positive electrode produces water and lead sulfate. The water dilutes the electrolyte, making it a weaker solution, and the lead sulfate that is produced at both positive and negative plates tends to fill the pores of the active material. Both these effects (dilution of the electrolyte and formation of lead sulfate) render each cell less efficient and eventually cause the battery output voltage to fall. When the battery is recharged, a current (conventional direction) is made to flow into the positive electrode of each cell. This current causes the lead sulfate at the negative electrode to recombine with hydrogen ions, thus re-forming sulfuric acid in the electrolyte and spongy lead on the negative plates. Also, the lead sulfate on the positive electrode recombines with water to regenerate lead peroxide on the positive plates and sulfuric acid in the electrolyte. The final result of charging the cell is that the electrodes are re-formed and the electrolyte is returned to its original strength. With proper care a lead–acid battery is capable of sustaining a great many cycles of charge and discharge, giving satisfactory service for a period of several years.

AMPERE-HOUR RATING OF LEAD–ACID BATTERY. Typical ampere-hour ratings for 12-V lead–acid automobile batteries range from

100 Ah to 300 Ah. This is usually specified for an 8-h discharge time, and it defines the amount of energy that can be drawn from the battery until the voltage drops to about 1.7 V per cell.

For a 240-Ah rating, the battery could be expected to supply 30 A for an 8-h period. With greater load currents, the discharge time is obviously shorter. However, the ampere-hour rating is also likely to be reduced for a shorter discharge time, because the battery is less efficient when supplying larger currents. Another method of rating a lead–acid battery is to define what its terminal voltage will be after about 5 s of supplying perhaps 250 A. This corresponds with the kind of load that a battery experiences in starting an automobile.

The rating of a battery is typically stated for temperatures around 25°C, and this must be revised for operation at lower temperatures. Since the chemical reactions occur more slowly at reduced temperatures, the available output current and voltage is less than that at 25°C. Around −18°C a fully charged battery may be capable of delivering only 60% of its normal ampere-hour rating. As the cell is discharged and the electrolyte becomes weaker, freezing of the electrolyte becomes more likely. A fully charged cell is less susceptible to freezing, but even a fully charged cell may fail when its temperature falls to about −21°C.

EXAMPLE 6-6

A lead–acid battery has a rating of 300 Ah.

a. Determine how long the battery might be employed to supply 25 A.

b. If the battery rating is reduced to 100 Ah when supplying large currents, calculate how long it could be expected to supply 250 A.

c. Under very cold conditions the battery supplies only 60% of its normal rating. Find the length of time that it might continue to supply 250 A to the starter motor of an automobile.

SOLUTION

a.

$$\text{Ah } rating = I \times t$$

Therefore,

$$t = \frac{\text{Ah}}{I} = \frac{300 \text{ Ah}}{25 \text{ A}}$$

$$\boldsymbol{t = 12 \text{ h}}$$

b.

$$t = \frac{\text{Ah}}{I} = \frac{100 \text{ Ah}}{250 \text{ A}} = 0.4 \text{ h}$$

$$\boldsymbol{t = 24 \text{ min}}$$

c.

$$Ah = 60\% \text{ of } 100\ Ah$$

$$\mathbf{Ah = 60\ Ah}$$

$$t = \frac{60\ Ah}{250\ A} = 0.24$$

$$\mathbf{t = 14.4\ min}$$

CHARGING. When a battery is to be charged, a dc charging voltage must be applied to its terminals. The polarity of the charging voltage must be such that it causes current to flow into the battery, in opposition to the normal direction of discharge current. This means that the positive output terminal of the *battery charger* must be connected to the positive terminal of the battery, and the charger negative terminal must be connected to the battery negative terminal. The arrangement is shown in Figure 6-18. The battery charger normally has a voltmeter and ammeter to monitor the charging voltage and current, and a control to adjust the rate of charge.

The output voltage of a battery charger must be greater than the battery voltage in order to cause current to flow into the battery positive

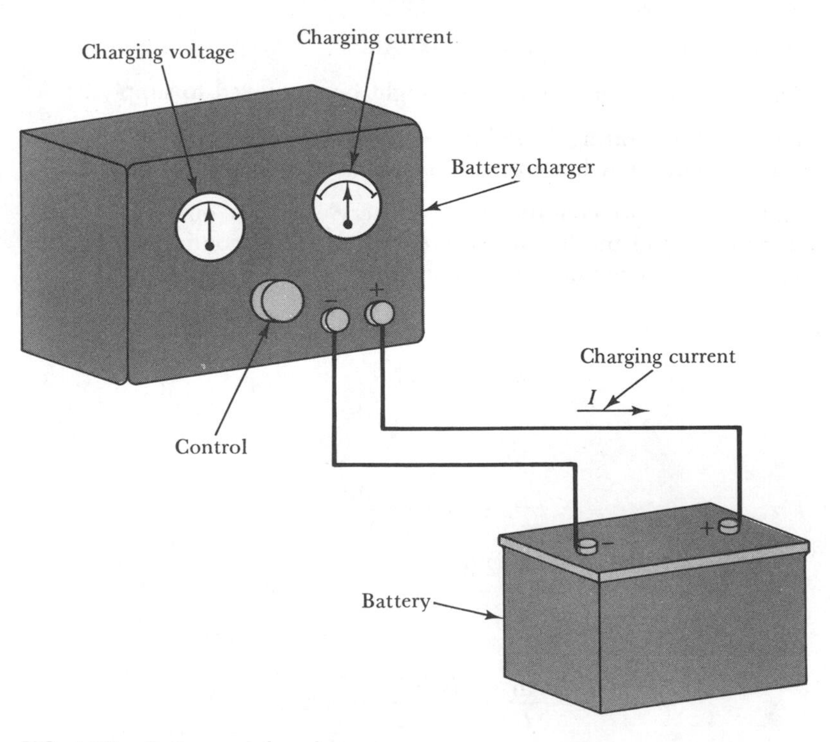

FIG. 6-18. Battery under charge

terminal. The charging current depends upon the difference between the battery voltage and the charging voltage and upon the internal resistances of the battery. A very large charging current is to be avoided because it could cause the battery to overheat, possibly resulting in warping of the lead plates. The maximum safe charging current is frequently taken as the maximum output current from the battery when discharging at its 8-h rate.

EXAMPLE 6-7

A battery with a rating of 300 Ah is to be charged.

a. Determine a safe maximum charging current.

b. If the internal resistance of the battery is 0.008 Ω and its (discharged) terminal voltage is 11.5 V, calculate the initial output voltage level for the battery charger.

SOLUTION

a. *Safe rate of charge at the* 8-h *discharge rate:*

$$I = \frac{\text{Ah}}{t} = \frac{300\ \text{Ah}}{8}$$

Therefore,

$$\mathbf{I = 37.5\ A}$$

b. *Charging current:*

$$I = \frac{\textit{charger voltage} - \textit{battery voltage}}{r_i}$$

Therefore,

$$\textit{charger voltage} = (I \times r_i) + \textit{battery voltage}$$

$$= (37.5\ \text{A} \times 0.008\ \Omega) + 11.5\ \text{V}$$

$$\textbf{\textit{charger voltage}} = \mathbf{11.8\ V}$$

SPECIFIC GRAVITY. When a lead–acid battery is in a nearly discharged condition, the electrolyte is in its weakest state. Conversely, the electrolyte is at its strongest (or greatest density) when the battery is fully charged. The density of electrolyte related to the density of water is termed its *specific gravity*. The specific gravity of the electrolyte is used as a measure of the state of charge of a lead–acid battery. Electrolyte with a specific gravity of 1100 to 1150 is 11 to 11.5 times as dense as water. At 1100 to 1150 the cell is completely discharged. When the specific gravity is 1280 to 1300, the cell may be assumed to be fully charged.

Specific gravity is measured by means of a *hydrometer*. This instrument consists of a glass tube with a rubber bulb on one end, a thin rubber tube on the other end, and a calibrated float inside the tube (see Figure 6-19).

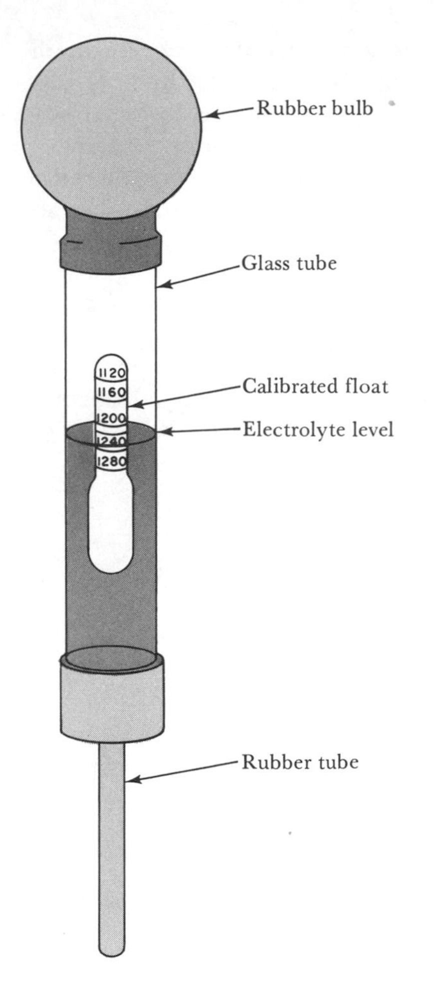

FIG. 6-19. Hydrometer used for testing the specific gravity of an electrolyte

Each cell of a battery is tested by inserting the hose into the electrolyte, then squeezing the rubber bulb and releasing it to suck some electrolyte into the glass tube. The calibrated float becomes partially submerged in the electrolyte inside the glass tube, sinking deep into weak electrolyte, and not so deep into more dense electrolyte. As illustrated in Figure 6-19, the specific gravity of the liquid is indicated by the level of electrolyte on the float.

CARE OF LEAD–ACID BATTERIES.

- The level of the electrolyte in each cell should be checked regularly, and distilled water added as necessary to keep the top of the plates covered by about 1 cm of liquid.

- Battery terminals should be kept clean and lightly coated with Vaseline to avoid corrosion.
- In cold weather, batteries should always be maintained in a nearly fully charged condition to avoid freezing.
- Lead–acid batteries should never be allowed to remain for long periods in a discharged state, because lead sulfate could harden and permanently clog the pores of the electrodes.
- Before storing for a long time the battery should be completely charged, then the electrolyte should be drained out so that the battery is stored dry.

6-7 STANDARD CELL

The *mercury–cadmium cell*, also known as the *Weston cadmium cell*, is applied exclusively as a standard reference voltage source. When such cells are correctly constructed by different people working to the same specification, the cell terminal voltage differences are on the order of microvolts.

The cell illustrated in Figure 6-20 consists of an H-shaped tube containing electrolyte with the electrodes at the bottom of each leg of the tube. The electrolyte is cadmium sulfate in sulfuric acid. The positive electrode is a pool of mercury, and the negative electrode is a pool of cadmium amalgam. The positive electrode also has a layer of mercurous sulfate which acts as a depolarizer. Two cork washers are included, as illustrated, to retain the liquid electrodes in place.

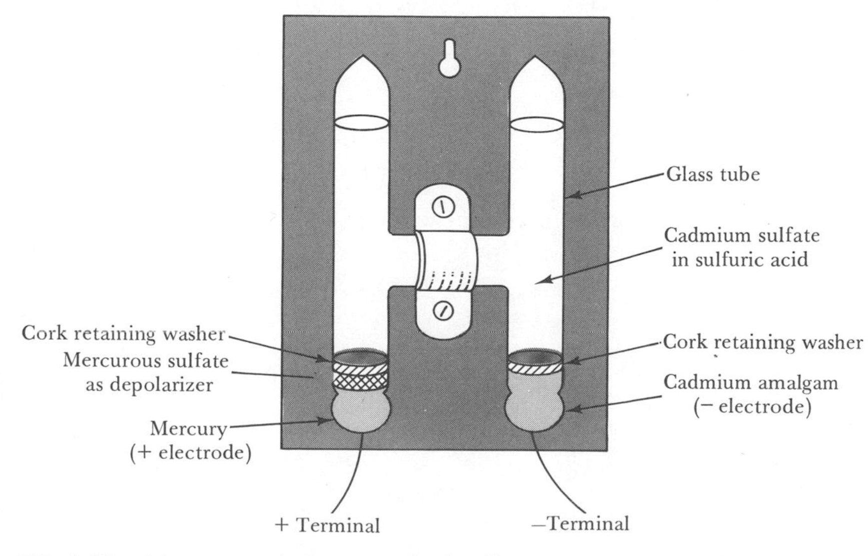

FIG. 6-20. Mercury–cadmium standard cell

In another more accurate version of the standard cell, the cork washers are omitted. This, of course, necessitates extreme care in handling.

For the type of standard cell illustrated in Figure 6-20, the terminal voltage is typically 1.0190 V to 1.0194 V. The actual value is specified by the manufacturer. Obviously, great care must be exercised in using this device. It is very important to remember that this cell is not for supplying current; nothing greater than 10 to 20 μA should be allowed to flow. If the cell is ever short-circuited it is rendered useless. In use, a resistor is normally connected in series with a standard cell to protect it from overload.

EXAMPLE 6-8

A standard cell is to be connected to a piece of electrical equipment which could have a terminal voltage of 2 V. Calculate the value of series resistance that must be employed to limit the cell current to a maximum of 10 μA.

SOLUTION

In the worst circumstances the equipment terminal voltage and the cell voltage might be connected series-aiding. Therefore, the voltage acting around the circuit is

$$E = \textit{cell voltage} + \textit{equipment terminal voltage}$$

$$= 1.019\ 4\ \text{V} + 2\ \text{V}$$

$$= 3.019\ 4\ \text{V}$$

$$R = \frac{E}{I}$$

Therefore,

$$\textit{series resistance} = \frac{3.019\ 4\ \text{V}}{10\ \mu\text{A}}$$

$$\textbf{\textit{series resistance}} \cong \mathbf{300\ k\Omega}$$

GLOSSARY OF FORMULAS

Cell output current:

$$I_o = \frac{E_c}{r_i + R_L}$$

Cell output voltage:

$$E_o = E_c - I_o r_i$$

For cells in series:

$$E = E_1 + E_2 + E_3 + \cdots$$

$$r_i = r_1 + r_2 + r_3 + \cdots$$

$$I = \frac{E_1 + E_2 + E_3 + \cdots}{R_L + r_1 + r_2 + r_3 + \cdots}$$

For cells in parallel:

$$E = E_1 = E_2 = E_3 = \cdots$$

$$r_i = r_1 \| r_2 \| r_3 \cdots$$

$$I = \frac{E}{R_L + (r_1 \| r_2 \| r_3 \cdots)}$$

REVIEW QUESTIONS

6-1 Draw a sketch to show the basic construction of a simple voltage cell. Explain the operation of the cell.

6-2 Discuss the output voltage and current that can be supplied by a simple voltage cell and define ampere-hour rating.

6-3 Define primary cell, secondary cell, open-circuit output voltage, no-load output voltage, internal resistance, and ideal cell.

6-4 Sketch the equivalent circuit for a voltage cell with an external load resistance (R_L). Write equations for the cell terminal voltage and output current. Explain briefly.

6-5 Draw a sketch to show the basic construction of a typical zinc–carbon dry cell. Explain the operation of the cell and discuss its ampere-hour rating.

6-6 Discuss the construction, terminal voltages, ampere-hour ratings, and application for the following: manganese–alkaline cell, mercury cell, and nickel–cadmium cell.

6-7 Sketch the circuit diagram for four voltage cells connected in series. Draw the equivalent circuit for the series combination of cells and write the equations for output voltage and current.

6-8 Draw circuit diagrams to show series-aiding and series-opposing connected cells and briefly explain. Also, show how a ±10-V supply could be constructed using 2-V cells.

6-9 Sketch the circuit diagram for four voltage cells connected in parallel. Draw the equivalent circuit for the parallel combination of cells and write the equations for output voltage and current.

6-10 Explain what occurs when cells having different terminals voltages are connected in parallel. Also, discuss the result of incorrectly connecting voltage cells in parallel.

6-11 Describe the construction of a 12-V lead–acid battery and explain the chemical process involved in charge and discharge within the individual cells.

6-12 Discuss the typical ampere-hour rating of a lead–acid battery with respect to supplying small currents, large currents, and low-temperature performance.

6-13 Explain how a lead–acid battery should be connected to a battery charger and discuss charging voltage and current levels.

6-14 Define specific gravity and explain how the specific gravity of a lead–acid cell is tested. Also, state typical specific-gravity values for discharged and fully charged batteries.

6-15 Discuss the care of lead–acid batteries.

6-16 Define a standard cell. Draw a sketch to show the construction of a mercury–cadmium cell. Explain briefly. Discuss briefly the care that should be exercised in using a standard cell.

PROBLEMS

6-1 A voltage cell supplies a current of 0.75 A for 10 h. Then its terminal voltage drops to a low level. Calculate the Ah rating of the cell and determine how long it might be expected to supply a current of 250 mA.

6-2 The open-circuit voltage from a certain voltage cell is 1.65 V and its internal resistance is 0.05 Ω. Calculate the terminal voltage of the cell when the output current is 5 A. Also, determine the cell current when the terminals are short-circuited.

6-3 A battery is made up of five series-connected voltage cells, each of which has an open-circuit terminal voltage of 1.6 V and an internal resistance of 0.08 Ω. Calculate the battery terminal voltages for no-load condition and for $R_L = 6\ \Omega$.

6-4 A 12-V supply is to be constructed using 1.5-V cells each of which has $r_i = 0.05\ \Omega$. Determine the number of cells required and calculate the internal resistance of the battery. Also, calculate the battery terminal voltage when the load resistance is 10 Ω.

6-5 A parallel-connected battery of 1.5-V cells is to supply current to a 0.15-Ω resistor. If each cell can supply a maximum current of 1 A, determine the number of cells required. Neglect the internal resistances of each cell.

6-6 Eight voltage cells connected in parallel each have an open-circuit terminal voltage of 1.7 V and an internal resistance of 0.07 Ω. Each cell can normally supply a maximum load current of 750 mA for a time period of 12 h. Determine the output voltage of the battery of cells and the maximum load current that can be supplied. Also, calculate the Ah rating and internal resistance for the parallel combination of cells.

6-7 A power supply to be constructed from voltage cells is required to supply a maximum current of 20 A with a terminal voltage of approximately 6 V. Several voltage cells are available, each of which has a terminal voltage of 1.5 V and each of which can supply a maximum current of 5 A. Determine the number of cells required and draw a circuit diagram to show how they should be connected.

6-8 A certain lead–acid battery has a rating of 400 Ah.

a. Calculate how long the battery should be able to supply a current of 12 A.

b. When supplying large currents the battery rating is reduced to 150 Ah. Determine how long it can be expected to supply 240 A to an automobile starter motor.

c. If the battery rating is reduced to 60% of normal when very cold, calculate how long it might supply 240 A.

6-9 A 400-Ah lead–acid battery is to be charged.

a. Calculate a maximum safe level of charging current.

b. The internal resistance of the battery is 0.007 Ω and the terminal voltage when discharged is 11.4 V. Determine the initial output voltage level for the battery charger.

6-10 A standard cell with a terminal voltage of 1.019 4 V is connected to a piece of equipment that can have an output of 2.5 V. Calculate the series resistance value that should be used to limit the maximum cell current to 10 μA.

7

SERIES RESISTIVE CIRCUITS

INTRODUCTION

Resistors are said to be in *series* when they are connected in such a way that there is only one path through which current can flow. This means that the *current in a series circuit* is the same in all parts of the circuit.

The *voltage drop* across each component *in a series circuit* is dependent upon the current level and the component resistance. Two or more series-connected resistors can be used as a *voltage divider*. The *potentiometer* is an adjustable resistor used as a variable voltage divider.

The total power supplied to *a series circuit* is the sum of the powers dissipated in the individual components.

Resistors connected in series with electrical equipment may be used for the purpose of *voltage dropping and current limiting*. *Open circuits* and *short circuits* affect the current flow in a series circuit.

7-1 CURRENT IN A SERIES CIRCUIT

Three *series-connected* resistors are shown in Figure 7-1(a) together with a battery to supply emf and current. It is seen that the resistors are connected end to end in such a way that there is only one path through which current can flow. The current path is from the positive terminal of the battery (conventional current direction), through resistor R_1 from top to bottom, and similarly through R_2 and R_3 and then into the negative terminal of the battery. Clearly, all of the current that flows into one end of R_1 must flow out at the other end. This same current flows into one end

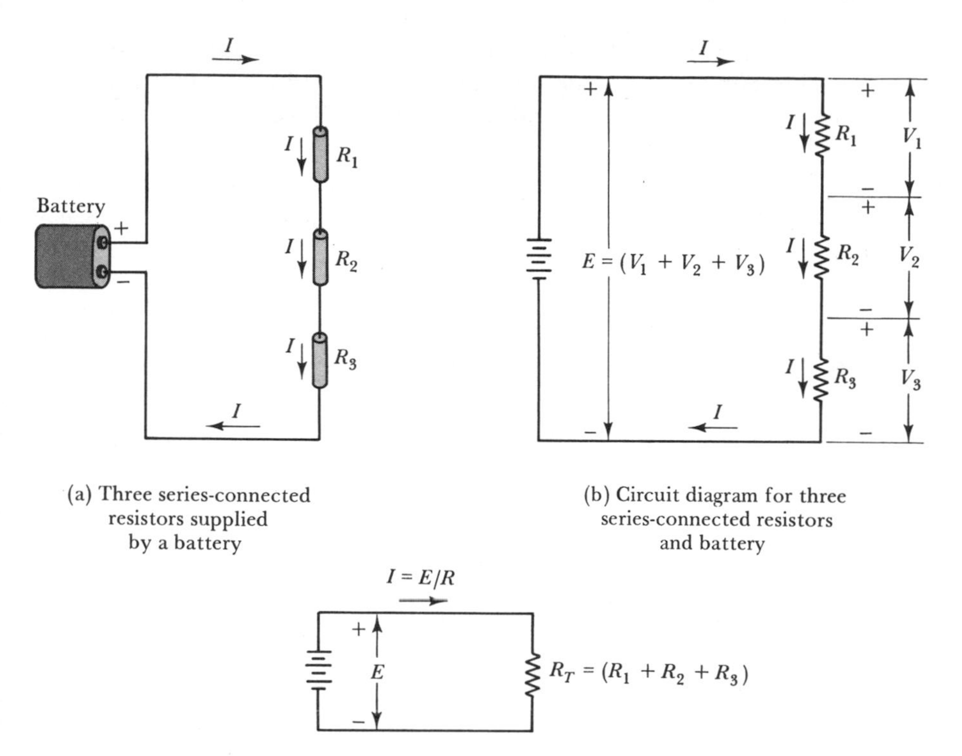

(a) Three series-connected resistors supplied by a battery

(b) Circuit diagram for three series-connected resistors and battery

(c) Equivalent circuit

FIG. 7-1. Series-connected resistors

of R_2 and out at the other end, and then through R_3. Thus, it is seen that

The current is the same in all parts of a series circuit.

Figure 7-1(b) shows the circuit diagram for the three resistors and battery. The total resistance connected across the battery terminals is obviously

$$R = R_1 + R_2 + R_3$$

For any series circuit with n resistors:

$$\boxed{R = R_1 + R_2 + R_3 + \cdots + R_n} \tag{7-1}$$

Using Ohm's law, the current through the series circuit is calculated as

$$\boxed{I = \frac{E}{R_1 + R_2 + R_3 + \cdots + R_n}} \tag{7-2}$$

An *equivalent circuit* can be drawn for the series resistance circuit as shown in Figure 7-1(c). The equivalent circuit consists simply of a voltage source and a resistance R equal to the sum of all resistors in series.

EXAMPLE 7-1

Calculate the current that flows through three resistors connected in series, as illustrated in Figure 7-1. The supply voltage is $E=9$ V, and the resistors are $R_1=15\ \Omega$, $R_2=25\ \Omega$, and $R_3=5\ \Omega$.

SOLUTION

Eq. (7-2):

$$I=\frac{E}{R_1+R_2+R_3}$$

$$=\frac{9\text{ V}}{15\ \Omega+25\ \Omega+5\ \Omega}$$

$$\boldsymbol{I=0.2\text{ A}}$$

7-2 VOLTAGE DROPS IN A SERIES CIRCUIT

Referring again to Figure 7-1(b), it is seen that the current flow causes a *voltage drop*, or potential difference, across each resistor. If there was no potential difference between the terminals of each resistor, there would be no current flow.

Across R_1:

$$V_1=IR_1$$

Across R_2:

$$V_2=IR_2$$

Across R_3:

$$V_3=IR_3$$

Note that the polarity of the resistor volt drops is always such that the (conventional) current direction is from positive to negative. Thus, for the circuit as shown, the polarity is + at the top of each resistor, − at the bottom. Also note that the most positive end of each resistor is the end nearest the positive terminal of the battery (i.e., via the current path). The most negative end of each resistor is nearest the battery negative terminal. The sum of the resistor voltage drops is $V_1+V_2+V_3$, and, as shown in Figure 7-1(b), these must be equal to the applied emf E. For any series circuit,

$$E=V_1+V_2+V_3+\cdots+V_n$$

or

$$\boxed{E = IR_1 + IR_2 + IR_3 + \cdots + IR_n} \qquad (7\text{-}3)$$

Therefore,

$$\boxed{E = I(R_1 + R_2 + R_3 + \cdots + R_n)} \qquad (7\text{-}4)$$

The relationship between the applied emf and the resistor voltage drops in a series circuit is defined by *Kirchhoff's voltage law**:

Kirchhoff's Voltage Law

In any closed electric circuit, the algebraic sum of the voltage drops must equal the algebraic sum of the applied emfs.

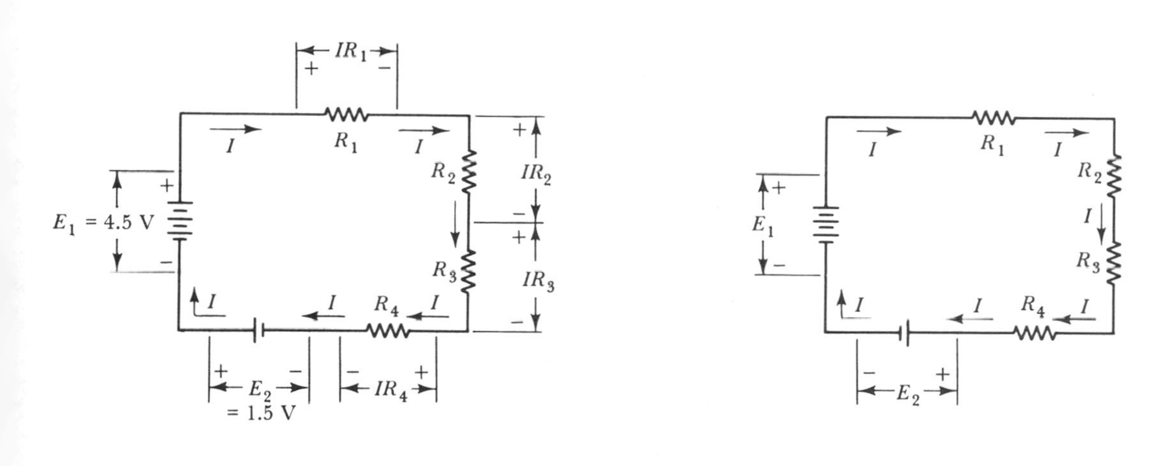

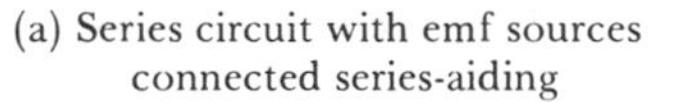
(a) Series circuit with emf sources connected series-aiding

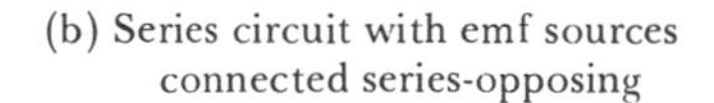
(b) Series circuit with emf sources connected series-opposing

FIG. 7-2. Series circuits with two sources of emf

Where more than one battery or other source of emf is involved, Kirchhoff's voltage law still applies. Consequently, for the circuit shown in Figure 7-2(a),

$$E_1 + E_2 = IR_1 + IR_2 + IR_3 + IR_4$$

When the sources of emf are connected *series-opposing*, as illustrated in

*Formulated by the German physicist Gustav Kirchhoff (1824–1887).

Figure 7-2(b), the voltage equation becomes

$$E_1 - E_2 = IR_1 + IR_2 + IR_3 + IR_4$$

EXAMPLE 7-2

The four resistors in Figure 7-2 have the following values: $R_1 = 5\ \Omega$, $R_2 = 13\ \Omega$, $R_3 = 25\ \Omega$, and $R_4 = 17\ \Omega$. The emfs are $E_1 = 4.5$ V and $E_2 = 1.5$ V. Determine the circuit current and resistor volts drops:

a. When the emf polarities are as in Figure 7-2(a).

b. When E_2 is reversed as in Figure 7-2(b).

SOLUTION

a.

$$E_1 + E_2 = I(R_1 + R_2 + R_3 + R_4)$$

Therefore,

$$I = \frac{E_1 + E_2}{R_1 + R_2 + R_3 + R_4} = \frac{4.5\text{ V} + 1.5\text{ V}}{5\ \Omega + 13\ \Omega + 25\ \Omega + 17\ \Omega}$$

$$\mathbf{I = 0.1\ A}$$

$$V_1 = IR_1 = 0.1\text{ A} \times 5\ \Omega$$

$$\mathbf{V_1 = 0.5\ V}$$

$$V_2 = IR_2 = 0.1\text{ A} \times 13\ \Omega$$

$$\mathbf{V_2 = 1.3\ V}$$

$$V_3 = IR_3 = 0.1\text{ A} \times 25\ \Omega$$

$$\mathbf{V_3 = 2.5\ V}$$

$$V_4 = IR_4 = 0.1\text{ A} \times 17\ \Omega$$

$$\mathbf{V_4 = 1.7\ V}$$

b.

$$I = \frac{E_1 + E_2}{R_1 + R_2 + R_3 + R_4} = \frac{4.5\text{ V} + (-1.5\text{ V})}{5\ \Omega + 13\ \Omega + 25\ \Omega + 17\ \Omega}$$

$$\mathbf{I = 50\ mA}$$

$$V_1 = IR_1 = 50\text{ mA} \times 5\ \Omega$$

$$\mathbf{V_1 = 250\ mV}$$

$$V_2 = IR_2 = 50\text{ mA} \times 13\ \Omega$$

$$\mathbf{V_2 = 650\ mV}$$

$$V_3 = IR_3 = 50\text{ mA} \times 25\ \Omega$$

$$\mathbf{V_3 = 1.25\ V}$$

$$V_4 = IR_4 = 50\text{ mA} \times 17\ \Omega$$

$$\mathbf{V_4 = 850\ mV}$$

7-3 VOLTAGE DIVIDER

It has been shown that the voltage drops across a *string* of series resistors add up to the value of the supply emf E. Another way of looking at this is that the applied emf is divided up between the series resistors.

Figure 7-3 shows two series-connected resistors used as a *voltage divider*. From previous studies,

$$I=\frac{E}{R_1+R_2}$$

Also,

$$V_1=IR_1$$

Therefore,

$$V_1=\frac{E}{R_1+R_2}\times R_1$$

or

$$\boxed{V_1=E\times\frac{R_1}{R_1+R_2}} \tag{7-5}$$

Similarly,

$$V_2=IR_2$$

or

$$\boxed{V_2=E\times\frac{R_2}{R_1+R_2}} \tag{7-6}$$

Equations (7-5) and (7-6) illustrate the *voltage divider theorem*:

Voltage Divider Theorem

In a series circuit, the portion of applied emf developed across each resistor is in the ratio of that resistor's value to the total series resistance.

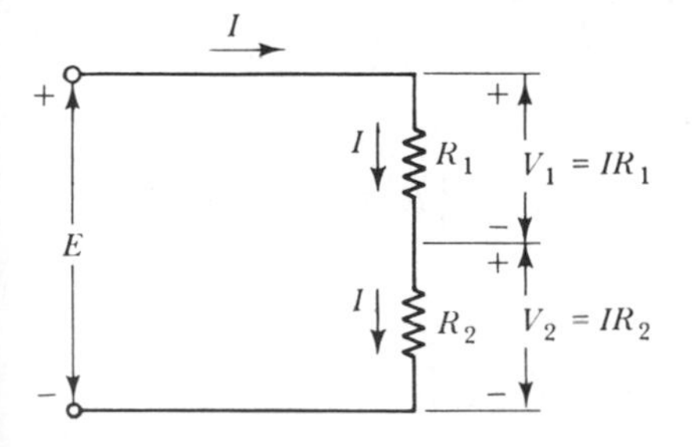

FIG. 7-3. Two series resistors as a voltage divider

EXAMPLE 7-3

For the voltage divider circuit in Figure 7-3, the applied emf is $E=100$ V, and the resistor values are $R_1=22\ \Omega$ and $R_2=28\ \Omega$. Calculate the values of V_1 and V_2.

SOLUTION

$$V_1=E\times\frac{R_1}{R_1+R_2}$$

$$=100\ \text{V}\times\frac{22\ \Omega}{28\ \Omega+22\ \Omega}$$

$$\mathbf{V_1=44\ V}$$

$$V_2=E\times\frac{R_2}{R_1+R_2}$$

$$=100\ \text{V}\times\frac{28\ \Omega}{28\ \Omega+22\ \Omega}$$

$$\mathbf{V_2=56\ V}$$

7-4 POTENTIOMETER

The circuit diagram of a variable resistor used as a *potentiometer* is illustrated in Figure 7-4. The construction of such a resistor is shown in Figure 5-8(a), and discussed in Section 5-6.

The potentiometer is essentially a single resistor with terminals at each end, and a moving contact that can be set to any point on the resistor. Thus, when the moving contact is exactly halfway between the two end terminals, the resistance from the moving contact to each terminal is exactly half the total resistance of the potentiometer (see Figure 7-4). In this situation the *output voltage between points C and B is*

$$V_o=E\times\frac{R_2}{R_1+R_2}$$

$$=E\times\frac{R/2}{R/2+R/2}$$

$$=\text{E}\times\tfrac{1}{2}$$

If the moving contact is adjusted until $R_2=\frac{1}{4}R$, then $R_1=\frac{3}{4}R$, and the output voltage becomes

$$V_o=E\times\frac{R/4}{3R/4+R/4}$$

$$=E\times\tfrac{1}{4}$$

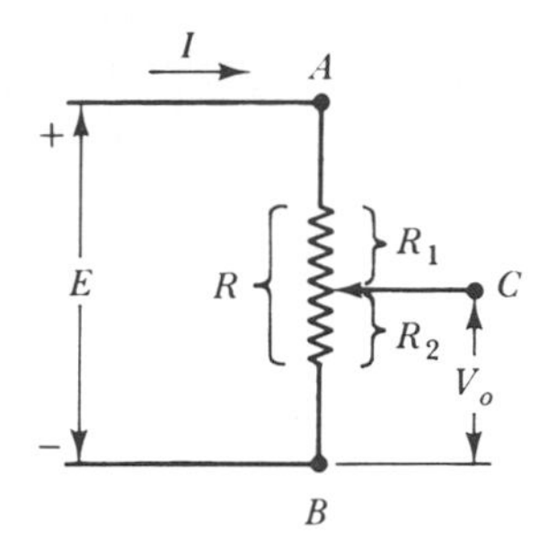

FIG. 7-4. The potentiometer

When $R_2 = \frac{3}{4}R$, the output voltage is

$$V_o = E \times \tfrac{3}{4}$$

and when $R_2 = R$,

$$V_o = E$$

Thus, it is seen that the potentiometer can be adjusted to give an output voltage ranging from 0 V to E.

Ideally, the resistance between the moving contact on a potentiometer and either one of the end terminals should increase or decrease smoothly as the device is adjusted. Because of the construction of most potentiometers, however, this resistance change normally occurs in small jumps. How smoothly any given potentiometer changes its resistance when adjusted is defined in terms of the potentiometer *resolution*. The resolution is usually defined in terms of a number of steps. Thus, the step changes in resistance of a 1-kΩ potentiometer with a resolution of 400 steps is

$$\text{resistance steps} = \frac{1\ \text{k}\Omega}{400} = 2.5\ \Omega$$

A resolution of 400 steps (or 1 in 400) also means that the step changes in the *output voltage* are

$$\Delta V = \frac{E}{400}$$

7-5 POWER IN A SERIES CIRCUIT

Using Eq. (3-5), (3-8), or (3-9), it is possible to calculate the power dissipated in a resistor from a knowledge of any two of the three quantities: current, voltage, and resistance. Thus, in Figure 7-3 power dissipated in R_1 is

$$P_1 = V_1 I$$

or

$$P_1 = \frac{V_1^2}{R_1}$$

or

$$P_1 = I^2 R_1$$

The power dissipated in R_2 is calculated in exactly the same way, and the total power dissipated in the circuit is the sum of the individual resistor power dissipations. For any series resistance circuit, the total power dissipated is

$$\boxed{\mathbf{P = P_1 + P_2 + P_3 + \cdots + P_n}} \qquad \textbf{(7-7)}$$

and

$$P = V_1 I + V_2 I + V_3 I + \cdots + V_n I$$
$$= I(V_1 + V_2 + V_3 + \cdots + V_n)$$
$$P = IE = \text{battery output power}$$

The total power can also be calculated as

$$\boxed{\mathbf{P = \frac{E^2}{R_1 + R_2 + R_3 + \cdots + R_n}}} \qquad \textbf{(7-8)}$$

or

$$\boxed{\mathbf{P = I^2 (R_1 + R_2 + R_3 + \cdots + R_n)}} \qquad \textbf{(7-9)}$$

EXAMPLE 7-4

Determine the total power dissipation and the power dissipated in each resistor in Figure 7-3 when the component values are as specified in Example 7-3.

SOLUTION

$$P_1 = \frac{V_1^2}{R_1} = \frac{(44\ \text{V})^2}{22}$$

$$\mathbf{P_1 = 88\ W}$$

$$P_2 = \frac{V_2^2}{R_2} = \frac{(56\ \text{V})^2}{28}$$

$$\mathbf{P_2 = 112\ W}$$

$$P = P_1 + P_2 = 88\ \text{W} + 112\ \text{W}$$

$$\mathbf{P = 200\ W}$$

or
$$P = \frac{E^2}{R_1 + R_2} = \frac{(100\ \text{V})^2}{22\ \Omega + 28\ \Omega}$$

$$P = 200\ \text{W}$$

EXAMPLE 7-5 A 2-kΩ potentiometer is connected across a supply voltage of $E = 100$ V. Determine the minimum power rating for the potentiometer.

SOLUTION

$$\textit{minimum power rating},\ P = \frac{E^2}{R}$$

$$= \frac{(100\ \text{V})^2}{2\ \text{k}\Omega}$$

$$P = 5\ \text{W}$$

7-6 VOLTAGE DROPPING AND CURRENT LIMITING

Sometimes a resistor is included in series with a circuit or electronic device to drop the supply voltage down to a required level. In other circumstances this same kind of series resistor arrangement can be thought of as a current-limiting resistor.

In the circuit shown in Figure 7-5(a), series resistor R_s provides voltage drop to operate an electronic circuit which requires a voltage level lower than the source voltage. In Figure 7-5(b) R_s limits the current that flows through the three series-connected lamps L_1, L_2, and L_3. It can also be shown that in Figure 7-5(b), R_s drops the supply voltage down to the level required by the three lamps.

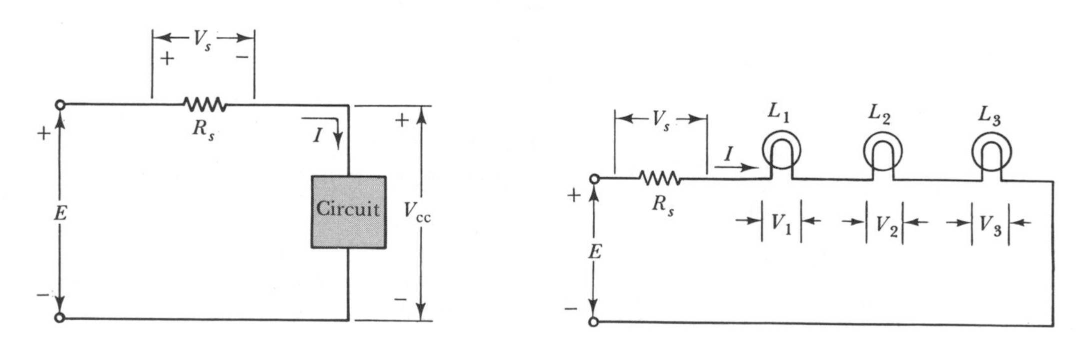

(a) Use of a series voltage-dropping resistor (b) Use of a series current-limiting resistor

FIG. 7-5. Series resistors used for voltage dropping and current limiting

EXAMPLE 7-6

A certain electronic circuit takes a current of $I = 20$ mA when supplied with $V_{cc} = 6$ V. If the circuit is to be supplied from a source of $E = 24$ V, as illustrated in Figure 7-5(a), determine the required value of series resistance. Also calculate the power rating of the series resistance.

SOLUTION

$$V_s = E - V_{cc} = 24\text{ V} - 6\text{ V}$$

$$\boldsymbol{V_s = 18\text{ V}}$$

$$R_s = \frac{V_s}{I_s} = \frac{18\text{ V}}{20\text{ mA}}$$

$$\boldsymbol{R_s = 900\ \Omega}$$

$$P_s = I^2 R_s = (20\text{ mA})^2 \times 900\ \Omega$$

$$\boldsymbol{P_s = 0.36\text{ W}}$$

EXAMPLE 7-7

Three 6-V 3-W lamps connected in series as shown in Figure 7-5(b) are to be supplied from a 50-V source. Calculate the resistance value and power rating for the series current-limiting resistor.

SOLUTION

For each lamp,

$$P = VI$$

Therefore,

$$I = \frac{P}{V} = \frac{3\text{ W}}{6\text{ V}}$$

$$\boldsymbol{I = 0.5\text{ A}}$$

Total voltage across the three lamps is

$$V = V_1 + V_2 + V_3$$

$$= 6\text{ V} + 6\text{ V} + 6\text{ V}$$

$$\boldsymbol{V = 18\text{ V}}$$

$$V_s = E - (V_1 + V_2 + V_3)$$

$$= 50\text{ V} - 18\text{ V}$$

$$\boldsymbol{V_s = 32\text{ V}}$$

$$R_s = \frac{V_s}{I} = \frac{32\text{ V}}{0.5\text{ A}}$$

$$\boldsymbol{R_s = 64\ \Omega}$$

$$P_s = V_s I = 32\text{ V} \times 0.5\text{ A}$$

$$\boldsymbol{P_s = 16\text{ W}}$$

7-7 OPEN CIRCUITS AND SHORT CIRCUITS IN A SERIES CIRCUIT

An *open circuit* occurs in a series resistance circuit when one of the resistors becomes disconnected from an adjacent resistor [see Figure 7-6(a)]. An open circuit can also occur when one of the resistors has been destroyed by excessive power dissipation.

In the circuit of Figure 7-6(a), the open circuit can be thought of as another resistance in series with R_1, R_2, and R_3. Thus, instead of the current being $I = E/(R_1 + R_2 + R_3)$, it becomes

$$I = \frac{E}{R_1 + R_2 + R_3 + R_{oc}}$$

Suppose that $R_{oc} = 100\ 000\ \text{M}\Omega$ and $E = 100$ V; then, assuming that $R_{oc} \gg (R_1 + R_2 + R_3)$,

$$I \cong \frac{100\text{ V}}{100\ 000\text{ M}}$$

$$= 1\text{ nA}$$

This small current causes an insignificant volts drop along R_1, R_2, and R_3. Consequently, the voltage across the open circuit is

$$V_{oc} \cong E$$

Figure 7-6(b) shows a series resistance circuit with resistor R_3 *short-circuited*. In this case the resistance between the terminals of R_3 is effectively zero. Consequently, instead of the current being $I = E/(R_1 + R_2 + R_3)$, it becomes

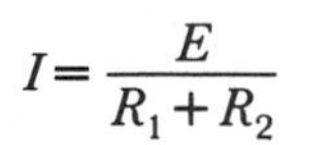

$$I = \frac{E}{R_1 + R_2}$$

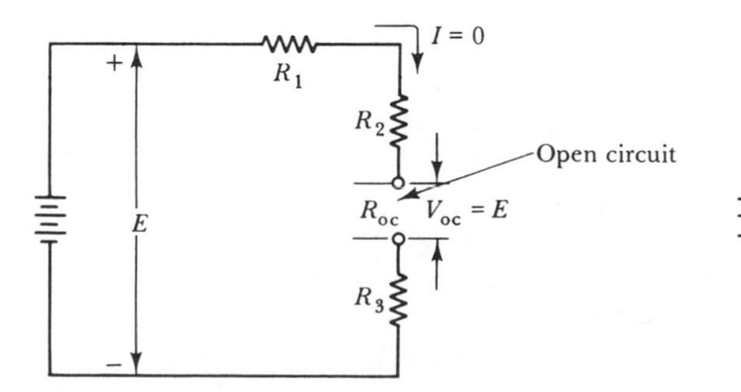

(a) Open-circuit condition in a series circuit

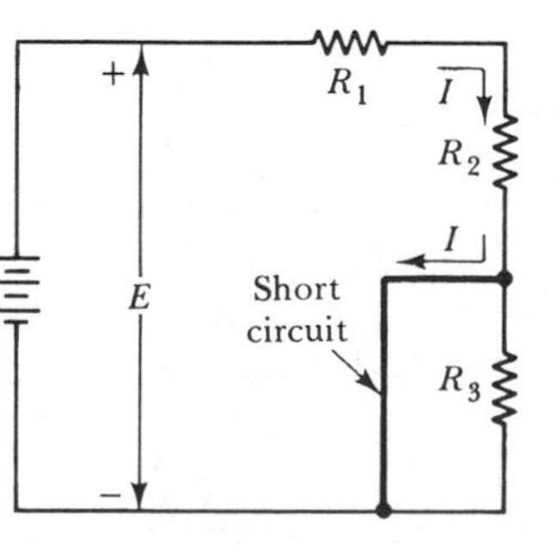

(b) Short-circuit condition in a series circuit

FIG. 7-6. Open-circuit and short-circuit conditions in a series resistance circuit

It is obvious that open circuits and short circuits considerably affect the current flow through a series resistance circuit.

GLOSSARY OF FORMULAS

Total series resistance:

$$R = R_1 + R_2 + R_3 + \cdots + R_n$$

Current in a series circuit:

$$I = \frac{E}{R_1 + R_2 + R_3 + \cdots + R_n}$$

Total voltage:

$$E = IR_1 + IR_2 + IR_3 + \cdots + IR_n$$

$$E = I(R_1 + R_2 + R_3 + \cdots + R_n)$$

For a voltage divider:

$$V_1 = E \times \frac{R_1}{R_1 + R_2}$$

$$V_2 = E \times \frac{R_2}{R_1 + R_2}$$

Total power:

$$P = P_1 + P_2 + P_3 + \cdots + P_n$$

$$P = \frac{E^2}{R_1 + R_2 + R_3 + \cdots + R_n}$$

$$P = I^2(R_1 + R_2 + R_3 + \cdots + R_n)$$

REVIEW QUESTIONS

7-1 Sketch the circuit diagram for four series-connected resistors supplied from a battery. Show the current direction and indicate the polarity of all voltage drops. Define a series circuit in terms of the current.

7-2 Write the equations for total resistance and current in a series resistance circuit.

7-3 Write the equations for the voltage drops across the resistors in a series circuit and for the total voltage. State Kirchhoff's voltage law.

7-4 Sketch the circuit of three series resistors supplied by two batteries which are connected series-opposing. Explain the effect that the series-opposing connection has on the circuit compared to the situation when the batteries are connected series-aiding.

7-5 Sketch the circuit of two resistors employed as a voltage divider. Derive equations for the voltage drop across each resistor in terms of the supply voltage. State the voltage divider theorem.

7-6 Sketch the circuit diagram of a potentiometer supplied with voltage from a battery. Explain the principle of the potentiometer and discuss what is meant by the resolution of a potentiometer.

7-7 Write equations for the total power supplied to a series circuit in terms of:

a. The power dissipated in each component.

b. The supply voltage and the resistor values.

c. The current and the resistor values.

7-8 Explain what is meant by a voltage dropping resistor. Sketch the diagram of a circuit using a voltage dropping resistor. Explain briefly.

7-9 Five 12-V lamps are to be supplied from a 120-V source. Sketch a circuit diagram to show how the lamps should be connected and explain briefly.

7-10 Using circuit diagrams, explain open circuits and short circuits, and discuss the effect that each has on a series resistance circuit.

PROBLEMS

7-1 Calculate the current that flows through four series-connected resistors when $R_1 = 150\ \Omega$, $R_2 = 250\ \Omega$, $R_3 = 125\ \Omega$, $R_4 = 75\ \Omega$, and the battery voltage is $E = 12$ V.

7-2 Calculate the voltage drops across each resistor in Problem 7-1 and show that they add up to equal the supply voltage.

7-3 For the circuit described in Review Question 7-4, take $R_1 = 12\ \Omega$, $R_2 = 22\ \Omega$, and $R_3 = 16\ \Omega$. Also, let the battery voltage be $E_1 = 9$ V and $E_2 = 6$ V. Determine the circuit current and resistor voltage drops:

a. When the batteries are connected series-aiding.

b. When they are connected series-opposing.

7-4 A voltage divider uses two resistors connected in series across a 75-V supply. If the resistor values are $R_1 = 37\ \Omega$ and $R_2 = 88\ \Omega$, calculate the output voltage that may be taken across each resistor.

7-5 Calculate the power dissipated in each resistor in the circuit of Problem 7-3 for both case a and b connections of the batteries as described. Also, calculate the total power dissipation in each case and the power output from each battery.

7-6 If the lamps described in Review Question 7-9 are each rated at 25 W, calculate the size and power rating of the required current-limiting resistor.

7-7 A small transistor radio normally takes 15 mA from a 9-V battery. Calculate the value of the resistor that must be connected in series with the radio if it is to be supplied from a 12-V battery. Also, calculate the power rating for the series resistor.

7-8 In the circuit described in Review Question 7-9 and Problem 7-6, one of the lamps becomes short-circuited. Calculate the new level of current and the power dissipated in each lamp.

8

PARALLEL RESISTIVE CIRCUITS

INTRODUCTION

Resistors are said to be connected in parallel when the same voltage appears across every component. With different resistance values, different currents flow through each resistor. The total current taken from the supply is the sum of all the individual resistor currents. The equivalent resistance of a parallel resistor circuit is most easily calculated by using the reciprocal of each individual resistance value (i.e., using conductance values).

Two resistors connected in parallel may be used as a *current divider*. In a parallel circuit, as in a series circuit, the total power supplied is the sum of the powers dissipated in the individual components. Open-circuit and short-circuit conditions in a parallel circuit have an effect on the total supply current.

8-1 VOLTAGE AND CURRENT IN A PARALLEL CIRCUIT

Resistors are connected in parallel when the circuit has two terminals which are common to every resistor. Figure 8-1 shows two resistors, R_1 and R_2, connected in parallel and supplied by a battery. It is seen that the battery voltage E appears across R_1 and across R_2. Thus, it can be stated:

Resistors are connected in parallel when the same voltage is applied across each resistor.

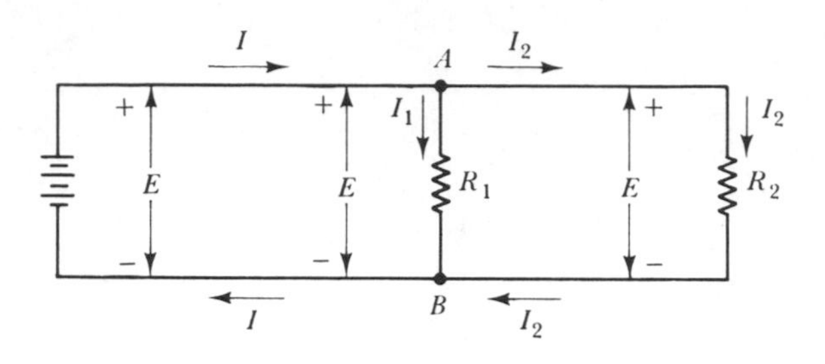

FIG. 8-1. Currents and voltages in a parallel resistance circuit

This should be contrasted to the situation in a series circuit where the same current flows through each resistor.

In a parallel circuit, *different* currents flow through each parallel component. From consideration of Figure 8-1, the current through each of the resistors is

$$I_1 = \frac{E}{R_1}$$

and

$$I_2 = \frac{E}{R_2}$$

Now look at the current directions with respect to junction A. I_1 flowing through R_1 is flowing *away* from junction A, and I_2 flowing through R_2 is also flowing away from A. The battery current I is flowing toward A, and I, I_1, and I_2 are the only currents entering or leaving the junction. Consequently,

$$I = I_1 + I_2$$

The same reasoning at junction B, where I_1 and I_2 are entering B and I is leaving B also gives

$$I = I_1 + I_2$$

In the case where there are n resistors in parallel, the battery current is

$$\boxed{I = I_1 + I_2 + I_3 + \cdots + I_n} \qquad \textbf{(8-1)}$$

The rule about currents entering and leaving a junction is defined in *Kirchhoff's current law*:

Kirchhoff's Current Law

The algebraic sum of the currents entering a point in an electric circuit must equal the algebraic sum of the currents leaving that point.

EXAMPLE 8-1

The parallel resistors shown in Figure 8-1 have values of $R_1 = 12\ \Omega$ and $R_2 = 15\ \Omega$. The battery voltage is $E = 9$ V. Calculate the current that flows through each resistor and the total current drawn from the battery.

SOLUTION

$$I_1 = \frac{E}{R_1} = \frac{9\text{ V}}{12\ \Omega}$$

$$\mathbf{I_1 = 0.75\ A}$$

$$I_2 = \frac{E}{R_2} = \frac{9\text{ V}}{15\ \Omega}$$

$$\mathbf{I_2 = 0.6\ A}$$

$$I = I_1 + I_2 = 0.75\text{ A} + 0.6\text{ A}$$

$$\mathbf{I = 1.35\ A}$$

EXAMPLE 8-2

Calculate the individual resistor currents in the circuit shown in Figure 8-2. Also, calculate the total current that must be supplied by the battery.

SOLUTION

$$I_1 = \frac{E}{R_1} = \frac{24\text{ V}}{2.2\text{ k}\Omega}$$

$$\mathbf{I_1 \cong 10.9\ mA}$$

$$I_2 = \frac{E}{R_2} = \frac{24\text{ V}}{5.6\text{ k}\Omega}$$

$$\mathbf{I_2 \cong 4.29\ mA}$$

$$I_3 = \frac{E}{R_3} = \frac{24\text{ V}}{3.3\text{ k}\Omega}$$

$$\mathbf{I_3 \cong 7.27\ mA}$$

$$I_4 = \frac{E}{R_4} = \frac{24\text{ V}}{4.7\text{ k}\Omega}$$

$$\mathbf{I_4 \cong 5.11\ mA}$$

$$I = I_1 + I_2 + I_3 + I_4$$

$$= 10.9\text{ mA} + 4.29\text{ mA} + 7.27\text{ mA} + 5.11\text{ mA}$$

$$\mathbf{I = 27.57\ mA}$$

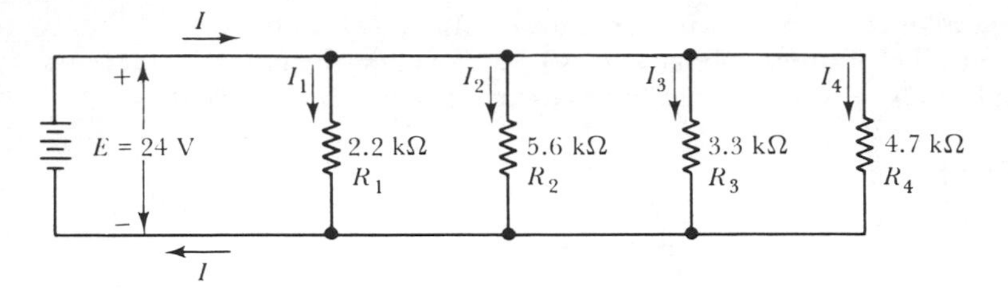

FIG. 8-2. Four resistors connected in parallel

8-2 PARALLEL EQUIVALENT CIRCUIT

Consider the case of four resistors in parallel, as reproduced in Figure 8-3(a). From Eq. (8-1), the battery current is

$$I = I_1 + I_2 + I_3 + I_4$$

which can be rewritten

$$I = \frac{E}{R_1} + \frac{E}{R_2} + \frac{E}{R_3} + \frac{E}{R_4}$$

or

$$I = E\left(\frac{1}{R_1} + \frac{1}{R_2} + \frac{1}{R_3} + \frac{1}{R_4}\right)$$

For n resistors in parallel, this becomes

$$\boxed{I = E\left(\frac{1}{R_1} + \frac{1}{R_2} + \frac{1}{R_3} + \cdots + \frac{1}{R_n}\right)} \tag{8-2}$$

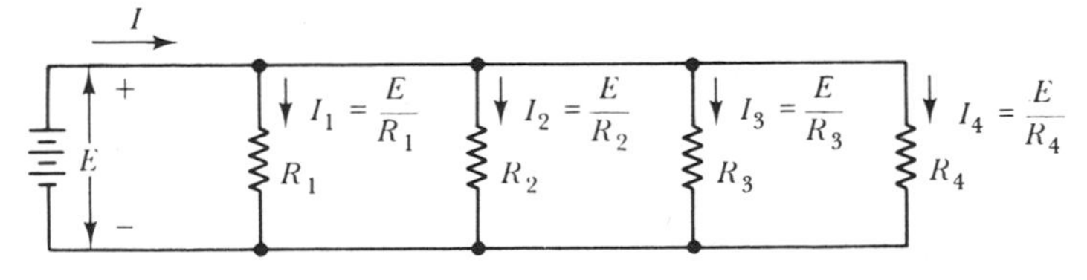

(a) Currents in parallel resistor circuit

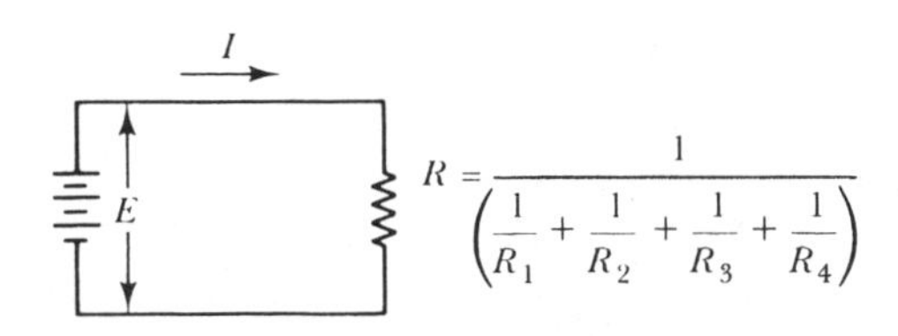

(b) Equivalent circuit for four resistors in parallel

FIG. 8-3. Equivalent circuit for four resistors in parallel

If all the resistors in parallel could be replaced by just one resistance which could draw the same current from the battery, the equation for current would be written

$$I=\frac{E}{R}$$

or

$$I=E\left(\frac{1}{R}\right)$$

where

$$\frac{1}{R}=\frac{1}{R_1}+\frac{1}{R_2}+\frac{1}{R_3}+\cdots+\frac{1}{R_n} \tag{8-3}$$

Thus, it is seen that:

The reciprocal of the equivalent resistance of several resistors in parallel is equal to the sum of the reciprocals of the individual resistances.

Equation (8-3) can be rearranged to give

$$R=\frac{1}{\frac{1}{R_1}+\frac{1}{R_2}+\frac{1}{R_3}+\cdots+\frac{1}{R_n}} \tag{8-4}$$

The equivalent circuit of the parallel resistors and battery can now be drawn as illustrated in Figure 8-3(b).

EXAMPLE 8-3

Determine the equivalent resistance of the four parallel resistors in Figure 8-2 and use it to calculate the total current drawn from the battery.

SOLUTION

Eq. (8-4):

$$R=\frac{1}{\frac{1}{R_1}+\frac{1}{R_2}+\frac{1}{R_3}+\frac{1}{R_4}}$$

$$=\frac{1}{\frac{1}{2.2\text{ k}\Omega}+\frac{1}{5.6\text{ k}\Omega}+\frac{1}{3.3\text{ k}\Omega}+\frac{1}{4.7\text{ k}\Omega}}$$

$$\mathbf{R\cong 870\ \Omega}$$

$$I=\frac{E}{R}=\frac{24\text{ V}}{870\ \Omega}$$

$$\mathbf{I=27.6\text{ mA}}$$

Note that because of approximations, this is slightly different from the result in Example 8-2.

8-3 CONDUCTANCES IN PARALLEL

As discussed in Section 3-4, conductance is the reciprocal of resistance, and its unit is the *siemens* (S). In the case of parallel circuits, it is sometimes more convenient to use the conductance values of the resistors involved instead of the resistance values.

When Eq. (8-3) is rewritten in terms of conductances, it becomes

$$\boldsymbol{G=(G_1+G_2+G_3+\cdots+G_n)} \tag{8-5}$$

It will be recalled that Ohm's law as applied to conductances is changed from

$$I=\frac{E}{R} \quad \text{to} \quad I=EG$$

Consequently, the current in each branch of the parallel circuit in Figure 8-2 can be calculated as

$$I_1=EG_1, \quad I_2=EG_2, \quad \text{etc.}$$

and the current drawn from the battery becomes

$$I=EG$$

EXAMPLE 8-4

For the circuit shown in Figure 8-2, express each resistance as a conductance, and using the conductance values, calculate the current through each resistor and the total current taken from the battery.

SOLUTION

$$G_1=\frac{1}{R_1}=\frac{1}{2.2\ \text{k}\Omega}$$

$$\cong 4.55\times 10^{-4}\ \text{S}$$

$$G_2=\frac{1}{R_2}=\frac{1}{5.6\ \text{k}\Omega}$$

$$\cong 1.79\times 10^{-4}\ \text{S}$$

$$G_3 = \frac{1}{R_3} = \frac{1}{3.3\ \text{k}\Omega}$$

$$\cong 3.03 \times 10^{-4}\ \text{S}$$

$$G_4 = \frac{1}{R_4} = \frac{1}{4.7\ \text{k}\Omega}$$

$$\cong 2.13 \times 10^{-4}\ \text{S}$$

$$I_1 = E \times G_1 = 24\ \text{V} \times 4.55 \times 10^{-4}\ \text{S}$$

$$\mathbf{I_1 \cong 10.9\ mA}$$

$$I_2 = E \times G_2 = 24\ \text{V} \times 1.79 \times 10^{-4}\ \text{S}$$

$$\mathbf{I_2 \cong 4.3\ mA}$$

$$I_3 = E \times G_3 = 24\ \text{V} \times 3.03 \times 10^{-4}\ \text{S}$$

$$\mathbf{I_3 \cong 7.3\ mA}$$

$$I_4 = E \times G_4 = 24\ \text{V} \times 2.13 \times 10^{-4}\ \text{S}$$

$$\mathbf{I_4 \cong 5.1\ mA}$$

$$G = G_1 + G_2 + G_3 + G_4$$

$$= (4.55 + 1.79 + 3.03 + 2.13) \times 10^{-4}\ \text{S}$$

$$= 11.5 \times 10^{-4}\ \text{S}$$

$$I = E \times G = 24\ \text{V} \times 11.5 \times 10^{-4}\ \text{S}$$

$$\mathbf{I = 27.6\ mA}$$

8-4 CURRENT DIVIDER

Refer again to the circuit in Figure 8-1. Such a parallel combination of two resistors is sometimes termed a *current divider*, because the battery current is divided between the two branches of the circuit. For this circuit

$$I_1 = \frac{E}{R_1} \quad \text{and} \quad I_2 = \frac{E}{R_2}$$

Also,

$$I = I_1 + I_2$$

$$= \frac{E}{R_1} + \frac{E}{R_2}$$

or

$$I = E\left(\frac{1}{R_1} + \frac{1}{R_2}\right)$$

Using $(R_1 \times R_2)$ as the common denominator for $1/R_1$ and $1/R_2$, the equation becomes

$$\boldsymbol{I = E\left(\frac{R_1 + R_2}{R_1 \times R_2}\right)} \tag{8-6}$$

or

$$I = E\Big/\left(\frac{R_1 \times R_2}{R_1 + R_2}\right)$$

Also,

$$I = \frac{E}{R}$$

where R is the equivalent resistance of R_1 and R_2 in parallel. Therefore, for two resistors in parallel the equivalent resistance is

$$\boldsymbol{R = \frac{R_1 \times R_2}{R_1 + R_2}} \tag{8-7}$$

From Eq. (8-6),

$$E = I\left(\frac{R_1 \times R_2}{R_1 + R_2}\right)$$

and substituting for E in

$$I_1 = \frac{E}{R_1}$$

gives

$$I_1 = \frac{I\left(\dfrac{R_1 \times R_2}{R_1 + R_2}\right)}{R_1}$$

or

$$I_1 = \frac{I}{R_1}\left(\frac{R_1 \times R_2}{R_1 + R_2}\right)$$

Therefore,

$$\boldsymbol{I_1 = I\left(\frac{R_2}{R_1 + R_2}\right)} \tag{8-8}$$

and similarly,

$$\boxed{I_2 = I\left(\frac{R_1}{R_1 + R_2}\right)} \tag{8-9}$$

Note that the expression for I_1 has R_2 on its top line, and that for I_2 has R_1 on its top line.

EXAMPLE 8-5

Calculate the equivalent resistance and the branch currents for the circuit in Figure 8-1 when $R_1 = 12\ \Omega$, $R_2 = 15\ \Omega$, and $E = 9$ V.

SOLUTION

Eq. (8-7):

$$R = \frac{R_1 \times R_2}{R_1 + R_2} = \frac{12\ \Omega \times 15\ \Omega}{12\ \Omega + 15\ \Omega}$$

$$\mathbf{R \cong 6.67\ \Omega}$$

$$I = \frac{E}{R} = \frac{9\ \text{V}}{6.67\ \Omega}$$

$$\mathbf{I = 1.35\ A}$$

Eq. (8-8):

$$I_1 = I\left(\frac{R_2}{R_1 + R_2}\right)$$

$$= 1.35\ \text{A} \times \frac{15\ \Omega}{12\ \Omega + 15\ \Omega}$$

$$\mathbf{I_1 = 0.75\ A}$$

Eq. (8-9):

$$I_2 = I\left(\frac{R_1}{R_1 + R_2}\right)$$

$$= 1.35\ \text{A} \times \frac{12\ \Omega}{12\ \Omega + 15\ \Omega}$$

$$\mathbf{I_2 = 0.6\ A}$$

This compares with the results of Example 8-1.

8-5 POWER IN PARALLEL CIRCUITS

Whether a resistor is connected in a series circuit or a parallel circuit, the power dissipated in the resistor is calculated in the same way. For the circuit in Figure 8-4,

$$P_1 = EI_1$$

or

$$P_1 = \frac{E^2}{R_1}$$

or

$$P_1 = I_1^2 R_1$$

The power dissipated in resistor R_2 is calculated in a similar way. The total power output from the battery is, of course,

$$P = EI$$

also,

$$P = P_1 + P_2$$

For any parallel (or series) combination of n resistors: Eq. (7-7):

$$P = P_1 + P_2 + P_3 + \cdots + P_n$$

EXAMPLE 8-6

For the circuit described in Example 8-5, calculate the power dissipations in R_1 and R_2 and the total power supplied from the battery.

SOLUTION

$$P_1 = \frac{E^2}{R_1} = \frac{(9\text{ V})^2}{12\ \Omega}$$

$$\mathbf{P_1 = 6.75\ W}$$

$$P_2 = \frac{E^2}{R_2} = \frac{(9\text{ V})^2}{15\ \Omega}$$

$$\mathbf{P_2 = 5.4\ W}$$

$$P = EI = 9\text{ V} \times 1.35\text{ A}$$

$$\mathbf{P = 12.15\ W}$$

Also,

$$P = P_1 + P_2 = 6.75\text{ W} + 5.4\text{ W}$$

$$\mathbf{P = 12.15\ W}$$

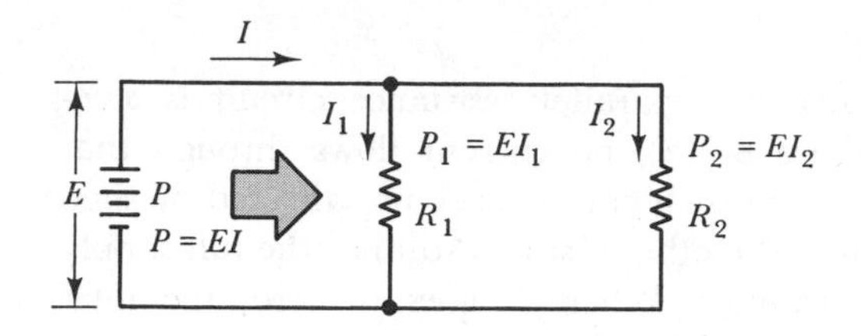

FIG. 8-4. Power dissipation in parallel resistors

EXAMPLE 8-7 Three 12-V lamps are connected in parallel to a 12-V supply, as illustrated in Figure 8-5. The lamp ratings are $L_1 = 3$ W, $L_2 = 5$ W, and $L_3 = 10$ W. Determine the current that flows through each lamp and the total power delivered by the battery.

SOLUTION

$$P_1 = EI_1$$

Therefore,

$$I_1 = \frac{P_1}{E} = \frac{3\text{ W}}{12\text{ V}}$$

$$\boldsymbol{I_1 = 0.25\text{ A}}$$

$$I_2 = \frac{P_2}{E} = \frac{5\text{ W}}{12\text{ V}}$$

$$\boldsymbol{I_2 \cong 0.42\text{ A}}$$

$$I_3 = \frac{P_3}{E} = \frac{10\text{ W}}{12\text{ V}}$$

$$\boldsymbol{I_3 \cong 0.83\text{ A}}$$

Total battery power output:
Eq. (7-7):

$$P = P_1 + P_2 + P_3$$

$$= 3\text{ W} + 5\text{ W} + 10\text{ W}$$

$$\boldsymbol{P = 18\text{ W}}$$

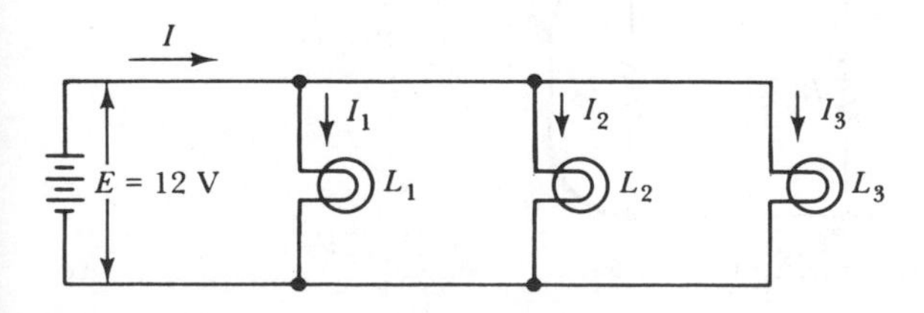

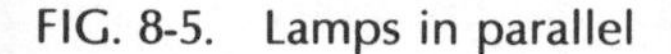
FIG. 8-5. Lamps in parallel

8-6 OPEN CIRCUITS AND SHORT CIRCUITS IN A PARALLEL CIRCUIT

When one of the components in a parallel resistance circuit is *open-circuited*, as illustrated in Figure 8-6(a), no current flows through that branch of the circuit. The other branch currents are not affected by such an open circuit, because each of the other resistors still has the full supply voltage applied across its terminals. When I_1 goes to zero, the total current drawn from the battery is reduced from

$$I = I_1 + I_2 + I_3$$

to

$$I = I_2 + I_3$$

An open circuit can also occur in the supply line to the parallel resistor combination, as shown in Figure 8-6(b). In this case the open circuit is effectively an infinite resistance in series with the battery and resistors. The result is that no supply current can flow, and consequently I_1, I_2, and I_3 are also zero.

Figure 8-6(c) shows a short circuit across resistor R_3. This has the same effect whether it is across R_1, R_2, or R_3, or the battery terminals. In this case, the current that flows through each resistor is effectively zero. However, the battery now has a short circuit across its terminals. Consequently, the battery short-circuit current flows:

$$I_{sc} = \frac{E}{r_i}$$

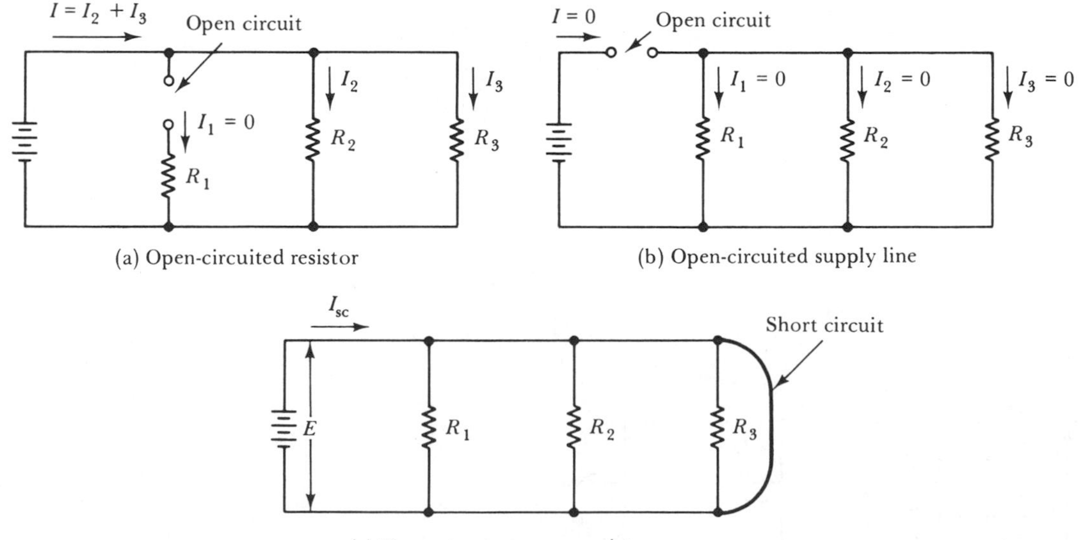

(a) Open-circuited resistor

(b) Open-circuited supply line

(c) Short circuit across a resistor

FIG. 8-6. Open-circuit and short-circuit conditions in a parallel circuit

where r_i is the battery internal resistance (see Section 6-2). In this situation an abnormally large current flows, and the battery could be seriously damaged.

GLOSSARY OF FORMULAS

Total supply current to parallel circuit:

$$I = I_1 + I_2 + I_3 + \cdots + I_n$$

$$I = E\left(\frac{1}{R_1} + \frac{1}{R_2} + \frac{1}{R_3} + \cdots + \frac{1}{R_n}\right)$$

Reciprocal of total parallel resistance:

$$\frac{1}{R} = \left(\frac{1}{R_1} + \frac{1}{R_2} + \frac{1}{R_3} + \cdots + \frac{1}{R_n}\right)$$

Total parallel resistance:

$$R = \frac{1}{\dfrac{1}{R_1} + \dfrac{1}{R_2} + \dfrac{1}{R_3} + \cdots + \dfrac{1}{R_n}}$$

Total parallel conductance:

$$G = G_1 + G_2 + G_3 + \cdots + G_n$$

Two resistors in parallel:

$$R = \frac{R_1 \times R_2}{R_1 + R_2}$$

Current through R_1:

$$I_1 = I\left(\frac{R_2}{R_1 + R_2}\right)$$

Current through R_2:

$$I_2 = I\left(\frac{R_1}{R_1 + R_2}\right)$$

Total power supplied:

$$P = P_1 + P_2 + P_3 + \cdots + P_n$$

REVIEW QUESTIONS

8-1 Sketch the circuit diagram for two parallel-connected resistors supplied by a battery. Show all voltage polarities and current directions and write expressions for the current through each resistor and for the total current drawn from the battery.

8-2 State Kirchhoff's current law. Also, explain the component voltage relationships in a parallel circuit.

8-3 Derive an equation for the equivalent resistance of n resistors connected in parallel. Repeat the derivation using conductances instead of resistances.

8-4 Derive the equations for the current through each of two parallel connected resistors, in terms of the total supply current. Also, show that for two parallel resistors,

$$R_1 \| R_2 = \frac{R_1 \times R_2}{R_1 + R_2}$$

8-5 Discuss the effect that open-circuit and short-circuit conditions can have on parallel resistance circuits.

PROBLEMS

8-1 For the circuit shown in Figure 8-7, determine the individual resistor currents and the current taken from the battery.

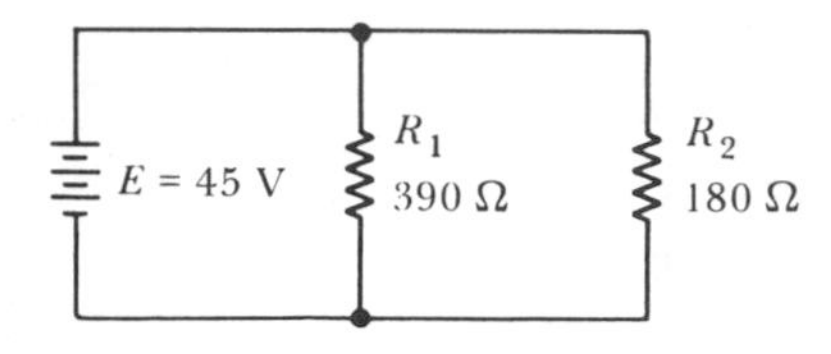

FIG. 8-7.

8-2 Calculate the current through each resistor and the total battery current for the circuit shown in Figure 8-8.

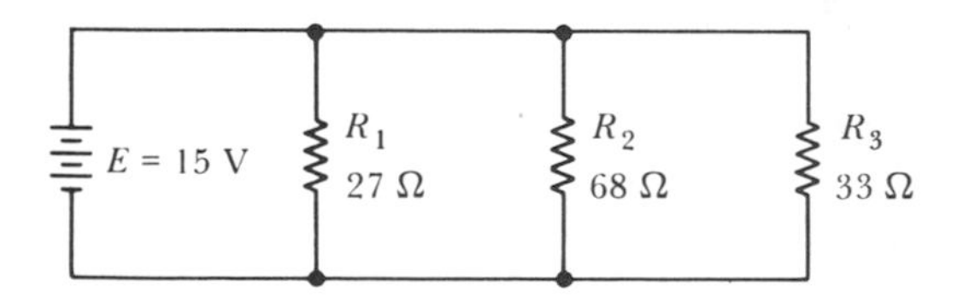

FIG. 8-8.

8-3 Four parallel-connected resistors are supplied from a 25-V battery. Three of the resistors have values of $R_1 = 1$ kΩ, $R_2 = 12$ kΩ, and $R_3 = 8.2$ kΩ. If the battery current is measured as 36.5 mA, determine the value of the fourth resistor.

8-4 Determine the equivalent resistance for the parallel resistors shown in Figure 8-7. Also, draw an equivalent circuit and calculate the total current supplied by the battery.

8-5 Repeat Problem 8-4 for the circuit in Figure 8-8.

8-6 For the circuit in Figure 8-7, express each resistance as a conductance. Using the conductance values, calculate all current levels.

8-7 Repeat Problem 8-6 for the circuit of Figure 8-8.

8-8 Use the current divider rule to determine each branch current in the circuit of Figure 8-7.

8-9 For the circuit of Figure 8-7, calculate the power dissipated in each resistor and the total power supplied by the battery.

8-10 Repeat Problem 8-9 for the circuit shown in Figure 8-8.

8-11 Four 115-V volt lamps are connected in parallel to a 115-V supply. The lamp ratings are $L_1 = 100$ W, $L_2 = 40$ W, $L_3 = 60$ W, and $L_4 = 25$ W. Determine the current that flows through each lamp and the total power delivered from the supply.

9

SERIES–PARALLEL CIRCUITS

INTRODUCTION

Not all circuits are simple series or parallel arrangements. A great many are combinations of parallel resistors connected in series with other resistors, or combined with other parallel groups. These can only be described as series–parallel circuits.

The simplest approach to analyzing a series–parallel circuit is to first resolve each series group into its single equivalent resistance. Then resolve each parallel group of resistors into its equivalent resistance. The resulting circuit is a simple series or parallel circuit which can be treated in the usual way.

As in all types of circuits, open-circuit and short-circuit conditions affect the currents and voltage drops throughout the circuit.

9-1 EQUIVALENT CIRCUIT OF A SERIES–PARALLEL CIRCUIT

In the circuit shown in Figure 9-1(a), resistors R_2 and R_3 are in parallel, and together they are in series with R_1. The level of current taken from the supply is easily calculated if R_2 and R_3 are first replaced with their equivalent resistance as illustrated in Figure 9-1(b). The circuit now becomes a simple two-resistor series circuit.

In Figure 9-2(a), another series–parallel resistor combination is shown. In this case the circuit is reduced to a simple parallel circuit when R_2 and R_3 are replaced by their equivalent resistance [Figure 9-2(b)].

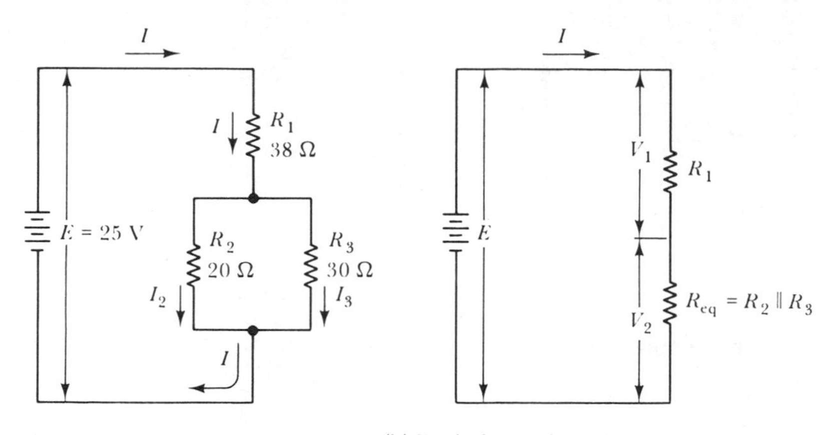

(a) Series-parallel resistance circuit (b) Equivalent series resistance circuit

FIG. 9-1. Simple series-parallel resistance circuit and its equivalent circuit

EXAMPLE 9-1 Calculate the current drawn from the supply in the circuit shown in Figure 9-1(a).

SOLUTION

For R_2 and R_3:

$$R_{eq} = R_2 \| R_3$$

$$= \frac{R_2 \times R_3}{R_2 + R_3}$$

$$= \frac{20\ \Omega \times 30\ \Omega}{20\ \Omega + 30\ \Omega}$$

$$\mathbf{R_{eq} = 12\ \Omega}$$

$$I = \frac{E}{R_1 + R_{eq}}$$

$$= \frac{25\ \text{V}}{38\ \Omega + 12\ \Omega}$$

$$\mathbf{I = 0.5\ A}$$

EXAMPLE 9-2 Determine the level of battery current for the circuit shown in Figure 9-2(a).

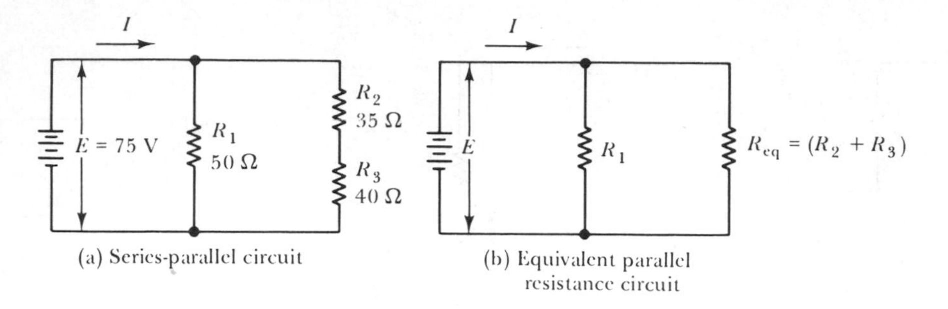

(a) Series-parallel circuit

(b) Equivalent parallel resistance circuit

FIG. 9-2. Series-parallel circuit and equivalent circuit

SOLUTION

For R_2 and R_3:

$$R_{eq} = R_2 + R_3$$
$$= 35\ \Omega + 40\ \Omega$$
$$\mathbf{R_{eq} = 75\ \Omega}$$

For R_1 and R_{eq} in parallel:

$$R = R_1 \| R_{eq}$$
$$= \frac{R_1 \times R_{eq}}{R_1 + R_{eq}}$$
$$= \frac{50\ \Omega \times 75\ \Omega}{50\ \Omega + 75\ \Omega}$$
$$\mathbf{R = 30\ \Omega}$$

$$I = \frac{E}{R} = \frac{75\ \text{V}}{30\ \Omega}$$
$$\mathbf{I = 2.5\ A}$$

9-2 CURRENTS IN A SERIES–PARALLEL CIRCUIT

The circuit of Figure 9-1 is reproduced again in Figure 9-3 with the branch currents and voltages identified. It is seen that the battery current flows through resistor R_1 but that it splits up into I_2 and I_3 in order to flow through R_2 and R_3. Returning to the battery negative terminal, the current is once again I. Obviously,

$$I = I_2 + I_3$$

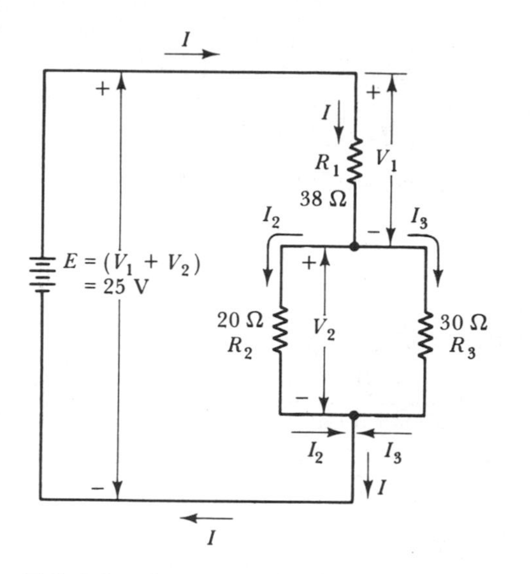

FIG. 9-3. Currents and voltages in the series-parallel circuit of Fig. 9-1(a)

Similarly, the battery current splits up between the resistors in Figure 9-4, which is a reproduction of the circuit shown in Figure 9-2(a). Here, I_1 flows through R_1, while I_2 flows through R_2 and R_3. Also, the battery current is

$$I = I_1 + I_2$$

In each of the cases above, the current through the individual resistors can be calculated easily using the current divider rule.

EXAMPLE 9-3 Calculate the branch currents in the circuit of Figure 9-3, using the information available from Example 9-1.

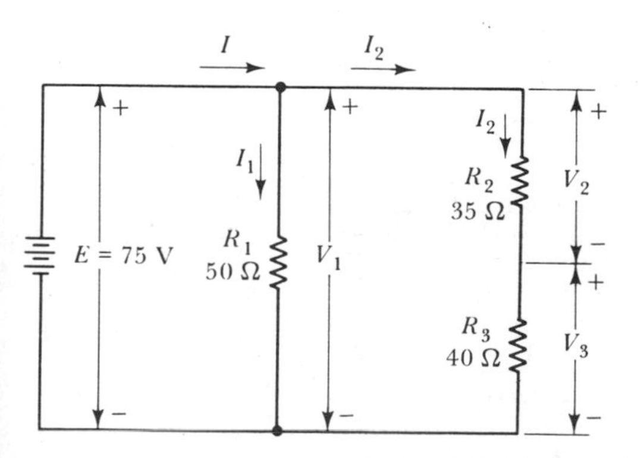

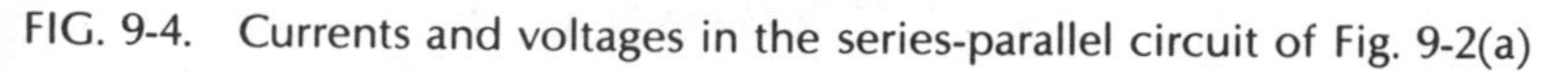
FIG. 9-4. Currents and voltages in the series-parallel circuit of Fig. 9-2(a)

SOLUTION

From Example 9-1,

$$I = 0.5\text{ A}$$

Therefore, *current through* $R_1 = I = 0.5\text{ A}$

from Eq. (8-8):

$$I_2 = I\left(\frac{R_3}{R_2 + R_3}\right)$$

$$= 0.5\text{ A} \times \frac{30\ \Omega}{20\ \Omega + 30\ \Omega}$$

$$\mathbf{I_2 = 0.3\text{ A}}$$

and

$$I_3 = I\left(\frac{R_2}{R_2 + R_3}\right)$$

$$= 0.5\text{ A} \times \frac{20\ \Omega}{20\ \Omega + 30\ \Omega}$$

$$\mathbf{I_3 = 0.2\text{ A}}$$

EXAMPLE 9-4 Find the individual branch current for the circuit of Figure 9-4, using the information given in Example 9-2.

SOLUTION

From Example 9-2:

$$I = 2.5\text{ A}$$

and

$$R_{eq} = 75\ \Omega$$

from Eq. (8-8):

$$I_1 = I\left(\frac{R_{eq}}{R_{eq} + R_1}\right)$$

$$= 2.5\text{ A} \times \frac{75\ \Omega}{75\ \Omega + 50\ \Omega}$$

$$\mathbf{I_1 = 1.5\text{ A}}$$

and

$$I_2 = I\frac{R_1}{R_{eq} + R_1}$$

$$= 2.5\text{ A} \times \frac{50\ \Omega}{75\ \Omega + 50\ \Omega}$$

$$\mathbf{I_2 = 1\text{ A}}$$

9-3 VOLTAGE DROPS IN A SERIES–PARALLEL CIRCUIT

As always, the voltage drop across any resistor is the product of the resistance value and the current through the resistor. In Figure 9-3,

$$V_1 = IR_1$$

and

$$V_2 = I_2R_2 = I_3R_3$$

Also,

$$E = V_1 + V_2$$

Similarly, in Figure 9-4,

$$V_1 = I_1R_1$$

$$V_2 = I_2R_2$$

$$V_3 = I_2R_3$$

$$E = V_1 = V_2 + V_3$$

Once the branch currents are known, the voltages across each resistor can be readily calculated. In some circumstances it may be more convenient to determine the resistor voltages first, then use these voltages to calculate the branch currents.

EXAMPLE 9-5

Using the information available from Example 9-3, calculate the voltage drop across each resistor in Figure 9-3.

SOLUTION

From Examples 9-1 *and* 9-3,

$$I = 0.5 \text{ A}$$

$$I_2 = 0.3 \text{ A}$$

and

$$I_3 = 0.2 \text{ A}$$

$$V_1 = IR_1 = 0.5 \text{ A} \times 38 \ \Omega$$

$$\mathbf{V_1 = 19 \ V}$$

$$V_2 = I_2R_2 = 0.3 \text{ A} \times 20 \ \Omega$$

$$\mathbf{V_2 = 6 \ V}$$

$$V_2 = I_3R_3 = 0.2 \text{ A} \times 30 \ \Omega$$

$$\mathbf{V_2 = 6 \ V}$$

EXAMPLE 9-6 Analyze the circuit in Figure 9-1(a) to determine the resistor voltage drops and the branch currents.

SOLUTION

$$R_{eq} = R_2 \| R_3$$

$$= \frac{20\ \Omega \times 30\ \Omega}{20\ \Omega + 30\ \Omega}$$

$$\mathbf{R_{eq} = 12\ \Omega}$$

For voltage divider R_1 *and* R_{eq}, *as shown in* Figure 9-1(b):

$$V_2 = E \frac{R_{eq}}{R_1 + R_{eq}}$$

$$= 25\ \text{V} \times \frac{12\ \Omega}{38\ \Omega + 12\ \Omega}$$

$$\mathbf{V_2 = 6\ V}$$

$$V_1 = E \frac{R_1}{R_1 + R_{eq}}$$

$$= 25\ \text{V} \times \frac{38\ \Omega}{38\ \Omega + 12\ \Omega}$$

$$\mathbf{V_2 = 19\ V}$$

$$I = \frac{V_1}{R_1} = \frac{19\ \text{V}}{38\ \Omega}$$

$$\mathbf{I = 0.5\ A}$$

$$I_2 = \frac{V_2}{R_2} = \frac{6\ \text{V}}{20\ \Omega}$$

$$\mathbf{I_2 = 0.3\ A}$$

$$I_3 = \frac{V_3}{R_3} = \frac{6\ \text{V}}{30\ \Omega}$$

$$\mathbf{I_3 = 0.2\ A}$$

EXAMPLE 9-7

Analyze the circuit in Figure 9-4 for resistor voltages and branch currents.

SOLUTION

$$V_1 = E = 75\text{ V}$$

$$I_1 = \frac{V_1}{R_1} = \frac{75\text{ V}}{50\ \Omega}$$

$$\mathbf{I_1 = 1.5\ A}$$

$$V_3 = E\frac{R_3}{R_2 + R_3}$$

$$= 75\text{ V} \times \frac{40\ \Omega}{40\ \Omega + 35\ \Omega}$$

$$\mathbf{V_3 = 40\ V}$$

$$V_2 = E\frac{R_2}{R_2 + R_3}$$

$$= 75\text{ V} \times \frac{35\ \Omega}{40\ \Omega + 35\ \Omega}$$

$$\mathbf{V_2 = 35\ V}$$

$$I_2 = \frac{V_2}{R_2} = \frac{35\text{ V}}{35\ \Omega}$$

$$\mathbf{I_2 = 1\ A}$$

$$I_2 = \frac{V_3}{R_3} = \frac{40\text{ V}}{40\ \Omega}$$

$$\mathbf{I_2 = 1\ A}$$

9-4 OPEN CIRCUITS AND SHORT CIRCUITS IN SERIES–PARALLEL CIRCUITS

The effect that an open-circuit or short-circuit condition has on a series–parallel circuit depends upon just where in the circuit the fault occurs. Consider Figure 9-5(a), where an open circuit is shown at one end of R_1. This has the same effect as an open circuit in the supply line, so that all current levels are zero. Also, because the currents are zero, there are no volt drops across the resistors, and consequently all of the supply voltage E appears across the open circuit.

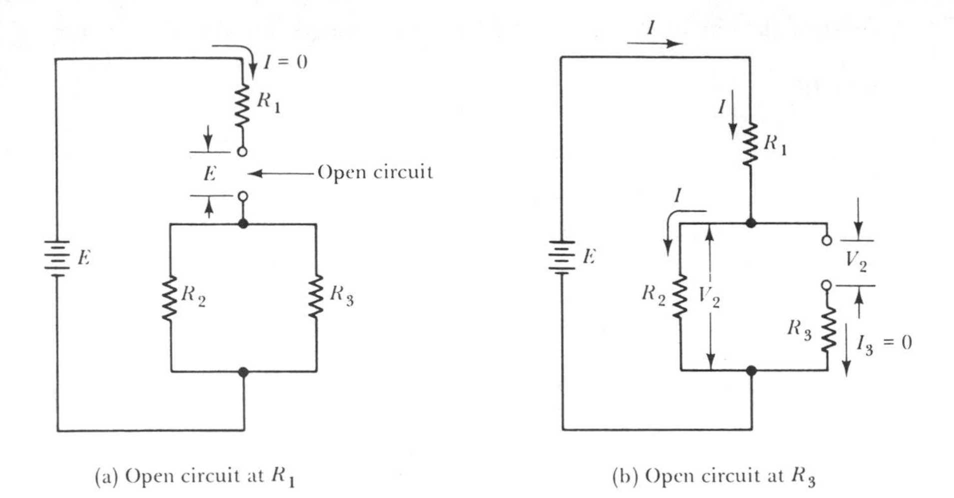

(a) Open circuit at R_1 (b) Open circuit at R_3

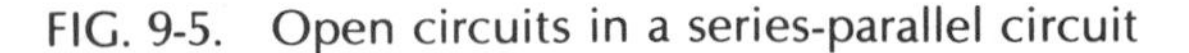

FIG. 9-5. Open circuits in a series-parallel circuit

In the case of an open circuit at one end of one of the parallel resistors, as shown in Figure 9-5(b), I_3 goes to zero. The current through R_1 and R_2 is now equal to the battery current and is calculated as

$$I = \frac{E}{R_1 + R_2}$$

Also, since there is no current through R_3, there is no voltage drop across it, and the voltage at the open circuit is equal to V_2.

For the short-circuit condition shown in Figure 9-6(a), the resistance between the terminals of R_1 is effectively zero. Therefore, the battery

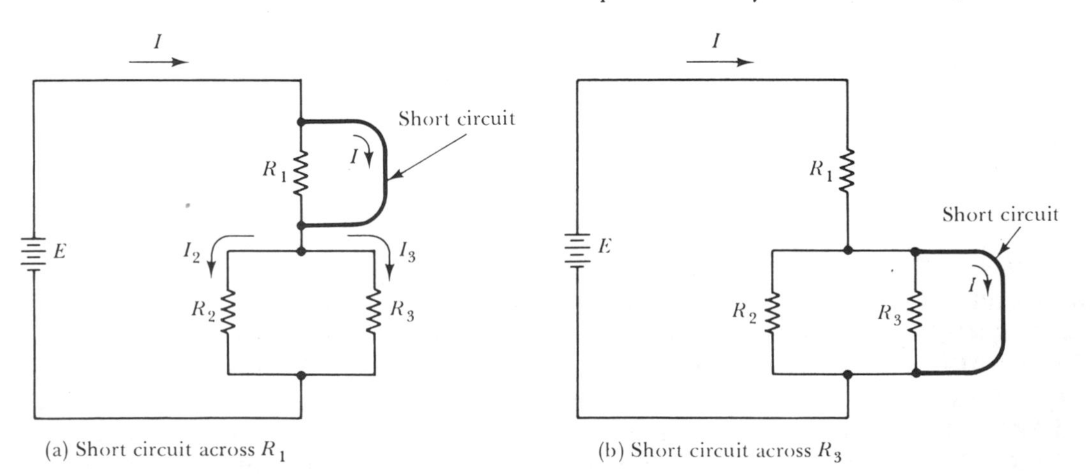

(a) Short circuit across R_1 (b) Short circuit across R_3

FIG. 9-6. Short circuits in a series-parallel circuit

voltage appears across R_2 and R_3 in parallel. This gives a battery current of

$$I = \frac{E}{R_2 \| R_3}$$

and the branch currents are

$$I_2 = \frac{E}{R_2}$$

and

$$I_3 = \frac{E}{R_3}$$

It is seen that the levels of current through R_2 and R_3 have been increased from the normal (i.e., before the short circuit) condition. This could cause excessive power dissipation in the components if they have previously been operating near their maximum rating.

The short-circuit condition illustrated in Figure 9-6(b) effectively reduces I_2 and I_3 to zero and increases the battery current to

$$I = \frac{E}{R_1}$$

Obviously, the current through R_1 is now greater than normal, and again power dissipation might present a problem.

9-5 ANALYSIS OF SERIES–PARALLEL CIRCUITS

The procedure for analyzing *any* series–parallel resistor circuit is as follows:

1. Draw a circuit diagram identifying all components by number and showing all currents and resistor voltage drops.
2. Convert all series branches of two or more resistors into a single equivalent resistance.
3. Convert all parallel combinations of two or more resistors into a single equivalent resistance.

The final circuit should be a straightforward series or parallel circuit, which can be analyzed in the normal way. Once the current through each equivalent resistance, or the voltage across it, is known, the original circuit can be used to determine individual resistor currents and voltages.

EXAMPLE 9-8 Analyze the series–parallel circuit in Figure 9-7(a) to determine all resistor currents and voltages.

SOLUTION

$$R_{e1} = R_1 + R_2$$
$$= 10\ \text{k}\Omega + 15\ \text{k}\Omega$$
$$\mathbf{R_{e1} = 25\ k\Omega}$$

The circuit is now modified as shown in Figure 9-7(b), *with* R_{e1} *replacing* R_1 *and* R_2.

$$R_{e2} = R_{e1} \| R_3$$
$$= \frac{R_{e1} \times R_3}{R_{e1} + R_3} = \frac{25\ \text{k}\Omega \times 27\ \text{k}\Omega}{25\ \text{k}\Omega + 27\ \text{k}\Omega}$$
$$\mathbf{R_{e2} = 12.98\ k\Omega}$$

$$R_{e3} = R_4 \| R_5$$
$$= \frac{R_4 \times R_5}{R_4 + R_5} = \frac{39\ \text{k}\Omega \times 22\ \text{k}\Omega}{39\ \text{k}\Omega + 22\ \text{k}\Omega}$$
$$\mathbf{R_{e3} = 14.07\ k\Omega}$$

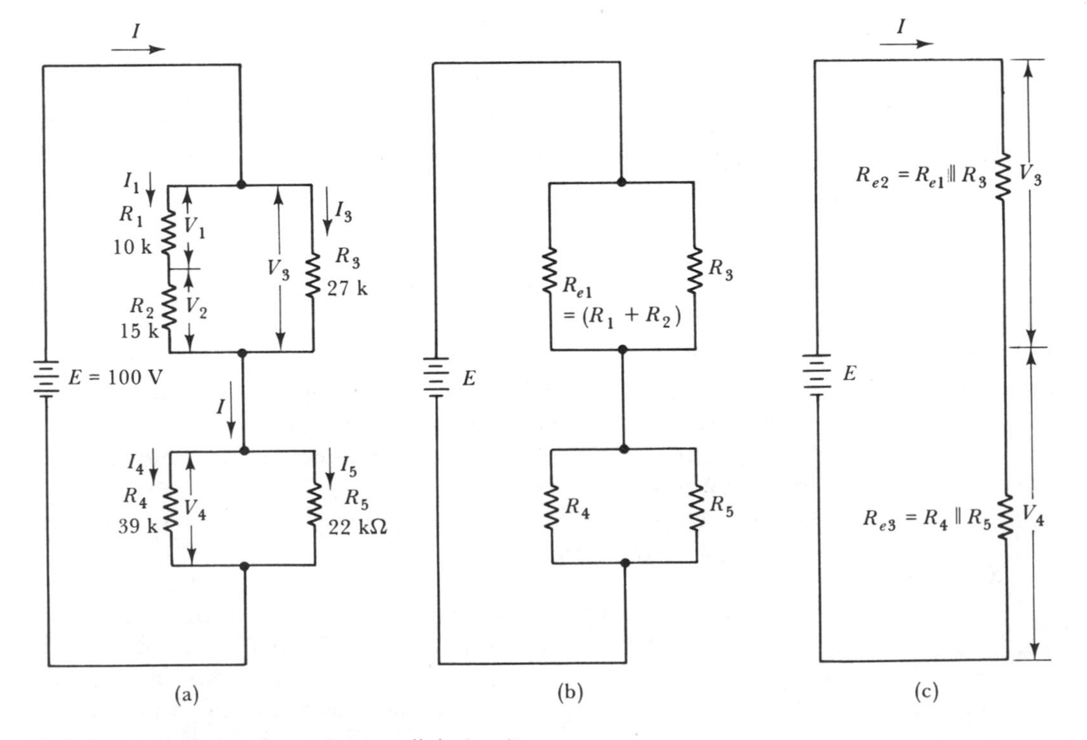

FIG. 9-7. Analysis of a series-parallel circuit

The circuit is now further modified as shown in Figure 9-7(c), *with* R_{e2} *replacing* $R_{e1} || R_3$ *and* R_{e3} *replacing* $R_4 || R_5$. *Now,*

$$I = \frac{E}{R_{e2} + R_{e3}} = \frac{100 \text{ V}}{12.98 \text{ k}\Omega + 14.07 \text{ k}\Omega}$$

$$\mathbf{I \cong 3.7 \text{ mA}}$$

$$V_3 = IR_{e2} = 3.7 \text{ mA} \times 12.98 \text{ k}\Omega$$

$$\mathbf{V_3 \cong 48 \text{ V}}$$

$$V_4 = IR_{e3} = 3.7 \text{ mA} \times 14.07 \text{ k}\Omega$$

$$\mathbf{V_4 \cong 52 \text{ V}}$$

$$I_1 = \frac{V_3}{R_1 + R_2} = \frac{48 \text{ V}}{25 \text{ k}\Omega}$$

$$\mathbf{I_1 \cong 1.92 \text{ mA}}$$

$$V_1 = I_1 R_1 = 1.92 \text{ mA} \times 10 \text{ k}\Omega$$

$$\mathbf{V_1 \cong 19.2 \text{ V}}$$

$$V_2 = I_1 \times R_2 = 1.92 \text{ mA} \times 15 \text{ k}\Omega$$

$$\mathbf{V_2 \cong 28.8 \text{ V}}$$

$$I_3 = \frac{V_3}{R_3} = \frac{48 \text{ V}}{27 \text{ k}\Omega}$$

$$\mathbf{I_3 \cong 1.78 \text{ mA}}$$

$$I_4 = \frac{V_4}{R_4} = \frac{52 \text{ V}}{39 \text{ k}\Omega}$$

$$\mathbf{I_4 \cong 1.33 \text{ mA}}$$

$$I_5 = \frac{V_4}{R_5} = \frac{52 \text{ V}}{22 \text{ k}\Omega}$$

$$\mathbf{I_5 \cong 2.36 \text{ mA}}$$

PROBLEMS

9-1 The circuit shown in Figure 9-1(a) has its components changed to $R_1 = 750\ \Omega$, $R_2 = 330\ \Omega$, and $R_3 = 560\ \Omega$. The supply voltage is also changed to 80 V. Analyze the circuit to determine all resistor currents and voltages.

9-2 The series–parallel circuit in Figure 9-2(a) is changed as follows: $R_1 = 2.7\ k\Omega$, $R_2 = 8.2\ k\Omega$, $R_3 = 15\ k\Omega$, and $E = 50\ V$. Analyze the circuit to determine all resistor currents and voltages.

9-3 Analyze the circuit in Figure 9-8 to determine all resistor currents and voltages and the total current drawn from the supply.

9-4 Repeat Problem 9-3 for the circuit in Figure 9-9.

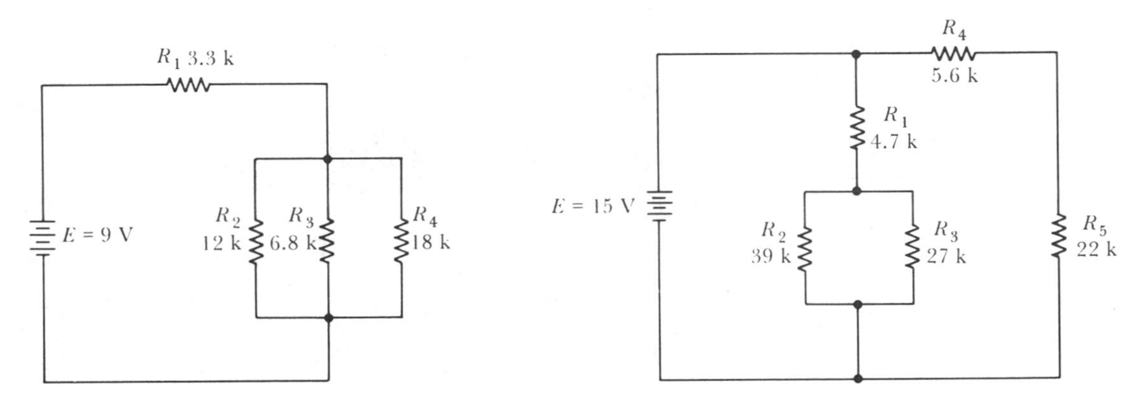

FIG. 9-8.

FIG. 9-9.

9-5 Repeat Problem 9-3 for the circuit in Figure 9-10.

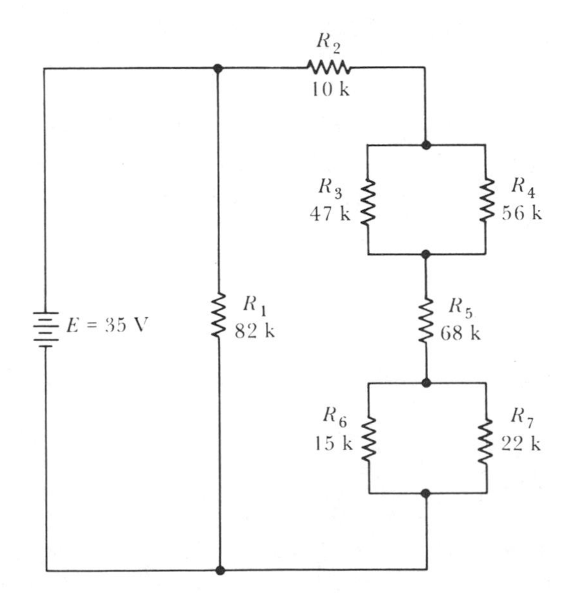

FIG. 9-10.

9-6 If resistor R_3 in Figure 9-10 becomes short-circuited, calculate the new levels of current and voltage throughout the circuit.

9-7 In the circuit of Figure 9-10, resistor R_7 becomes open-circuited. Determine the new levels of current and voltage throughout the circuit.

9-8 Determine the new level of battery current in the circuit of Figure 9-9 when R_2 is short-circuited.

9-9 Repeat Problem 9-8 for the case of R_3 open-circuited.

10

NETWORK ANALYSIS TECHNIQUES

INTRODUCTION

Some resistance networks are so complex that they cannot be analyzed by the simple techniques described in Chapter 9. One example is a circuit that has more than one voltage source. Also, some circuits have current sources as well as voltage sources.

A network can sometimes be simplified by converting voltage sources to current sources, or vice versa. By direct application of Kirchhoff's voltage and current laws, equations can be derived which afford analysis of any circuit. *Loop equations*, *nodal analysis*, and *delta–wye transformation* are useful techniques that simplify the analysis of complex circuits.

10-1 VOLTAGE SOURCES AND CURRENT SOURCES

CONSTANT VOLTAGE SOURCE. In Chapter 6 it was shown that voltage cells and batteries can be represented by an *ideal cell* in series with the internal resistance of the cell or battery. The ideal cell is assumed to be a *constant voltage source*, and the output current produces a voltage drop across the internal resistance. Figure 10-1(a) shows such a constant voltage source, with voltage E, source resistance R_S, and an external load resistance R_L. Sometimes, instead of a cell, the symbol for a dc voltage generator is used to represent a constant voltage source [see Figure 10-1(b)]. Using the voltage divider rule, the output voltage developed

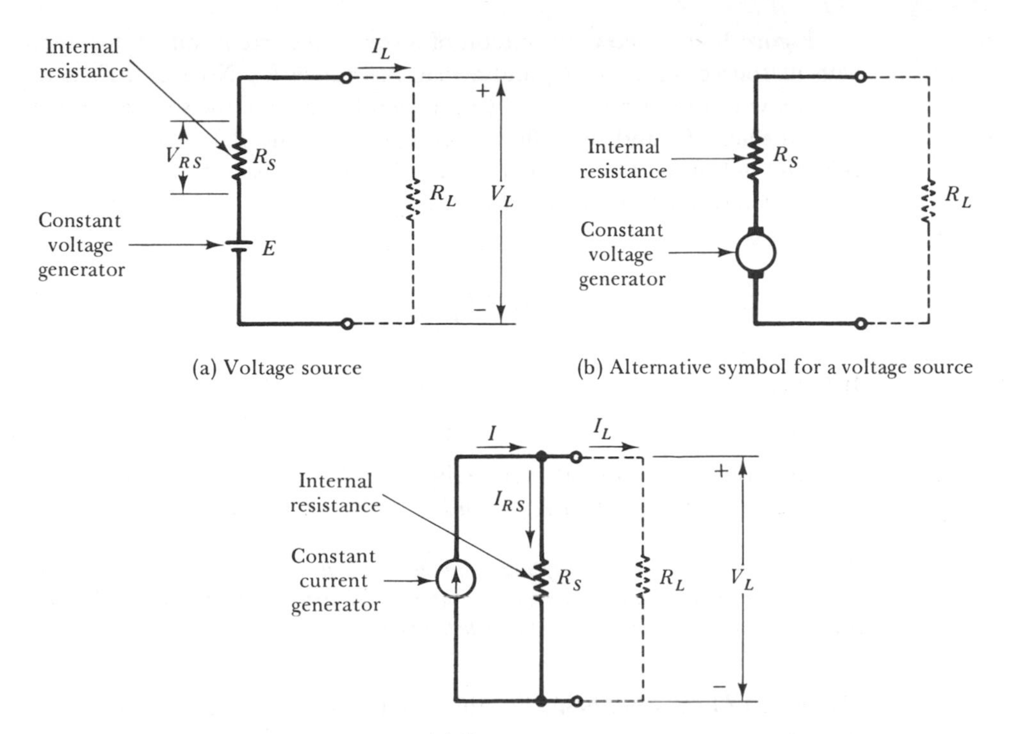

FIG. 10-1. Voltage and current sources

across R_L from the constant voltage source can be quickly determined:

$$\boxed{V_L = E\frac{R_L}{R_S + R_L}} \qquad (10\text{-}1)$$

If $R_S \ll R_L$,

$$V_L \cong E$$

Where the load resistance is very much larger than the source resistance, the constant voltage source is assumed to have zero source resistance, and all of the source voltage is assumed to be applied to the load.

CONSTANT CURRENT SOURCE. As its name implies, the constant voltage source is thought of as a source of voltage that does not change when the load current increases or decreases. Certain electronic devices can produce a current that tends to remain constant no matter how the load resistance varies. Thus, it is possible to have a *constant current source*.

Figure 10-1(c) shows the circuit of a constant current source, together with its source resistance R_S and a load resistance R_L. Note that R_S is in parallel with the current source. Consequently, some of the source current flows through R_S and some flows through R_L. Using the current divider rule, the output current (or load current) from a constant current source can be determined in terms of R_L and R_S:

$$\boxed{I_L = I\frac{R_S}{R_L + R_S}} \tag{10-2}$$

If $R_L \ll R_S$,

$$I_L \cong I$$

Referring to Figure 10-1(c) again, it is obvious that if R_S were very much larger than R_L, then I_{RS} would be very much smaller than I_L.

Where the load resistance is very much smaller than the source resistance, a constant current source is assumed to have an infinite parallel source resistance, and all of the source current is assumed to flow through the load.

SOURCE CONVERSIONS. For the voltage source in Figure 10-1(a), the load current is calculated as

$$I_L = \frac{E}{R_S + R_L}$$

If E in this equation is multiplied by R_S/R_S, nothing is changed, because it is the same as multiplying by 1:

$$I_L = \frac{E\left(\dfrac{R_S}{R_S}\right)}{R_S + R_L}$$

or

$$\boxed{I_L = \frac{E}{R_S}\left(\frac{R_S}{R_S + R_L}\right)} \tag{10-3}$$

In Eq. (10-3), E/R_S is a current and $R_S/(R_S + R_L)$ is exactly the same as in the current divider equation, Eq. (10-2). Thus, Eq. (10-3) represents the output from a current source.

A voltage source having a voltage E and a source resistance R_S can be replaced by a current source with a current E/R_S and a source resistance R_S.

For the current source in Figure 10-1(c), the load voltage is

$$V_L = I_L R_L$$

I_L can be replaced by Eq. (10-2). Thus,

$$V_L = (I \frac{R_S}{R_L + R_S}) R_L$$

or

$$\boldsymbol{V_L = IR_S \left(\frac{R_L}{R_L + R_S} \right)} \qquad \textbf{(10-4)}$$

In Eq. (10-4), IR_S is a voltage and $R_L/(R_L + R_S)$ is exactly the same as in the voltage divider equation, Eq. (10-1). Therefore, Eq. (10-4) represents the output from a voltage source.

A current source having a current I and a source resistance R_S can be replaced by a voltage source with a voltage IR_S and a source resistance R_S.

EXAMPLE 10-1

The voltage source in Figure 10-1(a) has $E = 10$ V, $R_S = 1\ \Omega$, and $R_L = 1\ \Omega$. Determine I and R_S for the equivalent current source. Then calculate V_L and I_L for each type of source.

SOLUTION

For the current source,

$$I = \frac{E}{R_S}$$

$$= \frac{10 \text{ V}}{1} = 10 \text{ A}$$

$$\text{current source } R_S = \text{voltage source } R_S = 1\ \Omega$$

Using the voltage source,

$$V_L = E \frac{R_S}{R_S + R_L}$$

$$= 10 \text{ V} \times \frac{1\ \Omega}{1\ \Omega + 1\ \Omega}$$

$$\boldsymbol{V_L = 5 \text{ V}}$$

$$I_L = \frac{E}{R_S + R_L}$$

$$= \frac{10 \text{ V}}{1\ \Omega + 1\ \Omega}$$

$$\boldsymbol{I_L = 5 \text{ A}}$$

Using the current source:

$$I_L = I \times \frac{R_S}{R_L + R_S}$$

$$= 10\ \text{A} \times \frac{1\ \Omega}{1\ \Omega + 1\ \Omega}$$

$$\mathbf{I_L = 5\ A}$$

$$V_L = I \times (R_L \| R_S)$$

$$= I \frac{R_L \times R_S}{R_L + R_S}$$

$$= 10\ \text{A} \times \frac{1\ \Omega \times 1\ \Omega}{1\ \Omega + 1\ \Omega}$$

$$\mathbf{V_L = 5\ V}$$

PARALLEL AND SERIES OPERATION. In Chapter 6 it was shown that *voltage sources can be operated in series without difficulty*. It was also shown that for parallel operation, voltage sources must have closely equal terminal voltages. *Parallel operation of voltage sources with differing terminal voltages is not possible*, because one source will tend to discharge the other.

Similar but converse rules apply to current sources. Consider Figure 10-2(a), in which two current sources are shown connected in parallel. The output current from the two is

$$I = I_{S1} + I_{S2}$$

$$= 2\ \text{A} + 3\ \text{A}$$

$$= 5\ \text{A}$$

and the internal resistance of the combination is

$$R = R_1 \| R_2$$

$$= 2\ \Omega \| 2\ \Omega$$

$$= 1\ \Omega$$

The same two current sources are shown in Figure 10-2(b) with the direction of I_{S1} altered. In this case

$$I = I_{S2} - I_{S1}$$

$$= 3\ \text{A} - 2\ \text{A}$$

$$= 1\ \text{A}$$

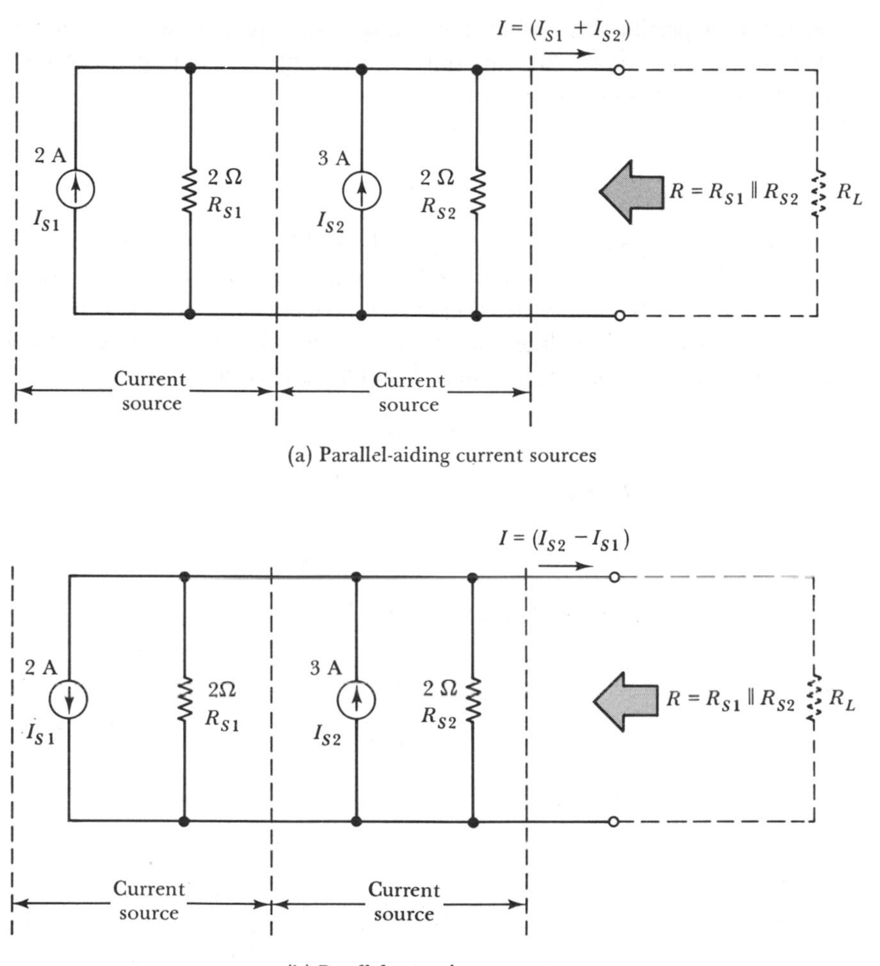

(a) Parallel-aiding current sources

(b) Parallel-opposing current sources

FIG. 10-2. Current sources in parallel

It is seen that *current sources may be operated in parallel without difficulty.*

Now consider two current sources connected in series. Where they both generate exactly the same level of current, there is no problem. But suppose that one generates an output of 1 A and the other has an output of 2 A. The 2-A current would flow through the 1-A source, thus changing its output level. Or the 2-A current would be unable to flow and its output level would become altered. Consequently, *series operation of current sources with differing output currents is not possible.*

A current source may be connected in series or in parallel with a voltage source. However, the rules discussed above still apply. When a conversion is made, so that they are both voltage sources, parallel operation is possible only when the voltages are equal. When the conversion makes them both current sources, series operation is possible only when the currents are equal.

10-2 NETWORK ANALYSIS USING KIRCHHOFF'S LAWS

Kirchhoff's voltage law (Section 7-2) and Kirchhoff's current law (Section 8-1) readily lend themselves to the derivation of equations for solving complex circuits. Consider the simple series parallel circuit in Figure 10-3. Applying Kirchhoff's voltage law to the closed path starting at point *A* and going through points *B*, *E*, and *F* and back to *A* again gives

$$E = V_1 + V_2$$

and since, $V_1 = I_1R_1$ and $V_2 = I_2R_2$,

$$E = I_1R_1 + I_2R_2 \tag{1}$$

Summing the voltage drops around path *BCDEB*,

$$V_2 = V_3$$

or

$$I_3R_3 = I_2R_2 \tag{2}$$

For the path *ABCDEFA*,

$$E = V_1 + V_3$$

Therefore,

$$E = I_1R_1 + I_3R_3 \tag{3}$$

And from Kirchhoff's current law, at point *B*,

$$I_1 = I_2 + I_3 \tag{4}$$

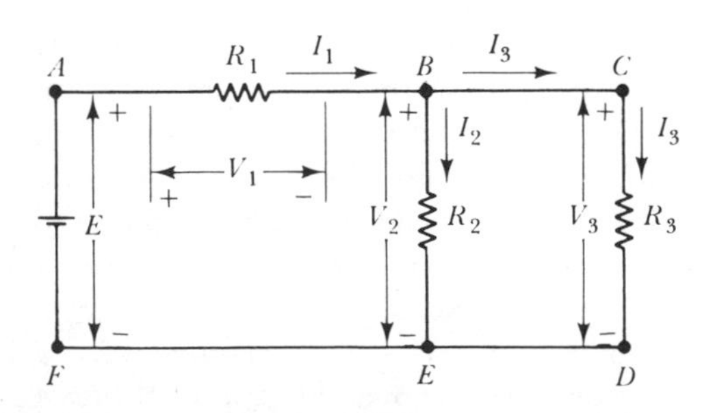

FIG. 10-3. Use of Kirchhoff's Laws to analyze a resistance network

Substituting the expression for I_1 from Eq. (4) into Eq. (1),

$$E=(I_2+I_3)R_1+I_2R_2$$

Therefore,

$$E=I_2R_1+I_3R_1+I_2R_2 \tag{5}$$

and, from Eq. (2),

$$I_3R_3=I_2R_2$$

$$I_3=\frac{I_2R_2}{R_3} \tag{6}$$

Substituting for I_3 from Eq. (6) into Eq. (5),

$$E=I_2R_1+\frac{I_2R_2}{R_3}R_1+I_2R_2$$

or

$$E=I_2\left(R_1+\frac{R_2R_1}{R_3}+R_2\right)$$

$$I_2=\frac{E}{R_1+\dfrac{R_2R_1}{R_3}+R_2} \tag{7}$$

From Eq. (7) the level of current I_2 can be calculated. Then I_2 can be substituted into Eq. (6) to find I_3, and I_1 can be determined by substituting I_2 and I_3 into Eq. (4). Once all the current levels are known, the voltage levels are easily calculated.

The procedure for network analysis using Kirchhoff's laws is as follows:

1. Letter all junctions on the network A, B, C, etc.
2. Identify current directions and voltage polarities, and number them according to the resistor involved.
3. Identify each current path according to the lettered junctions and, applying Kirchhoff's voltage law, write the voltage equations for the paths.
4. Applying Kirchhoff's current law, write the equations for the currents entering and leaving all junctions where more than one current is involved.
5. Solve the equations by substitution to find the unknown currents.

Note that in some circumstances currents and voltage polarities will turn out to be negative when the circuit is analyzed. This simply means that the assumed current directions and voltage polarities were incorrect.

EXAMPLE 10-2 Using Kirchhoff's laws, analyze the circuit in Figure 10-4 to determine I_1, I_2, and I_3.

SOLUTION

For the path ABEFA,

$$E_1 = V_1 + V_3$$

$$E_1 = I_1R_1 + I_3R_3$$

Therefore,

$$6\,V = 120I_1 + 200I_3 \qquad (1)$$

For CBEDC,

$$E_2 = V_2 + V_3$$

$$E_2 = I_2R_2 + I_3R_3$$

$$12\,V = 240I_2 + 200I_3 \qquad (2)$$

For ABCDEFA,

$$E_1 = V_1 - V_2 + E_2$$

$$E_1 - E_2 = I_1R_1 - I_2R_2$$

$$6\,V - 12\,V = 120I_1 - 240I_2$$

$$-6\,V = 120I_1 - 240I_2 \qquad (3)$$

For point B,

$$I_3 = I_1 + I_2 \qquad (4)$$

Substituting for I_3 from Eq. (4) *into* Eq. (1),

$$6\,V = 120I_1 + 200I_1 + 200I_2$$

$$6\,V = 320I_1 + 200I_2 \qquad (5)$$

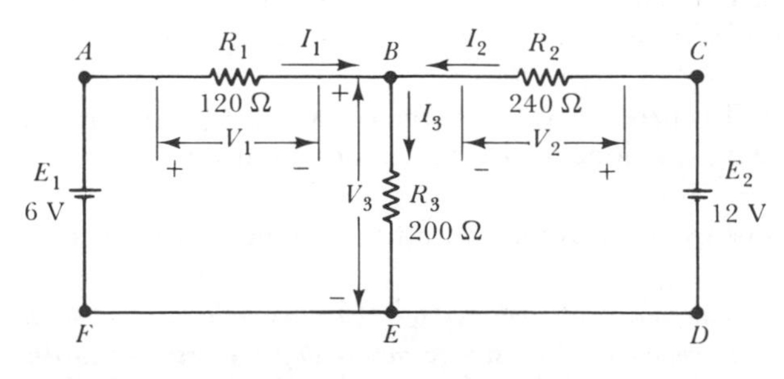

FIG. 10-4. Resistor network with two voltage sources

From Eq. (3),

$$240I_2 = 120I_1 + 6V$$

Therefore,

$$I_2 = \frac{120}{240}I_1 + \frac{6V}{240}$$

$$I_2 = \frac{1}{2}I_1 + \frac{1}{40} \qquad (6)$$

Substituting from Eq. (6) *into* Eq. (5),

$$6V = 320I_1 + \frac{200}{2}I_1 + \frac{200}{40}$$

Therefore,

$$6V = 420I_1 + 5$$

$$I_1 = \frac{6V - 5V}{420}$$

$$\mathbf{I_1 = 2.38\ mA}$$

Substituting for I_1 *in* Eq. (1),

$$6V = (120 \times 2.38\ \text{mA}) + 200I_3$$

$$\mathbf{I_3 = 28.57\ mA}$$

Substituting for I_1 *and* I_3 *in* Eq. (4),

$$28.57\ \text{mA} = 2.38\ \text{mA} + I_2$$

$$\mathbf{I_2 = 26.19\ mA}$$

10-3 LOOP EQUATIONS

By the use of *loop equations*, also termed *mesh equations*, complex circuits can be solved with greater ease than by direct application of Kirchhoff's laws.

The circuit of Figure 10-4 is reproduced in Figure 10-5, and *loop currents* are shown in a clockwise direction. The loop current is simply the current that circulates in the closed current path. In Figure 10-5, loop current I_1 is exactly the same as the branch current that flows through R_1. However, I_1 is *not* the branch current in R_3, because loop current I_2 also flows through R_3 in an opposite direction to I_1. The branch current in R_3 is $I_3 = I_1 - I_2$. Loop current I_2 is the same as the branch current flowing in resistor R_2.

The loop currents could actually be assigned an arbitrary direction, but assigning a clockwise direction simplifies the process of writing

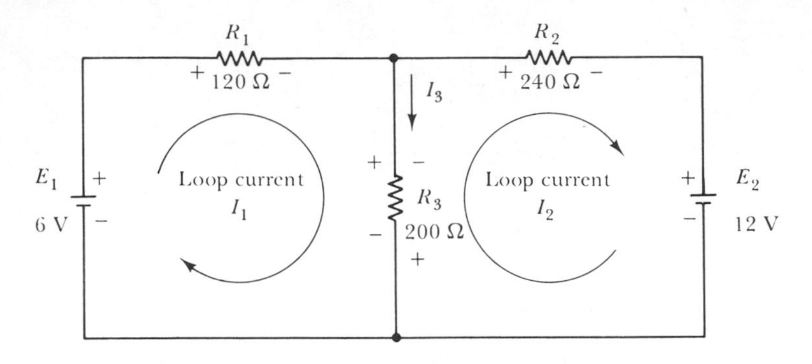

FIG. 10-5. Loop equation method of network analysis

equations for the volt drops around each loop. When the circuit analysis is complete, those branch currents which are in the same direction as the (clockwise) loop currents have a + sign. Those which are in an opposite direction come out with a − sign.

For very complex circuits, the process of combining the loop equation to determine the unknown quantities can sometimes become very lengthy and tedious. In this case it is advantageous to use the mathematical *method of determinants*.

The procedure for dc circuit analysis by loop equations is as follows:

1. Draw all loop currents in a clockwise direction and identify them by number.
2. Identify all resistor voltage drops as + to − in the direction of the loop current. (In Figure 10-5, note that for I_1, V_{R3} is + at the top of R_3, while for I_2, V_{R3} is + at the bottom.)
3. Identify all voltage sources according to their correct polarity. (*Note that in Figure* 10-5, E_1 *is in the opposite direction to the resistor voltage drops in loop* 1, *while in loop* 2, E_2 *is in the same direction as the resistor voltage drops*.)
4. Write the equations for the voltage drops around each loop in turn, by equating the sum of the voltage drops to zero.
5. Solve the equations to find the unknown currents.

EXAMPLE 10-3 Using loop equations, solve the resistor network shown in Figure 10-5 to determine the current through R_3.

SOLUTION

For loop 1,

$$0 = I_1R_1 + I_1R_3 - I_2R_3 - E_1$$

Therefore,

$$E_1 = I_1(R_1 + R_3) - I_2R_3$$

$$6\,V = I_1(120\ \Omega + 200\ \Omega) - I_2 \cdot 200\ \Omega$$

$$6\,V = 320I_1 - 200I_2 \qquad (1)$$

For loop 2,

$$0 = I_2R_2 + E_2 + I_2R_3 - I_1R_3$$

Therefore,

$$-E_2 = I_2(R_2 + R_3) - I_1R_3$$

$$-12\,V = I_2(240\ \Omega + 200\ \Omega) - I_1 \cdot 200\ \Omega$$

$$-12\,V = 440I_2 - 200I_1 \qquad (2)$$

Equations (1) *and* (2) *both contain two unknown quantities,* I_1 *and* I_2. *One of these two quantities must be eliminated before the other can be determined. Examining* Eqs. (1) *and* (2) *it is seen that if* Eq. (1) *is multiplied by* 440/200, *then the multiple of* I_2 *becomes* 440, *as in* Eq. (2). *This facilitates elimination of* I_2.

Eq. (1) $\times \dfrac{440}{200}$:

$$13.2\,V = \quad 704I_1 - 440I_2$$

Eq. (2):

$$-12\,V = -200I_1 + 440I_2$$

Adding,

$$1.2\,V = \quad 504I_1 + 0I_2$$

Therefore,

$$I_1 = \frac{1.2\,V}{504}$$

$$\mathbf{I_1 = 2.38\ mA}$$

Substituting for I_1 *in* Eq. (2),

$$-12\,V = 440I_2 - (200 \times 2.38\ \text{mA})$$

Therefore,

$$I_2 = \frac{-12\,V + (200 \times 2.38\ \text{mA})}{440}$$

$$\mathbf{I_2 = -26.19\ mA}$$

The negative sign indicates that the actual current direction is opposite to the (clockwise) loop current.

$$I_3 = I_1 - I_2$$
$$= 2.38 \text{ mA} - (-26.19 \text{ mA})$$
$$\mathbf{I_3 = 28.57 \text{ mA}}$$

Compare the current levels to the answers obtained in Example 10-2.

10-4 NODAL ANALYSIS

A *voltage node* is a junction in an electrical circuit at which a voltage can be measured with respect to another (reference) node. If one point in the circuit is grounded [see Figure 10-6(a)], that point is usually selected as the reference node. Otherwise, any convenient junction can be treated as a reference node.

From Figure 10-6(a), the current equation at node 1 is

$$I_3 = I_2 + I_1 \qquad (1)$$

An equation for each current can also be written in terms of the battery voltages and the node voltage:

$$I_1 = \frac{E_1 - V_1}{R_1} \qquad (2)$$

$$I_2 = \frac{E_2 - V_1}{R_2} \qquad (3)$$

$$I_3 = \frac{V_1}{R_3} \qquad (4)$$

Substitution of Eqs. (2), (3), and (4) into Eq. (1) gives

$$\frac{V_1}{R_3} = \frac{E_2 - V_1}{R_2} + \frac{E_1 - V_1}{R_1}$$

$$= \frac{E_2}{R_2} - \frac{V_1}{R_2} + \frac{E_1}{R_1} - \frac{V_1}{R_1}$$

or

$$V_1\left(\frac{1}{R_3} + \frac{1}{R_2} + \frac{1}{R_1}\right) = \frac{E_2}{R_2} + \frac{E_1}{R_1}$$

$$V_1 = \frac{\dfrac{E_1}{R_1} + \dfrac{E_2}{R_2}}{\dfrac{1}{R_3} + \dfrac{1}{R_1} + \dfrac{1}{R_2}} \qquad (5)$$

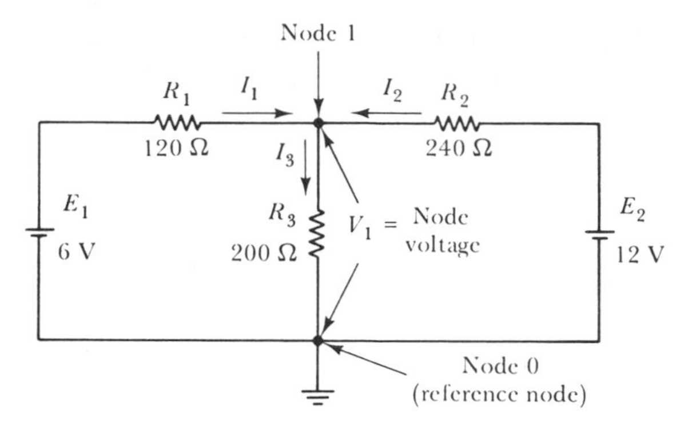

(a) Nodal analysis using voltage sources

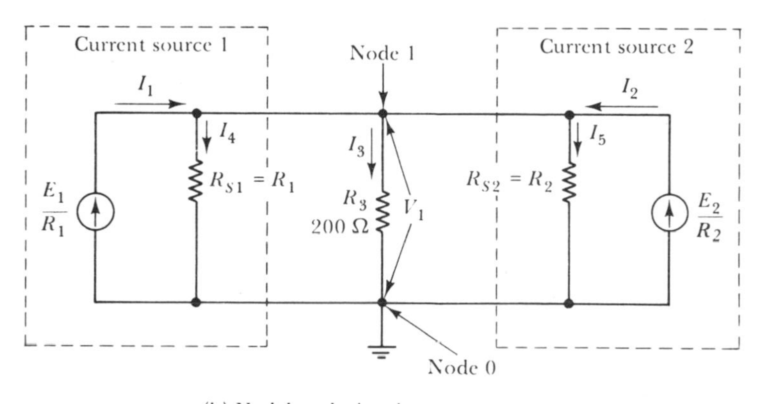

(b) Nodal analysis using current sources

FIG. 10-6. Nodal analysis method of solving complex networks

Since V_1 is the only unknown quantity in Eq. (5), the node voltage can be quickly determined. Then, using Eqs. (2), (3), and (4), the branch currents are readily calculated.

The method above is obviously simpler than either the Kirchhoff's law or loop-equation approach to network analysis. However, it can be simplified even further if the voltage sources and their series resistances are first converted to current sources, as illustrated in Figure 10-6(b).

The procedure for nodal analysis is as follows:

1. Convert all voltage sources to current sources, and redraw the circuit.
2. Identify all nodes and choose a reference node.
3. Write the equations for the currents flowing into and out of each node, with the exception of the reference node.
4. Solve the equations to determine the node voltages and the required branch currents.

EXAMPLE 10-4 For the circuit shown in Figure 10-6(a), convert the voltage sources to current sources, and using nodal analysis determine the current through R_3.

SOLUTION

The circuit with voltage sources converted to current sources is shown in Figure 10-6(b). *Voltage source* E_1 *and series resistance* R_1 *convert to current source 1, where from* Eq. (10-3),

$$I_1 = \frac{E_1}{R_1} = \frac{6\text{ V}}{120\ \Omega}$$

and

$$R_{S1} = R_1 = 120\ \Omega$$

For voltage source E_2 *and series resistance* R_2, *current source 2 is*

$$I_2 = \frac{E_2}{R_2} = \frac{12\text{ V}}{240\ \Omega}$$

and

$$R_{S2} = R_2 = 240\ \Omega$$

The modified circuit now shows three current paths, through R_{S1}, R_3, *and* R_{S2}, *as well as the current generators* I_1 *and* I_2.

Writing the current equation for node 1,

$$I_1 + I_2 = I_3 + I_4 + I_5 \qquad (1)$$

where

$$I_3 = \frac{V_1}{R_3}$$

$$I_4 = \frac{V_1}{R_{S1}}$$

$$I_5 = \frac{V_1}{R_{S2}}$$

Substituting into Eq. (1),

$$\frac{E_1}{R_1} + \frac{E_2}{R_2} = \frac{V_1}{R_3} + \frac{V_1}{R_{S1}} + \frac{V_1}{R_{S2}} \qquad (2)$$

or

$$\frac{E_1}{R_1} + \frac{E_2}{R_2} = V_1\left(\frac{1}{R_3} + \frac{1}{R_{S1}} + \frac{1}{R_{S2}}\right)$$

and

$$V_1 = \frac{\frac{E_1}{R_1} + \frac{E_2}{R_2}}{\frac{1}{R_3} + \frac{1}{R_{S1}} + \frac{1}{R_{S2}}}$$

$$V_1 = \frac{\frac{6\text{ V}}{120\ \Omega} + \frac{12\text{ V}}{240\ \Omega}}{\frac{1}{200\ \Omega} + \frac{1}{120\ \Omega} + \frac{1}{240\ \Omega}}$$

$$\mathbf{V_1 = 5.714\ V}$$

$$I_3 = \frac{V_1}{R_3} = \frac{5.714\text{ V}}{200\ \Omega}$$

$$\mathbf{I_3 = 28.57\ mA}$$

EXAMPLE 10-5

Using nodal analysis, determine the current that flows through resistor R_3 in the circuit shown in Figure 10-7(a).

SOLUTION

Voltage source E_1 *with series resistance* R_1 *becomes current source 1 in* Figure 10-7(b), *with*

$$I_1 = \frac{E_1}{R_1} \quad \text{and} \quad R_S = R_1$$

Voltage source E_2 *with series resistance* R_2 *becomes current source 2 in* Figure 10-7(b), *with*

$$I_4 = \frac{E_2}{R_2} \quad \text{and} \quad R_S = R_2$$

The nodes are identified and numbered as 0, 1, and 2 [see Figure 10-7(b)].
The circuit is redrawn in Figure 10-7(c) *to more easily identify which currents flow into the nodes and which flow out.*
At node 1:

$$I_1 + I_4 = I_2 + I_3 + I_5 \qquad (1)$$

Therefore

$$\frac{E_1}{R_1} + \frac{E_2}{R_2} = \frac{V_1}{R_1} + \frac{V_1}{R_2} + \frac{V_1 - V_2}{R_3}$$

$$\frac{50\text{ V}}{1.2\text{ k}\Omega} + \frac{20\text{ V}}{2.2\text{ k}\Omega} = V_1\left(\frac{1}{1.2\text{ k}\Omega} + \frac{1}{2.2\text{ k}\Omega} + \frac{1}{1.5\text{ k}\Omega}\right) - \frac{V_2}{1.5\text{ k}\Omega}$$

$$50.76 \times 10^{-3} = 1.95 \times 10^{-3} V_1 - 0.667 \times 10^{-3} V_2 \qquad (2)$$

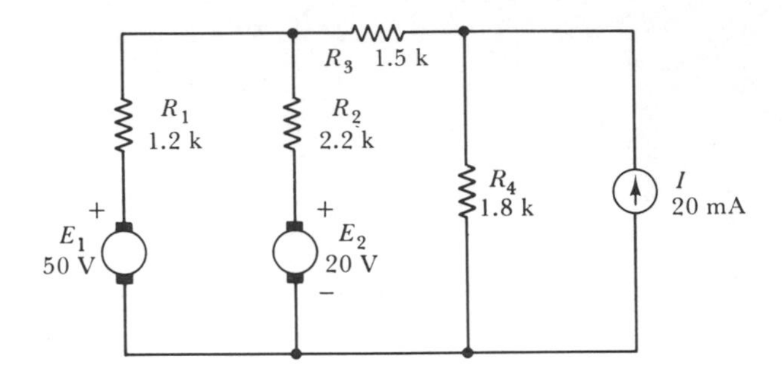

(a) Circuit to be analyzed

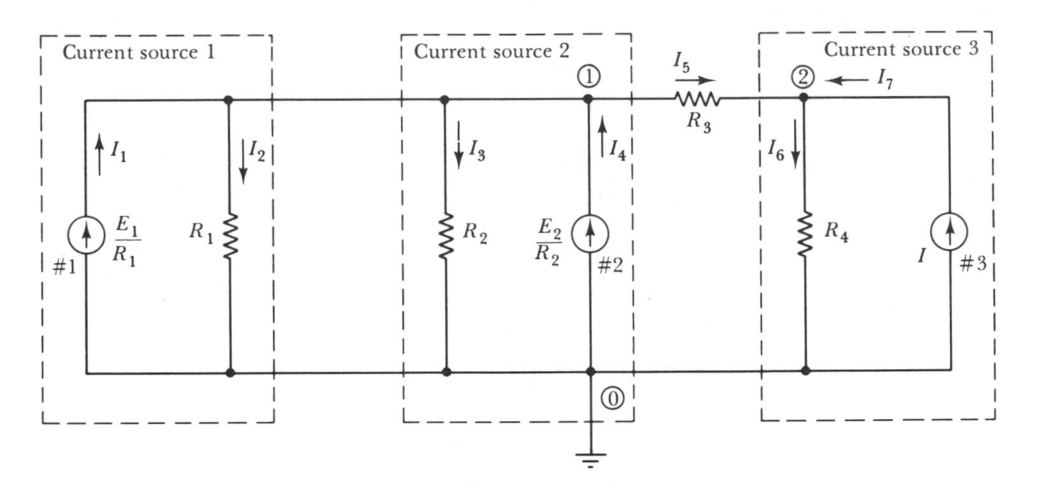

(b) Voltage sources replaced by current sources

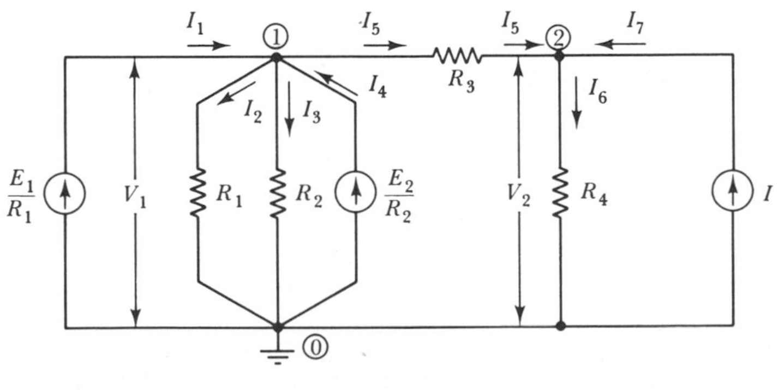

(c) Circuit redrawn

FIG. 10-7. Nodal analysis circuit for Example 10-5

At node 2:

$$I_6 = I_5 + I_7 \qquad (3)$$

$$\frac{V_2}{R_4} = \frac{V_1 - V_2}{R_3} + I_7$$

$$\frac{V_2}{1.8\text{ k}\Omega} = \frac{V_1}{1.5\text{ k}\Omega} - \frac{V_2}{1.5\text{ k}\Omega} + 20\times10^{-3}$$

$$-20\times10^{-3} = 0.667\times10^{-3}V_1 - 1.22\times10^{-3}V_2 \qquad (4)$$

Eq. (2):

$$50.76\times10^{-3} = 1.95\times\ 10^{-3}V_1 - 0.667\times10^{-3}V_2$$

Eq. (4) $\times \dfrac{0.667}{1.22}$:

$$-10.93\times10^{-3} = 0.365\times10^{-3}V_1 - 0.667\times10^{-3}V_2$$

Subtracting,

$$61.69\times10^{-3} = 1.585\times10^{-3}V_1$$

$$\mathbf{V_1 = 38.92\ V}$$

From Eq. (2),

$$50.76\times10^{-3} = (1.95\times10^{-3}\times38.92\text{ V}) - 0.667\times10^{-3}V_2$$

$$\mathbf{V_2 = 37.68\ V}$$

$$I_3 = \frac{V_1 - V_2}{R_3}$$

$$= \frac{38.92\text{ V} - 37.68\text{ V}}{1.5\text{ k}\Omega}$$

$$\mathbf{I_3 = 0.83\ mA}$$

10-5 DELTA–WYE TRANSFORMATION

Consider the circuit shown in Figure 10-8(a), in which resistors R_{ab}, R_{ac}, and R_{bc} are in the form of a *delta* (Δ) network. The circuit, as shown, could be analyzed by loop equations or by nodal analysis. However, the circuit can be considerably simplified if the Δ network is replaced by the *wye* (Y) network shown in Figure 10-8(b). If the Δ network is to be replaced by the Y network, both networks must have exactly the same effect on the rest of the circuit. Consequently, the resistance measured between any two of terminals A, B, and C must be exactly the same in each case.

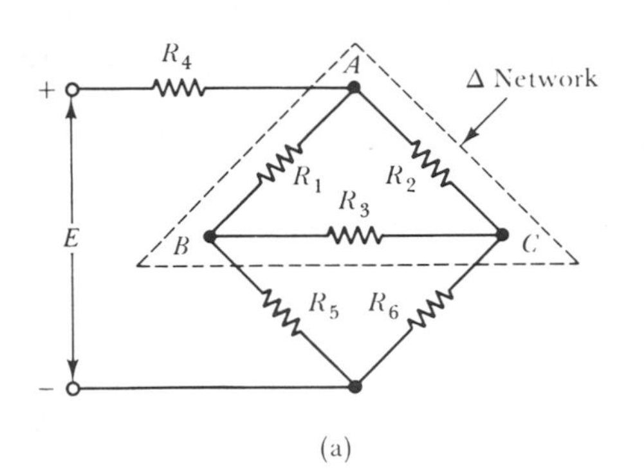

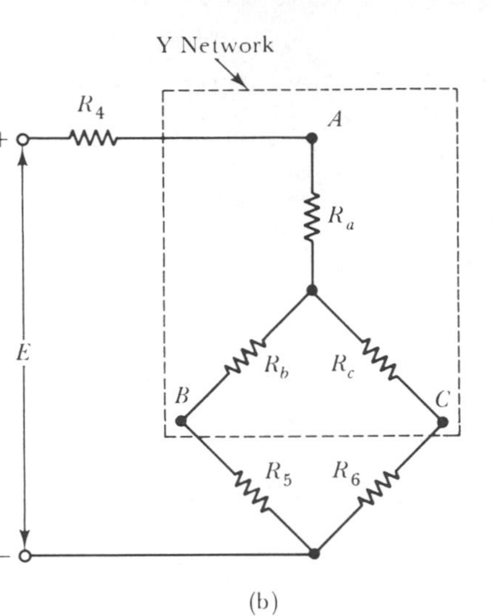

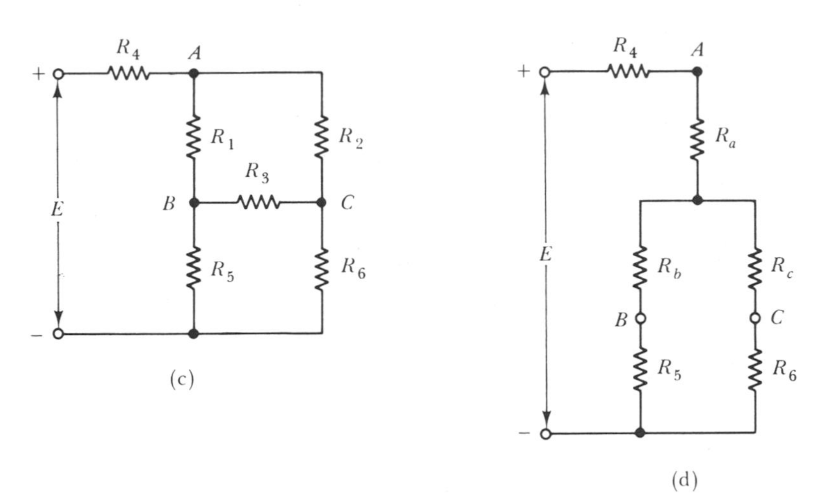

FIG. 10-8. Delta–wye transformation

Figure 10-8(c) and (d) represent exactly the same circuits as in Fig. 10-8(a) and (b), respectively.

For Δ–Y and Y–Δ transformation, formulas must be derived to relate the resistors in the two networks. Consider Figure 10-9(a) and (b). *For the Y network, the resistance between terminals A and B is*

$$R_{AB} = R_a + R_b$$

and for the Δ network,

$$R_{AB} = R_{ab} \| (R_{ac} + R_{bc})$$

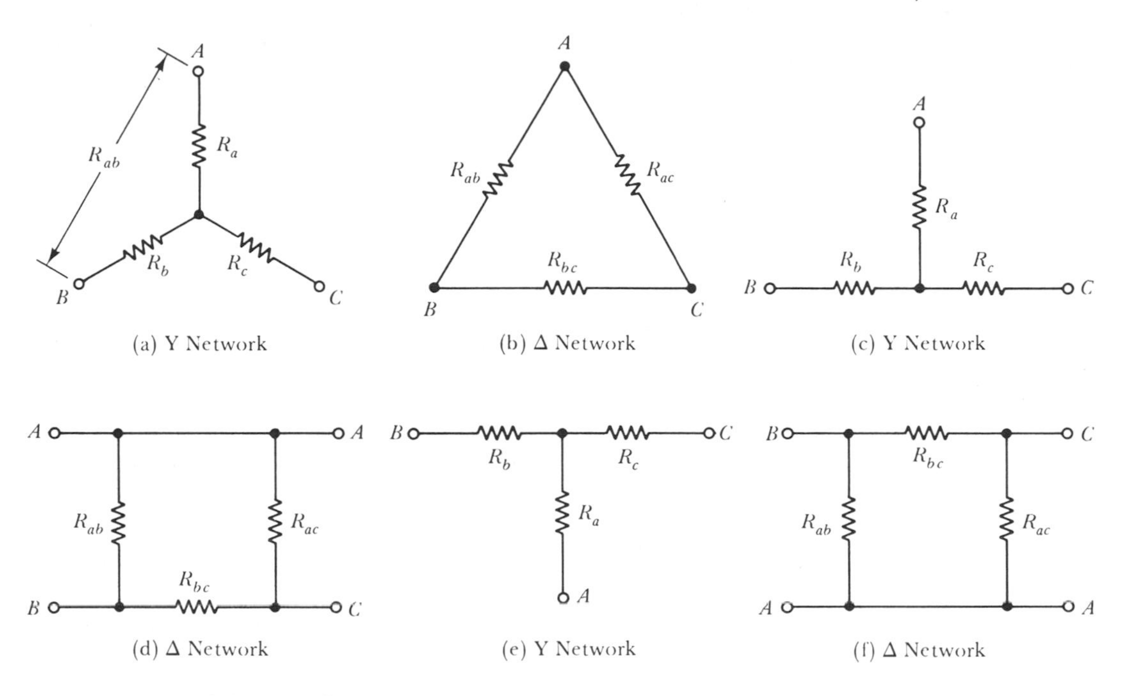

FIG. 10-9. Y and Δ networks

Therefore,

$$R_{AB} = \frac{R_{ab}(R_{ac} + R_{bc})}{R_{ab} + R_{ac} + R_{bc}}$$

$$R_{AB} = \frac{R_{ab}R_{ac} + R_{ab}R_{bc}}{R_{ab} + R_{ac} + R_{bc}}$$

For the same resistance value between terminals A and B on both networks,

$$R_a + R_b = \frac{R_{ab}R_{ac} + R_{ab}R_{bc}}{R_{ab} + R_{ac} + R_{bc}} \qquad (1)$$

Similarly, it can be shown that

$$R_a + R_c = \frac{R_{ab}R_{ac} + R_{ac}R_{bc}}{R_{ab} + R_{ac} + R_{bc}} \qquad (2)$$

and

$$R_b + R_c = \frac{R_{ab}R_{bc} + R_{ac}R_{bc}}{R_{ab} + R_{ac} + R_{bc}} \qquad (3)$$

Adding Eqs. (1) *and* (2) *and subtracting Eq.* (3) *gives*

$$\boxed{R_a = \frac{R_{ab}R_{ac}}{R_{ab} + R_{ac} + R_{bc}}} \qquad \textbf{(10-5)}$$

Also, Eq. (1) + *Eq.* (3) − *Eq.* (2) *gives*

$$R_b = \frac{R_{ab}R_{bc}}{R_{ab}+R_{ac}+R_{bc}} \tag{10-6}$$

and Eq. (2) + *Eq.* (3) − *Eq.* (1) *gives*

$$R_c = \frac{R_{ac}R_{bc}}{R_{ab}+R_{ac}+R_{bc}} \tag{10-7}$$

Equations (10-5), (10-6), and (10-7) can be used to convert from Δ to Y. Examination of Eq. (10-5) shows that *to obtain the Y network resistor* (R_a) *connected to terminal A, the two Δ network resistors connected to A must be multiplied together, and the product divided by the sum of the three Δ network resistors.* A similar procedure applies for obtaining the Y network resistors connected to terminals *B* and *C*.

By further manipulation of Eqs. (1), (2), and (3), equations for conversion from Y to Δ can be obtained:

$$R_{ab} = \frac{R_aR_b + R_aR_c + R_bR_c}{R_c} \tag{10-8}$$

$$R_{ac} = \frac{R_aR_b + R_aR_c + R_bR_c}{R_b} \tag{10-9}$$

$$R_{bc} = \frac{R_aR_b + R_aR_c + R_bR_c}{R_a} \tag{10-10}$$

Once again, examination of the three equations reveals a definite pattern. *To obtain the Δ resistor connected between any two terminals, divide the Y resistor connected to the terminal opposite those two (terminal C for the resistor between A and B) into the sum of the products of each pair of Y resistors.*

The Y network is also known as a *star network* or a *T network*. Figure 10-9(c) and (e) shows Y networks redrawn in a slightly different form. Delta networks are also termed *mesh networks* or *π networks*. Figure 10-9(d) and (f) shows delta networks redrawn in a different form.

EXAMPLE 10-6

Convert the Δ network shown in Figure 10-9 (b) to a Y network. Then convert back again to prove the formula. Take $R_{ab}=500\ \Omega$, $R_{ac}=400\ \Omega$, and $R_{bc}=300\ \Omega$.

SOLUTION

Eq. (10-5):

$$R_a=\frac{R_{ab}R_{ac}}{R_{ab}+R_{ac}+R_{bc}}$$

$$=\frac{500\ \Omega\times 400\ \Omega}{500\ \Omega+400\ \Omega+300\ \Omega}$$

$$\mathbf{R_a=166.\dot{7}\ \Omega}$$

Eq. (10-6):

$$R_b=\frac{R_{ab}R_{bc}}{R_{ab}+R_{ac}+R_{bc}}$$

$$=\frac{500\ \Omega\times 300\ \Omega}{500\ \Omega+400}$$

$$\mathbf{R_b=125\ \Omega}$$

Eq. (10-7):

$$R_c=\frac{R_{ac}R_{bc}}{R_{ab}+R_{ac}+R_{bc}}$$

$$=\frac{400\ \Omega\times 300\ \Omega}{500\ \Omega+400\ \Omega+300\ \Omega}$$

$$\mathbf{R_c=100\ \Omega}$$

Converting back from Y to Δ:
Eq. (10-8):

$$R_{ab}=\frac{R_aR_b+R_aR_c+R_bR_c}{R_c}$$

$$=\frac{(166.\dot{7}\ \Omega\times 125\ \Omega)+(166.\dot{7}\ \Omega\times 100\ \Omega)+(125\ \Omega\times 100\ \Omega)}{100}$$

$$=\frac{50\ 000\ \Omega}{100\ \Omega}$$

$$\mathbf{R_{ab}=500\ \Omega}$$

Eq. (10-9):

$$R_{ac} = \frac{R_aR_b + R_aR_c + R_bR_c}{R_b}$$

$$= \frac{50\ 000\ \Omega}{125\ \Omega}$$

$$\mathbf{R_{ac} = 400\ \Omega}$$

Eq. (10-10):

$$R_{bc} = \frac{R_aR_b + R_aR_c + R_bR_c}{R_a}$$

$$= \frac{50\ 000\ \Omega}{166.\dot{6}\ \Omega}$$

$$\mathbf{R_{bc} = 300\ \Omega}$$

GLOSSARY OF FORMULAS

Voltage source-to-current source conversion:

$$I = \frac{E}{R_s} \qquad R_s = R_s$$

Current source-to-voltage source conversion:

$$V = IR_s \qquad R_s = R_s$$

Δ-to-Y conversion:

$$R_a = \frac{R_{ab}R_{ac}}{R_{ab} + R_{ac} + R_{bc}}$$

$$R_b = \frac{R_{ab}R_{bc}}{R_{ab} + R_{ac} + R_{bc}}$$

$$R_c = \frac{R_{ac}R_{bc}}{R_{ab} + R_{ac} + R_{bc}}$$

Y-to-Δ conversion:

$$R_{ab} = \frac{R_aR_b + R_aR_c + R_bR_c}{R_c}$$

$$R_{ac} = \frac{R_aR_b + R_aR_c + R_bR_c}{R_b}$$

$$R_{bc} = \frac{R_aR_b + R_aR_c + R_bR_c}{R_a}$$

REVIEW QUESTIONS

10-1 Define constant voltage source, constant current source, loop equations, loop currents, and mesh questions.

10-2 Define voltage node, nodal analysis, delta network, π network, Y network, and T network.

10-3 Sketch the circuit diagram for a constant voltage source and for a constant current source. Identify all voltage polarities and current directions.

10-4 Derive the necessary equation for converting a constant current source to a constant voltage source, and vice versa. Also, discuss the effects of operating each type of source in series and in parallel.

10-5 List the procedure for analyzing a complex network by use of Kirchhoff's laws.

10-6 List the procedure for analyzing a complex network using loop equations.

10-7 List the procedure for nodal analysis.

10-8 Write the equations for Δ–Y conversion and for Y–Δ conversion.

PROBLEMS

10-1 A constant current source has $I = 15$ A and $R_s = 5\ \Omega$. Determine E and R_s for the equivalent constant voltage source. If $R_L = 25\ \Omega$, calculate the output voltage and current in each case.

10-2 Using direct application of Kirchhoff's laws, analyze the circuit shown in Figure 10-10 to determine the voltage across R_5.

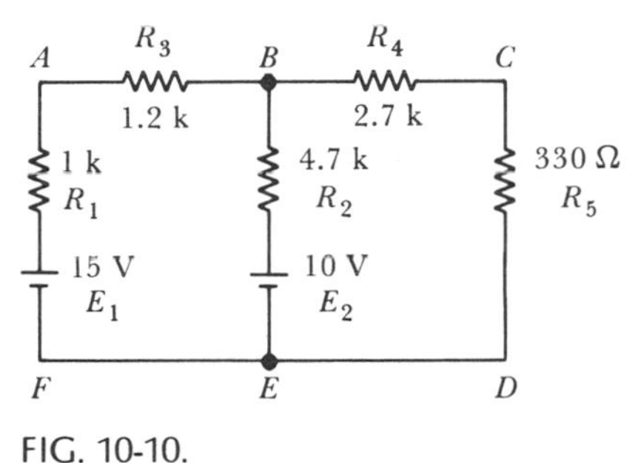

FIG. 10-10.

10-3 Using direct application of Kirchhoff's laws, analyze the circuit in Figure 10-11 to determine the current through R_5.

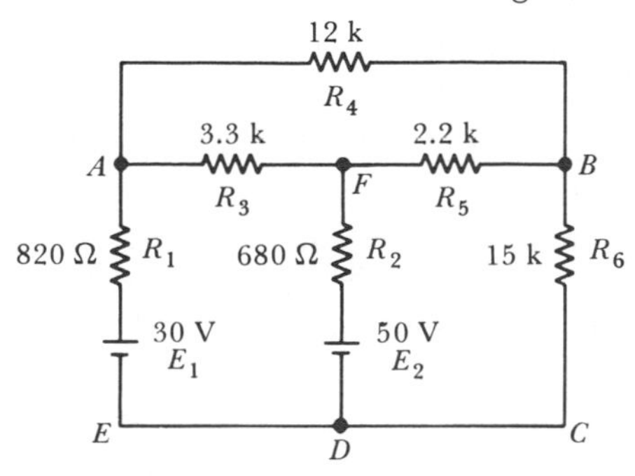

FIG. 10-11.

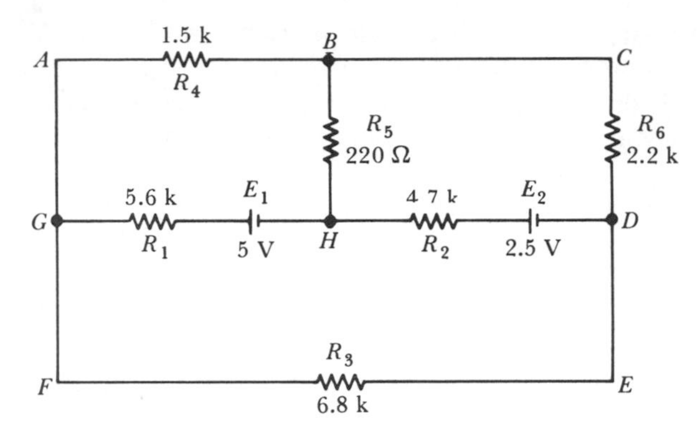

FIG. 10-12.

10-4 Using direct application of Kirchhoff's laws, analyze the circuit in Figure 10-12 to determine the voltage across R_6.

10-5 Using direct application of Kirchhoff's laws, analyze the circuit in Figure 10-13 to determine the current that flows through R_3.

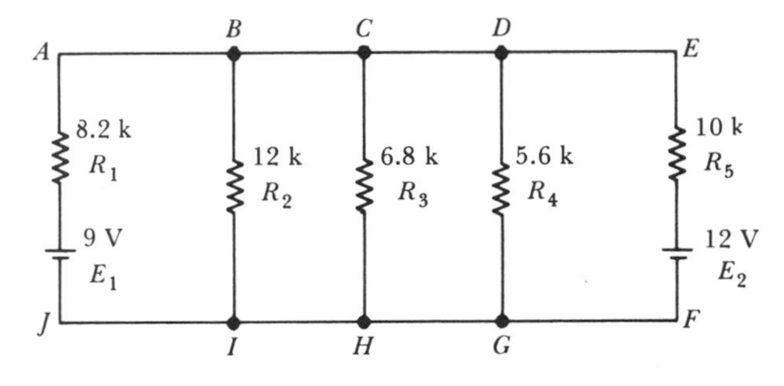

FIG. 10-13.

10-6 Repeat Problem 10-2 using loop equations.

10-7 Repeat Problem 10-3 using loop equations.

10-8 Repeat Problem 10-4 using loop equations.

10-9 Repeat Problem 10-5 using loop equations.

10-10 Repeat Problem 10-2 using nodal analysis.

10-11 Repeat Problem 10-5 using nodal analysis.

10-12 Apply Δ–Y transformation to analyze the circuit of Figure 10-11 to determine the voltage across R_6.

10-13 Apply Y–Δ transformation to resistors R_1, R_2, and R_3 in the circuit shown in Figure 10-6(a). Then determine the current from E_1.

10-14 Using nodal analysis determine the current through R_4 in the circuit of Figure 10-7(a).

11

NETWORK THEOREMS

INTRODUCTION

Network analysis can be simplified by the use of *network theorems*, which state certain rules that may be applied in particular circumstances. The *superposition theorem*, for example, enables a circuit with several voltage and/or current sources to be treated as several circuits that each have only one source. *Thévenin's theorem* permits complex networks to be reduced to a single voltage source in series with a resistance. Thus simplified, the load current and voltage can be very easily determined for many different values of load resistance. *Norton's theorem* is just as powerful as Thévenin's theorem. In this case the network is reduced to a single current source and parallel resistance. How a circuit consisting of many sources in parallel may be treated as just one source is defined by *Millman's theorem*, while the *maximum power transfer theorem* predicts optimum load conditions.

11-1 THE SUPERPOSITION THEOREM

In a network containing more than one source of voltage or current, the current through any branch is the algebraic sum of the currents produced by each source acting independently.

Application of the superposition theorem simplifies the analysis of a network which has more than one source. The procedure is as follows:

1. Select one source, and replace all other sources with their internal impedances.

2. Determine the level and direction of the current that flows through the desired branch as a result of the single source acting alone.
3. Repeat steps 1 and 2 using each source in turn until the branch current components have been calculated for all sources.
4. Algebraically sum the component currents to obtain the actual branch current.

EXAMPLE 11-1

Use the superposition theorem to calculate the current I_3 in the circuit shown in Figure 10-4.

SOLUTION

Figure 10-4 *is reproduced in* Figure 11-1(a). *The voltage source* E_2 *is replaced with its internal resistance* (R_2) *in* Figure 11-1(b), *and in* Figure 11-1(c) *voltage source* E_1 *is replaced with its internal resistance* (R_1).

For Figure 11-1(b):

The equivalent resistance of the circuit is

$$R_{eq1} = R_1 + R_2 \| R_3$$

$$= 120\ \Omega + 240\ \Omega \| 200\ \Omega$$

$$= 229.09\ \Omega$$

and

$$I_a = \frac{E_1}{R_{eq1}} = \frac{6\ \text{V}}{229.09\ \Omega}$$

$$= 26.19\ \text{mA}$$

Using the current divider rule (*see* Section 8-4),

$$i_1 = I_a \times \frac{R_2}{R_3 + R_2}$$

$$= 26.19\ \text{mA} \times \frac{240\ \Omega}{200\ \Omega + 240\ \Omega}$$

Therefore,

$$i_1 = 14.29\ \text{mA}$$

For Figure 11-1(c):

$$R_{eq2} = R_2 + R_1 \| R_3$$

$$= 240\ \Omega + 120\ \Omega \| 200\ \Omega$$

$$= 315\ \Omega$$

$$I_b = \frac{E_2}{R_{eq2}} = \frac{12\ \text{V}}{315\ \Omega}$$

$$= 38.10\ \text{mA}$$

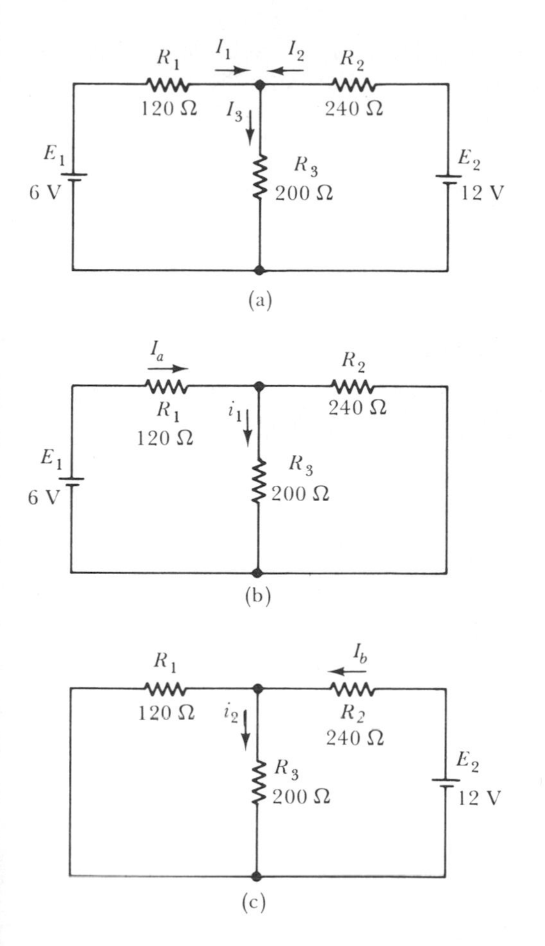

FIG. 11-1. Application of the superposition theorem

Using the current divider rule,

$$i_2 = I_b \times \frac{R_1}{R_1 + R_3}$$

$$= 38.1\text{ mA} \times \frac{120\ \Omega}{120\ \Omega + 200\ \Omega}$$

$$= 14.29\text{ mA}$$

$$I_3 = i_1 + i_2$$

$$= 14.29\text{ mA} + 14.29\text{ mA}$$

$$\mathbf{I_3 = 28.58\ mA}$$

Note: The fact that i_1 and i_2 happen to be equal is peculiar to this particular circuit.

EXAMPLE 11-2 For the circuit of Figure 10-7(a), reproduced in Figure 11-2(a), determine the current through resistor R_3, using the superposition theorem.

SOLUTION

The circuits for each individual source acting independently are shown in Figure 11-2(b), (c), and (d).

For Figure 11-2(b),

$$I_a = \frac{E_1}{R_1 + R_2 \| (R_3 + R_4)}$$

$$= \frac{50 \text{ V}}{1.2 \text{ k}\Omega + 2.2 \text{ k}\Omega \| (1.5 \text{ k}\Omega + 1.8 \text{ k}\Omega)}$$

$$= 19.84 \text{ mA}$$

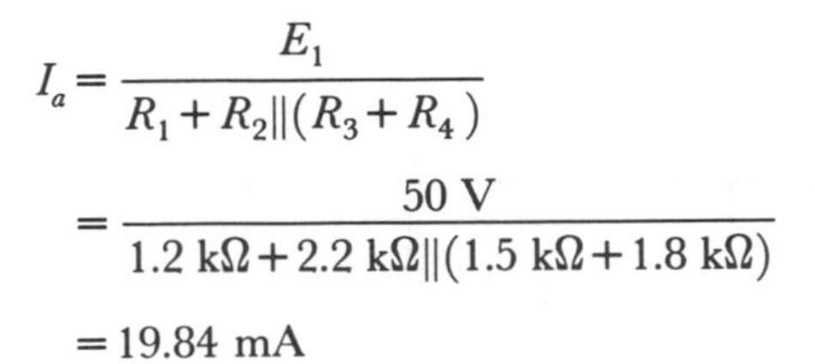

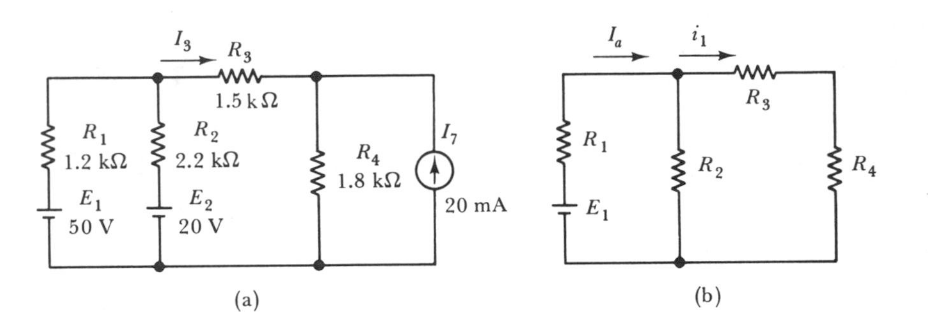

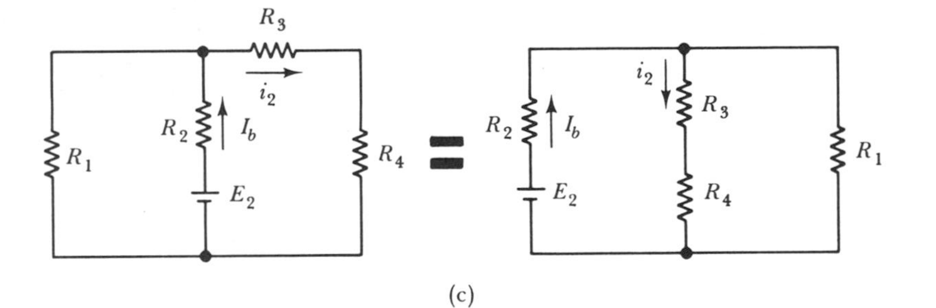

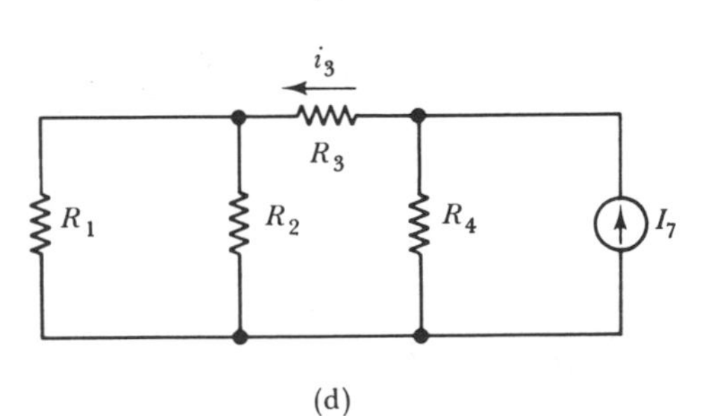

FIG. 11-2. Circuit for Example 11-2

and using the current divider rule,

$$i_1 = I_a \times \frac{R_2}{R_2 + (R_3 + R_4)}$$

$$= 19.83 \text{ mA} \times \frac{2.2 \text{ k}\Omega}{2.2 \text{ k}\Omega + (1.5 \text{ k}\Omega + 1.8 \text{ k}\Omega)}$$

$$= 7.94 \text{ mA}$$

For Figure 11-2(c),

$$I_b = \frac{E}{R_2 + R_1 \| (R_3 + R_4)}$$

$$= \frac{20 \text{ V}}{2.2 \text{ k}\Omega + 1.2 \text{ k}\Omega \| (1.5 \text{ k}\Omega + 1.8 \text{ k}\Omega)}$$

$$= 6.49 \text{ mA}$$

Using the current divider rule,

$$i_2 = I_b \times \frac{R_1}{R_1 + (R_3 + R_4)}$$

$$= 6.49 \text{ mA} \times \frac{1.2 \text{ k}\Omega}{1.2 \text{ k}\Omega + (1.5 \text{ k}\Omega + 1.8 \text{ k}\Omega)}$$

$$= 1.73 \text{ mA}$$

For Figure 11-2(d):
Using the current divider rule,

$$i_3 = I_7 \times \frac{R_4}{R_4 + R_3 + (R_1 \| R_2)}$$

$$= 20 \text{ mA} \times \frac{1.8 \text{ k}\Omega}{1.8 \text{ k}\Omega + 1.5 \text{ k}\Omega + (1.2 \text{ k}\Omega \| 2.2 \text{ k}\Omega)}$$

$$= 8.83 \text{ mA}$$

$$I_3 = i_1 + i_2 - i_3$$

$$= 7.94 \text{ mA} + 1.73 \text{ mA} - 8.83 \text{ mA}$$

$$\mathbf{I_3 = 0.84 \text{ mA}}$$

This compares with the result of Example 10-5.

11-2 THÉVENIN'S THEOREM

Any two-terminal network containing resistances and voltage sources and/or current sources may be replaced by a single voltage source in series with a single resistance. The emf of the voltage source is the open-circuit emf at the network terminals, and the series resistance is the resistance between the network terminals when all sources are replaced by their internal impedances.

By means of Thévenin's theorem, any one resistor in a network can be isolated. The entire remaining portion of the network can be replaced by a single source of emf and a single resistor. Then the current through the isolated resistor may be easily calculated for any value of resistance.

Consider the resistor network shown in Figure 11-3(a). Resistor R_L is variable. Consequently, a complete circuit analysis would seem to be necessary for each value of R_L. The circuit is simplified by replacing all of the network to the left of terminals A and B by its *Thévenin equivalent circuit*. This results in the circuit shown in Figure 11-3(b). Clearly, the calculation of load voltage and current can now be easily repeated for any number of values of R_L.

Determining the Thévenin equivalent circuit for a network is sometimes termed *thevenizing the circuit*. The procedure for thevenizing is as

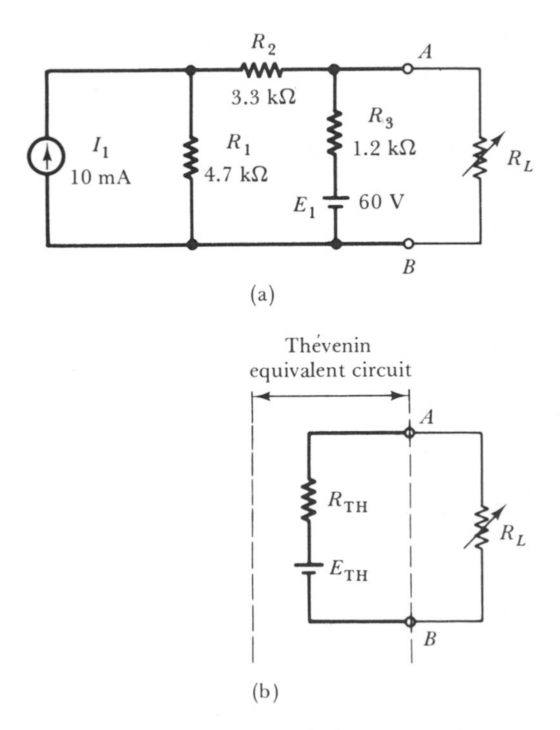

FIG. 11-3. Application of Thévenin's theorem

follows:

1. Calculate the open-circuit terminal voltage of the network.
2. Redraw the network with each voltage source replaced by a short circuit in series with its internal resistance, and each current source replaced by an open circuit in parallel with its internal resistance.
3. Calculate the resistance of the redrawn network as *seen* from output terminals.

EXAMPLE 11-3

For the circuit shown in Figure 10-4, an external load R_L is to be connected across resistor R_3, as illustrated in Figure 11-4(a). Determine the Thévenin equivalent circuit for the network and calculate the load current when R_L is 330 Ω.

SOLUTION

The open-circuit voltage at terminals B and E must first be calculated.
From Example 10-2,

$$I_3 = 28.57 \text{ mA}$$

$$V_{BE} = V_{R3} = I_3 R_3$$

$$= 28.57 \text{ mA} \times 200 \ \Omega$$

$$\mathbf{E_{TH} = V_{BE} = 5.71 \ V}$$

Replacing E_1 *and* E_2 *by short circuits gives the circuit of* Figure 11-4(b).
The resistance seen looking into terminals B and E is

$$R_{TH} = R_3 \| R_1 \| R_2$$

$$= 200 \ \Omega \| 120 \ \Omega \| 240 \ \Omega$$

$$\mathbf{R_{TH} = 57.1 \ \Omega}$$

The Thévenin equivalent circuit is now as shown in Figure 11-4(c).
The load current is calculated as

$$I_L = \frac{E_{TH}}{R_{TH} + R_L}$$

$$= \frac{5.71 \text{ V}}{57.1 \ \Omega + 330 \ \Omega}$$

$$\mathbf{I_L = 14.75 \ mA}$$

Obviously, the load current can be readily recalculated for any number of values of R_L.

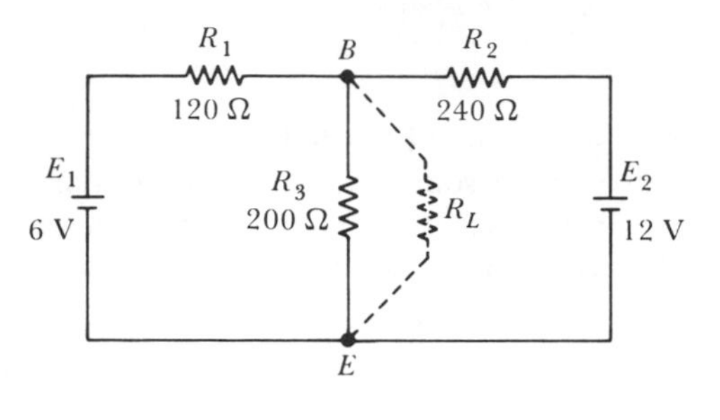

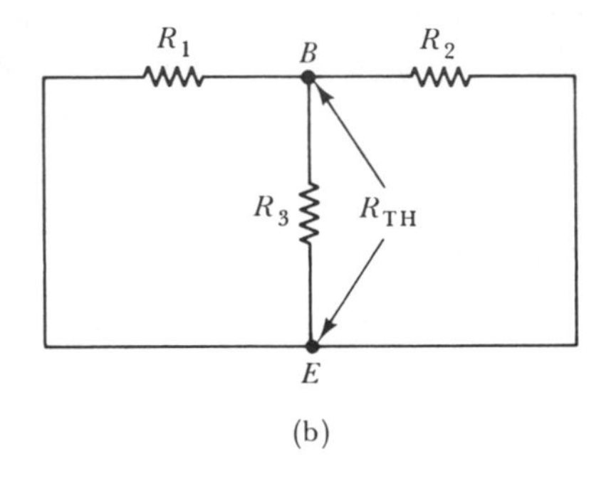

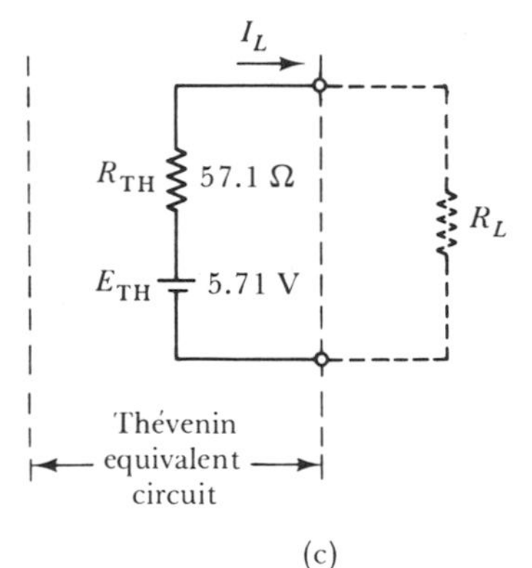

FIG. 11-4. Circuit for Example 11-3

EXAMPLE 11-4 Derive the Thévenin equivalent circuit for the network shown in Figure 11-3(a), and calculate the output voltage for $R_{L1} = 12$ kΩ and for $R_{L2} = 5.6$ kΩ.

SOLUTION

The circuit of Figure 11-3(a) *is reproduced in* Figure 11-5(a) *with the load resistance removed. The open-circuit voltage at terminals A and B can now be calculated by any of the methods already discussed.*

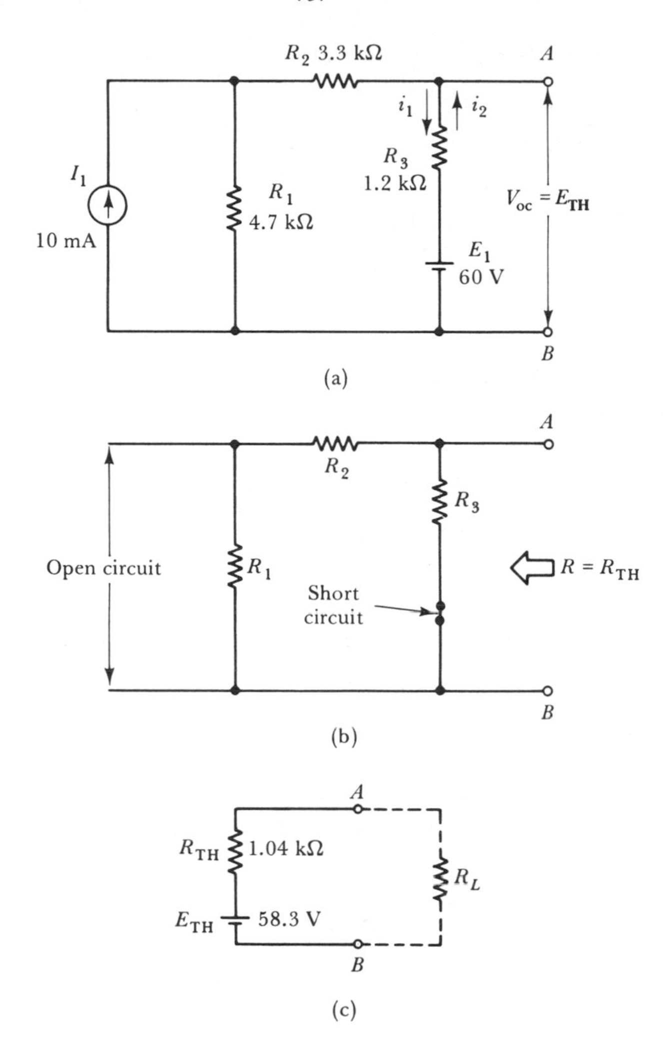

FIG. 11-5. Circuit for Example 11-4

Using the superposition theorem, a current i_1 flows through R_3 due to current source I_1, and a current i_2 flows in the opposite direction due to voltage source E_1. For I_1 acting alone (E_1 short-circuited), using the current divider rule:

$$i_1 = I_1 \times \frac{R_1}{R_1 + (R_2 + R_3)}$$

$$= 10 \text{ mA} \times \frac{4.7 \text{ k}\Omega}{4.7 \text{ k}\Omega + 3.3 \text{ k}\Omega + 1.2 \text{ k}\Omega}$$

$$\mathbf{i_1 = 5.11 \text{ mA}}$$

For E_1 acting alone (I_1 open-circuited):

$$i_2 = \frac{E}{R_3 + R_2 + R_1}$$

$$= \frac{60\ \text{V}}{1.2\ \text{k}\Omega + 3.3\ \text{k}\Omega + 4.7\ \text{k}\Omega}$$

$$\mathbf{i_2 = 6.52\ mA}$$

Therefore,

$$I_3 = i_2 - i_1$$

$$= 6.52\ \text{mA} - 5.11\ \text{mA}$$

$$\mathbf{I_3 = 1.41\ mA}$$

and

$$V_{AB} = E_1 - I_3 R_3 \qquad [\textit{see}\ \text{Figure 11-5(a)}]$$

$$= 60\ \text{V} - (1.41\ \text{mA} \times 1.2\ \text{k}\Omega)$$

$$\mathbf{E_{TH} = V_{AB} = 58.3\ V}$$

$$R_{TH} = R_3 \| (R_2 + R_1) \qquad [\textit{see}\ \text{Figure 11-5(b)}]$$

$$= 1.2\ \text{k}\Omega \| (3.3\ \text{k}\Omega + 4.7\ \text{k}\Omega)$$

$$\mathbf{R_{TH} = 1.04\ k\Omega}$$

The Thévenin equivalent circuit is now as shown in Figure 11-5(c).

The output voltage across R_L can be calculated using the voltage divider rule:

$$V_L = E_{TH} \times \frac{R_L}{R_L + R_{TH}}$$

For $R_L = 12\ \text{k}\Omega$,

$$V_{L1} = 58.3\ \text{V} \times \frac{12\ \text{k}\Omega}{12\ \text{k}\Omega + 1.04\ \text{k}\Omega}$$

$$\mathbf{V_{L1} = 53.65\ V}$$

For $R_L = 5.6\ \text{k}\Omega$,

$$V_{L2} = 58.3\ \text{V} \times \frac{5.6\ \text{k}\Omega}{5.6\ \text{k}\Omega + 1.04\ \text{k}\Omega}$$

$$\mathbf{V_{L2} = 49.17\ V}$$

Example 11-4 illustrates the circuit simplification afforded by Thévenin's theorem. There would obviously be much more work involved if the circuit had to be completely analyzed each time a new value of R_L

is substituted. Now look at Example 11-5, which further demonstrates how this useful circuit theorem tremendously cuts down on calculations. Were Thévenin's theorem not available, the circuit would have to be completely analyzed nine times for the nine different values of R_L listed.

EXAMPLE 11-5

For the circuit shown in Figure 11-5(a), calculate the load current for the following value of R_L: 12 kΩ, 5.6 kΩ, 3.3 kΩ, 10 kΩ, 6.8 kΩ, 8.2 kΩ, 4.7 kΩ, 13 kΩ, and 15 kΩ.

SOLUTION

Using the Thévenin equivalent circuit in Figure 11-5(c):

$$I_L = \frac{E_{TH}}{R_{TH} + R_L}$$

$$I_{L1} = \frac{58.3 \text{ V}}{1.04 \text{ k}\Omega + 12 \text{ k}\Omega} = \mathbf{4.47 \text{ mA}}$$

$$I_{L2} = \frac{58.3 \text{ V}}{1.04 \text{ k}\Omega + 5.6 \text{ k}\Omega} = \mathbf{8.78 \text{ mA}}$$

$$I_{L3} = \frac{58.3 \text{ V}}{1.04 \text{ k}\Omega + 3.3 \text{ k}\Omega} = \mathbf{13.43 \text{ mA}}$$

$$I_{L4} = \frac{58.3 \text{ V}}{1.04 \text{ k}\Omega + 10 \text{ k}\Omega} = \mathbf{5.28 \text{ mA}}$$

$$I_{L5} = \frac{58.3 \text{ V}}{1.04 \text{ k}\Omega + 6.8 \text{ k}\Omega} = \mathbf{7.44 \text{ mA}}$$

$$I_{L6} = \frac{58.3 \text{ V}}{1.04 \text{ k}\Omega + 8.2 \text{ k}\Omega} = \mathbf{6.31 \text{ mA}}$$

$$I_{L7} = \frac{58.3 \text{ V}}{1.04 \text{ k}\Omega + 4.7 \text{ k}\Omega} = \mathbf{10.16 \text{ mA}}$$

$$I_{L8} = \frac{58.3 \text{ V}}{1.04 \text{ k}\Omega + 13 \text{ k}\Omega} = \mathbf{4.15 \text{ mA}}$$

$$I_{L9} = \frac{58.3 \text{ V}}{1.04 \text{ k}\Omega + 15 \text{ k}\Omega} = \mathbf{3.63 \text{ mA}}$$

11-3 NORTON'S THEOREM

In Section 10-1 it was shown that constant voltage sources can be converted to constant current sources, and vice versa. Therefore, a Thévenin equivalent circuit which is a constant voltage source can be converted to a constant current source. Alternatively, instead of deriving the Thévenin constant voltage equivalent circuit for a complex network, a constant current equivalent circuit can be derived directly for the network. In this case the equivalent circuit is termed a *Norton equivalent circuit*, and the theory for its application is stated in *Norton's theorem*:

Norton's Theorem

Any two-terminal network containing resistances and voltage source and/or current sources may be replaced by a single current source in parallel with a single resistance. The output from the current source is the short-circuit current at the network terminals, and the parallel resistance is the resistance between the network terminals when all sources are replaced by their internal impedances.

As with thevenizing, the determination of the Norton equivalent circuit is termed *nortonizing*. The procedure for nortonizing is as follows:

1. Calculate the short-circuit current at the network terminals.
2. Redraw the network with each voltage source replaced by a short circuit in series with its internal resistance, and each current source replaced by an open circuit in parallel with its internal resistance.
3. Calculate the resistance of the redrawn network as *seen* from the output terminals.

Like the Thevenin theorem, application of the Norton theorem can save a lot of calculations when several values of load resistor are involved.

EXAMPLE 11-6

Derive the Norton equivalent circuit for the network shown in Figure 11-3(a) and calculate the load current for $R_L = 12\ \text{k}\Omega$ and for $R_L = 5.6\ \text{k}\Omega$.

SOLUTION

The circuit of Figure 11-3(a) *is reproduced in* Figure 11-6(a) *with* R_L *replaced by a short circuit. Using the superposition theorem to determine* I_{sc}:

$$I_{sc} = i_1 + i_2 \qquad [\textit{see}\ \text{Figure 11-6(b)}\ \textit{and}\ \text{(c)}]$$

$$i_1 = I_1 \times \frac{R_1}{R_1 + R_2} \qquad \textit{by current divider rule}$$

$$= 10\ \text{mA} \times \frac{4.7\ \text{k}\Omega}{4.7\ \text{k}\Omega + 3.3\ \text{k}\Omega}$$

$$= 5.875\ \text{mA}$$

$$i_2 = \frac{E_1}{R_3}$$

$$= \frac{60\ \text{V}}{1.2\ \text{k}\Omega}$$

$$= 50\ \text{mA}$$

$$I_{sc} = 5.875\ \text{mA} + 50\ \text{mA}$$

$$\boldsymbol{I_N = I_{sc} = 55.875\ \textbf{mA}}$$

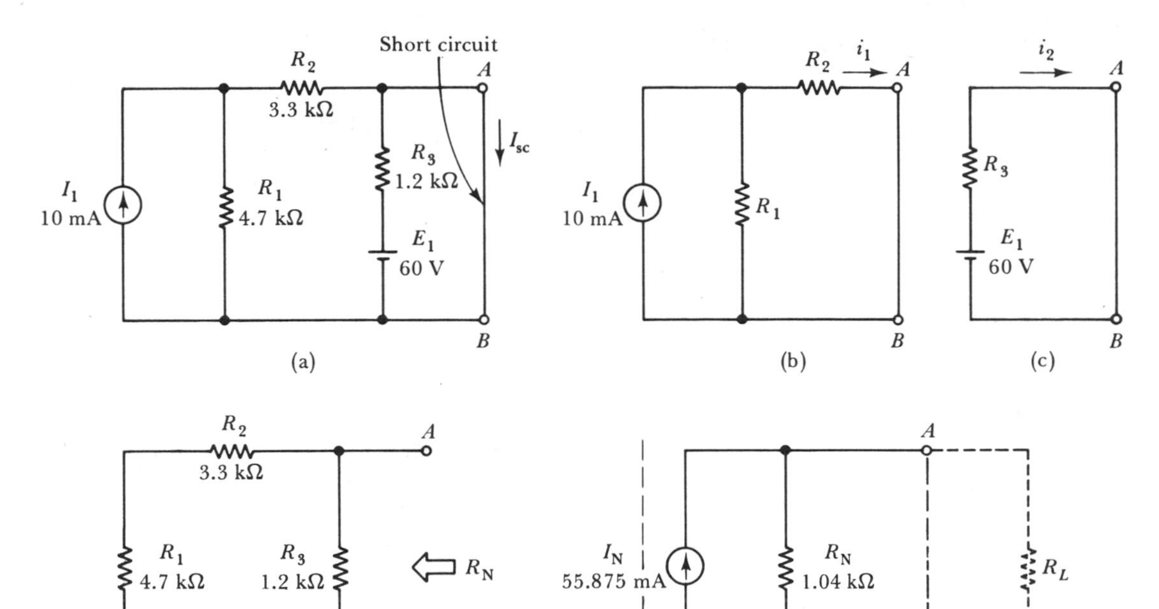

FIG. 11-6. Application of Norton's theorem

From Figure 11-6(d),

$$R_N = R_3 \| (R_2 + R_1)$$
$$= 1.2\ \text{k}\Omega \| (3.3\ \text{k}\Omega + 4.7\ \text{k}\Omega)$$
$$\mathbf{R_N = 1.04\ k\Omega}$$

The Norton equivalent circuit is now as shown in Figure 11-6(e).

The output current through R_L *can now be calculated using the current divider rule:*

$$I_L = I_N \times \frac{R_N}{R_N + R_L}$$

For $R_L = 12$ kΩ,

$$I_{L1} = 55.875\ \text{mA} \times \frac{1.04\ \text{k}\Omega}{1.04\ \text{k}\Omega + 12\ \text{k}\Omega}$$

$$\mathbf{I_{L1} = 4.46\ mA}$$

For $R_L = 5.6\ \text{k}\Omega$,

$$I_{L2} = 55.875\ \text{mA} \times \frac{1.04\ \text{k}\Omega}{1.04\ \text{k}\Omega + 5.6\ \text{k}\Omega}$$

$$\mathbf{I_{L2} = 8.75\ mA}$$

Note that because of approximations, I_{L1} *and* I_{L2} *as calculated above are slightly different from the values obtained in* Example 11-5.

11-4 MILLMAN'S THEOREM

Any number of current sources in parallel can be represented by a single current generator in which the generator current is the algebraic sum of the individual source currents, and the generator resistance is the parallel combination of the individual source resistances.

This definition does not give a complete idea of the applications of Millman's theorem, because it refers only to current sources. In Section 10-1 it is shown that voltage sources can be converted to current sources, and vice versa. Consequently, Millman's theorem can also be applied to voltage sources in parallel, or to a combination of voltage and current sources. Also, instead of ending up with a single current generator, the complete network can be represented as a single voltage generator.

One very important point is that *this theorem applies only to sources connected directly in parallel*; it does not apply where there are resistors between the sources. For example, Millman's theorem could not be applied directly to the two-generator circuit shown in Figure 11-3(a) because of the presence of resistor R_2.

The procedure for application of Millman's theorem is illustrated in Figure 11-7 and is stated as follows:

1. Convert all voltage sources to current sources.
2. Algebraically add all source currents to obtain the final generator current.
3. Determine the resistance value of the source resistances in parallel to obtain the final generator resistance.
4. If required, convert the current generator to a voltage generator.

EXAMPLE 11-7

Apply Millman's theorem to the circuit shown in Figure 11-7(a), to obtain the equivalent current generator circuit and the equivalent voltage generator circuit.

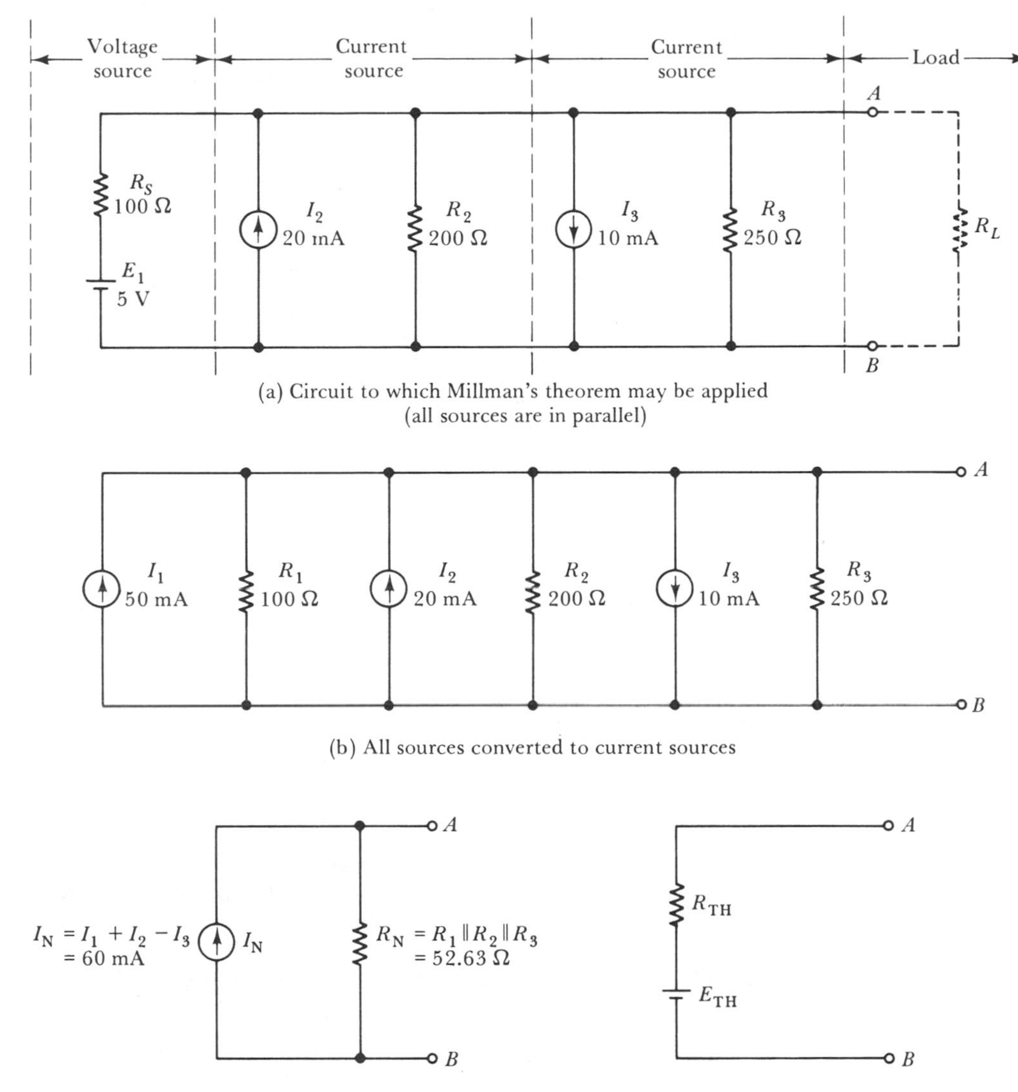

FIG. 11-7. Application of Millman's theorem

SOLUTION

Converting the voltage generator,

$$I_1 = \frac{E_1}{R_s} = \frac{5\ \text{V}}{100\ \Omega}$$

$$\mathbf{I_1 = 50\ mA} \qquad [\textit{see}\ \text{Figure 11-7(b)}]$$

Current source resistance,

$$R_1 = R_s = \mathbf{100\ \Omega} \qquad \left[\textit{see}\ \text{Figure 11-7(b)}\right]$$

For the final equivalent current generator,

$$I_N = I_1 + I_2 + I_3$$
$$= 50\ \text{mA} + 20\ \text{mA} + (-10\ \text{mA})$$
$$\mathbf{I_N = 60\ mA} \qquad [\textit{see}\ \text{Figure 11-7(c)}]$$

$$R_N = R_1 \| R_2 \| R_3$$
$$= 100\ \Omega \| 200\ \Omega \| 250\ \Omega$$
$$= \frac{1}{\dfrac{1}{100\ \Omega} + \dfrac{1}{200\ \Omega} + \dfrac{1}{250\ \Omega}}$$
$$\mathbf{R_N = 52.63\ \Omega} \qquad [\textit{see}\ \text{Figure 11-7(c)}]$$

For the equivalent voltage generator,

$$E_{TH} = I_N R_N = 60\ \text{mA} \times 52.63\ \Omega$$
$$\mathbf{E_{TH} = 3.16\ V}$$

and

$$\mathbf{R_{TH} = R_N = 52.63\ \Omega} \qquad [\textit{see}\ \text{Figure 11-7(d)}]$$

11-5 MAXIMUM POWER TRANSFER THEOREM

Maximum output power is obtained from a network or source when the load resistance is equal to the output resistance of the network or source as seen from the terminals of the load.

The truth of this theorem is easily tested by considering the Thévenin equivalent circuit in Figure 11-8(a), and calculating the power output for various values of load resistance. For the circuit in Figure 11-8(a), $R_{TH} = 50\ \Omega$ and $E_{TH} = 100$ V.

For $R_L = 50\ \Omega$:

$$I_L = \frac{E_{TH}}{R_{TH} + R_L} = \frac{100\ \text{V}}{50\ \Omega + 50\ \Omega}$$
$$\mathbf{I_L = 1\ A}$$

$$V_L = I_L \times R_L = 1\ \text{A} \times 50\ \Omega$$
$$\mathbf{V_L = 50\ V}$$

$$P_L = V_L \times I_L = 50\ \text{V} \times 1\ \text{A}$$
$$\mathbf{P_L = 50\ W}$$

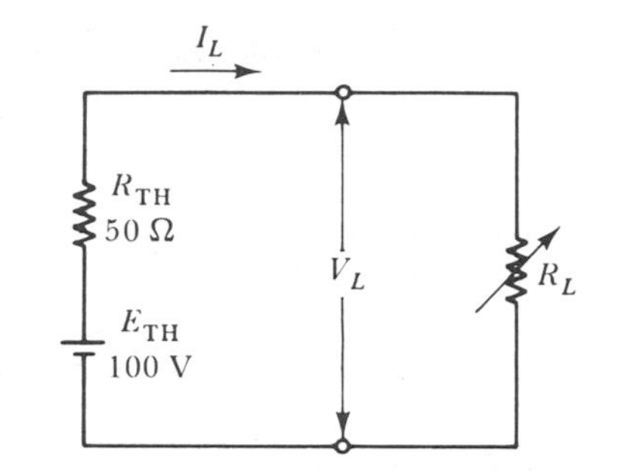

(a) Thévenin equivalent circuit with a variable load

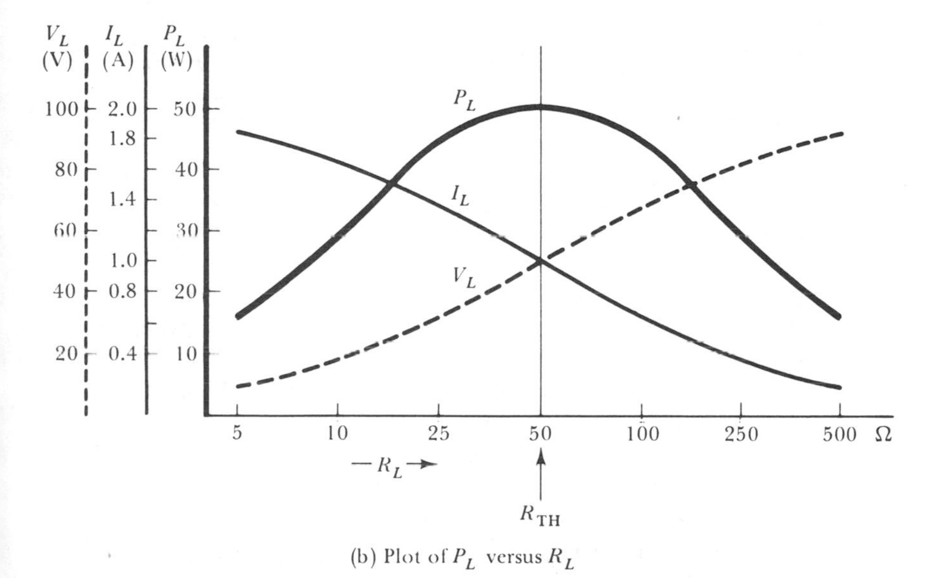

(b) Plot of P_L versus R_L

FIG. 11-8. Illustrating the maximum power transfer theorem

Table 11-1 gives the values of load current, voltage, and power for values of R_L ranging from 5 Ω to 500 Ω. The values listed in Table 11-1 are plotted in Figure 11-8(b), showing clearly that the output power is at a peak value when $R_L = R_{TH}$. Also, note that the output current is greatest

TABLE 11-1. I_L, V_L and P_L for the Circuit of Figure 11-8(a) with Various Values of R_L

R_L(Ω)	5	10	25	50	100	250	500
V_L(V)	9.09	16.7	33.3	50	66.7	83.3	90.9
I_L(A)	1.82	1.67	1.33	1	0.67	0.33	0.18
P_L(W)	16.5	27.8	44.4	50	44.4	27.8	16.5

when R_L is very small, and the output voltage is greatest when R_L is very large.

REVIEW QUESTIONS

11-1 State the superposition theorem and list the steps involved in applying it to the analysis of a resistor network.

11-2 State Thévenin's theorem and give the procedure for thevenizing a circuit. Explain the major advantages afforded by use of this theorem.

11-3 State Norton's theorem and list the steps for nortonizing a circuit. Compare the Norton equivalent circuit to the Thévenin equivalent circuit.

11-4 State Millman's theorem and list the procedure for its application to circuit analysis.

11-5 State the maximum power transfer theorem.

PROBLEMS

11-1 Using the superposition theorem, analyze the circuit shown in Figure 10-11 to determine the current through R_6, when R_5 is open-circuited.

11-2 Applying the superposition theorem, determine the current through resistor R_6 in the circuit of Figure 10-12, when R_5 is open-circuited.

11-3 Apply the superposition theorem to the circuit of Figure 10-13 to determine the current that flows through resistor R_3.

11-4 Use Thévenin's theorem to determine the current through resistor R_6 in the circuit of Figure 10-11, when R_5 is open-circuited.

11-5 Apply Thévenin's theorem to determine the voltage across resistor R_6 in the circuit of Figure 10-12, when R_5 is open-circuited.

11-6 Using Thévenin's theorem, calculate the voltage across resistor R_3 in Figure 10-13.

11-7 For the circuit of Figure 10-11, use Norton's theorem to determine the current that flows through resistor R_3, when R_4 is open-circuited.

11-8 Apply Norton's theorem to the circuit in Figure 9-8 to determine the current flowing in resistor R_4.

11-9 Using Norton's theorem, determine the voltage across resistor R_4 in the circuit shown in Figure 9-9.

11-10 Apply Millman's theorem to the circuit of Figure 10-13 to determine the current flowing in resistor R_4.

11-11 Apply Millman's theorem to the circuit of Figure 10-4 to determine the voltage across resistor R_3.

11-12 Using Millman's theorem, reduce the voltage and current generators in Figure 11-9 to a single current generator. Then calculate the current through the load for $R_L = 3.3\ \text{k}\Omega$ and for $R_L = 4.7\ \text{k}\Omega$.

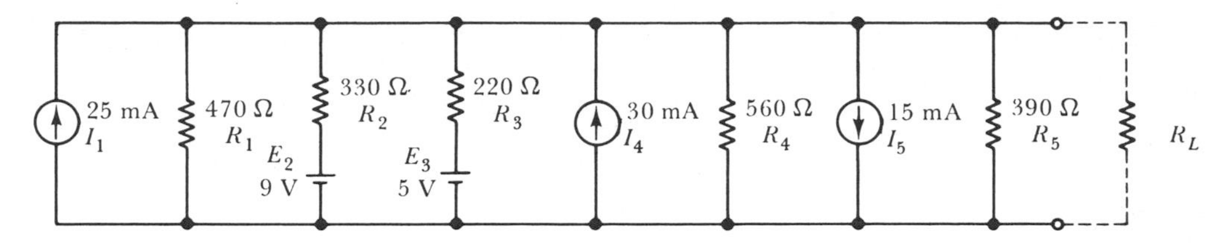

FIG. 11-9.

11-13 Demonstrate the maximum power transfer theorem by calculating load voltage, current, and power for various values of load resistance connected to the current generator derived in Problem 11-12. Also plot a graph of the quantities versus R_L.

12
MAGNETISM

INTRODUCTION

Some two thousand years ago, Chinese mariners used a suspended bar of mineral known as *lodestone* as a crude compass. The mineral had magnetic properties, so its ends tended to point to the earth's magnetic poles. Today's magnetic compasses are much more sensitive than the lodestone, but they fulfill essentially the same purpose. As well as direction finding, there are a great many other applications of magnetism.

The magnetic force field around a bar magnet is easily plotted by means of iron filings or by use of compasses. It is found that magnetic flux passes through nonmagnetic material, including air and vacuum. However, iron and steel are by far the best conductors of magnetic flux, and in the presence of a magnetic field pieces of iron and steel become magnetized.

A magnetic flux also exists around a current-carrying conductor, and a current-carrying coil generates a field similar to that of a bar magnet.

12-1 MAGNETIC FIELD

A bar magnet suspended on thread tends to align itself in relation to the earth in a north–south direction. This is also the case with the magnetic compass, which is of course, simply a pivoted magnetized needle. The two ends of the bar magnet and the compass are identified as *north pole* (N) and *south pole* (S), in accordance with the direction that they tend to point.

Consider the effect of bringing the N pole of another bar magnet close to the N pole of the suspended magnet. As illustrated in Figure 12-1(a), the result is repulsion between the two magnets. A similar result occurs when the S pole of the hand-held magnet is brought close to the S pole of the suspended magnet. However, when N and S poles are brought together, the magnets are attracted to each other [see Figure 12-1(b)]. The same effects are demonstrated when one pole of a bar magnet is

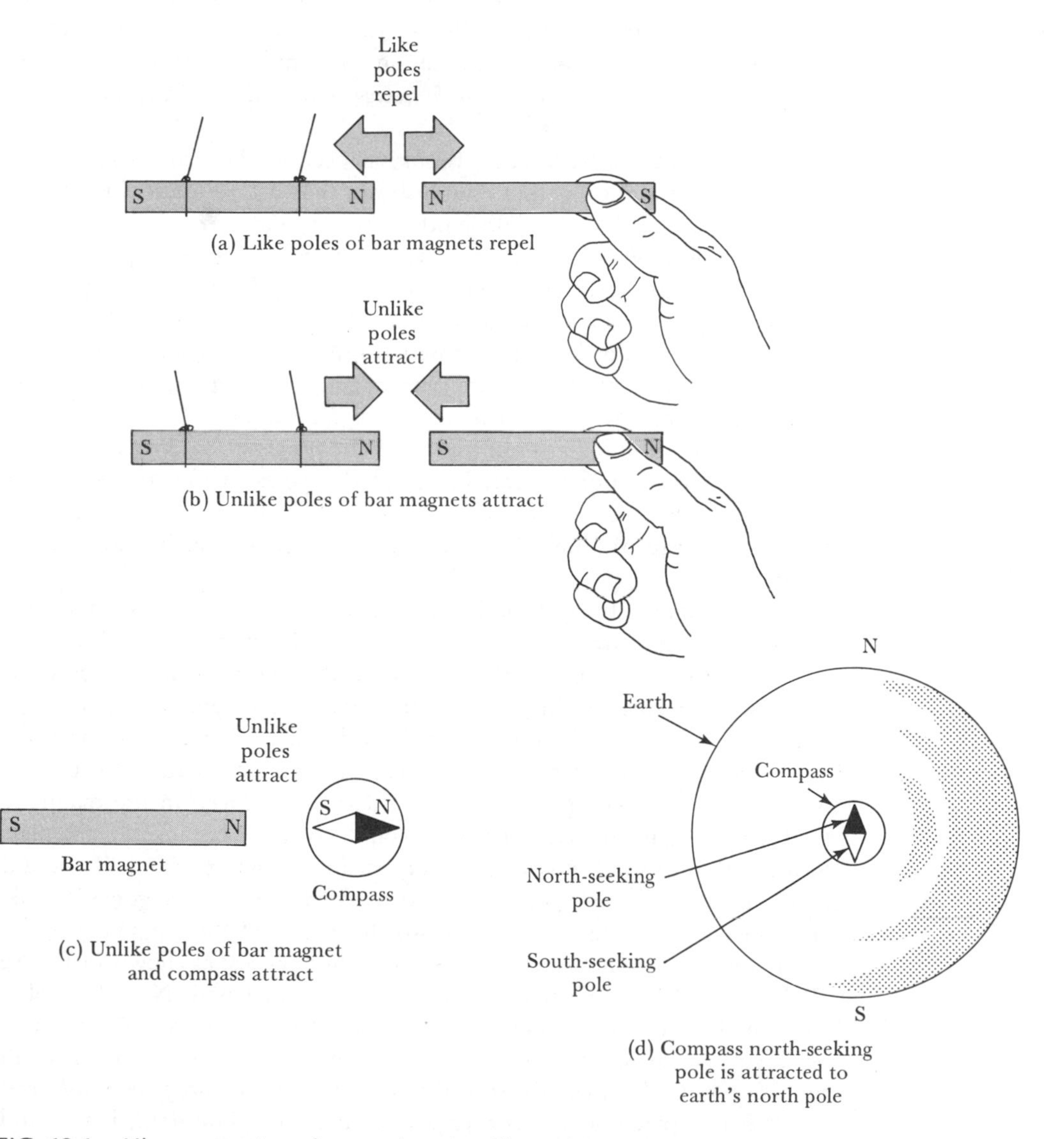
Like poles repel
S
N
N
S
(a) Like poles of bar magnets repel
Unlike poles attract
S
N
S
N
(b) Unlike poles of bar magnets attract
Unlike poles attract
S
N
Bar magnet
S
N
Compass
(c) Unlike poles of bar magnet and compass attract
N
Earth
Compass
North-seeking pole
South-seeking pole
S
(d) Compass north-seeking pole is attracted to earth's north pole

FIG. 12-1. Like magnetic poles repel and unlike poles attract

brought close to a magnetic compass [Figure 12-1(c)]. The needle of the compass always turns so that its S pole points to the N pole of the magnet, or so that its N pole points to the magnet's S pole. These results give rise to the *fundamental law of magnetism*:

Like poles repel; unlike poles attract.

Because unlike poles attract, the pole of the magnetic compass that points to the earth's north pole, Figure 12-1(d), must be *unlike* the earth's north pole. Therefore, the north poles on a compass and on a bar magnet are more correctly described as *north-seeking poles*, and similarly their south poles are *south-seeking poles*.

The fact that attraction or repulsion occurs when the poles of two magnets are brought together demonstrates that a *force field* exists around the magnets, and that the field tends to be concentrated at the magnet's poles. This force field around a magnet is termed a *magnetic field*, and it can be investigated further by means of iron filings sprinkled on a piece of cardboard placed on top of a magnet. When the cardboard is gently tapped, the iron filings align themselves into chains, as shown in Figure 12-2(a). The lines plotted by the pattern of iron filings stretch from one pole to the other and are termed *magnetic lines of force*. Collectively, the magnetic lines of force are referred to as the *magnetic flux*.

The concentration of iron filings at the poles of the magnet shows that the magnetic force field has its greatest strength close to the magnet's poles. Obviously, the field becomes weaker at points farther away from the poles.

The magnetic lines of force can also be plotted by the use of several compasses, as illustrated in Figure 12-2(b). The compass needles point along the lines of force in a direction from the magnet's north pole to its south pole. Because of this, it is assumed that the magnetic lines of force exit from the north pole of the magnet and enter the magnet at its south pole. The direction of the lines of force is defined as the direction in which an *isolated north pole* would move if it were placed in the magnetic field. In fact, an isolated north pole could not exist because all magnets have north and south poles. However, it can be imagined that if one did exist, an isolated north pole would be repelled from the magnet's N pole and move along a line of force toward the S pole of the magnet.

If the bar magnet were to be cut into two pieces, as shown in Figure 12-2(c), it is found that each half is still a magnet with N and S poles. Also, the lines of force cross the gap between the two pieces of the magnet, as illustrated, demonstrating that the magnetic lines of force pass through the magnet itself. Thus, it is seen that *magnetic lines of force form closed loops*.

When a piece of soft iron is placed in a magnetic field, it is found that the lines of force bend away from their usual paths in order to pass

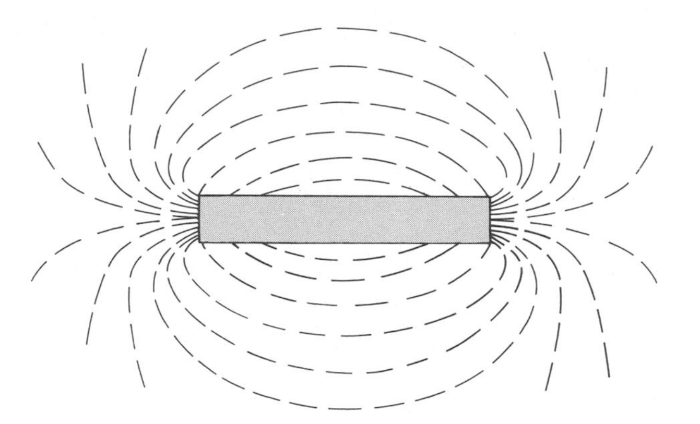

(a) Force field around a magnet, plotted by iron filings

(b) Force field plotting using compasses

(c) Each half of a magnet has N and S poles

FIG. 12-2. A magnetic force field exists around a magnet

through the iron [see Figure 12-3(a)]. It is also found that the iron has become magnetized temporarily and that a S pole can be identified at the end of the soft iron nearest the magnet's N pole, and a N pole can be identified at the other end. This effect is known as *induced magnetism*. No similar effect occurs when pieces of wood or glass or rubber or other such *nonmagnetic materials* are placed in the magnetic field. Thus, it can be stated that *magnetic lines of force pass most easily through iron*; that is, iron is a better conductor of magnetic lines of force than such materials as air or wood or glass.

It can now be seen that the individual iron filings used to plot the magnetic lines of force must become magnetized, so that each has its own N and S poles. Figure 12-3(b) illustrates the situation and shows that the poles of each little magnet or group of magnets are likely to be adjacent to like poles of other magnetized iron filings alongside them. Thus, repulsion occurs, and this demonstrates another piece of information about the magnetic lines of force. That is, *magnetic lines of force tend to repel each other*. This fact is also shown by the field pattern at the poles of the

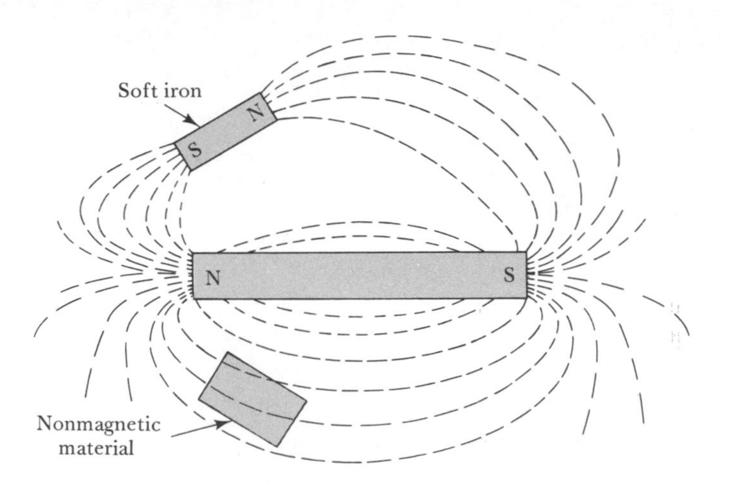

(a) Effects of magnetic and nonmagnetic material on shape of magnetic force field

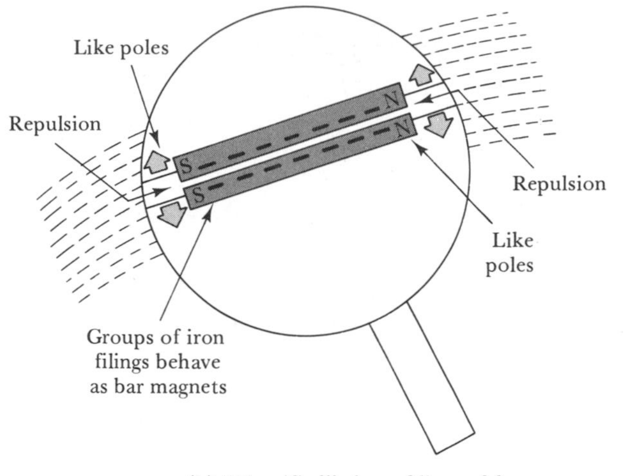

(b) "Magnified" view of lines of force plotted by iron filings

FIG. 12-3. Some characteristics of magnetic lines of force

magnet in Figure 12-2(a). Since lines of force repel each other, then *two lines of force cannot intersect.*

Figure 12-4(a) shows the pattern of the magnetic field when two N poles (or two S poles) are brought close together. It is clearly seen that the lines of force from each pole repel one another. Conversely, when a N and S pole are placed close together [Figure 12-4(b)], the lines of force run out of one pole into the other. In Figure 12-4(c), the field pattern at the poles of a horseshoe-shaped magnet is shown. Note that the magnetic lines of

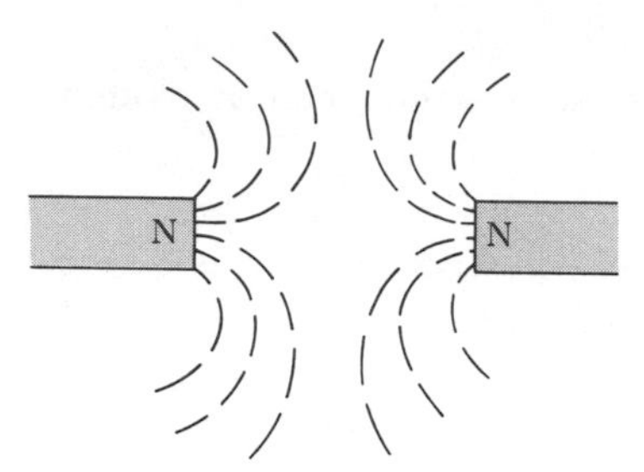

(a) Magnetic field at like poles

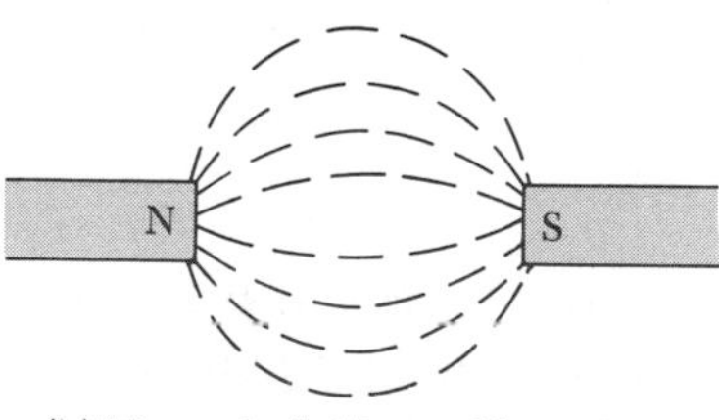

(b) Magnetic field at unlike poles

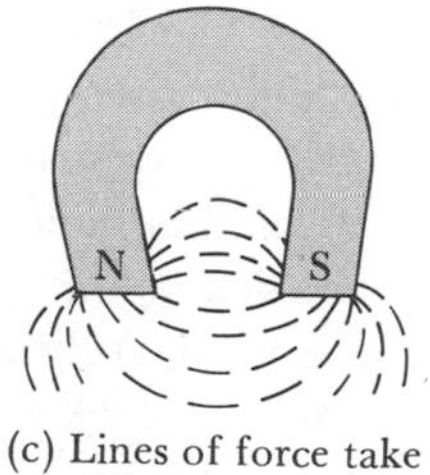

(c) Lines of force take the shortest path

FIG. 12-4. Magnetic lines of force at like and unlike poles

force take the shortest path between the poles, and this shows that *the lines of force are always in a state of tension.*

Summarizing:

- Every magnet has two poles identified as a north pole (N) and a south pole (S).
- The N pole is actually a *north-seeking pole*, and the S pole is a *south-seeking pole*, with respect to the earth.
- The force field that exists around a magnet is concentrated at its poles.

- Like poles repel; unlike poles attract.
- Magnetic lines of force in a field around a magnet can be plotted by the use of iron filings.
- Lines of force form closed loops.
- Lines of force exit from a magnet's N pole and enter at its S pole.
- Lines of force travel most easily through soft iron.
- Lines of force repel each other.
- Lines of force cannot intersect.
- Lines of force from two like poles repel each other.
- Lines of force from two unlike poles run into each other.
- Lines of force are always in a state of tension.
- A piece of soft iron placed in a magnetic field is temporarily magnetized by induction.

12-2 ELECTROMAGNETISM

When an electric current flows in a conductor, a magnetic field is set up around the conductor. This is easily demonstrated by again using iron filings, as illustrated in Figure 12-5(a). A conductor is passed vertically through a hole in a horizontal piece of cardboard. Iron filings are sprinkled on the cardboard, and when the current through the conductor is zero, there is no evidence of a magnetic field. However, when a current flows through the conductor, gentle tapping of the cardboard causes the iron filings to settle in concentric rings around the conductor.

As with other magnetic fields, the field around a current-carrying conductor can also be investigated by the use of magnetic compasses. Figure 12-5(b) shows the result of placing four compasses on the horizontal cardboard. For a current flowing in a (conventional) direction down through the cardboard, the compass needles point in a clockwise direction around the conductor. Since the direction of magnetic lines of force are defined in terms of the direction in which a free north pole would move, it can be stated that for the conditions illustrated, the magnetic lines of force are in a clockwise direction. When the current direction is reversed so that it flows up through the cardboard, the compass needles reverse and point in a counterclockwise direction.

Figure 12-5(c) and (d) further illustrates the direction of a magnetic field around a current-carrying conductor. The conductor is assumed to be directly opposite the viewer, so only its cross-sectional area is shown. In Figure 12-5(c), current is assumed to be flowing away from the viewer (i.e., into the page). This is indicated by the + sign, which represents the tail of an arrow. In this case the magnetic lines of force around the conductor have a clockwise direction. In Figure 12-5(d) the dot at the center of the conductor cross section represents the point of an arrow. Therefore,

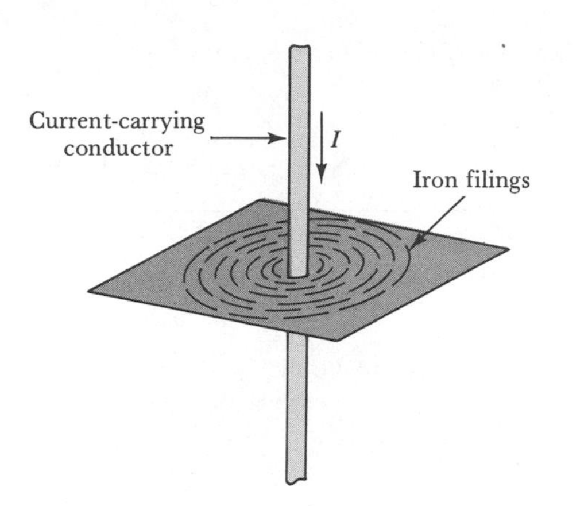

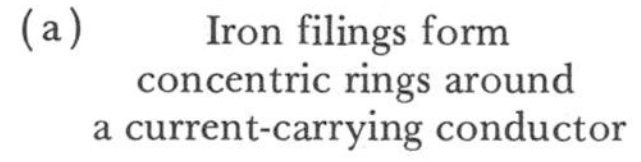
(a) Iron filings form concentric rings around a current-carrying conductor

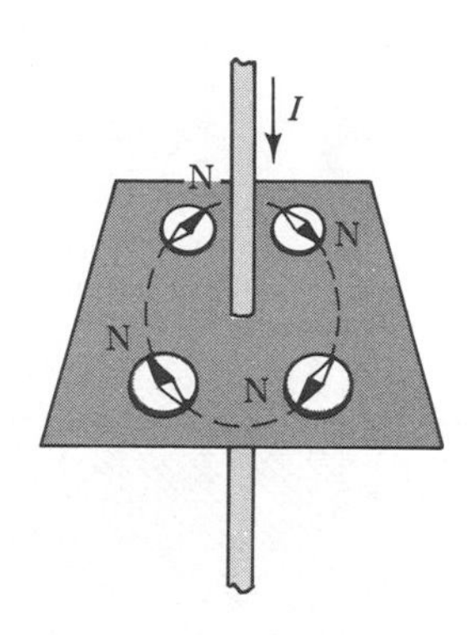

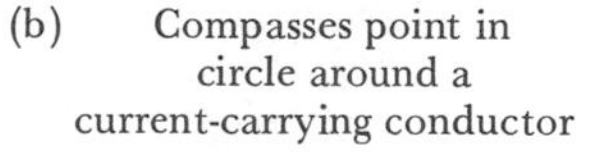
(b) Compasses point in circle around a current-carrying conductor

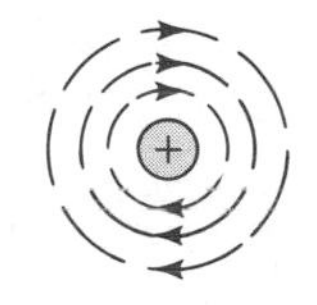

(c) Lines of force are in a clockwise direction around a conductor which is carrying current away from the viewer

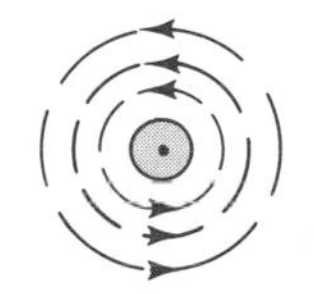

(d) Lines of force are in a counterclockwise direction around a conductor which is carrying current toward the viewer

FIG. 12-5. Magnetic lines of force around a current-carrying conductor

current is flowing toward the viewer (i.e., out of the page). The magnetic lines of force now have a counterclockwise direction.

Two memory aids for determining the direction of the magnetic flux around a current-carrying conductor are shown in Figure 12-6. The *right-hand-screw rule* as illustrated in Figure 12-6(a) shows a wood screw being turned clockwise and progressing into a piece of wood. The horizontal direction of the screw is analogous to the direction of current in a conductor, and the circular motion of the screw shows the direction of magnetic flux around the conductor. In the *right-hand rule*, illustrated in Figure 12-6(b), when the thumb points in the (conventional) direction of current flow, the fingers show the direction of the magnetic lines of force around the conductor.

Since a current-carrying conductor has a magnetic field around it, when two current-carrying conductors are brought close together there will be interaction between the fields. Figure 12-7(a) shows the effect on

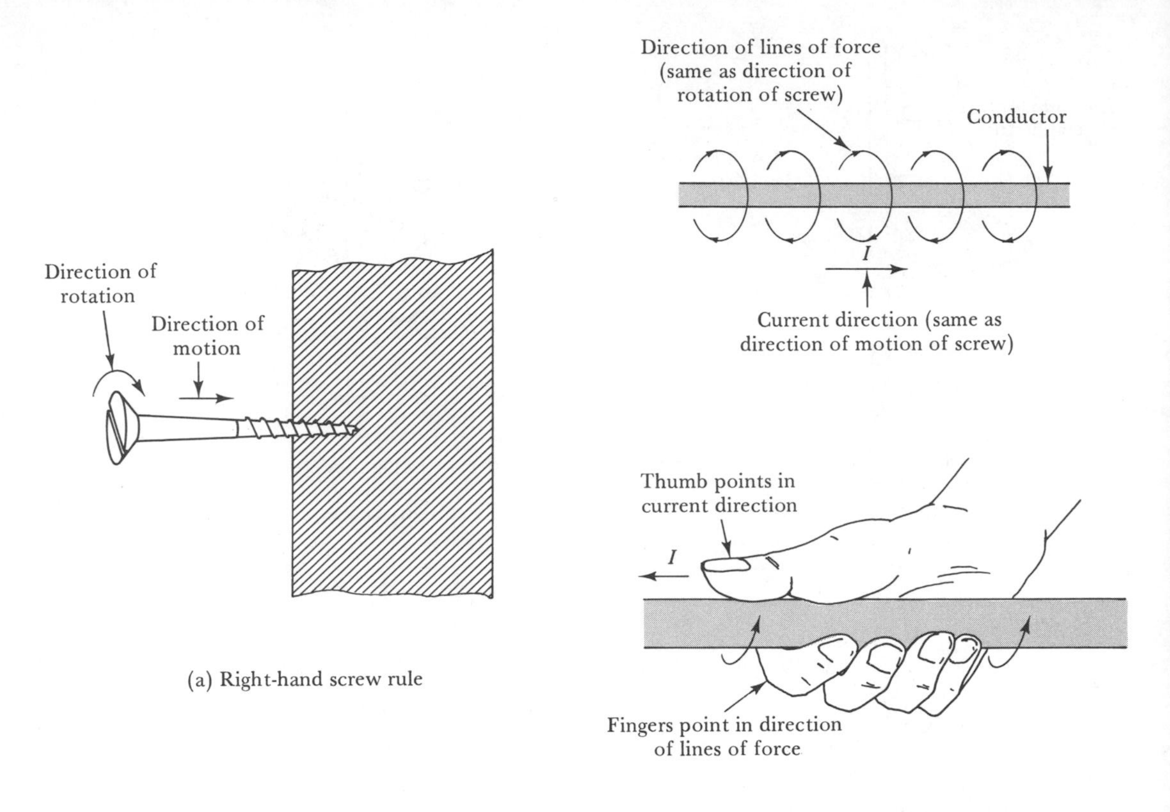

FIG. 12-6. Right-hand-screw rule and right-hand rule for determining the magnetic flux direction around a current-carrying conductor

the fields when two conductors carrying currents in opposite directions are adjacent. The directions of the magnetic fields around the conductors are in opposition, and this is similar to the situation when two like magnetic poles are brought close together. The fields exert a force of repulsion on each other, tending to push the conductors apart. When the adjacent conductors have currents flowing in the same direction [Figure 12-7(b) and (c)], the magnetic fields assist each other. Since the lines of force are always in tension, they are always trying to find the shortest path. Consequently, the fields exert a force that tends to pull the conductors together, as illustrated.

Now consider the effect of passing a current through a one-turn coil of wire. Figure 12-8(a) and (b) shows that all the magnetic flux generated by the electric current passes through the center of the coil. Therefore, the one-turn coil acts like a little magnet, and has a magnetic field with an identifiable N pole and S pole. Instead of a single turn, the coil may have

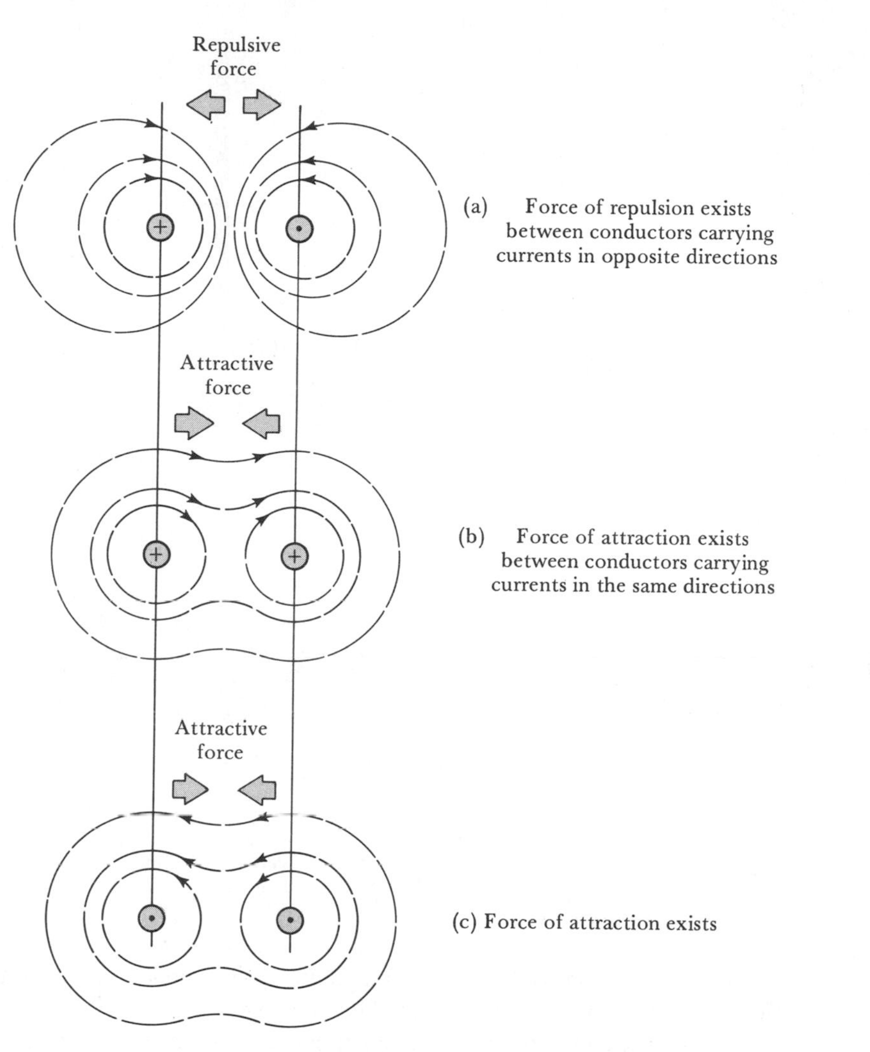
Repulsive force
(a) Force of repulsion exists between conductors carrying currents in opposite directions
Attractive force
(b) Force of attraction exists between conductors carrying currents in the same directions
Attractive force
(c) Force of attraction exists

FIG. 12-7. Force between parallel current-carrying conductors

many turns, as illustrated in Figure 12-8(c). In this case the flux generated by each of the individual current-carrying turns tends to link up and pass out of one end of the coil and back into the other end. This type of coil, known as a *solenoid*, obviously has a magnetic field pattern very similar to that of a bar magnet.

The right-hand rule for determining the direction of flux from a solenoid is illustrated in Figure 12-8(d). When the solenoid is gripped with the right hand such that the fingers are pointing in the direction of current flow in the coils, the thumb points in the direction of the flux (i.e., toward the N-pole end of the solenoid).

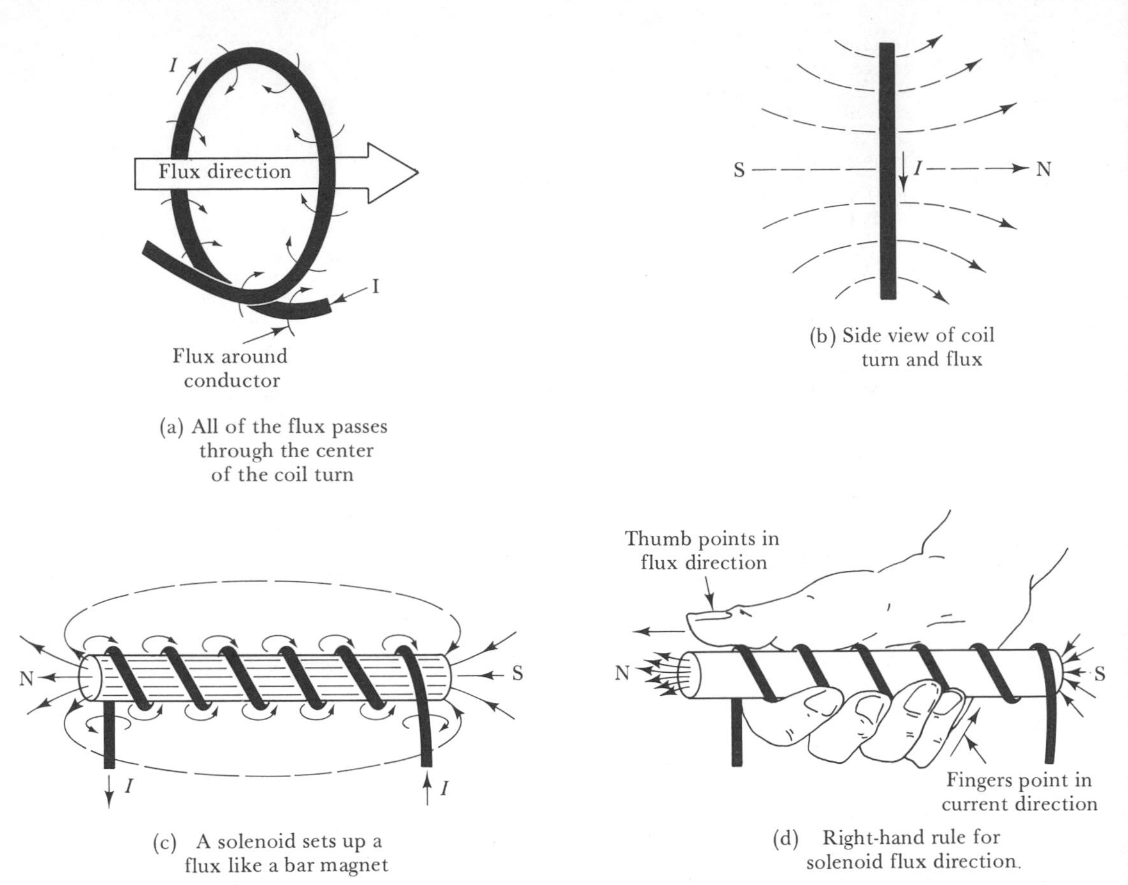

FIG. 12-8. Flux generated by current-carrying coils

It has been demonstrated that a magnetic flux is generated by an electric current flowing in a conductor. The converse is also possible; that is, a magnetic flux can produce a current flow in a conductor. Consider Figure 12-9(a), in which a hand-held bar magnet is shown being brought close to a coil of wire. As the bar magnet approaches the coil, the flux from the magnet brushes across the coil conductors or *cuts* the conductors. This produces a current flow in the conductors proportional to the actual flux that *cuts* the coil. If the coil circuit is closed by a resistor (as shown dashed in the figure), a current flows. Whether or not the circuit is closed, an emf can be measured at the coil terminals. This effect is known as *electromagnetic induction*.

It is important to note that the emf within the coil (which produces the current flow) is generated only when the magnetic field is in motion with respect to the coil. When both the field and the coil are stationary, no emf is produced.

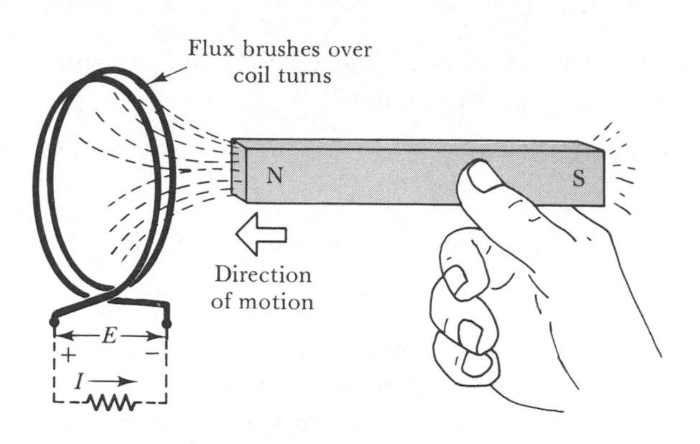

(a) emf induced in a coil by the motion of the flux from the bar magnet

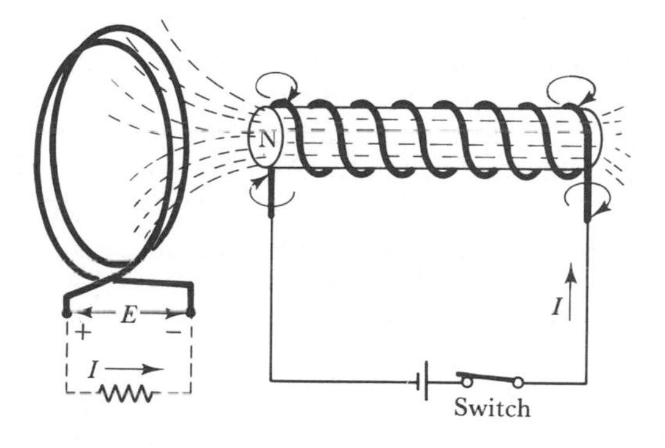

(b) emf induced in a coil by the motion of the flux from the solenoid when the current is switched on or off

FIG. 12-9. Electromagnetic induction

Now consider Figure 12-9(b), which shows a solenoid placed close to another coil. Both the solenoid and the other coil are stationary. The solenoid also has a battery and a switch that can be closed to pass current through the turns of the solenoid.

With the switch open, there is obviously no current flow and therefore no flux to generate an emf in the second coil. However, when the switch is closed, current flows through the solenoid, and the flux grows from zero to its maximum level. While the flux is growing, it is moving out from the solenoid and *brushing over* the coil. Consequently, an emf is generated in the coil. When the flux has reached its maximum level, it becomes stationary and is no longer able to generate the emf in the coil. If the switch is now opened, the flux falls to zero again, and in doing so it once more brushes over the conductors of the coil. Again, an emf is generated in the coil while the flux is in motion.

12-3 THEORY OF MAGNETISM

In relation to magnetism all materials can be categorized into four groups:

- *Nonmagnetic* materials have no more effect on a magnetic field than air or vacuum. This group includes wood and rubber.
- *Diamagnetic* materials exhibit a very slight opposition to magnetic lines of force. They tend to be repelled from both poles of a magnet, and if a bar of diamagnetic substance is suspended in a magnetic field, it tends to align itself at right angles to the field. The effect is so slight that these materials are usually classified as nonmagnetic. Copper and silver are diamagnetic.
- *Paramagnetic* materials assist the passage of magnetic lines of force. Sometimes this term is used to include ferromagnetic materials. In general, however, it is applied only to material that shows a very slight magnetic effect. Aluminum and platinum are paramagnetic.
- *Ferromagnetic* materials tremendously assist the passage of magnetic lines of force. These are the materials used as permanent magnets and as electromagnets. Iron, nickel, cobalt, and a certain type of ceramic known as *ferrite* are all ferromagnetic materials.

It was shown in Section 12-2 that a current flowing through a conductor produces a magnetic field around the conductor, and that when the current-carrying conductor is bent in a circle, a magnetic flux having a definite direction is generated. Now recall that current flow is in fact electron motion. Thus, it follows that an electron in orbit around an atom can be thought of as a circular current flow, and that a magnetic flux is produced by the orbital motion of the electrons. This effect is referred to as the *magnetic moment* of the electron.

Figure 12-10(a) shows two electrons orbiting in the same plane and direction around adjacent atoms. Note that conventional current direction is opposite to that of the electron motion. The orbital motion of the two electrons produces magnetic fluxes in the same direction, and these fluxes reinforce each other. Now look at Figure 12-10(b). In this case the two electrons are shown orbiting in opposite directions. Thus, they generate magnetic fluxes that cancel each other, and the net result is zero flux.

Within nonmagnetic material, it can be assumed that all the magnetic effects of orbiting electrons add up to zero, and that the condition is unchanged by the influence of a magnetic field. Diamagnetic materials tend to be affected by a magnetic field, to the extent that they set up their

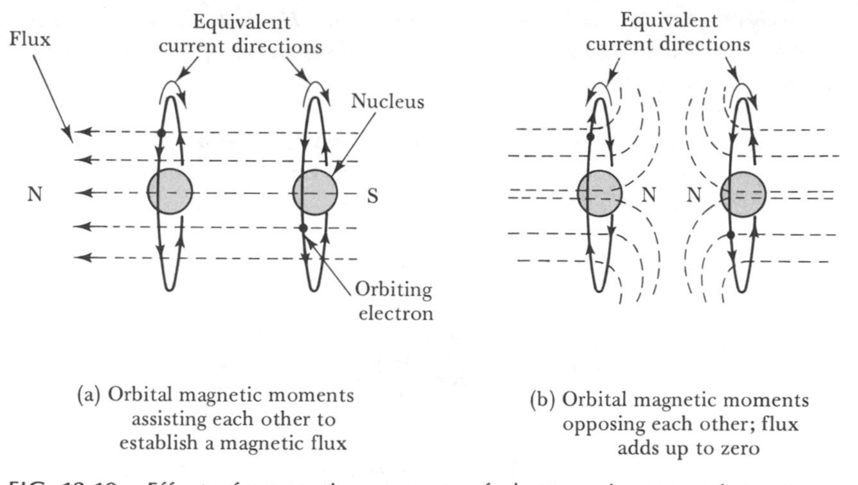

(a) Orbital magnetic moments assisting each other to establish a magnetic flux

(b) Orbital magnetic moments opposing each other; flux adds up to zero

FIG. 12-10. Effect of magnetic moments of electrons in generating a magnetic flux

own very weak magnetic flux which opposes the external magnetic field. The effect is almost undetectable, but it is fair to assume that the external magnetic field causes an alignment of some orbital electrons, and that the combined effect of these produces the opposing flux. With paramagnetic materials that exhibit very weak magnetization, it may be assumed that some orbiting electrons are aligned in such a way that they produce a magnetic flux in the same direction as the external flux.

In *ferromagnetic* materials, groups of atoms seem to act together. Within these groups, large numbers of orbital electrons are aligned in such a way that their magnetic fields reinforce each other. The groups of atoms are termed *magnetic domains*, and while the material is unmagnetized, the net effect of the magnetic domains is to cancel each other completely. When an external magnetic field is applied to ferromagnetic material, the resulting alignment of the magnetic domains produces a tremendous increase in the magnetic flux. In the case of soft iron, the flux reduces almost to zero when the external field is removed. Thus, the magnetic domains appear to return (almost) to their original state. With hard steel and certain alloys of iron and nickel, the magnetic domains apparently remain realigned, and the steel continues to produce a magnetic flux even when the external field is removed. In this case the metal has become a permanent magnet.

12-4 MAGNETIC FLUX AND FLUX DENSITY

The total lines of force in a magnetic field are referred to as the *magnetic flux*, and the flux per unit cross-sectional area of the field is termed the *flux density*.

*The **weber*** **(Wb)** is the SI unit of magnetic flux. The **weber** is defined as the magnetic flux which, linking a single turn coil, produces an emf of 1 V when the flux is reduced to zero at a uniform rate in 1 s.*

*The **tesla**† **(T)** is the SI unit of magnetic flux density. The **tesla** is the flux density in a magnetic field when 1 Wb of flux occurs in a plane of 1 m^2; that is, the **tesla** can be described as 1 Wb/m^2.*

The relationship between flux and flux density is stated by the equation

$$\boxed{B = \frac{\Phi}{A}} \tag{12-1}$$

where B is the flux density in teslas, Φ is the total field flux in webers, and A is the cross-sectional area in m^2.

The flux density is frequently different for different parts of a magnetic field. For example, in the case of the bar magnet in Figure 12-2(a), the flux density is obviously greatest at points close to the poles of the magnet, or within the metal.

EXAMPLE 12-1

The total flux emitted from the pole of a bar magnet is 2×10^{-4} Wb.

a. If the magnet has a cross-sectional area of 1 cm^2, determine the flux density within the metal.

b. If the flux spreads out so that at a certain distance from a pole it is distributed over an area of 2 cm by 2 cm, find the flux density at that point.

SOLUTION

a. *Flux density within the metal*:

Eq. (12-1):

$$B = \frac{\Phi}{A}$$

$$= \frac{2 \times 10^{-4}\ \text{Wb}}{1 \times 10^{-4}\ \text{m}^2}$$

$$\mathbf{B = 2\ T}$$

---

*Named after the German physicist Wilhelm Weber (1804–1890).

†Named for the Croatian–American researcher and inventor Nikola Tesla (1857–1943).

b. *Flux density away from the pole*:

$$B = \frac{\Phi}{A}$$

$$= \frac{2\times10^{-4}\ \text{Wb}}{(2\times10^{-2})^2\ \text{m}^2}$$

$$\mathbf{B = 0.5\ T}$$

12-5 MAGNETOMOTIVE FORCE AND MAGNETIC FIELD STRENGTH

Just as an electric current is the result of an electromotive force (emf) acting on the electric circuit, so a magnetic flux is produced by a *magnetomotive force* (mmf), symbol $\mathcal{F}$, acting upon the magnetic circuit. In the case of a solenoid or any other current-carrying coil, the magnetomotive force is the product of the current and the number of turns on the coil. Figure 12-11(a) shows a one-turn coil carrying a current of 1 A. All of the flux set up by the 1 A passes through the center of the coil. Therefore, the magnetomotive force is

$$\mathcal{F} = 1\ \text{A} \times 1\ \text{turn}$$
$$= 1\ \text{A}$$

In the case of the four-turn coil with a current of 1 A [Figure 12-11(b)], the effect of the current is multiplied four times, because each turn of the coil generates a flux that passes through the center of the coil. Therefore,

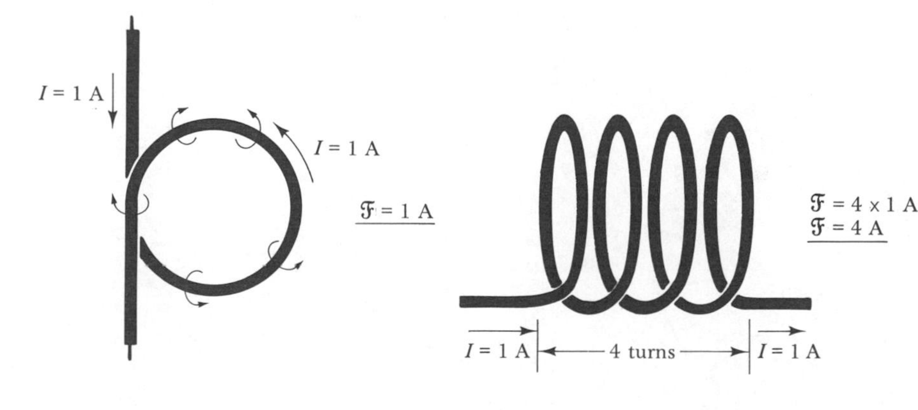

(a) Magnetomotive force of 1 A from a one-turn coil with $I = 1$ A

(b) Magnetomotive force of 4 A from a 4-turn coil with $I = 1$ A

FIG. 12-11. Magnetomotive force

the magnetomotive force is

$$\mathcal{F} = 1\ \text{A} \times 4\ \text{turns}$$
$$= 4\ \text{A}$$

In general, for a coil of N turns carrying a current of I amperes, the magnetomotive force is

$$\boxed{\textit{magnetomotive force},\ \mathcal{F} = NI \qquad \textbf{amperes}} \tag{12-2}$$

The more descriptive term *ampere-turns* is sometimes used as the units of mmf. However, the approved SI unit of mmf is the ampere.

The quantity of flux that can be set up in a magnetic circuit is, of course, proportional to the mmf. It is also inversely proportional to the length of the magnetic circuit. Consider the *toroidal* coil shown in Figure 12-12. The length of the magnetic path is l, as illustrated. If l is very large, the mmf has to act over a long distance, and if l is very small, the mmf acts over a short distance. Obviously, the greatest intensity of magnetic flux occurs for the shortest magnetic path. This gives rise to the expression for *magnetic field strength*:

$$\textit{magnetic field strength},\ H = \frac{\mathcal{F}}{l}$$

or

$$\boxed{H = \frac{NI}{l} \qquad \textbf{amperes/meter}} \tag{12-3}$$

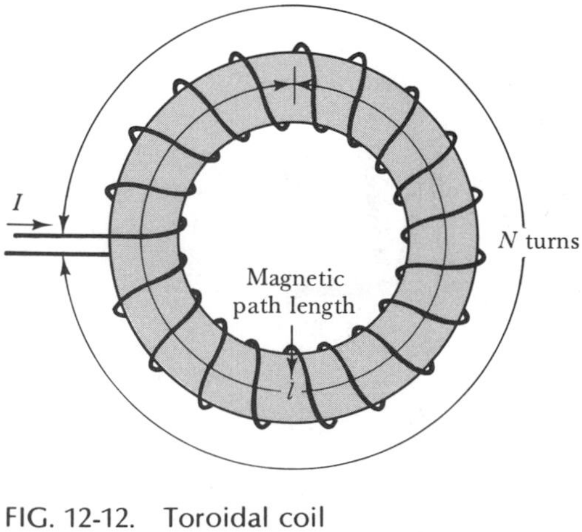

FIG. 12-12. Toroidal coil

When I is in amperes, N is number of turns, and l is in meters, the units of magnetic field strength are amperes per meter (A/m).

EXAMPLE 12-2

The toroidal coil shown in Figure 12-12 has 100 turns and carries a current of 0.5 A. If the length of the magnetic circuit is 10 cm, determine the mmf and the magnetic field strength.

SOLUTION

Eq. (12-2):

$$\mathcal{F} = NI$$
$$= 100 \times 0.5 \text{ A}$$
$$\boldsymbol{\mathcal{F} = 50 \text{ A}}$$

Eq. (12-3):

$$H = \frac{NI}{l}$$
$$= \frac{100 \times 0.5 \text{ A}}{10 \times 10^{-2} \text{ m}}$$
$$\boldsymbol{H = 500 \text{ A/m}}$$

12-6 FORCE ON A CURRENT-CARRYING CONDUCTOR IN A MAGNETIC FIELD

Figure 12-13(a) shows an end view of a current-carrying conductor in a magnetic field. With the current direction as indicated by the + sign (i.e., into the page), the flux set up by the current is in a clockwise direction around the conductor. When the field flux has the vertically down direction indicated, the flux from the conductor tends to assist the field flux on the right-hand side of the conductor. The conductor flux also opposes the field flux on the left-hand side of the conductor. Since the magnetic lines of force always tend to be in tension, those lines of force on the right-hand side of the conductor are tending to straighten out. The effect of this is to produce a force that pushes the conductor to the left.

When the current direction is reversed, as illustrated in Figure 12-13(b), the flux around the conductor is in a counterclockwise direction. Thus, the field flux is assisted on the left-hand side of the conductor and weakened on its right side. The result is a force that pushes the conductor to the right.

Note that the conductor must be at right angles to the direction of the magnetic flux; otherwise, maximum force is not generated.

The actual force acting upon a current-carrying conductor in a magnetic field is porportional to the flux density of the field and to the

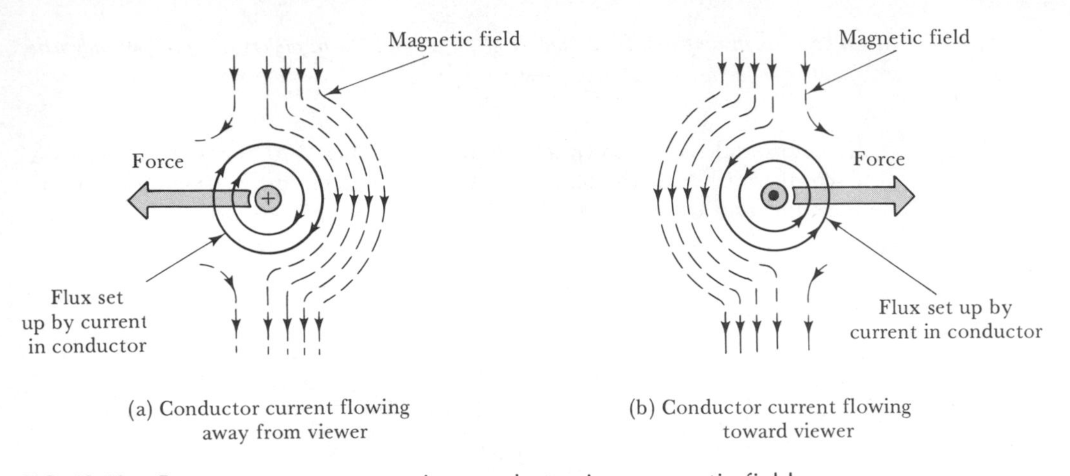

FIG. 12-13. Force on a current-carrying conductor in a magnetic field

density of the flux set up by the conductor. Since the flux set up around the conductor is directly proportional to the current, the force on the conductor is proportional to the current and to the flux density of the magnetic field. If a 2-m length of conductor is within the magnetic field, the force exerted upon it is obviously twice the force that would be exerted upon a 1-m length. The force on the conductor is thus seen to be proportional to the length of the conductor within the field, as well as the field flux density and the current carried by the conductor. Therefore,

$$(\text{force on conductor}) \propto (\text{flux density}) \times (\text{current}) \times (\text{length})$$

or

$$\boxed{F = BIl \qquad \text{newtons}} \tag{12-4}$$

When B is in teslas, I is in amperes, and l is in meters, the force F is given in ***newtons***. Thus:

A force of 1 newton is exerted upon a 1 meter length of conductor carrying a current of 1 ampere and situated in a magnetic field having a flux density of 1 tesla.

EXAMPLE 12-3 A conductor carrying a current of 15 A is situated at right angles to a magnetic field with a flux density of 0.7 T. If the length of the conductor within the field is 20 cm, determine the force on the conductor. Also calculate the new current level required to increase the force to 25 N.

SOLUTION

Eq. (12-4):

$$F = B\,I\,l \qquad \text{newtons}$$

$$= 0.7\ \text{T} \times 15\ \text{A} \times 20 \times 10^{-2}\ \text{m}$$

$$\mathbf{F = 2.1\ N}$$

$$I = \frac{F}{Bl}$$

$$= \frac{25\ \text{N}}{0.7\ \text{T} \times 20 \times 10^{-2}}$$

$$\mathbf{I = 179\ A}$$

In Figure 12-14(a), a single-turn coil is shown pivoted in the magnetic field set up by the N and S poles of a magnet. Consideration of the current direction through the coil shows that a downward force is exerted on its left-hand side, while an upward force is exerted on its right-hand side [see Figure 12-14(b)]. Since there are two lengths (l) of conductor within the magnetic field, the force exerted on the coil is

$$F = 2\,B\,I\,l \qquad \text{newtons}$$

If the coil has N turns, the equation for force is

$$F = 2\,B\,I\,l\,N \qquad \text{newtons}$$

and if the force acts at a radius r [see Figure 12-14(a)], the torque acting on the coil is

$$\boxed{\textit{torque} = B\,I\,l\,N\,r \qquad \text{newton meters}} \qquad \textbf{(12-5)}$$

EXAMPLE 12-4

Calculate the torque acting on a 100-turn coil pivoted in a magnetic field having a flux density of 0.5 T. The current through the coil is 100 mA, its radius is 1 cm, and its axial length is 2 cm.

SOLUTION

Eq. (12-5):

$$\textit{torque} = 2\,B\,I\,l\,N\,r \qquad \text{newtons}$$

$$= 2 \times 0.5 \times 100 \times 10^{-3} \times 2 \times 10^{-2} \times 100 \times 1 \times 10^{-2}$$

$$\textbf{\textit{torque}} \mathbf{= 2 \times 10^{-3}\ Nm}$$

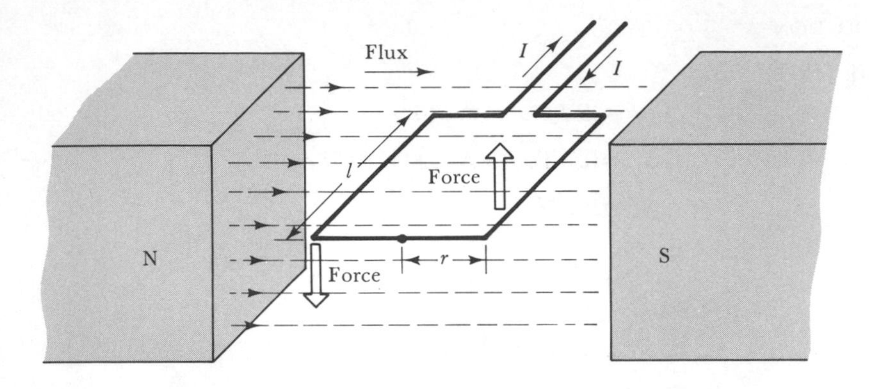

(a) Single-turn coil pivoted in a magnetic field

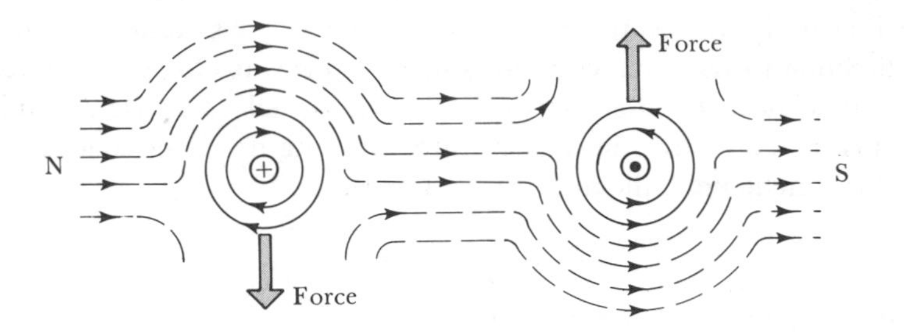

(b) Showing the force on each side of a single-turn coil pivoted in a magnetic field

FIG. 12-14. A force is exerted on each side of a current-carrying coil pivoted in a magnetic field

The simple arrangement illustrated in Figure 12-14 is the basic moving part of an electric measuring instrument. It also shows the basic principle behind the electric motor.

GLOSSARY OF FORMULAS

Flux density:

$$B = \frac{\Phi}{A}$$

Magnetomotive force:

$$\mathcal{F} = NI$$

Magnetic field strength:

$$H = \frac{NI}{l}$$

Mechanical force:

$$F = B\,I\,l$$

Mechanical torque:

$$\text{torque} = 2\,B\,I\,l\,N\,r$$

REVIEW QUESTIONS

12-1 Discuss the terms north-seeking pole and south-seeking pole in relation to a bar magnet and the earth's magnetic field. Also, explain what occurs when the ends of two bar magnets are brought close together.

12-2 Draw sketches to show the force field set up around a bar magnet and describe two methods of mapping the force field.

12-3 Define magnetic lines of force and magnetic flux, and list the characteristics of magnetic lines of force. Also, explain how their direction is defined.

12-4 Sketch the magnetic fields that occur at:

a. Two adjacent N poles of bar magnets.

b. Two adjacent S poles.

c. Adjacent N and S poles.

d. A bar magnet with a piece of soft iron in its field.

12-5 Sketch the shape of the magnetic field around a current-carrying conductor and show how the direction of the field is related to the direction of the current in the conductor.

12-6 Sketch the magnetic field that occurs around two adjacent current-carrying conductors:

a. For currents in opposite directions.

b. For currents in the same direction.

Explain how a force occurs between the conductors and indicate the direction of the force in each case.

12-7 Draw sketches to show the form of magnetic field that occurs with:

a. A single-turn coil with current flowing through it.

b. A long current-carrying solenoid.

Also, sketch a toroidal-shaped coil and show what its magnetic field should look like.

12-8 Discuss *electromagnetic induction* and draw sketches to explain the principle.

12-9 Define the following classifications of materials: nonmagnetic, diamagnetic, paramagnetic, and ferromagnetic.

12-10 Using illustrations, explain the theory of magnetism.

12-11 Define the weber and the tesla and state their relationship.

12-12 Define magnetomotive force and magnetic field strength and explain how the two are related.

12-13 Draw illustrations to show that a force is exerted on a current-carrying conductor situated in a magnetic field. Show how the direction of the force is related to the directions of the current and the magnetic field.

12-14 Discuss the relationships between the force on a current-carrying conductor in a magnetic field and the factors responsible for the force. Also, evolve the equation for the force and state the units for each factor.

12-15 Derive the expression for the torque on a pivoted coil situated within a magnetic field.

PROBLEMS

12-1 The total flux emitted from the pole of a magnet is 0.5×10^{-6} Wb.

a. If the magnet's cross-sectional area is 1.5 cm^2, determine the flux density within the metal.

b. Find the flux density at a short distance from the pole, if all the flux is contained in an area of 5 cm by 5 cm.

12-2 The flux density in an air gap between two N and S poles is 2.5 T. The poles have a circular cross section with a diameter of 5.6 cm. Calculate the total flux that crosses the air gap.

12-3 The cross-sectional area of a solenoid is 15 cm^2, and the flux density within it is 30×10^{-3} T. Calculate the total flux.

12-4 A toroidal-shaped coil has 280 turns and carries a current of 7.5 A. The length of the magnetic circuit within the coil is 50 cm. Calculate the mmf and the magnetic field strength.

12-5 A conductor situated at right angles to a magnetic field is carrying a current of 5 A. The field has a flux density of 1.5 T, and the length of the conductor within the field is 12 cm. Calculate the force on the conductor and the new value of flux density required if the force is to remain the same when the current is reduced to 500 mA.

12-6 A magnetic field with a flux density of 500×10^{-3} T has two conductors situated at right angles to the field. One of the conductors has a current of 100 A, and the current in the other is 3.9 A. Both conductors are 25 cm long. Calculate the force exerted on each of the conductors.

12-7 A 600-turn coil is pivoted within a magnetic field having a flux density of 2.2 T. The current through the coil is 50 μA, its radius is 0.75 cm, and its axial length is 1 cm. Calculate the torque acting upon the coil.

12-8 A 320-turn coil has a radius of 1.5 cm and an axial length of 3 cm. The coil is situated within a magnetic field having a flux density of 5 T. If the torque that acts on the coil is to be 1.5×10^{-3} Nm, determine the current that must flow through the coil.

12-9 A 500-turn coil with a radius of 1 cm and a length of 2.5 cm is pivoted between the poles of a magnet at right angles to the magnetic flux. The coil current is 100 μA and the torque acting on the coil is 0.5×10^{-4} Nm. Calculate the flux density of the magnetic field.

12-10 For Problem 12-9, the magnetic pole area is 9 cm^2. Calculate the total flux emitted from the poles.

13

MAGNETIC CIRCUITS

INTRODUCTION

Magnetic circuits are in many ways analogous to electrical circuits. *Magnetomotive force* (mmf), *flux*, and *reluctance* are the counterparts of electromotive force, current, and resistance, respectively. Electrical conductivity also has its magnetic analog in *permeability*. For all nonmagnetic materials, the permeability has a very small fixed value, known as the *permeability of free space*. *Ferromagnetic* materials have relative permeability values which relate their magnetic properties to that of free space.

Magnetic circuits made up of cores of different thickness, and perhaps an air gap, are termed *composite magnetic circuits*. The mmfs required to set up a given flux are determined for each section of the circuit and then added together to find the total circuit mmf. Each type of ferromagnetic material has its own particular *magnetization curve*, the shape of which can be explained in terms of *magnetic domains*. Because of the *hysteresis* effect, the magnetization curves change to loops when the flux within a sample of ferromagnetic material is reversed several times. This gives rise to a loss of energy in magnetic cores.

13-1 RELUCTANCE AND PERMEABILITY

Recall from Section 5-4 that the resistance of a conductor can be calculated from Eq. (5-1):

$$\text{resistance,} \qquad R = \frac{\rho l}{A}$$

where ρ is the resistivity of the conducting material, l is the length of the conductor, and A is its cross-sectional area.

The magnetic circuit analogy to the resistance of the electrical circuit is termed *reluctance*. Therefore, *reluctance is a measure of the opposition offered by a magnetic circuit to the setting up of flux*, just as resistance is opposition to current flow in an electrical circuit. The corresponding equation for reluctance is

$$\textbf{reluctance, } \mathfrak{R} = \frac{1}{\mu} \times \frac{l}{A} \tag{13-1}$$

Here, l and A are length and cross-sectional area, respectively, of the magnetic circuit, and μ is the *permeability* of the material in the magnetic circuit. Obviously, the reciprocal of permeability corresponds to the resistivity (ρ) of the electrical circuit. This means that magnetic permeability is the analog of electrical conductivity. Just as ρ is resistance of a cubic meter of electrical material, so $1/\mu$ can be termed *reluctance of a cubic meter of magnetic material.*

Ohm's law for an electrical circuit can be written:

$$\text{resistance, } R = \frac{\text{emf}}{\text{current}} = \frac{E}{I}$$

and

$$\text{conductance} = \frac{1}{R} = \frac{I}{E}$$

Similarly, for the magnetic circuit:

$$\text{reluctance} = \frac{\text{mmf}}{\text{flux}}$$

or

$$\mathfrak{R} = \frac{\mathfrak{F}}{\Phi} \tag{13-2}$$

The magnetic circuit analog of conductance is *permeance*, and since conductance is the reciprocal of resistance, so permeance is the reciprocal of reluctance:

$$\textbf{permeance} = \frac{1}{\mathfrak{R}} = \frac{\Phi}{\mathfrak{F}} \tag{13-3}$$

Substituting for $\mathcal{R}$ from Eq. (13-1) into Eq. (13-3) gives

$$\frac{1}{\frac{1}{\mu}\times\frac{l}{A}}=\frac{\Phi}{\mathcal{F}}$$

or

$$\frac{\mu A}{l}=\frac{\Phi}{\mathcal{F}}$$

Therefore,

$$\mu=\frac{\Phi}{A}\times\frac{l}{\mathcal{F}}$$

from Eq. (12-1),

$$\frac{\Phi}{A}=\text{flux density, } B$$

and from Eqs. (12-2) and (12-3),

$$\frac{\mathcal{F}}{l}=\frac{NI}{l}=\text{magnetic field strength, } H$$

Therefore,

$$\text{permeability}=\frac{\text{flux density}}{\text{magnetic field strength}}$$

or

$$\boxed{\mu=\frac{B}{H}} \qquad \textbf{(13-4)}$$

The SI units for permeability are henrys/meter (H/m). The *henry* is the unit of inductance and is discussed in Chapter 15. However, as will be seen, μ is invariably used as a ratio.

13-2 PERMEABILITY OF FREE SPACE

A current-carrying conductor (A) is shown in Figure 13-1 with the magnetic flux around it in concentric circles. Another current-carrying conductor (B) is situated within the magnetic field generated by A, and therefore a force is exerted upon conductor B. The magnitude of the force can be determined using Eq. (12-4):

$$F=B\,I\,l$$

or

$$B=\frac{F}{I\,l}$$

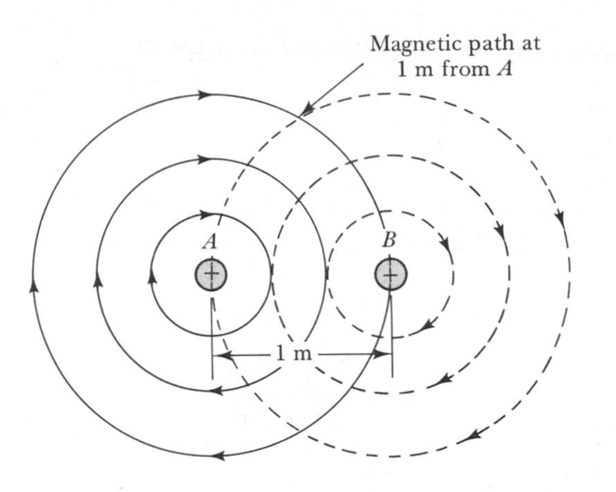

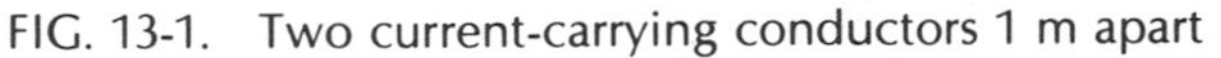
FIG. 13-1. Two current-carrying conductors 1 m apart

Now recall that the definition of the ampere (see Section 3-1) states that the force per meter length exerted between two conductors 1 m apart when they are each carrying a current of 1 A is 2×10^{-7} N. Therefore, substituting: $F=2\times10^{-7}$ N, $I=1$ A, and $l=1$ m:

$$B=2\times10^{-7}\text{ teslas}$$

Thus, it is clear that *the flux density in air (or any nonmagnetic material) at 1 m from a long conductor carrying 1 A is 2×10^{-7} T.*

Returning to Figure 13-1, the length of the magnetic path at 1 m from conductor A is obviously the circumference of a circle with a radius of 1 m. Therefore, the length of the magnetic path is

$$\begin{aligned} l&=2\pi\times1\text{ m}\\ &=2\pi\text{ meters}\end{aligned}$$

and the mmf is

$$\begin{aligned}\mathscr{F}&=I\times N\\ &=1\text{ A}\times1\\ &=1\text{ A}\end{aligned}$$

Therefore, the magnetic field strength is calculated as

$$\begin{aligned} H&=\frac{\mathscr{F}}{l}\\ &=\frac{1\text{ A}}{2\pi\text{ m}}\\ &=\frac{1}{2\pi}\text{ A/m}\end{aligned}$$

From Eq. (13-4), the *permeability of the material around the conductor* is

$$\mu = \frac{B}{H}$$

The material around the conductor was assumed to be air; however, it could be a vacuum or any nonmagnetic material. In this case, the permeability is a constant which is designated μ_0 and is referred to as the *permeability of free space*. Substituting for B and H gives

$$\mu_o = \frac{2 \times 10^{-7}\ \text{T}}{1/2\pi\ \text{A/m}}$$

or *permeability of free space*, $\mu_o = 4\pi \times 10^{-7}$

The term μ_o may now be employed in calculations relating to flux density and magnetic field strength.

EXAMPLE 13-1

An air-cored toroidal coil of the type shown in Figure 12-12 has 3000 turns and carries a current of 0.1 A. The cross-sectional area of the coil is 4 cm^2, and the length of the magnetic circuit is 15 cm. Determine the magnetic field strength, the flux density, and the total flux within the coil.

SOLUTION

Eq. (12-3):

$$H = \frac{NI}{l}$$

$$= \frac{3000 \times 0.1\ \text{A}}{15 \times 10^{-2}\ \text{m}}$$

magnetic field strength, $\boldsymbol{H = 2000\ \text{A/m}}$

Eq. (13-4):

$$\mu = \frac{B}{H}$$

Therefore,

$$B = \mu_o H$$

$$= 4\pi \times 10^{-7} \times 2000\ \text{A/m}$$

flux density, $\boldsymbol{B = 2.5 \times 10^{-3}\ \text{T}}$

Eq. (12-1):

$$B = \frac{\Phi}{A}$$

Therefore,

$$\Phi = BA$$

$$= 2.5 \times 10^{-3}\,\text{T} \times 4 \times 10^{-4}\,\text{m}^2$$

total flux, $\boldsymbol{\Phi = 1\ \mu\text{Wb}}$

EXAMPLE 13-2 Determine the mmf required to generate a total flux of 100μWb in an air gap that is 0.2 cm long. The cross-sectional area of the air gap is 25 cm^2.

SOLUTION

Eq. (12-1):

$$B = \frac{\Phi}{A}$$

$$= \frac{100 \times 10^{-6}\,\text{Wb}}{25 \times 10^{-4}\,\text{m}^2}$$

$$\boldsymbol{B = 4 \times 10^{-2}\,\text{T}}$$

Eq. (13-4):

$$\mu_o = \frac{B}{H}$$

or

$$H = \frac{B}{\mu_o}$$

$$= \frac{4 \times 10^{-2}\,\text{T}}{4\pi \times 10^{-7}}$$

$$\boldsymbol{H = 3.18 \times 10^4\,\text{A/m}}$$

Eq. (12-3):

$$H = \frac{\mathcal{F}}{l}$$

Therefore,

$$\mathcal{F} = H \times l$$

$$= 3.18 \times 10^4 \times 0.2 \times 10^{-2}$$

$$\boldsymbol{\mathcal{F} = 63.7\,\text{A}}$$

13-3 SOLENOID

A *solenoid* is a long, thin air-cored coil, and when a current is passed through the coil, a magnetic field is set up, as shown in Figure 13-2(a). The path of the magnetic lines of flux is made up of two components: the length of the path within the coil (l_1), and the length of the path followed by the flux outside the coil (l_2). The length l_2 is longer than l_1, but not very much longer. Also, as illustrated in the figure, the cross-sectional area of the path outside the coil is very much larger than that inside the coil. Consequently, since they are both the same material (i.e., air), the reluctance of path l_2 is very much smaller than the reluctance of path l_1. Thus, the mmf required to set up a given total flux along path l_1 is very much larger than that required to set up the same flux along l_2. The total mmf required for the solenoid is the sum of the two components:

$$\text{total mmf} = (\text{mmf for } l_1) + (\text{mmf for } l_2)$$

but $$(\text{mmf for } l_1) \gg (\text{mmf for } l_2)$$

Therefore, $$\text{total mmf} \cong \text{mmf for } l_1$$

In Chapter 12 it was shown that magnetic lines of force are always in a state of tension, and always seeking to flow through the path of least reluctance. It was also seen that the reluctance of ferromagnetic materials is very much smaller than that of air. When a bar of ferromagnetic material is brought close to one end of a solenoid, the bar becomes magnetized by induction [see Figure 13-2(b)]. The induced magnetism produces poles as illustrated, so that the bar is attracted to the solenoid. If the bar is of suitable dimensions, it will be *sucked* right into the solenoid, and remain there. When this happens, the magnetic path inside the (iron-cored) solenoid has a very small reluctance, and it is the air path outside the solenoid that requires the greatest mmf to set up a given total flux.

EXAMPLE 13-3

An air-cored solenoid has length $l = 15$ cm and inside diameter $D = 1.5$ cm. If the coil has 900 turns, determine the total flux within the solenoid when the coil current is 100 mA.

SOLUTION

$$\begin{aligned}\mathcal{F} &= NI \\ &= 900 \times 100 \text{ mA} \\ &= 90 \text{ A}\end{aligned}$$

$$H = \frac{\mathcal{F}}{l}$$

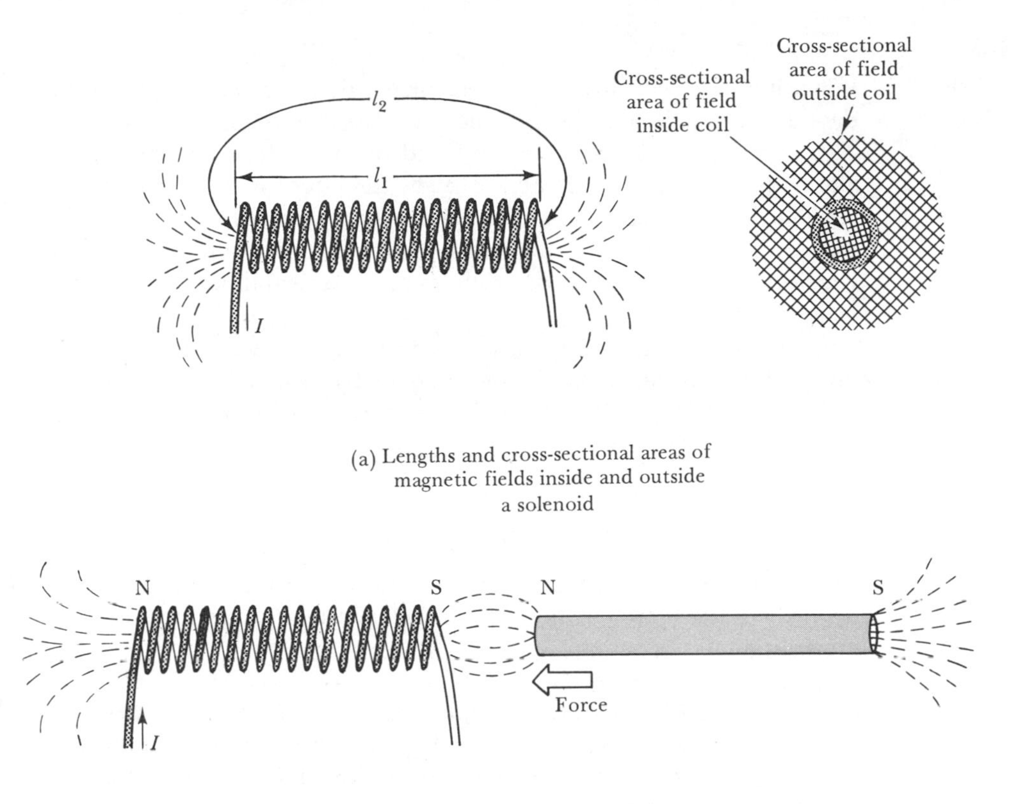

(a) Lengths and cross-sectional areas of magnetic fields inside and outside a solenoid

(b) Effect of bringing a soft iron core close to a solenoid

FIG. 13-2. The magnetic field of a solenoid and its effect on an iron core

For the solenoid, $l \cong$ coil length. Therefore,

$$H \cong \frac{90 \text{ A}}{15 \times 10^{-2} \text{ m}}$$

$$H = 600 \text{ A/m}$$

$$B = \mu_o \times H$$

$$= 4\pi \times 10^{-7} \times 600 \text{ A/m}$$

$$\mathbf{B = 24\pi \times 10^{-5} \text{ T}}$$

$$\Phi = B \times A$$

$$= 24\pi \times 10^{-5} \times \pi \left(\frac{D}{2}\right)^2$$

$$= 24\pi \times 10^{-5} \times \pi \times \left(\frac{1.5 \times 10^{-2}}{2}\right)^2$$

$$\mathbf{\Phi = 1.33 \times 10^{-7} \text{ Wb}}$$

13-4 RELATIVE PERMEABILITY

When iron or steel is inserted into a current-carrying coil, it is found that the total magnetic flux through the coil is increased a great many times. This occurs even through the mmf and magnetic field strength are not altered in any way. The reason (already discussed) is that iron and steel are much better conductors of magnetic flux than any other materials. Since permeability is the magnetic analog of conductivity, it is obvious that iron and steel have very much greater permeabilities than nonmagnetic materials.

The improvement in total flux and flux density when iron or steel is involved in a circuit is taken into account by assigning each material a *relative permeability* μ_r. Equation (13-4) is then modified to

$$\boxed{\mu_r\mu_o = \frac{B}{H}} \qquad (13\text{-}5)$$

In the case of air and other nonmagnetic materials, $\mu_r = 1$. Depending upon the particular type of iron or steel, the relative permeability may range from 400 to perhaps 2500.

Figure 13-3 shows the plot of flux density versus magnetic field strength (or B/H *curves*) for various specimens of iron and steel. The measurements were made on ring-shaped specimens, so the magnetic circuit was a closed iron circuit with no air gaps. Thus, maximum possible flux densities were achieved. The magnetization curves can be used for direct determinations of flux density for a given value of magnetic field strength. Alternatively, the relative permeabilities of each specimen can be calculated and plotted versus H. The dashed line in Figure 13-3 shows the plot of μ_r for sheet steel. This is derived simply by taking corresponding values of B and H from the permeability curve for sheet steel, and substituting them into Eq. (13-5) to calculate the values of μ_r. Note that the relative permeability is by no means a constant quantity; instead, it is very much dependent upon the magnetic field strength H. From the plot of μ_r versus H for sheet steel, the greatest permeability is obviously achieved at a magnetic field strength of approximately 250 A/m.

It should be noted that all the B/H curves shown in Figure 13-3 are typical, and that for any given specimen of magnetic material, the actual B/H relationships may be different from those illustrated. This is because slight differences in the manufacturing process can significantly affect the material's magnetic properties. The shape of the B/H curve is considered further in Section 13-7.

EXAMPLE 13-4 Calculate the permeability of sheet steel at a magnetic field strength of 250 A/m.

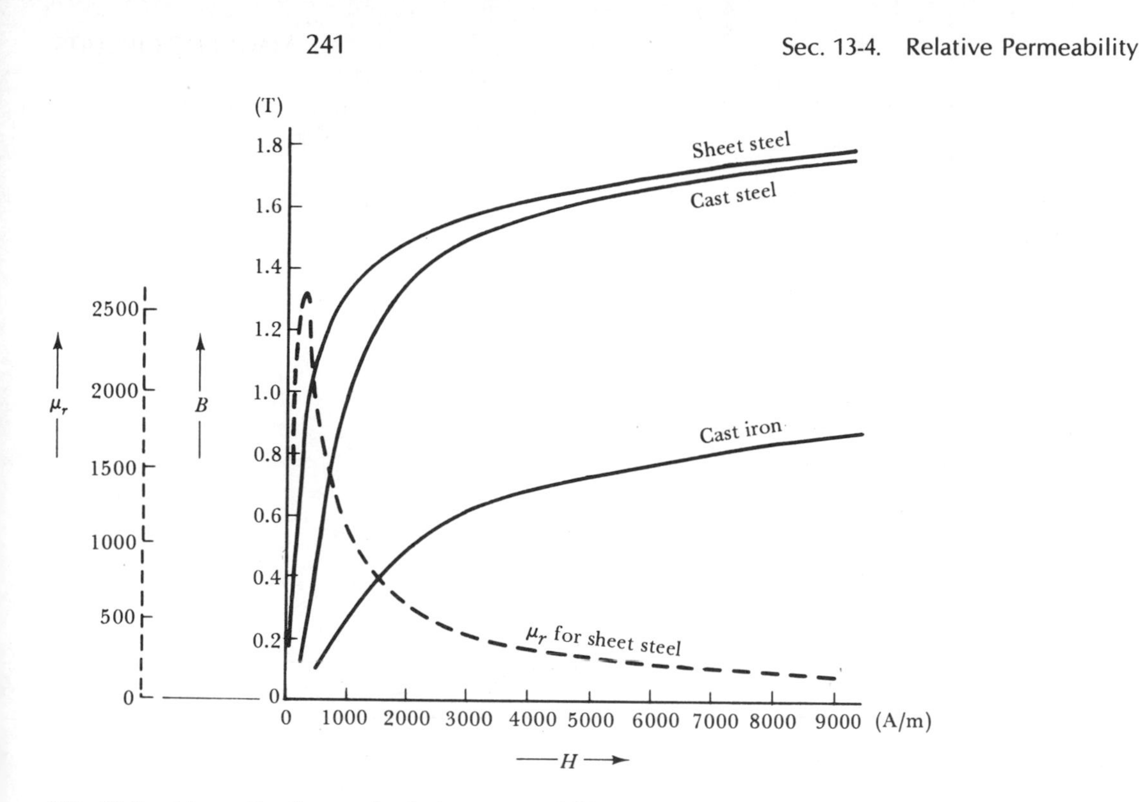

FIG. 13-3. Magnetization and relative permeability curves

SOLUTION

From Figure 13-3, *at* $H = 250$ A/m, $B = 0.8$ T:
Eq. (13-5):

$$\mu_r \mu_o = \frac{B}{H}$$

Therefore,

$$\mu_r = \frac{B}{\mu_o H}$$

$$= \frac{0.8 \text{ T}}{4\pi \times 10^{-7} \times 250 \text{ A/m}}$$

$$\mu_r = 2546$$

EXAMPLE 13-5 The cast iron ring illustrated in Figure 13-4 has 3000 turns and carries a current of 0.1 A. The cross-sectional area of the ring is 4 cm^2, and the length of the magnetic path is 15 cm. Determine the flux density and the total flux in the ring.

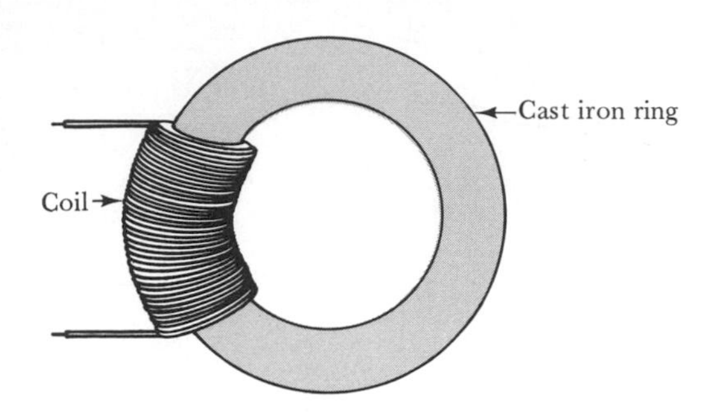

FIG. 13-4. Cast iron ring with winding

SOLUTION

Eq. (12-3):

$$H=\frac{NI}{l}$$

$$=\frac{3000\times 0.1\ \text{A}}{15\times 10^{-2}\text{m}}$$

$$\boldsymbol{H=2000\ \text{A/m}}$$

From Figure 13-3, *for cast iron at* $H=2000$ A/m:

flux density, B **= 0.5 T**

Eq. (12-1):

$$B=\frac{\Phi}{A}$$

Therefore,

$$\Phi=BA$$

$$=0.5\ \text{T}\times 4\times 10^{-4}\ \text{m}^2$$

total flux, $\boldsymbol{\Phi=200\ \mu\text{Wb}}$

Compare this to the results of Example 13-1.

EXAMPLE 13-6 The cast steel core shown in Figure 13-5 has a relative permeability of 800, and the coil has 700 turns. If the total flux in the core is to be 4×10^{-4} Wb, determine the required current through the coil.

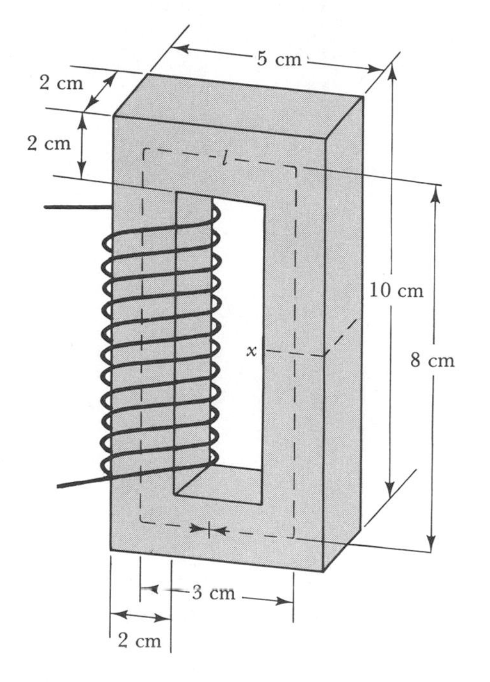

FIG. 13-5. Cast steel core

SOLUTION

$$B = \frac{\Phi}{A}$$

$$= \frac{4 \times 10^{-4}\ \text{Wb}}{2 \times 10^{-2}\ \text{m} \times 2 \times 10^{-2}\ \text{m}}$$

$$\boldsymbol{B = 1\ \text{T}}$$

From Eq. (13-5),

$$H = \frac{B}{\mu_r \mu_o}$$

$$= \frac{1\ \text{T}}{800 \times 4\pi \times 10^{-7}}$$

$$\boldsymbol{H = 994.7\ \text{A/m}}$$

The length of the magnetic circuit is l. As shown in Figure 13-5,

$$l = 2(3\ \text{cm} + 8\ \text{cm})$$

$$= 22\ \text{cm}$$

From Eq. (12-3),

$$\mathcal{F} = H \times l$$

$$= 994.7\ \text{A/m} \times 22 \times 10^{-2}\ \text{m}$$

$$\cong 219\ \text{A}$$

and $$\mathcal{F} = NI$$

Therefore, $$I = \frac{219\ \text{A}}{700\ \text{turns}}$$

$$\boldsymbol{I \cong 313\ \text{mA}}$$

13-5 COMPOSITE MAGNETIC CIRCUITS

Where a magnetic circuit has an air gap in its path, the air gap is *in series* with the rest of the magnetic circuit. The necessary magnetomotive forces must be calculated independently for the air gap and for the magnetic path, and then the two are added to determine the total required mmf. The same procedure applies when the cross-sectional area of a core is not constant along its entire length.

EXAMPLE 13-7

The core shown in Figure 13-5 has a 1-mm air gap at point x. $\mu_r = 800$ and $N = 700$, as described for Example 13-6. Calculate the new value of current through the coil to give a total core flux of 4×10^{-4} Wb.

SOLUTION

From Example 13-6,

$$B = 1\ \text{T}$$

and for the iron path,

$$\mathcal{F} = 219\ A$$

For the air gap,

$$H = \frac{B}{\mu_o} = \frac{1\ \text{T}}{4\pi \times 10^{-7}}$$

$$\boldsymbol{H \cong 7.96 \times 10^5\ \text{A/m}}$$

and $$l = 1\text{ mm} = 1 \times 10^{-3}\text{ m}$$

$$\mathscr{F} = H \times l$$

$$= 7.96 \times 10^5 \times 1 \times 10^{-3}$$

$$\boldsymbol{\mathscr{F} = 7.96 \times 10^2\text{ A}}$$

$$\textit{total } \mathscr{F} = 219\text{ A} + (7.96 \times 10^2)$$

$$\boldsymbol{\mathscr{F} = 1015\text{ A}}$$

$$I = \frac{\mathscr{F}}{N} = \frac{1015\text{ A}}{700\text{ turns}}$$

$$\boldsymbol{I \cong 1.45\text{ A}}$$

EXAMPLE 13-8 The ring-shaped core shown in Figure 13-6 has a relative permeability of 1000, and the flux density in the thickest section is to be 0.75 T. If the current through a coil wound on the core is to be 500 mA, determine the number of coil turns required.

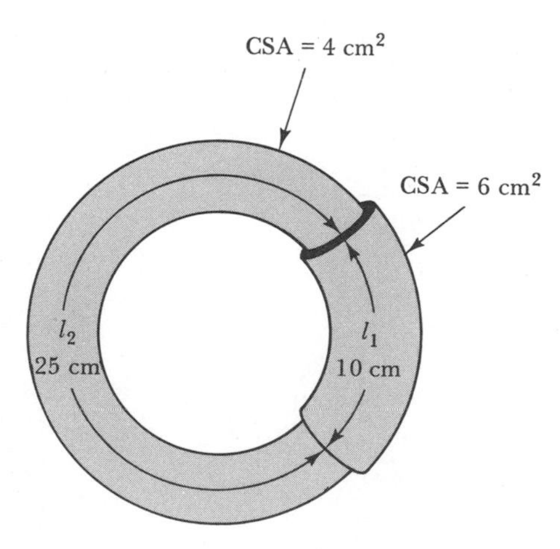

FIG. 13-6. An uneven ring-shaped core which must be treated as a composite magnetic circuit

SOLUTION

For the thick section,

$$H = \frac{B}{\mu_r \mu_o}$$

$$= \frac{0.75}{1000 \times 4\pi \times 10^{-7}}$$

$$\mathbf{H_1 = 597\ A/m}$$

$$\mathcal{F} = H \times l$$

$$= 597\ \text{A/m} \times 10 \times 10^{-2}\ \text{m}$$

$$\mathbf{\mathcal{F}_1 = 59.7\ A}$$

total flux in core, $\Phi = B \times A$

$$= 0.75\ \text{T} \times (A \text{ of thick section})$$

$$= 0.75\ \text{T} \times 6 \times 10^{-4}\ \text{m}$$

$$\mathbf{\Phi = 4.5 \times 10^{-4}\ Wb}$$

For the thin section,

$$B = \frac{\Phi}{A}$$

$$= \frac{4.5 \times 10^{-4}\ \text{Wb}}{4 \times 10^{-4}}$$

$$\mathbf{B = 1.125\ T}$$

$$H = \frac{B}{\mu_r \mu_o}$$

$$= \frac{1.125\ \text{T}}{1000 \times 4\pi \times 10^{-7}}$$

$$\mathbf{H_2 \cong 895\ A/m}$$

$$\mathcal{F} = H \times l$$

$$= 895\ \text{A/m} \times 25 \times 10^{-2}\ \text{m}$$

$$\mathbf{\mathcal{F}_2 \cong 224\ A}$$

$$\textit{total}\ \text{mmf} = \mathcal{F}_1 + \mathcal{F}_2$$

$$= 59.7 + 224$$

$$\mathbf{\mathcal{F} = 283.7\ A}$$

$$\mathcal{F} = N \times I$$

Therefore,
$$N = \frac{\mathcal{F}}{I}$$
$$= \frac{283.7\ \text{A}}{500\ \text{mA}}$$
$$\mathbf{N \cong 567\ turns}$$

When crossing an air gap, the magnetic lines of force tend to bulge out as shown in Figure 13-7(a). This is because lines of force repel each other when passing through nonmagnetic material. The effect of this bulging, or *fringing* as it is termed, is to increase the cross-sectional area of the magnetic field at the air gap and thus decrease its flux density. In a short air gap with a large cross-sectional area, the fringing may be insignificant. In other situations, 10% is typically added to the air gap's cross-sectional area to allow for fringing.

In Figure 13-7(b) an electromagnet is shown lifting an iron bar. In this case there are air gaps at the junction between the magnet and the iron bar, and fringing occurs as illustrated. Some lines of force cross the gap between the magnet's poles without passing through the iron bar. This is termed *leakage flux*, and unlike the fringing flux, it does not assist in holding the iron bar to the magnet. Thus the leakage flux is wasted. However, the iron path around the magnetic circuit has a much lower reluctance than the air path taken by the leakage flux; consequently, leakage flux is usually small enough to be neglected.

EXAMPLE 13-9 The magnetic core shown in Figure 13-8 has the following dimensions: $l_1 = 10$ cm, $l_2 = l_3 = 18$ cm, cross-sectional area of l_1 path $= 6.25 \times 10^{-4}$ m^2,

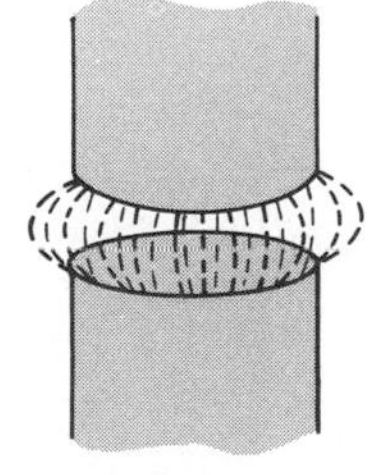

(a) Fringing flux at an air gap

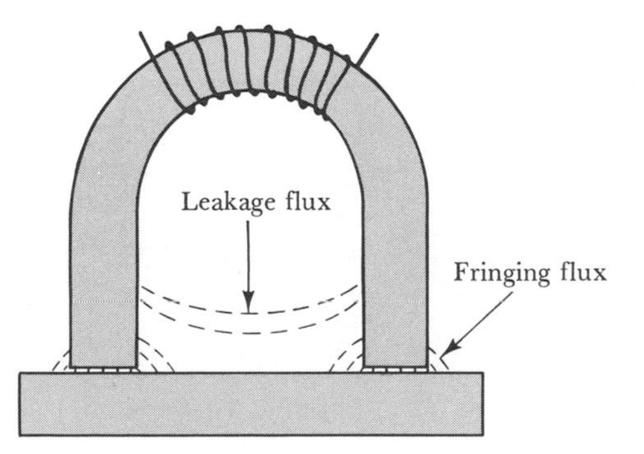

(b) Electromagnet and iron bar

FIG. 13-7. Fringing flux and leakage flux

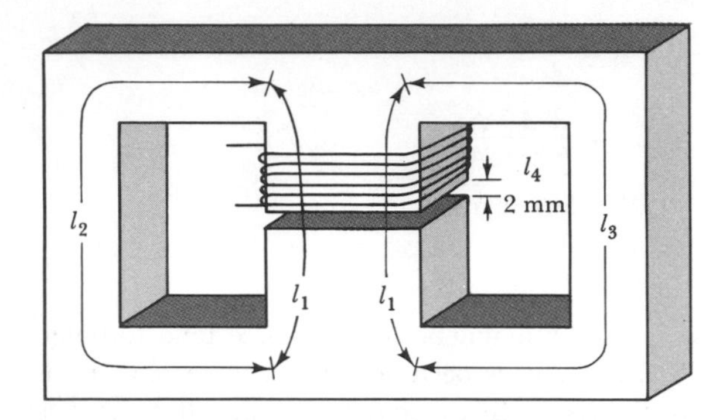

FIG. 13-8. Composite magnetic core

cross-sectional area of l_2 and l_3 paths$=3\times10^{-4}$ m^2, length of air gap (l_4)$=2$ mm. Determine the current that must be passed through the 600-turn coil to produce a total flux of 100 μWb in the air gap. Assume that the metal has a permeability of 800.

SOLUTION

For the air gap:

$$B=\frac{\Phi}{A}=\frac{100\times10^{-6}\text{ Wb}}{6.25\times10^{-4}\text{ m}^2}$$

$$=0.16\text{ T}$$

$$H=\frac{B}{\mu_o}=\frac{0.16\text{ T}}{4\pi\times10^{-7}}$$

$$\boldsymbol{H=1.27\times10^5\text{ A/m}}$$

and $$H=\frac{\mathscr{F}}{l}$$

Therefore, $$\mathscr{F}=H\times l_4$$

$$=1.27\times10^5\text{ A/m}\times2\times10^{-3}\text{ m}$$

$$\boldsymbol{\mathscr{F}_4=254\text{ A}}$$

For path l_1:

$$B=0.16\text{ T}$$

$$H=\frac{B}{\mu_r\mu_o}=\frac{0.16\text{ T}}{800\times4\pi\times10^{-7}}$$

$$\boldsymbol{H=159\text{ A/m}}$$

and $$\mathscr{F}=H\times l_1=159\text{ A/m}\times10\times10^{-2}\text{ m}$$

$$\boldsymbol{\mathscr{F}_1=15.9\text{ A}}$$

Since l_2 and l_3 are in parallel, and each has a cross-sectional area of 3×10^{-4} m^2, they can be treated as a single path with a cross-sectional area of 6×10^{-4} m^2.

For path l_2 and l_3:

$$B=\frac{\Phi}{A}=\frac{100\times10^{-6}\ \text{Wb}}{6\times10^{-4}\ \text{m}^2}$$

$$\cong 0.167\ \text{T}$$

$$H=\frac{B}{\mu_r\mu_o}=\frac{0.167\ \text{T}}{800\times4\pi\times10^{-7}}$$

$$\mathbf{H=166\ A/m}$$

and $$\mathcal{F}=H\times l_2=166\ \text{A/m}\times18\times10^{-2}\ \text{m}$$

$$\mathbf{\mathcal{F}_2=29.9\ A}$$

$$\textit{total mmf}=254\ \text{A}+15.9\ \text{A}+29.9\ \text{A}$$

$$=300\ \text{A}$$

and $$\mathcal{F}=NI$$

Therefore, $$I=\frac{\mathcal{F}}{N}=\frac{300}{600}$$

$$\mathbf{I\cong500\ mA}$$

13-6 FORCE BETWEEN TWO MAGNETIC SURFACES

The magnetic field strength required to establish a given flux density in an air gap is

$$H=\frac{IN}{l}$$

Therefore, $$I=\frac{H\times l}{N}$$

If I is increased from zero to its maximum level during time Δt, the core flux changes from zero to a maximum of Φ. This flux change links with the coil and induced a voltage V at the coil terminals. From the definition of the weber (Section 12-4)—that a flux change of 1 Wb in a time of 1 s induces 1 V in a one-turn coil:

$$\text{induced voltage } V=\frac{\Phi N}{\Delta t}$$

The supply voltage is equal in magnitude to the induced voltage, and the power input to the coil can be calculated as:

$$P = VI$$

However, P increased from zero to its maximum level during time Δt. Assuming that the power increased linearly, the average power input can be shown to be:

$$P_{av} = \frac{VI}{2}$$

Also, the energy supplied to the coil in a time of Δt is

$$W = P_{av} \times \Delta t$$

$$= \frac{V \times I \times \Delta t}{2} \quad \text{joules}$$

Substituting for V and I,

$$W = \frac{\Phi \times N}{\Delta t} \times \frac{H \times l}{2N} \times \Delta t$$

Therefore,

$$W = \frac{\Phi \times H \times l}{2}$$

Since

$$\Phi = B \times A \text{ and } H = B/\mu_o,$$

$$W = \frac{(B \times A) \times \left(\dfrac{B}{\mu_o}\right) \times l}{2}$$

or

$$\textbf{\textit{energy stored in air gap, }} W = \frac{B^2 A l}{2\mu_o} \quad \text{joules} \qquad (13\text{-}6)$$

When two magnetic surfaces are separated by a short distance, the mechanical energy involved in pulling them apart is

$$W = F \times d$$

Assuming that the surfaces are still attracted to each other, the electrical energy supplied to the air gap is equal to the mechanical energy supplied to pull the surfaces apart. Therefore,

$$F \times d = \frac{B^2 A l}{2\mu_o}$$

In this case d and l both represent the thickness of the air gap, and they cancel out, giving

$$\boxed{F = \frac{B^2 A}{2\mu_o}} \qquad \textbf{(13-7)}$$

This expression includes the cross-sectional area (A) of the air gap, but does not include its length. When B is in teslas and A is in square meters, the force F is in newtons. Equation (13-7) can now be used to calculate the mechanical pull exerted by an electromagnet.

EXAMPLE 13-10 The electromagnet shown in Figure 13-7(b) has pole pieces with cross-sectional areas of 25 cm^2. The total flux crossing each pole is 250 μWb. Determine the maximum weight of iron plate that can be lifted by the magnet.

SOLUTION

$$B = \frac{\Phi}{A}$$

$$= \frac{250\ \mu\text{Wb}}{25 \times 10^{-4}\ \text{m}^2}$$

$$\boldsymbol{B = 0.1\ \text{T}}$$

Eq. (13-7):

$$F = \frac{B^2 A}{2\mu_o}$$

$$A = \textit{total}\ \text{cross-sectional area} = 2 \times 25 \times 10^{-4}\ \text{m}^2$$

$$= 50 \times 10^{-4}\ \text{m}^2$$

Therefore,

$$F = \frac{(0.1\ \text{T})^2 \times 50 \times 10^{-4}\ \text{m}^2}{2 \times 4\pi \times 10^{-7}}\ \text{newtons}$$

$$\boldsymbol{F = 19.9\ \text{N}}$$

and

$$F = m \times a$$

Therefore, $\quad \text{mass } m = \dfrac{F}{a}$ *where a is the acceleration due to gravity*

$$= \frac{19.9 \text{ N}}{9.81 \text{ m/s}^2}$$

weight that can be lifted $\cong$ **2.03 kg**

13-7 MAGNETIZATION CURVES

In Figure 13-9 a typical B/H curve is shown to a slightly larger scale than those in Figure 13-3. Examination of the curve shows that the material is initially unmagnetized, because at zero magnetic field strength the flux density is zero. As the field strength is increased from zero, the flux density increases very slowly at first, giving a shallow slope, section *o–a* of the curve. Further increase in H causes B to increase progressively more rapidly, until there is a near-linear relationship between B and H over section *a–b* of the curve. With continued increase in H past point *b*, changes in B become less and less until the curve flattens out again to a very shallow slope beyond point *c*.

The shape of the curve can be explained in terms of the magnetic domains in ferromagnetic material (see Section 12-3). During portion *o–a* of the curve, the magnetic field strength is too weak to produce much realignment of the magnetic domains. Consequently, the increase in flux density is relatively small for these levels of field strength. Over portion *a–b* the increasing field strength is obviously causing more and more of the domains to be aligned in the same direction. Thus, B increases in a

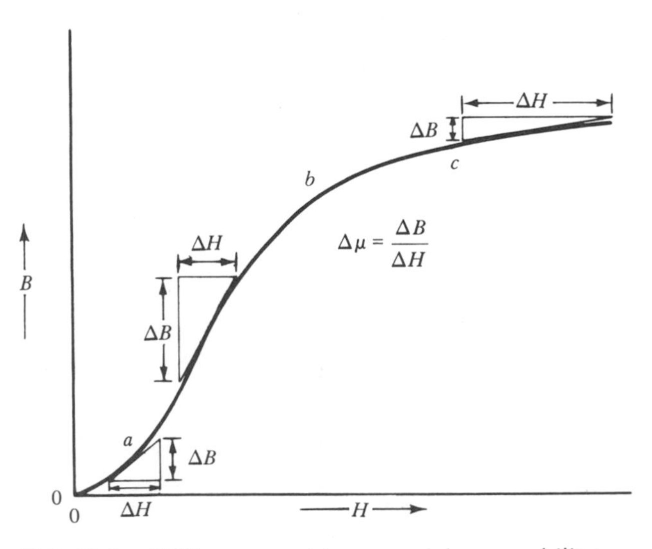

FIG. 13-9. B/H curve and incremental permeability

near-linear relationship to H. From point b to c, all of the domains are becoming completely aligned in the required direction for maximum magnetic effect. So, during the b–c portion of the curve, there are fewer and fewer magnetic domains available to be realigned. Therefore, once again, the increases in flux density in relation to increases in H are becoming smaller. Finally, from point c on, it can be assumed that effectively all available magnetic domains are completely aligned in the direction enforced by the magnetic field strength. The magnetic material is said to be *saturated* at this point. The slope of the B/H curve beyond point c shows the same increase in B with respect to H as would be obtained with nonmagnetic material.

From Eq. (13-5), the relationship between B and H is

$$\textit{permeability},\ \mu_r\mu_o = \frac{B}{H}$$

Ideally, it is desirable to have the largest possible value of permeability, in order to achieve the greatest flux density with the smallest possible magnetic field strength. The plot of μ_r for sheet steel (Figure 13-3) shows that the largest possible value of μ_r is achieved around the center of the linear portion of the B/H curve. This direct relationship between constant values of B and H can be referred to as the *normal permeability* of the material. Another important relationship is that between changing values of B and H. Referring to Figure 13-9 again, the slope of the B/H curve at any point is known as the *incremental permeability*. This parameter defines the flux density change (ΔB) for a given magnetic field strength change (ΔH). Thus, the incremental permeability is

$$\Delta\mu = \frac{\Delta B}{\Delta H}$$

Reference to the B/H curve in Figure 13-9 reveals that (as with μ_r) $\Delta\mu$ is greatest at the center of the linear portion of the magnetization curve. Both the initial and final values of incremental permeability (see the illustration) are considerably smaller than that at the center of the curve.

The incremental permeability is important in applications where only small changes occur in the magnetic field strength and the largest possible changes in flux density are desired. One such application is the telephone earphone (or headphone) (see Figure 13-10), where audio signals are represented by small changes in coil current (i.e., small changes in magnetic field strength). The resulting changes in flux density cause the metal disc (or *diaphragm*) to vibrate and re-create the original sound wave. In order to produce the largest possible changes in flux density (i.e., largest $\Delta\mu$), the magnetic field strength is usually set (or biased) at the center of the magnetization curve.

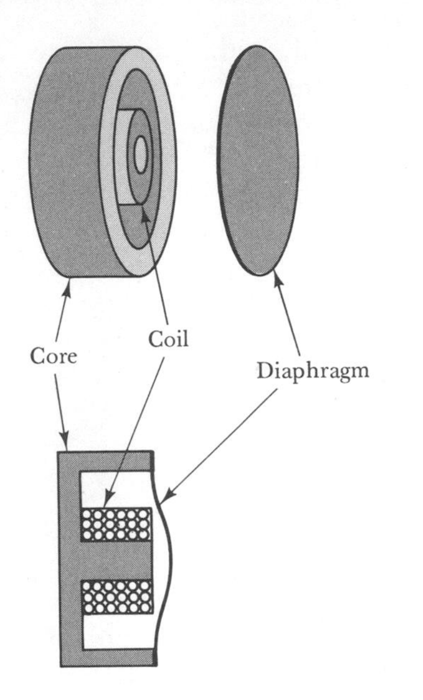

FIG. 13-10. Construction of earphone

13-8 HYSTERESIS

The magnetization curve of Figure 13-9 is reproduced in Figure 13-11(a), in order to investigate what occurs when the magnetic field strength is reduced from its maximum level. As already discussed, portion *o–a–b–c* of the curve occurs when the material is initially unmagnetized and the field strength is increased from zero. Commencing at point *c*, when the field strength is decreased, the flux density does not move down the *c–b–a–o* curve, as might be expected. Instead, it falls along the curve *c–d*, as illustrated. The effect can be explained in terms of a kind of friction-force-resisting movement of the magnetic domains. Thus, instead of returning to their original unmagnetized state, the magnetic domains remain partially aligned, and the material retains some magnetic flux, even though the field strength has been reduced to zero. The material has, in fact, been magnetized, and the retained flux density (*o–d*) is referred to as *remanence*, or *residual magnetism*. The ability of a given ferromagnetic material to retain residual magnetism is termed its *retentivity*.

To reduce the remanence to zero, a negative or reverse magnetic field strength must be applied (i.e., the current in the magnetizing coil must be reversed). As the reversed value of H is increased, the flux density moves down the *d–e* portion of the curve to a zero level of B. The magnetic field strength (*o–e*) required to reduce the remanence to zero is termed the *coercive force*.

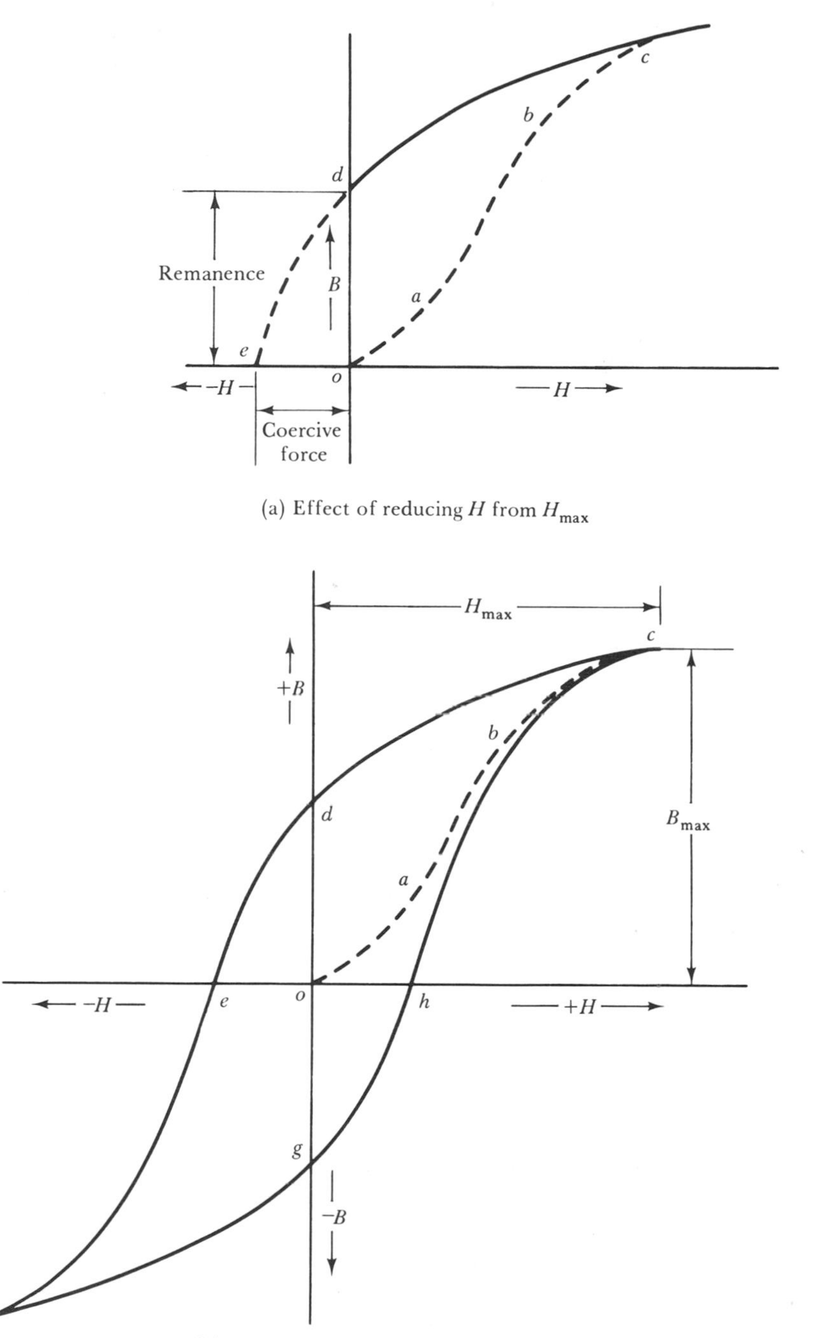

(a) Effect of reducing H from H_{max}

(b) Effect of reversing H and then increasing it in a positive direction once again

FIG. 13-11. Generation of a hysteresis loop

When the reversed value of H is increased beyond the level needed to reduce B to zero, it is found that the flux density now increases in a negative direction. A new magnetization curve is generated, giving *e–f* in Figure 13-11(b). At point *f*, the reversed flux density has reached its saturation point, and this corresponds to point *c* on the positive half of the curve. Once again, when the reversed field strength is reduced to zero, the reversed flux density does not become zero. Instead, portion *f–g* of the B/H graph is generated, and *o–g* represents the retained flux density. Section *o–g* is exactly equal to *o–d*, although in the opposite direction. To reduce the flux density to zero once more, the field strength must be increased in a positive direction, giving portion *g–h* of the graph. This again represents a coercive force, and *o–h* is found to be equal in magnitude to *o–e*. Finally, when the magnetic field strength is further increased in a positive direction, the B/H relationship traces out the *h–c* section of the graph.

It is seen that the B/H graph now forms a closed loop. The loop is symmetrical, with maximum positive and negative values of B and H being equal in magnitude. When a positive magnetic field strength (*o–h*) is applied to the specimen to which the graph refers, the flux density B is zero. Only when H is increased to H_{max} does the flux density become B_{max}. Similarly, when H becomes zero, B remains equal to *o–d*. Continuing around the loop, it is seen that the changing levels of flux density lag behind the changes in magnetic field strength. This lagging effect is termed *hysteresis*, and the B/H graph is then referred to as a *hysteresis loop*.

The B/H loop demonstrates that some energy is absorbed into a magnetic core to overcome the *friction* involved in changing the alignment of the magnetic domains. Thus, a core that is subjected to repeated and rapid reversals of the magnetic field (as in the case of alternating currents) may absorb a lot of energy in this way. This energy results in heating of the core, and it is obviously wasted or lost energy. In fact, the area enclosed by the hysteresis loop can be shown to be proportional to the lost energy.

Figure 13-12 shows three typical hysteresis loops for three different types of ferromagnetic material. Loop (*a*) is the type of hysteresis loop obtained for soft iron. The fact that it is narrow means that the area enclosed by the loop is relatively small, and consequently the hysteresis losses in the core would be a minimum. For greatest efficiency, therefore, a soft iron core should be employed in situations where the magnetic field has to undergo a large number of reversals each second. Also, note that the residual magnetism is small for the soft iron core; consequently, soft iron is not suitable for permanent magnets.

The hysteresis loop shown in Figure 13-12(b) is typical of hard steel, and its large area obviously results in a large core loss. Thus, hard steel is not suitable where the magnetic field is being rapidly reversed. However,

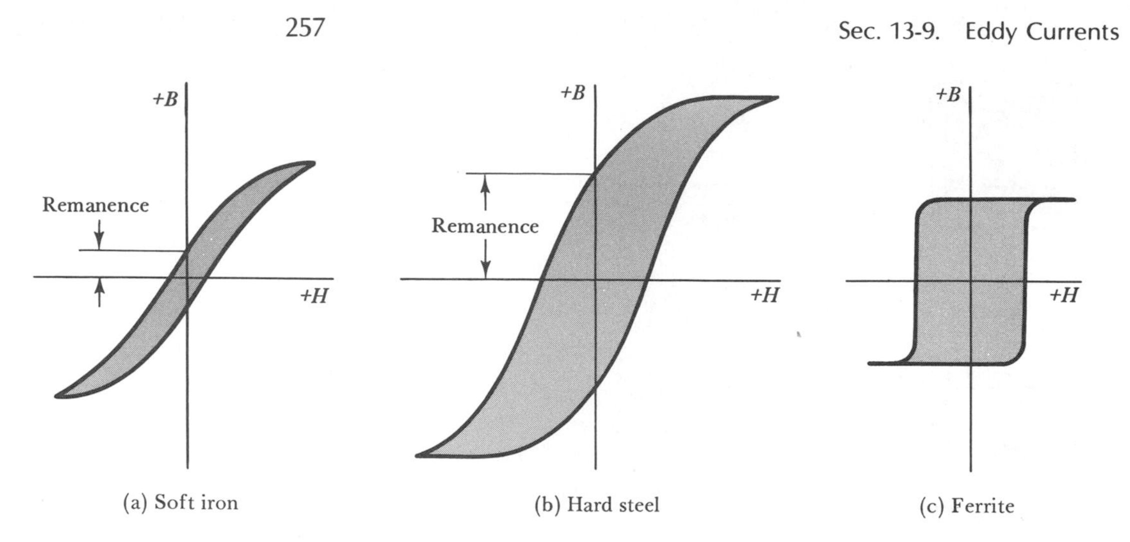

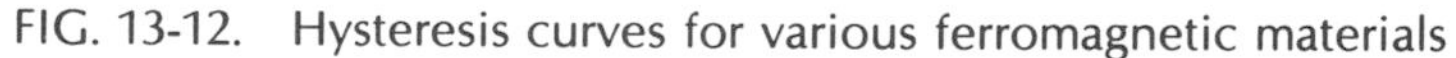
FIG. 13-12. Hysteresis curves for various ferromagnetic materials

because its residual magnetism is large, hard steel is very suitable for permanent magnets.

The third hysteresis loop, Figure 13-12(c), is that of a material known as *ferrite*. This is a *ceramic core* made up of iron oxides. The shape of the loop suggests a large hysteresis loss. However, it also shows that B tends to remain constant in one direction until the value of H is increased almost to its maximum level in the opposite direction; then B rapidly reverses. This is exactly the characteristic required for a *magnetic memory*, and consequently memories are one area of ferrite application.

13-9 EDDY CURRENTS

In Section 12-2 it was explained that a changing magnetic field induces a voltage in a conductor which is situated within the field. When the current direction in a coil is continually reversing, a ferromagnetic core in that coil constitutes a conductor in a changing magnetic field. Consequently, voltages are induced in the core, and circulating currents (or *eddy currents*) are caused to flow, as illustrated in Figure 13-13(a).

Eddy currents cause heating of the core and add considerably to the total core losses. Even when the core is nonmagnetic, eddy currents are generated if the material is an electrical conductor. To combat the eddy current losses, magnetic cores used with alternating current are always made up of thin sheets, termed *laminations* [see Figure 13-13(b)]. The surfaces of the laminations are varnished or otherwise thinly insulated on either side so that they offer a high resistance to the flow of circulating eddy currents. By this method the eddy current core losses are rendered negligible without affecting the magnetic performance of the core. Note

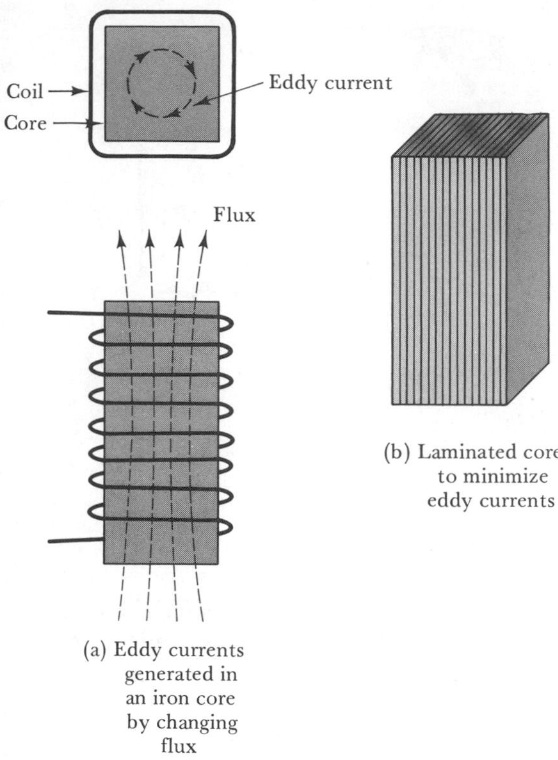

FIG. 13-13. Eddy currents and a laminated core

that the orientation of the laminations is such that they allow the setting up of the magnetic lines of force without difficulty. The varnished surface interface between laminations does not constitute an air gap in the magnetic path. The single negative effect of laminated cores is that the total cross-sectional area of the magnetic material is reduced by the total thickness of the insulation. This is usually taken into account by allowing an approximately 10% reduction in thickness of the magnetic core when making the magnetic calculations.

GLOSSARY OF FORMULAS

Reluctance:

$$\mathcal{R} = \frac{1}{\mu} \times \frac{l}{A}$$

$$\mathcal{R} = \frac{\mathcal{F}}{\Phi}$$

Permeability:

$$\mu = \frac{B}{H}$$

$$\mu_o = 4\pi \times 10^{-7}$$

$$\mu_r \mu_o = \frac{B}{H}$$

$$\Delta\mu = \frac{\Delta B}{\Delta H}$$

Energy stored in air gap:

$$W = \frac{B^2 A l}{2\mu_o}$$

Force between two magnetic surfaces:

$$F = \frac{B^2 A}{2\mu_o}$$

REVIEW QUESTIONS

13-1 Define reluctance and permeability and derive the equation relating the two quantities.

13-2 Show that the permeability of free space is $4\pi \times 10^{-7}$.

13-3 Using illustrations, explain why the length of a solenoid can be taken as the total length of the magnetic path in calculations of mmf, flux density, and so on. Also, discuss what occurs when a thin cylindrical iron core is brought close to one end of the solenoid.

13-4 Explain the term relative permeability and write the equation relating relative permeability to flux density and magnetic field strength.

13-5 Define fringing and leakage flux and discuss their effects on magnetic calculations.

13-6 Derive the equation for energy stored in an air gap and the equation for force between two magnetic surfaces.

13-7 Sketch a typical magnetization curve for an initially unmagnetized sample of ferromagnetic material and explain its shape. Also, show how the incremental permeability can be determined from the curve and discuss its importance in some applications.

13-8 Sketch a typical hysteresis loop for a sample of ferromagnetic material and explain its shape. Also, identify and define remanence and coercive force.

13-9 Sketch typical hysteresis loops for:

a. Soft iron.

b. Hard steel.

c. Ferrite.

In each case state the applications that the material is most suitable for and explain why.

13-10 Explain the origin of eddy currents in a magnetic core and discuss their effects and the method employed to combat them.

PROBLEMS

13-1 A toroidal coil has 1750 turns and carries a current of 5 A. The cross-sectional area of the coil is 2.25 cm^2 and the length of the magnetic path is 11 cm. Calculate the magnetic field strength, total flux, and flux density within the coil.

13-2 If the coil described in Problem 13-1 is to have a total flux of 0.2 μWb, determine the new level of current that must flow through the coil.

13-3 A 0.5-mm air gap has a cross-sectional area of 7 cm^2. Calculate the mmf required to generate a total flux of 50 μWb in the air gap.

13-4 An air-cored solenoid is 18 cm long and has an inside diameter of 1.7 cm. The coil has 1400 turns. Calculate the current that must flow to give a total flux of 0.1 μWb within the coil. Also, determine the flux density when the current is 150 mA.

13-5 Referring to the magnetization curves shown in Figure 13-3, calculate the permeability of:

a. Cast iron at a magnetic field strength of 2000 A/m.

b. Cast steel at 1000 A/m.

13-6 If the coil described in Problem 13-1 has a cast steel core, determine the value of total flux within the core when the current is 500 mA.

13-7 A cast iron ring has a cross-sectional area of 9 cm^2, and the length of the magnetic path is 25 cm. A coil of 5000 turns is wound on the ring and a current of 75 mA is passed through the coil. Calculate the magnetic field strength, the total flux, and the flux density within the core.

13-8 If the cast iron ring described in Problem 13-7 has a 2-mm air gap, calculate the new level of coil current required to establish the previously calculated level of flux density in the core.

13-9 The horseshoe-shaped magnet in Figure 13-7(b) has a magnetic path length of 50 cm and a cross-sectional area of 25 cm^2. The bar attached to the magnet's poles has a cross-sectional area of 35 cm^2 and the magnetic path through it is 15 cm long. Each air gap is approximately 5 mm thick. The magnet is cast iron and the bar is cast steel. Determine the current that must be passed through the 3000-turn coil to establish a total flux of 500 μWb around the circuit.

13-10 The magnetic core shown in Figure 13-5 has four 0.5-mm air gaps. If the core material has $\mu_r = 800$ and the coil has 1400 turns, determine the current required to give a total core flux of 700 μWb.

13-11 Assume that the magnetic core shown in Figure 13-6 has a 1-mm air gap halfway along the thickest section. The relative permeability of the material is 1000 and the flux density in the air gap is to be 0.75 T. Determine the number of coil turns required if the coil current is to be 500 mA.

13-12 In Problem 13-8, fringing at the 2-mm air gap tends to increase the cross-sectional area of the air gap by 10%. Recalculate the coil current required to establish the previously calculated level of flux density in the air gap.

13-13 The magnetic core shown in Figure 13-14 is constructed of sheet steel, and the coil has 900 turns. Calculate the coil current required to produce a flux of 330 μWb in the air gap.

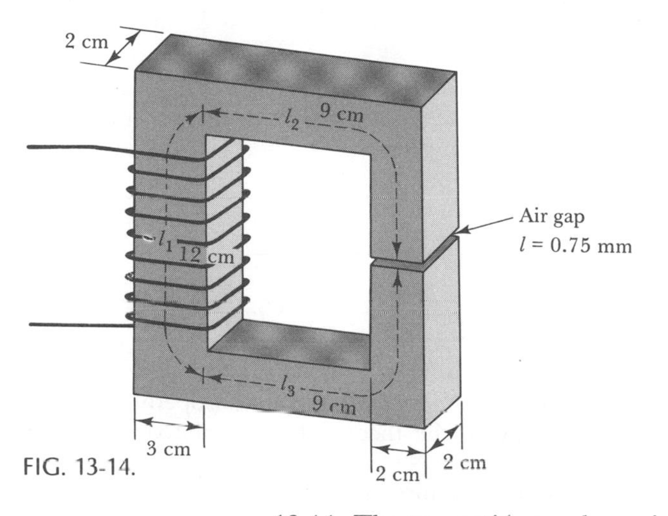

FIG. 13-14.

13-14 The magnetic core shown in Figure 13-8 is made of sheet steel and has the following dimensions: $l_1 = 8$ cm, $l_2 = l_3 = 15$ cm, cross-sectional area of l_1 path $= 5 \times 10^{-4}$ m^2, cross-sectional area of l_2 and l_3 paths $= 2 \times 10^{-4}$ m^2, and length of air gap $= 1.75$ mm. Calculate the number of coil turns required if the coil current is to be 5 mA and the flux density in the air gap is to be 1.2 T.

13-15 An electromagnet has two poles, each with a cross-sectional area of 35 cm^2. When the magnet is lifting a soft iron plate that completely covers the pole faces, the total flux around the magnetic circuit is 620 μWb. Determine the maximum weight of the iron plate that can be lifted.

13-16 For the electromagnet and steel bar described in Problem 13-9, calculate the maximum weight of steel bar that can be lifted.

14

DC MEASURING INSTRUMENTS

INTRODUCTION

The *permanent magnet moving coil* (PMMC) instrument is a basic deflection meter that may be connected to function as a *moving coil ammeter*, a *moving coil voltmeter*, or an *ohmmeter*. An understanding of the operation of the PMMC instrument is essential to the theory of more complex measurement methods.

It is possible to measure resistance by the use of an ammeter and voltmeter. However, certain errors can occur, and these dictate how the instruments should be connected. The *ohmmeter* provides a means of measuring medium-range resistance quickly, but not very accurately. For the measurement of very high resistance the *megger* is applicable, and precise resistance measurement is made on a *Wheatstone bridge*.

The electrodynamic instrument, which has similarities to a PMMC instrument, is most frequently applied for *power measurement*.

All dc instruments can be accurately calibrated by means of a *dc potentiometer*. The correct procedures for using electrical measuring instruments are described in Chapter 4.

14-1 PERMANENT MAGNET MOVING COIL INSTRUMENT

A moving coil instrument consists basically of a permanent magnet to provide a magnetic field, and a small lightweight coil pivoted within the field. When a current is passed through the winding of the coil, a torque is exerted on the coil by the interaction of the magnet's field and the field set up by the current in the coil. The resulting deflection of the coil is indicated by a pointer which moves over a calibrated scale. Figure 14-1 shows the basic construction of a *permanent magnet moving coil* (PMMC) instrument.

As well as a *deflecting force* provided by the coil current and the field from the permanent magnet, there is need for a *controlling force*. This is the force that returns the coil and pointer to the zero position when no current is flowing through the coil. The controlling force also *balances* the deflecting force, so that the pointer remains stationary for any constant value of current through the coil. The controlling force is usually provided by *spiral springs*, as illustrated in Figure 14-1. The springs are also employed as connecting leads for conducting current through the coil.

One other force, known as a *damping force*, is required for correct operation of a deflection instrument. When no damping force is present, the pointer swings above and below its final position on the scale for some time before settling down. In the case of the PMMC instrument, the damping force uses *eddy currents* (see Section 13-9). To facilitate this, the coil is wound on an aluminum former in which eddy currents are generated by any rapid movement of the coil in the magnetic field. The eddy currents set up a magnetic flux which opposes the original movement that generates them. Thus, oscillations of the meter pointer are *damped out*.

The amount of deflection of the pointer of a moving coil instrument is directly proportional to the current flowing through the coil. As shown in the sketch, the scale might be calibrated to indicate a maximum

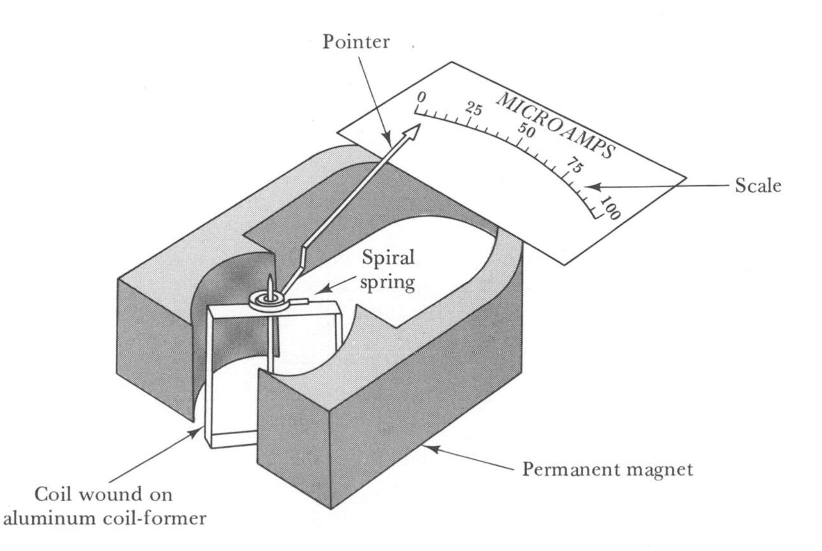

FIG. 14-1. Basic construction of a permanent magnet moving coil instrument

current of 100 μA. When the coil current is 50 μA, the pointer is at the halfway mark between zero and 100 μA. Therefore, the scale is calibrated to read 50 μA at that point. Similarly, at $\frac{1}{4}$ of full-scale deflection (FSD), the scale reads 25 μA, and at $\frac{3}{4}$ FSD the reading is 75 μA. The scale divisions for a given change of current are equal at all points on the scale, and therefore the scale is said to be *linear*. In certain other types of deflection instruments the scale is not linearly divided.

In the PMMC instrument it is important that the current flows through the coil in the correct direction. When the current is in the wrong direction, the magnetic field set up around the coil reacts with the field from the magnet in a way that tends to cause the pointer to deflect to the left of the zero position. Therefore, for correct (i.e., positive) deflection over the scale of the instrument, there is only one direction for coil current flow. As already discussed in Chapter 4, the terminals of the instrument are identified as + and − to show the polarity for correctly connecting the meter into a circuit.

14-2 MOVING COIL AMMETER

Since the deflection of a PMMC instrument is directly proportional to the current through its coil, the instrument is essentially an *ammeter*. However, the meter can be employed directly as an ammeter for only very small current levels. Consequently, some modifications must be made where it is desired to have an ammeter that measures larger currents.

The modified circuit for the ammeter is illustrated in Figure 14-2(a), where a resistor known as a *shunt* is shown connected in parallel with the PMMC instrument. Figure 14-2(b) shows the equivalent circuit of the complete instrument, including the coil resistance r_m. Some of the current

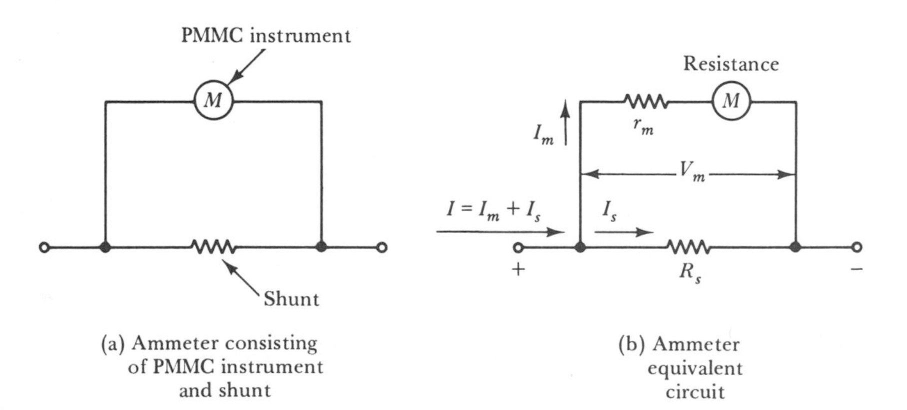

FIG. 14-2. An ammeter using a PMMC instrument and shunting resistor

to be measured passes through the instrument, and the rest of it passes through the shunt R_s. From a knowledge of the coil resistance and the FSD current of the instrument, the resistance of the shunt can be determined for any desired level of current to be measured. When the ammeter is designed to indicate a current of (for example) 100 A, the scale is recalibrated to read 100 A at FSD, and proportional levels at other points. The procedure for determining the shunt resistance value is demonstrated by Example 14-1.

EXAMPLE 14-1

A PMMC instrument has a coil resistance of 200 Ω and gives full-scale deflection for a current of 750 μA. Determine the value of shunt resistance required if the instrument is to be employed as an ammeter with a FSD of 1 A.

SOLUTION

From Figure 14-2(b):

$$\textit{meter voltage} = V_m = I_m \times r_m$$

at FSD,

$$I_m = 750\ \mu\text{A}$$

Therefore,

$$V_m = 750\ \mu\text{A} \times 200\ \Omega$$
$$= 150\ \text{mV}$$

and

$$\textit{meter voltage} = \textit{shunt voltage} = 150\ \text{mV}$$

Since

$$I = I_s + I_m$$
$$\textit{shunt current, } I_s = I - I_m$$
$$= 1\ \text{A} - 750\ \mu\text{A}$$
$$= 999.25\ \text{mA}$$

and

$$\textit{shunt resistance, } R_s = \frac{V_m}{I_s}$$
$$= \frac{150\ \text{mV}}{999.25\ \text{mA}}$$

Therefore,

$$\mathbf{R_s \cong 0.15\ \Omega}$$

EXAMPLE 14-2

The instrument described in Example 14-1 is to have its range changed to 100 mA at FSD. Calculate the new value of shunt resistor.

SOLUTION

$$V_m = I_m \times r_m$$
$$= 750\ \mu\text{A} \times 200\ \Omega$$
$$= 150\ \text{mV}$$

$$I_s = I - I_m$$
$$= 100\ \text{mA} - 750\ \mu\text{A}$$
$$= 99.25\ \text{mA}$$

$$R_s = \frac{V_m}{I_s}$$
$$= \frac{150\ \text{mV}}{99.25\ \text{mA}}$$
$$\boldsymbol{R_s \cong 1.51\ \Omega}$$

Note that *it is very important for an ammeter to have a very low resistance.* This is because the ammeter is always connected in series with the load which is to have its current measured (see Figure 4-1). If the ammeter resistance is not very much smaller than the load resistance, the load current can be substantially altered by the inclusion of the ammeter in the circuit. This is illustrated by Figure 14-3 and Example 14-3.

EXAMPLE 14-3

An ammeter having a resistance of 1 Ω is to be used to measure the current supplied to a 4-Ω resistor from a 100-V source. Calculate the current through the resistor before the ammeter is connected, and after it is included in the circuit.

SOLUTION

Without the ammeter:

$$I = \frac{E}{R_L} = \frac{100\ \text{V}}{4\ \Omega}$$

$$\boldsymbol{I = 25\ \text{A}}$$

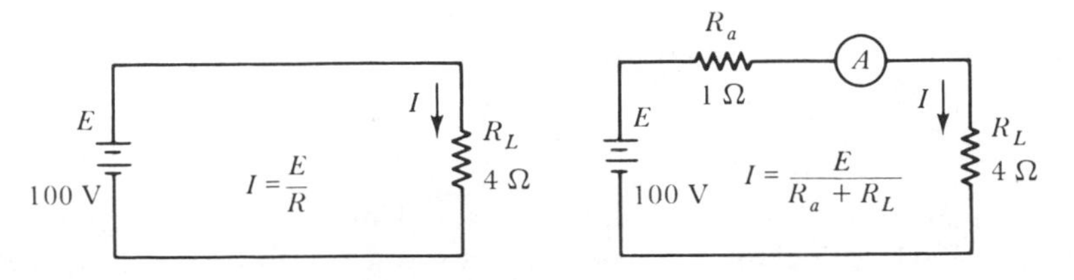

FIG. 14-3. Effect of ammeter resistance on circuit current

With the ammeter in circuit:

$$I = \frac{E}{R_L + R_a} = \frac{100\text{ V}}{4\ \Omega + 1\ \Omega}$$

$$\boldsymbol{I = 20\text{ A}}$$

A multirange ammeter can be constructed simply by employing several values of shunt resistor, with a rotary switch to select the desired range. Figure 14-4(a) shows the circuit arrangement. When an instrument is used in this fashion, care must be taken to ensure that the shunt does not become open-circuited, even for a brief instant. An ammeter is connected in series with the circuit in which current is to be measured. Consequently, if the shunt is open-circuited, a very large current may flow through the deflection instrument, possibly resulting in its destruction.

The *make-before-break* switch illustrated in Figure 14-4(b) protects an instrument from the possibility of the shunts becoming open-circuited in a multirange ammeter. The wide-ended moving contact connects to the next terminal to which it is being moved before it loses contact with the previous terminal. Thus, during the switching time there are two shunts in parallel with the instrument, and an open-circuited shunt is avoided.

The *Ayrton shunt* shown in Figure 14-5(a) is another method employed to protect the deflection instrument in an ammeter from the possibility of excessive current flow. When the moving contact of the rotary switch is connected to point B, the resistance of the shunt in parallel with the instrument is $(R_1 + R_2 + R_3)$. This is illustrated in Figure 14-5(b). When the moving contact is switched to point C [Figure 14-5(c)] the shunt becomes $(R_1 + R_2)$, and resistor R_3 is now in series with the meter. Finally, with the moving contact at point D [Figure 14-5(d)], R_1 is the shunt and

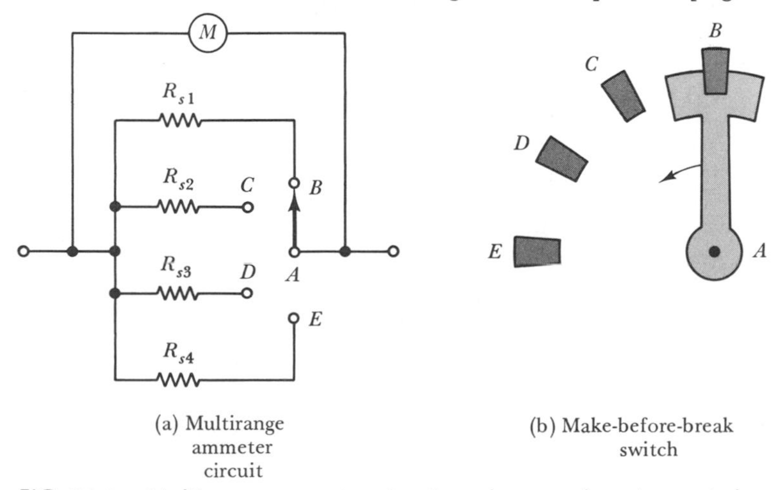

(a) Multirange ammeter circuit

(b) Make-before-break switch

FIG. 14-4. Multirange ammeter circuit and range-changing switch

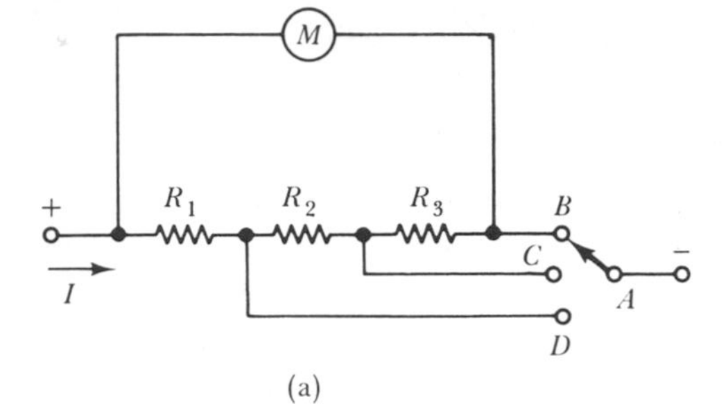

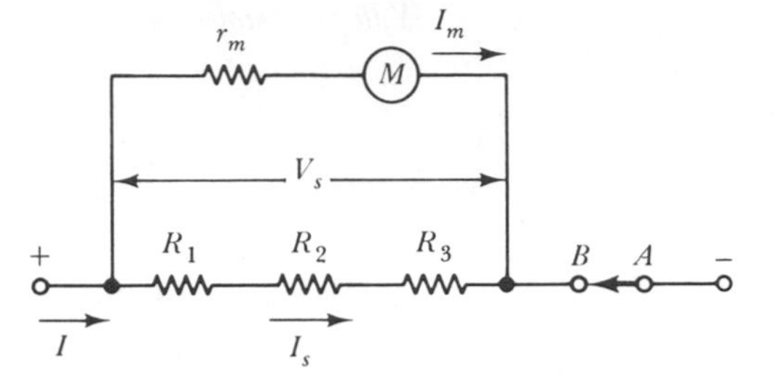

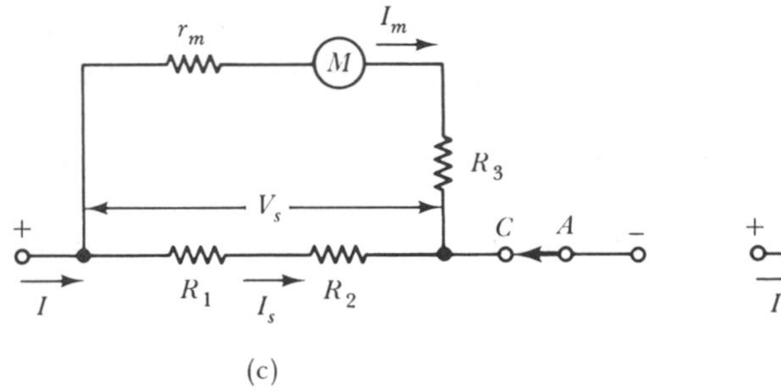

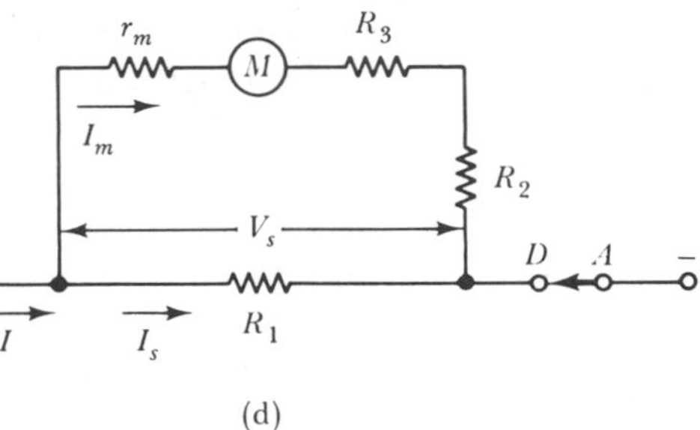

FIG. 14-5. Multirange ammeter using an Ayrton shunt

(R_2+R_3) is in series with the meter. Note that there is a shunt in parallel with the instrument at all times.

EXAMPLE 14-4 An ammeter using an Ayrton shunt as illustrated in Figure 14-5 has $R_1=R_2=R_3=0.0334\ \Omega$, and the meter resistance is 50 Ω. The instrument gives FSD when 200 μA passes through the moving coil. Calculate the ammeter full-scale range for each of the switched positions.

SOLUTION

Switch at point B:

$$V_s = I_m \times r_m$$
$$= 200\ \mu\text{A} \times 50\ \Omega$$
$$= 10\ \text{mV}$$

$$I_s = \frac{V_s}{R_1+R_2+R_3}$$
$$= \frac{10\ \text{mV}}{0.0334\ \Omega + 0.0334\ \Omega + 0.0334\ \Omega}$$
$$= 99.8\ \text{mA}$$

$$I = I_s + I_m$$
$$= 99.8\ \text{mA} + 200\ \mu\text{A}$$
$$\boldsymbol{I = 100\ \text{mA} = \textit{range of meter}}$$

Switch at point C:

$$V_s = I_m \times (r_m + R_3)$$
$$= 200\ \mu\text{A} \times (50\ \Omega + 0.0334\ \Omega)$$
$$= 10.007\ \text{mV}$$

$$I_s = \frac{V_s}{R_1 + R_2}$$
$$= \frac{10.007\ \text{mV}}{0.0334\ \Omega + 0.0334\ \Omega}$$
$$= 149.8\ \text{mA}$$
$$I = I_s + I_m$$
$$= 149.8\ \text{mA} + 200\ \mu\text{A}$$
$$\boldsymbol{I = 150\ \text{mA} = \textit{range of meter}}$$

Switch at point D:

$$V_s = I_m \times (r_m + R_3 + R_2)$$
$$= 200\ \mu\text{A} \times (50\ \Omega + 0.0334\ \Omega + 0.0334\ \Omega)$$
$$= 10.013\ \text{mV}$$

$$I_s = \frac{V_s}{R_1}$$
$$= \frac{10.013\ \text{mV}}{0.0334\ \Omega}$$
$$= 299.8\ \text{mA}$$
$$I = I_s + I_m$$
$$= 299.8\ \text{mA} + 200\ \mu\text{A}$$
$$\boldsymbol{I = 300\ \text{mA} = \textit{range of meter}}$$

14-3 MOVING COIL VOLTMETER

The deflection of a PMMC instrument is proportional to the current through its coil, and the current is, of course, proportional to the voltage across the coil resistance. Therefore, the scale of the instrument could be calibrated to indicate the applied voltage. Since the coil resistance is usually quite small, this arrangement would result in a voltmeter that could measure only very low voltage levels. To increase the voltage range

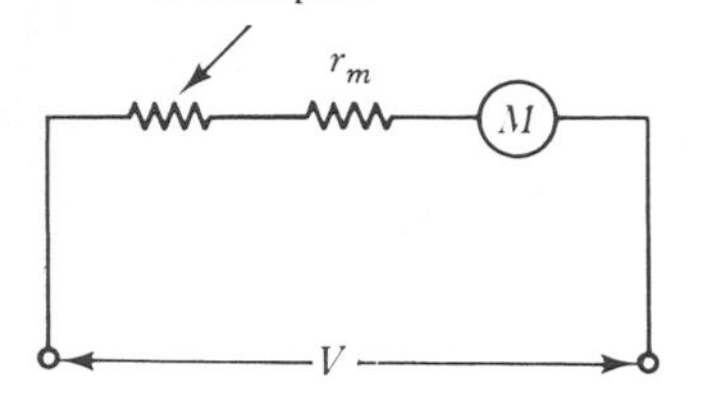

(a) Single-range voltmeter

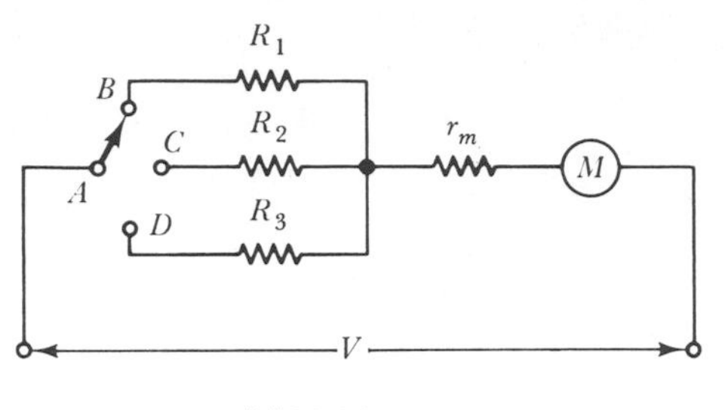

(b) Multirange voltmeter

FIG. 14-6. Voltmeter circuits

the voltmeter resistance must be increased, and this is easily done by connecting a resistor in series with the instrument [see Figure 14-6(a)]. Because the series resistor increases the range of the voltmeter, it is termed a *multiplier resistance.*

The circuit of a multirange voltmeter is shown in Figure 14-6(b). Any one of several multiplier resistors is selected by means of a rotary switch, as illustrated. Unlike the case of the ammeter, the rotary switch used with the voltmeter should be a *break-before-make* type; that is, the moving contact should disconnect from one terminal before connecting to the next terminal. Also, unlike an ammeter, a voltmeter should have a very high resistance. This is because it is normally connected *in parallel* with the circuit where the voltage is to be measured. To avoid altering the circuit conditions, the voltmeter should draw a very small current. Figure 14-7 and Example 14-6 illustrate the situation where a low-resistance voltmeter could substantially affect a circuit.

EXAMPLE 14-5

A PMMC meter having a coil resistance of 100 Ω and an FSD current of 50 μA is to be used as a voltmeter having ranges of 100 V, 50 V, and 10 V. Determine the required values of multiplier resistor for each range.

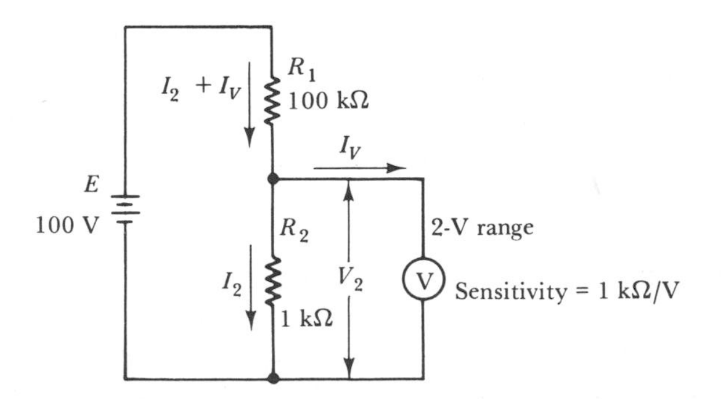

FIG. 14-7. Voltmeter effect on circuit conditions

SOLUTION

For the 100-V *range*:

$$I = \frac{E}{R_1 + r_m}$$

Therefore,

$$R_1 = \frac{E}{I} - r_m$$

$$= \frac{100\text{ V}}{50\ \mu\text{A}} - 100\ \Omega$$

$$\mathbf{R_1 = 1.999\ 9\ M\Omega}$$

For the 50-V *range*:

$$R_2 = \frac{50\text{ V}}{50\ \mu\text{A}} - 100\ \Omega$$

$$\mathbf{R_2 = 0.999\ 9\ M\Omega}$$

For the 10-V *range*:

$$R_3 = \frac{10\text{ V}}{50\ \mu\text{A}} - 100\ \Omega$$

$$\mathbf{R_3 = 199.9\ k\Omega}$$

In Example 14-5, the total voltmeter resistance is

$$R_V = R \text{ of multiplier} + r_m$$

For the 100-V range this adds up to 2 MΩ. Thus, the voltmeter *sensitivity* or *resistance per volt* is

$$\text{resistance per volt} = \frac{2\text{ M}\Omega}{100\text{ V}}$$

$$= 20\text{ k}\Omega/\text{V}$$

Calculating for the 50-V range and 10-V range gives the same 20 kΩ/V constant. The voltmeter sensitivity is an important constant and is usually printed on the face of all voltmeters. To find the total voltmeter resistance, the resistance per volt is multiplied by the voltmeter range. For example, on a 50-V range with 20 kΩ/V, the total voltmeter resistance is

$$R_V = (20\text{ k}\Omega/\text{V}) \times (50\text{ V})$$

$$= 1\text{ M}\Omega$$

EXAMPLE 14-6

For the circuit shown in Figure 14-7, determine the voltage across resistance R_2:

a. Without the voltmeter in the circuit.

b. With the voltmeter connected.

SOLUTION

a. *Without the voltmeter*:

$$V_2 = \frac{E \times R_2}{R_1 + R_2}$$

$$= \frac{100 \text{ V} \times 1 \text{ k}\Omega}{100 \text{ k}\Omega + 1 \text{ k}\Omega}$$

$$\mathbf{V_2 \cong 0.99\ V}$$

b. *With the voltmeter*:

$$\textit{voltmeter resistance } R_v = (\text{range}) \times (\text{sensitivity}) = (1 \text{ k}\Omega/\text{V}) \times (2 \text{ V})$$

$$= 2 \text{ k}\Omega$$

$$R_v \| R_2 = 2 \text{ k}\Omega \| 1 \text{ k}\Omega$$

$$= 666.7\ \Omega$$

$$V_2 = \frac{E \times (R_v \| R_2)}{R_1 + (R_v \| R_2)}$$

$$= \frac{100 \text{ V} \times 666.7\ \Omega}{100 \text{ k}\Omega + 666.7\ \Omega}$$

$$\mathbf{V_2 \cong 0.66\ V}$$

14-4 MEASURING RESISTANCE BY AMMETER AND VOLTMETER

If the voltage across a resistor and the current flowing through it are measured, the resistance value can be calculated by applying Ohm's law. However, an error occurs depending upon how the ammeter and voltmeter are connected, and this error may be insignificantly small or quite large.

Consider the arrangement shown in Figure 14-8(a). Since the voltmeter is connected directly across R_L, it measures the actual load voltage. However, the ammeter measures the load current I_L as well as the current I_V that flows through the voltmeter.

$$\text{measured resistance} = \frac{\text{voltmeter reading}}{\text{ammeter reading}}$$

$$\boxed{R = \frac{V_L}{I_L + I_V}} \qquad \textbf{(14-1)}$$

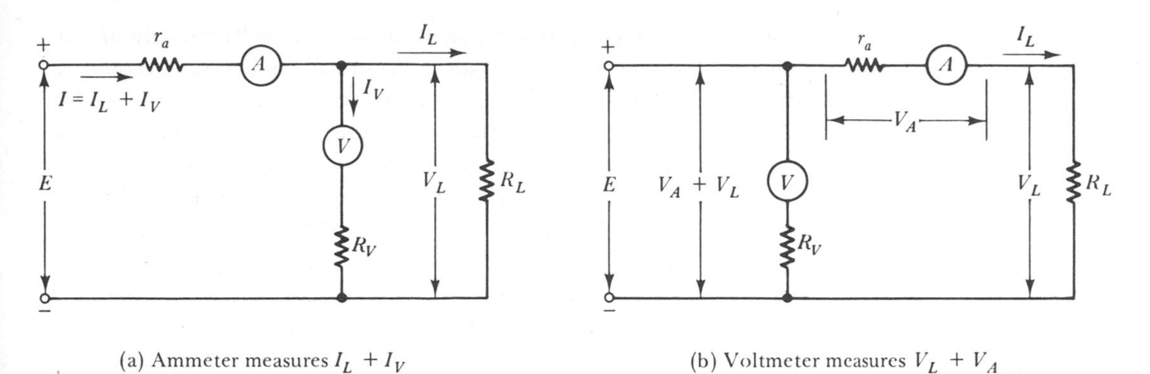

FIG. 14-8. Error sources when measuring resistance by ammeter and voltmeter

The actual resistance of R_L is

$$R_L = \frac{V_L}{I_L}$$

Therefore, the presence of I_V in the equation introduces an error into the result. If I_V is very much smaller than I_L, the error may be insignificant. Obviously, this requires that I_L be a large current, which is the case when R_L is a small resistance.

Now consider the arrangement in Figure 14-8(b), where R_L and the ammeter are in series and the voltmeter is connected in parallel with them. In this case the ammeter measures the actual current through R_L, but the voltmeter measures the voltage across the ammeter plus the voltage across R_L. Therefore,

$$\text{measured resistance} = \frac{\text{voltmeter reading}}{\text{ammeter reading}}$$

$$\boxed{R = \frac{V_L + V_A}{I_L}} \tag{14-2}$$

Again, the actual resistance of R_L is

$$R_L = \frac{V_L}{I_L}$$

Consequently, the presence of V_A in the equation produces an error in the result. If V_A is very much smaller than V_L, the error could be insignificant. This requires that V_L be a large voltage, which means that R_L should be a large resistance.

It is seen that where R_L is a small resistance, the voltmeter should be connected directly across the resistance. Where R_L is a large resistance, the ammeter should be connected directly in series with the resistance for greatest accuracy. The correct arrangement can be easily determined by first connecting the ammeter in series with R_L, then observing the ammeter reading with the voltmeter temporarily connected directly across R_L [i.e., connected as in Figure 14-8(a)]. If the ammeter reading is not noticeably altered when the voltmeter is connected, the readings will give an accurate result. When the ammeter reading is noticeably changed by the voltmeter, the voltmeter should be moved to the other side of the ammeter.

EXAMPLE 14-7

An ammeter and voltmeter when connected as in Figure 14-8(a) indicate 10 A and 99 V, respectively. When the voltmeter is changed to reconnect the circuit as in Figure 14-8(b), the readings become 10 A and 100 V. The ammeter has a resistance of 0.1 Ω. The voltmeter is on its 100-V range and its sensitivity is 20 kΩ/V. Calculate the value of measured resistance for each case and determine which arrangements give the most accurate result.

SOLUTION

For the circuit of Figure 14-8(a):

$$R=\frac{V_L}{I_L+I_V}=\frac{99\text{ V}}{10\text{ A}}$$

$$\mathbf{R=9.9\ \Omega}$$

and *voltmeter resistance*, $R_V=20\text{ k}\Omega/\text{V}\times(\textit{voltmeter range})$

$$=20\text{ k}\Omega/\text{V}\times(100\text{ V})$$

$$\mathbf{R_V=2\ M\Omega}$$

Therefore,

$$I_V=\frac{V_L}{R_V}=\frac{99\text{ V}}{2\text{ M}\Omega}$$

$$\mathbf{I_V=49.5\ \mu A}$$

$$I_V \ll I_L$$

For the circuit of Figure 14-8(b):

$$R=\frac{V_L+V_A}{I_L}=\frac{100\text{ V}}{10\text{ A}}$$

$$\mathbf{R=10\ \Omega}$$

and

$$V_a = I_L \times r_a$$
$$= 10\ A \times 0.1\ \Omega$$
$$V_a = 1\ \text{V}$$

In this case V_A *is* not *very much smaller than* V_L, *and connection as in* Figure 14-8(a) *gives the most accurate result.*

14-5 OHMMETER

The *ohmmeter* provides a quick, but not very accurate, means of resistance measurement. The simplest direct reading ohmmeter is the basic *series ohmmeter* circuit illustrated in Figure 14-9(a). The circuit consists of a moving coil instrument in series with a battery and adjustable resistor R_1, as shown.

With terminals A and B shorted together, R_1 is adjusted for full-scale deflection (FSD) on the instrument. A and B are the ohmmeter terminals and when they are short-circuited, the ohmmeter should read zero resistance. Thus, as shown in Figure 14-9(b), FSD is taken as an indication of zero ohms. When terminals A and B are open-circuited, the pointer should indicate infinity. Therefore, the zero deflection point on the dial is marked as infinite resistance [see Figure 14-9(b)].

When an unknown resistance R_x is connected to terminals A and B, some current flows through the meter, giving a reading between zero and infinity. As will be seen, for a given value of R_x the pointer position on the dial depends upon the value of R_1.

From Figure 14-9(a) it is seen that the meter current is

$$I_m = \frac{E_b}{R_1 + R_x + r_m} \tag{14-3}$$

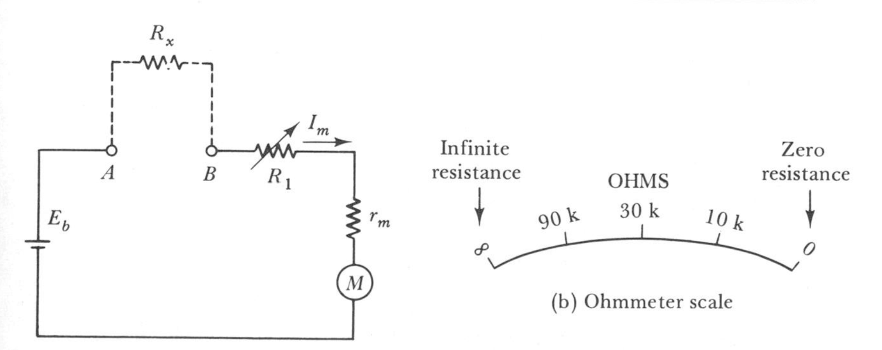

(a) Simple ohmmeter circuit

(b) Ohmmeter scale

FIG. 14-9. Simple ohmmeter circuit and scale

where r_m is the coil resistance of the instrument. If $r_m \ll R_1$, the equation is simplified to

$$\boxed{I_m \cong \frac{E_b}{R_1 + R_x}} \tag{14-4}$$

EXAMPLE 14-8

The series ohmmeter shown in Figure 14-9 is made up of a 3-V battery, a 100-μA meter, and a resistance R_1 which has a fixed value of 30 kΩ. Determine the value of the unknown resistance R_x when the pointer indicates:

a. $\frac{1}{2}$ FSD.

b. $\frac{1}{4}$ FSD.

c. $\frac{3}{4}$ FSD.

SOLUTION

Eq. (14-4):

$$I_m \cong \frac{E_b}{R_1 + R_x}$$

When $R_x = 0$:

$$I_m = \frac{3 \text{ V}}{30 \text{ k}\Omega}$$

$$\mathbf{I_m = 100\ \mu A = FSD}$$

a. *At* $\frac{1}{2}$ FSD:

$$I_m = \frac{100\ \mu\text{A}}{2} = 50\ \mu\text{A}$$

From Eq. (14-4):

$$R_1 + R_x = \frac{E_b}{I_m}$$

Therefore,

$$R_1 + R_x = \frac{3 \text{ V}}{50\ \mu\text{A}}$$

$$R_1 + R_x = 60 \text{ k}\Omega$$

and $$R_x = 60\ \text{k}\Omega - R_1$$
$$= 60\ \text{k}\Omega - 30\ \text{k}\Omega$$
$$\mathbf{R_x = 30\ k\Omega}$$

b. *At* $\frac{1}{4}$ FSD:

$$I_m = \frac{100\ \mu\text{A}}{4} = 25\ \mu\text{A}$$

Therefore, $$R_1 + R_x = \frac{3\ \text{V}}{25\ \mu\text{A}}$$
$$= 120\ \text{k}\Omega$$

and $$R_x = 120\ \text{k}\Omega - 30\ \text{k}\Omega$$
$$\mathbf{R_x = 90\ k\Omega}$$

c. *At* $\frac{3}{4}$ FSD:

$$I_m = \tfrac{3}{4} \times 100\ \mu\text{A} = 75\ \mu\text{A}$$

Therefore, $$R_1 + R_x = \frac{3\ \text{V}}{75\ \mu\text{A}}$$
$$= 40\ \text{k}\Omega$$

and $$R_x = 40\ \text{k}\Omega - 30\ \text{k}\Omega$$
$$\mathbf{R_x = 10\ k\Omega}$$

The ohmmeter scale is now as marked in Figure 14-9(b).

The results of Example 14-8 demonstrate that the ohmmeter scale is nonlinear. Also, the portion of the scale from $\frac{3}{4}$ FSD to FSD includes all resistance measurements from 10 kΩ to zero, while the portion from zero deflection to $\frac{1}{4}$ FSD includes all values from 90 kΩ to infinity. This shows that the useful range of the ohmmeter scale is approximately from $\frac{1}{4}$ FSD to $\frac{3}{4}$ FSD. The actual resistance values marked upon the scale depend upon the value of R_1, which (instead of being variable) should be a fixed-value precision resistor. Where $R_1 = 30$ kΩ, as in Example 14-8, the resistance value at $\frac{1}{2}$ FSD is 30 kΩ. If R_1 were changed to (for example) 100 kΩ, the scale must be marked 100 kΩ at $\frac{1}{2}$ FSD (E_b would also have to be changed).

The circuit of Figure 14-9(a) relies upon the battery voltage remaining absolutely constant. When the battery terminal voltage falls (as they all do with use), the instrument scale is no longer accurate. Thus, some

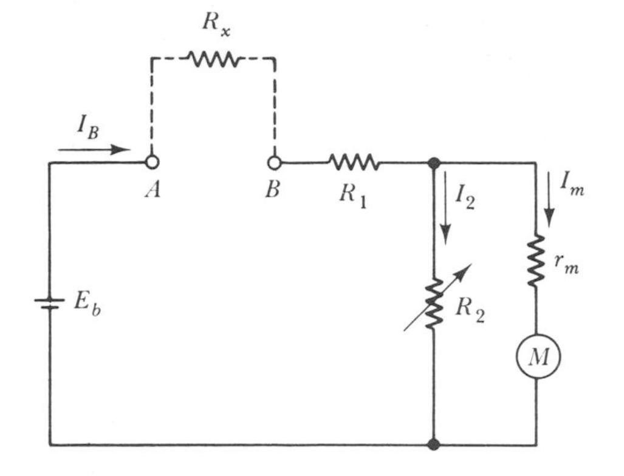

FIG. 14-10. Ohmmeter circuit with zero control

means of adjusting for battery voltage variations must be built into the circuit. Variable resistor R_2 shown in the circuit of Figure 14-10 provides the necessary adjustment.

The total current from the battery is

$$\boxed{I_B = \frac{E_b}{R_1 + R_x + (r_m \| R_2)}} \tag{14-5}$$

and if $(r_m \| R_2) \ll R_1$, then, once again,

$$\boxed{I_B \cong \frac{E_b}{R_1 + R_x}} \tag{14-6}$$

The total current supplied by the battery divides between R_2 and r_m,

$$\boxed{I_B = I_2 + I_m} \tag{14-7}$$

and this must be taken into account when determining the scale readings of the ohmmeter. This is best demonstrated by an example (see Example 14-9).

Figure 14-11 shows the circuit of a multirange ohmmeter. The range is set by selecting one of several different standard resistors, as illustrated.

EXAMPLE 14-9

The ohmmeter circuit in Figure 14-10 is made up of a 3-V battery, a 50-μA meter, a resistance $R_1 = 30$ kΩ, and an adjustable resistor R_2 which can be set to 20 Ω. The meter resistance r_m is also 20 Ω. Determine the meter indication when $R_x = 0$ and when $R_x = 30$ kΩ. Also determine the

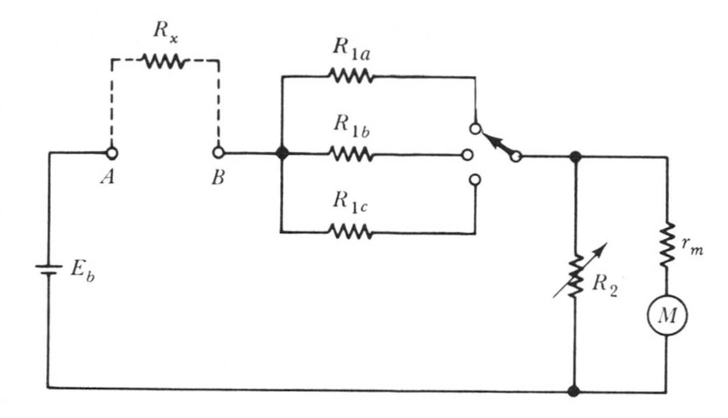

FIG. 14-11. Multirange ohmmeter circuit

new value for R_2 when E_B falls to 2.5 V, and show that the ohmmeter scale is unaffected.

SOLUTION

For $E_b = 3$ V:

Eq. (14-6):

$$I_B \simeq \frac{E_b}{R_1 + R_x}$$

When $R_x = 0$,

$$I_B = \frac{3\text{ V}}{30\text{ k}\Omega}$$

$$\mathbf{I_B = 100\ \mu A}$$

When $R_2 = R_m$, I_B *divides equally between the two. Therefore,*

$$I_m = I_2 = \frac{I_B}{2} = \frac{100\ \mu\text{A}}{2}$$

$$\mathbf{I_m = 50\ \mu A = FSD}$$

When $R_x = R_1 = 30$ kΩ,

$$I_B = \frac{E_b}{R_1 + R_x} = \frac{3\text{ V}}{30\text{ k}\Omega + 30\text{ k}\Omega}$$

$$\mathbf{I_B = 50\ \mu A}$$

and

$$I_m = I_2 = \frac{I_B}{2} = \frac{50\ \mu\text{A}}{2}$$

$$\mathbf{I_m = 25\ \mu A = \tfrac{1}{2}\ FSD}$$

When $E_b = 2.5$ V, terminals A and B are shorted, and R_2 is adjusted to give FSD (i.e., zero ohms) on the instrument.

$$\text{meter voltage, } V_m = I_m \times r_m$$
$$= 50\ \mu\text{A} \times 20\ \Omega$$
$$\mathbf{V_m = 1\ mV}$$

$$I_B \cong \frac{E_b}{R_1 + R_x} = \frac{2.5\ \text{V}}{30\ \text{k}\Omega}$$
$$\mathbf{I_B = 83.3\ \mu A}$$

Since

$$I_B = I_m + I_2$$
$$I_2 = I_B - I_m$$
$$= 83.3\ \mu\text{A} - 50\ \mu\text{A}$$
$$\mathbf{I_2 = 33.3\ \mu A}$$

$$R_2 = \frac{V_m}{I_2} = \frac{1\ \text{mV}}{33.3\ \mu\text{A}}$$
$$\mathbf{R_2 = 30\ \Omega}$$

When $E_b = 2.5$ V and $R_x = 30$ kΩ,

$$I_B \cong \frac{E_b}{R_1 + R_x} = \frac{2.5}{30\ \text{k}\Omega + 30\ \text{k}\Omega}$$
$$\mathbf{I_B = 41.7\ \mu A}$$

Using the current divider rule,

$$I_m = \frac{I_B \times R_2}{R_2 + R_m} = \frac{41.7\ \mu\text{A} \times 30\ \Omega}{30\ \Omega + 20\ \Omega}$$
$$\mathbf{I_m = 25\ \mu A = \tfrac{1}{2}\ FSD}$$

It is seen that when R_2 is adjusted from 20 Ω to 30 Ω, the ohmmeter scale remains correct when the battery voltage has fallen from 3 V to 2.5 V.

14-6 MEGGER

The *megger* (or *megohmmeter*) is an instrument for measuring very high resistances, such as the insulation resistance of electric cables. Before considering how a megger operates, it is useful to first look at high-resis-

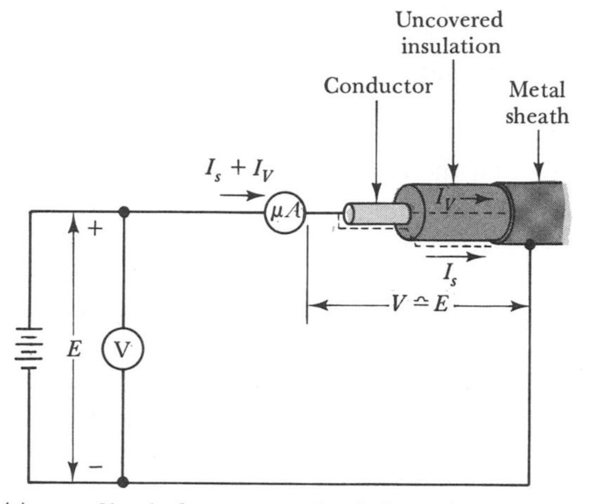

(a) Circuit that measures insulation volume resistance in parallel with surface leakage resistance

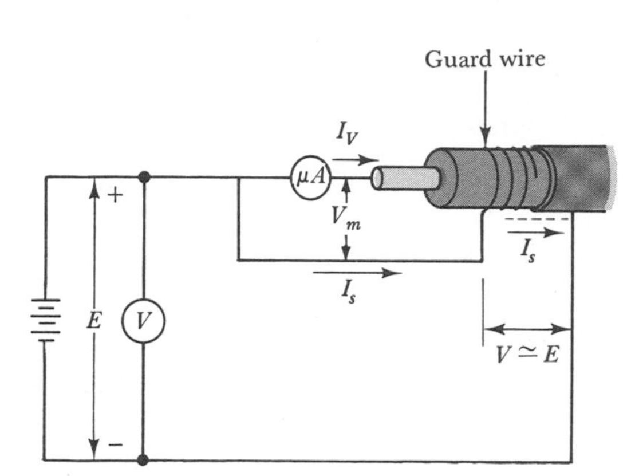

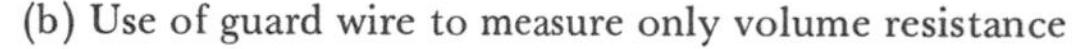

(b) Use of guard wire to measure only volume resistance

FIG. 14-12. Measurement of insulation resistance

tance measurement by means of an ammeter and voltmeter, as illustrated in Figure 14-12. The arrangement shown measures the resistance of the insulation between a conductor and an outer metal sheath on a cable.

Since insulation resistance is normally very high, a high-voltage source (perhaps 1000 V or greater) is required in order to pass a measurable current through the insulation. Despite the high voltage, only an extremely low current is likely to flow; consequently, an ammeter capable of measuring microamps is required. As illustrated in Figure 14-12(a), the voltmeter must be connected on the *supply side* of the ammeter so that the ammeter does not measure the voltmeter current as well as the insulation current. The resistance is simply calculated as

$$R = \frac{E}{I} = \frac{\text{voltmeter readings}}{\mu\text{A meter readings}}$$

Referring again to Figure 14-12(a), note that the current registered by the microammeter is $(I_v + I_s)$, where I_v is the current that passes through the volume of the insulation and I_s is the *surface leakage current*. Therefore, the resistance calculated is the combination of the *volume resistance* and the *surface leakage resisistance* of the insulation. For most practical purposes, this combined surface and volume resistance is the effective resistance of the insulation. However, in some circumstances the volume resistance and the surface leakage resistance must be separated. This is achieved by the use of the *guard-wire method* (also known as the *guard-ring method*), illustrated in Figure 14-12(b). The guard wire is simply a bare copper wire wrapped around the uncovered insulation of the cable and connected to the supply side of the microammeter. It is seen that the surface leakage current I_s now bypasses the microammeter, so the instrument registers only the insulation volume current. Note that there is no significant surface leakage between the cable conductor and the guard wire. This is because the potential difference between them is only the voltage drop across the meter, whereas the potential difference between the guard wire and the metal sheath is the applied voltage E.

The megger is a deflection instrument, provided with a source of high voltage, usually from a hand-driven generator built into the instrument (see Figure 14-13). The generated voltage may be anything from 100 V to 2.5 kV. The deflection movement is a permanent magnet moving coil system with *two coils*. There is no mechanical controlling force. Instead, the coils are connected to oppose each other, so that one becomes a *control coil* while the other is a *deflection coil*.

The control coil is supplied via a standard resistor (R), so the controlling force is proportional to the generator voltage divided by the standard resistor. The deflection coil is supplied via the resistance to be measured (R_x), so the deflecting force is proportional to the supply voltage divided by the unknown resistance. Thus, deflection is proportional to the difference between the standard and the unknown resistances, and the instrument can be calibrated to read directly in the resistance of the unknown.

When measuring the resistance of an open circuit, there is no current flowing in the deflection coil. Therefore, the force from the control coil

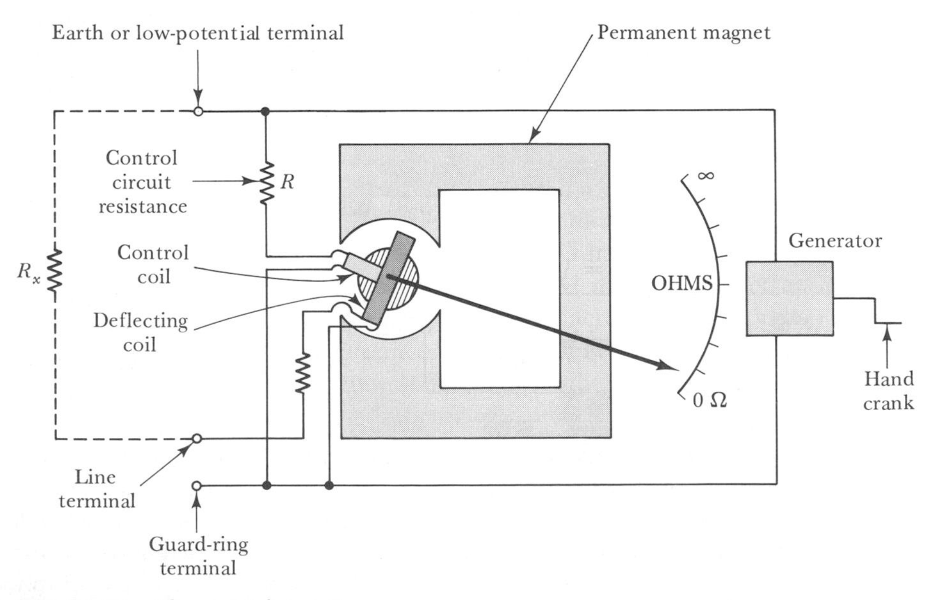

FIG. 14-13. The megohmmeter

causes the pointer to be deflected to one end of the scale, and this is marked *infinity* ∞. When measuring a short circuit, the force from the deflecting coil is very much greater than that from the control coil. Consequently, the pointer is deflected to the opposite end of the scale from infinity, and this (opposite) end is marked 0 Ω (see Figure 14-13). When the deflecting coil force and the control coil force are equal, the pointer becomes stationary at the center of the scale, and the unknown resistance is then equal to the resistance in the control circuit. The scale is thus marked accordingly, and the range of the instrument can be altered by switching different values of standard resistors into the control circuit. Note that the megger is provided with a *guard-ring* terminal, to facilitate measurement of volume resistance.

14-7 POWER MEASUREMENT

In a dc circuit, the power supplied to a load can be determined by measuring the load voltage and current and multiplying them together:

$$P = EI$$

However, it is much more convenient to have an instrument that indicates power directly. The meter used for this purpose is called a *wattmeter*, and the instrument that can be applied as a wattmeter is known as a *dynamometer*, or sometimes as an *electrodynamic instrument*. The construction of a dynamometer instrument to some extent resembles the permanent magnet moving coil instrument. The major change from the PMMC construction is that the permanent magnet is replaced by two coils, as illustrated in Figure 14-14(a). (Compare this to Figure 14-1.) The magnetic field in which the lightweight moving coil is situated is generated by passing current through the stationary *field coils*. Then, when a current is passed through the moving coil, the coil and the meter pointer are deflected.

The deflection of the pointer of a dynamometer instrument is proportional to the current through the moving coil, but it is also proportional to the flux density of the magnetic field set up by the stationary coils. This means, of course, that deflection is also proportional to the current through the field coils.

Consider the arrangement shown in Figure 14-14(b). The moving coil of the instrument has a series resistor and is connected in parallel with the load that is to have its power measured. Consequently, current I_V through the moving coil is directly proportional to the load voltage. The field coils are connected in series with the load, so the current flowing through them is $(I_V + I_L)$, as shown. If $I_V \ll I_L$, I_V can be neglected and the field coil current assumed to be approximately equal to I_L. Because the meter deflection is proportional to the field coil current and to the moving coil

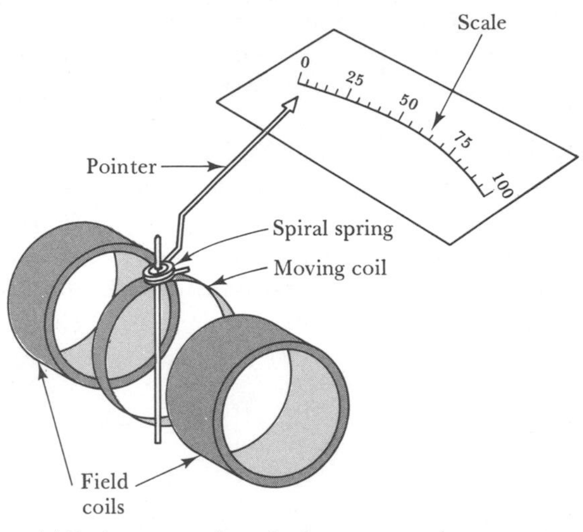

(a) Basic construction of a dynamometer instrument

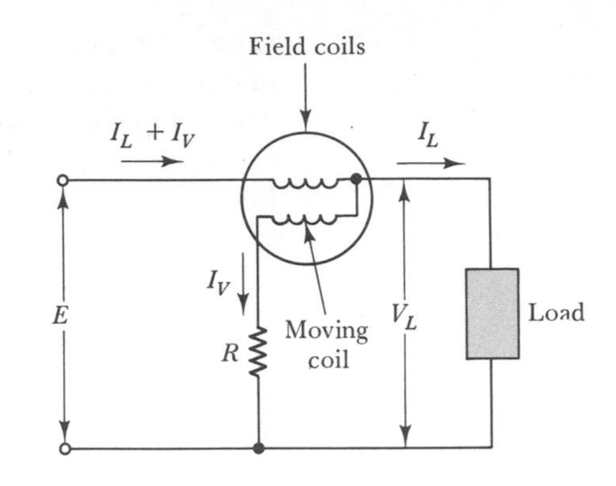

(b) Dynamometer wattmeter

FIG. 14-14. Dynamometer instrument construction and application as a wattmeter

current,

$$\text{deflection} \propto V_L \times I_L$$

or

$$\text{deflection} \propto P$$

The scale of the instrument can be calibrated to indicate watts, and thus it becomes a *dynamometer wattmeter*.

The dynamometer instrument can also be employed as a voltmeter or an ammeter. For use as a voltmeter, the field coils are connected in series with the moving coil and a multiplier resistor. For ammeter applications the field coils are connected in parallel with the moving coil, and in parallel with a shunt, as required.

Because fairly large currents are required to set up the necessary field flux, the dynamometer instrument is not as sensitive as a PMMC instrument. Consequently, its major application is as a wattmeter. One advantage that the dynamometer has over a PMMC instrument is that it can be used for both direct or alternating current/voltage measurements. This is discussed further in Chapter 26.

14-8 WHEATSTONE BRIDGE

Like the ohmmeter, the *Wheatstone bridge* is employed for measurement of resistance. But *unlike* the ohmmeter, the Wheatstone bridge measures resistance with a high degree of accuracy.

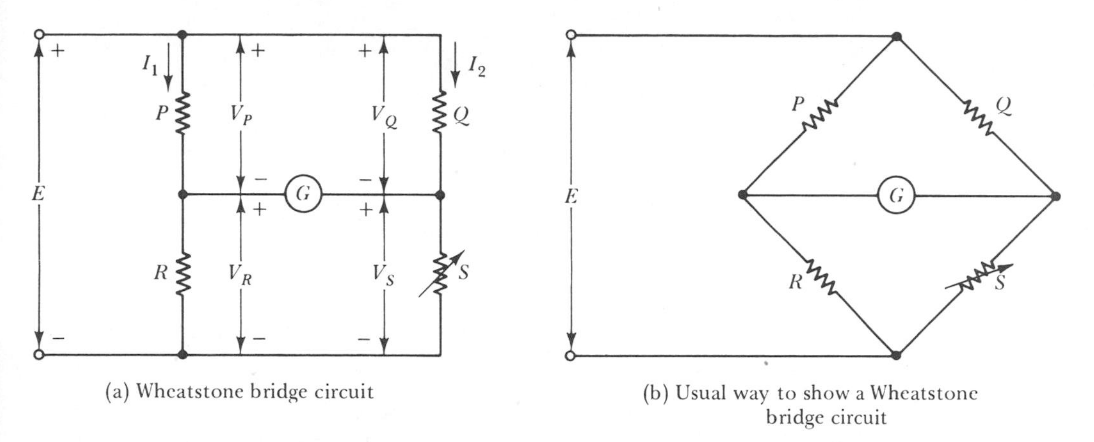

(a) Wheatstone bridge circuit

(b) Usual way to show a Wheatstone bridge circuit

FIG. 14-15. The Wheatstone bridge

The circuit of a Wheatstone bridge is shown in Figure 14-15(a). P and Q are precision resistors, S is an adjustable precision resistor, R is the unknown resistance that is to be measured, and G is a *galvanometer*. A galvanometer is essentially a PMMC instrument which has its pointer indicating zero at the center of its scale. It is also an extremely sensitive instrument, so very small currents through its coil will cause it to deflect either to the left or to the right of zero. As applied in the Wheatstone bridge, the galvanometer is used to detect the zero-current condition. Consequently, it is usually termed a *null detector*.

In Figure 14-15(b), the Wheatstone bridge circuit is drawn in a slightly different form. This is the way the circuit is most frequently represented; however, although it may look different, it does not differ in any way from the circuit shown in Figure 14-15(a).

Referring again to Figure 14-15(a), when the galvanometer indicates a null condition, the voltages on each side of the galvanometer must be equal. Also, since there is no current flowing through the galvanometer, current I_1 flows through P and R, and current I_2 flows through Q and S. Therefore,

$$V_R = V_S$$

or

$$I_1 R = I_2 S \tag{1}$$

also

$$V_P = V_Q$$

or

$$I_1 P = I_2 Q \tag{2}$$

Dividing Eq. (1) by Eq. (2) gives

$$\frac{I_1 R}{I_1 P} = \frac{I_2 S}{I_2 Q}$$

Therefore,

$$\frac{R}{P}=\frac{S}{Q}$$

or

$$\boxed{\textit{unknown resistance, } R=\frac{SP}{Q}} \tag{14-8}$$

Since the precise values of S, P and Q are known, the resistance R can be accurately calculated.

EXAMPLE 14-10 A Wheatstone bridge has $P=5\ \text{k}\Omega$, $Q=10\ \text{k}\Omega$, and the null condition is indicated on the galvanometer when $S=1994\ \Omega$. Determine the value of the unknown resistance R.

SOLUTION

Eq. (14-8):

$$R=\frac{SP}{Q}$$

$$=\frac{1994\ \Omega\times 5\ \text{k}\Omega}{10\ \text{k}\Omega}$$

$$\mathbf{R=997\ \Omega}$$

14-9 DC POTENTIOMETER

The *dc potentiometer* is used for the precise measurement of voltages, and its most important application is in checking the accuracy, or *calibrating*, voltmeters and ammeters.

As shown in Figure 14-16, the basic dc potentiometer consists of a resistance wire of uniform cross section AB stretched alongside a calibrated scale (e.g., a meter stick). A current is passed through the wire from a battery (B_1), and the current is controlled by a variable resistance (R_1). A sensitive galvanometer (G) is connected to the resistance wire via a moving contact (C). The other side of the galvanometer is connected to a switch (S), which allows contact to be made to either a *standard cell* (B_2), or to an unknown voltage (V_x). Resistor R_2 is included in series with the standard cell to protect it against excessive current flow (see Section 6-7).

Suppose that S is set to connect B_2 to the galvanometer, and assume that the standard cell voltage is known to be exactly 1.0190 V. Also, let the resistance wire (or *slide wire*) be exactly 100 cm in length, and assume

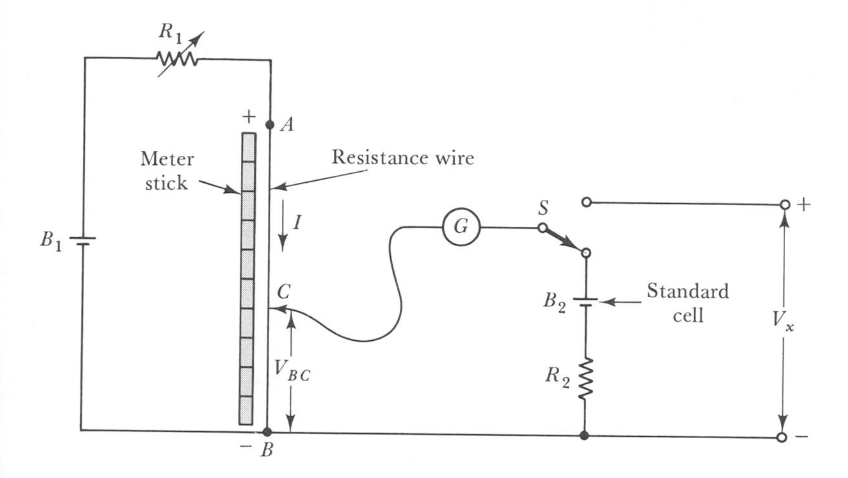

FIG. 14-16. Basic dc potentiometer

that the moving contact is set to 50.95 cm from point B. Now, when R_1 is adjusted until the galvanometer indicates null, the voltage between points B and C will be exactly equal to the standard cell voltage. Therefore,

$$V_{BC} = 1.0190 \text{ V}$$

and V_{BC} is the voltage drop across 50.95 cm of the slide wire. Therefore,

$$\begin{aligned} \text{voltage drop/cm} &= \frac{1.0190 \text{ V}}{50.95 \text{ cm}} \\ &= 20 \text{ mV/cm} \end{aligned}$$

The voltage at any point on the slide wire can now be determined by measuring its distance in cm from point B and multiplying by 20 mV/cm.

The procedure of setting S to B_2 and adjusting R_1 for null conditions on the galvanometer is termed *calibrating the potentiometer*. Once calibration is performed, S may be switched to connect the unknown voltage (V_x) to the galvanometer. Then, moving contact C is adjusted on the slide wire until the galvanometer indicates null once again. When this occurs, V_{BC} is equal to V_x. The new distance BC is now measured on the scale, and the value of V_x is calculated by multiplying the distance by 20 mV/cm.

Figure 14-17(a) shows how a potentiometer may be used to calibrate an ammeter. The ammeter is connected in series with a precision resistor R. The current through the ammeter also flows through the resistor, and the voltage drop across the resistor can be accurately measured by the potentiometer. Once V_x is known, the exact value of I can be calculated and compared to the scale reading on the ammeter.

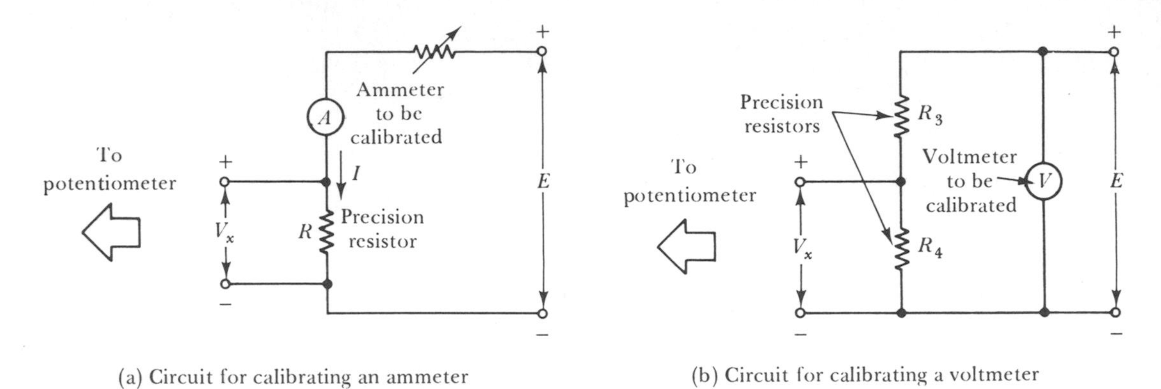

(a) Circuit for calibrating an ammeter (b) Circuit for calibrating a voltmeter

FIG. 14-17. Circuits for calibration of an ammeter and a voltmeter

Since the dc potentiometer measures voltages very accurately, it can be used to calibrate a voltmeter. However, it has the limitations that it can only measure small voltages. At 20 mV/cm for the potentiometer in Figure 14-16, the voltage between points A and B is

$$V_{AB} = 20 \text{ mV/cm} \times 100 \text{ cm}$$
$$= 2 \text{ V}$$

For calibrating voltmeters with larger ranges than 2 V, the meter terminal voltage must be divided down as illustrated in Figure 14-17(b). If R_3 and R_4 are known precision resistors, then when V_x is measured by the dc potentiometer, the actual terminal voltage of the voltmeter can be calculated and compared to its scale reading.

EXAMPLE 14-11

An ammeter and voltmeter are calibrated by means of the potentiometer shown in Figure 14-16. The ammeter is connected in series with a 1.5-Ω precision resistor, and the galvanometer null is found at 76.5 cm on the potentiometer scale. For the voltmeter, R_3 and R_4 [in Figure 14-17(b)] are precisely 99 kΩ and 1 kΩ, respectively, and null is found at 49.75 cm on the potentiometer scale. Calculate the exact level of current through the ammeter, and the exact voltage at the voltmeter terminals.

SOLUTION

For the ammeter:

$$V_x = 76.5 \text{ cm} \times 20 \text{ mV/cm}$$
$$= 1.53 \text{ V}$$
$$I = \frac{V}{R} = \frac{1.53 \text{ V}}{1.5\ \Omega}$$
$$\boldsymbol{I = 1.02 \text{ A}}$$

For the voltmeter:

$$V_x = 49.75\ \text{cm} \times 20\ \text{mV/cm}$$
$$= 0.995\ \text{V}$$
$$E = V_x \times \frac{R_3 + R_4}{R_4}$$
$$= 0.995\ \text{V} \times \frac{99\ \text{k}\Omega + 1\ \text{k}\Omega}{1\ \text{k}\Omega}$$
$$\boldsymbol{E = 99.5\ \text{V}}$$

Laboratory-type dc potentiometers are more complicated than the simple instrument illustrated in Figure 14-16. The slide wire is bent into a circle, so the sliding contact can be adjusted by means of a knob and its position indicated on a circular scale. Switched precision resistors are included in series with the slide wire. These are also usually knob-controlled, and their equivalent voltage drop is indicated on a scale.

GLOSSARY OF FORMULAS

For measuring resistance by ammeter and voltmeter:

$$R = \frac{V_L}{I_L + I_V}$$
$$R = \frac{V_L + V_A}{I_L}$$

Ohmmeter circuit:

$$I_m = \frac{E_b}{R_1 + R_x + r_m}$$
$$I_m \cong \frac{E_b}{R_1 + R_x}$$
$$I_B = \frac{E_b}{R_1 + R_x + (r_m \| R_2)}$$
$$I_B \cong \frac{E_b}{R_1 + R_x}$$
$$I_B = I_2 + I_m$$

Wheatstone bridge circuit:

$$R = \frac{SP}{Q}$$

REVIEW QUESTIONS

14-1 Draw a sketch to show the construction of a PMMC instrument. Define the three forces involved in operation of the instrument and explain briefly.

14-2 Sketch the circuit of an ammeter and briefly explain its operation. Comment on the resistance of an ammeter.

14-3 Draw a sketch to show the circuit of a multirange ammeter and discuss the type of switch required for range changing. Also, sketch the circuit of an ammeter using an Ayrton shunt and briefly explain.

14-4 Sketch the circuit of a multirange voltmeter and briefly explain its operation. Discuss any special requirement for the range-changing switch and comment on the resistance of a voltmeter in relation to its usual applications.

14-5 Sketch circuits to show the two possible arrangements for measuring resistance by use of an ammeter and voltmeter. Discuss the error sources with each arrangement.

14-6 Sketch the basic circuit of a series ohmmeter and explain its operation. Also, show how adjustments may be made for a falling battery voltage.

14-7 Draw a sketch to show the basic construction of a dynamometer instrument and explain its operation. Also, show how this instrument can be employed as a wattmeter, and explain how it can give a direct indication of power.

14-8 Sketch the circuit of a Wheatstone bridge, explain its operation, and derive an expression from which the unknown resistance can be calculated.

14-9 Sketch the circuit of a basic dc potentiometer, identify the function of every component, and explain the operation of the instrument.

14-10 Sketch circuits to show how to calibrate

a. An ammeter.

b. A voltmeter.

by means of a dc potentiometer. Briefly explain in each case.

14-11 Discuss the measurement of insulation resistance by means of an ammeter and voltmeter. Sketch the circuitry involved.

14-12 Draw a sketch to show the basic operation of a megger. Carefully explain the operation of the instrument.

PROBLEMS

14-1 A PMMC instrument has a coil resistance of 270 Ω and gives FSD for a current of 100 μA. Determine the value of shunt resistance required to convert the instrument into a 100-mA ammeter.

14-2 Determine the new values of shunt resistor for the ammeter described in Problem 14-1 to change its range to:

a. 1 A.

b. 10A.

14-3 An ammeter having a resistance of 0.5 Ω is connected in series with a 20-V supply and a load which normally takes a current of 20 A from the supply. Calculate the current indicated by the ammeter.

14-4 A PMMC instrument has a resistance of 100 Ω and a FSD for a current of 100 μA. An Ayrton shunt is connected to the instrument to convert it to an ammeter. The Ayrton shunt has four resistors, each of which is 0.001 Ω. Determine the various ranges to which the ammeter may be switched.

14-5 A PMMC instrument with a resistance of 75 Ω and FSD current of 100 μA is to be used as a voltmeter with 250-V, 100-V, and 50-V ranges. Determine the required values of multiplier resistor for each range. Also, calculate the sensitivity of the voltmeter.

14-6 Two resistors are connected in series across a 250-V supply. The resistor values are $R_1 = 330\ \Omega$ and $R_2 = 220\ \Omega$. The voltmeter described in Problem 14-5 is used to measure the voltage across each resistor. Determine the voltage indicated in each case.

14-7 An ammeter and voltmeter employed to measure resistance give readings of 196 μA and 240 V, respectively, when the voltmeter is connected directly in parallel with the resistor to be measured. When the ammeter is connected directly in series with the resistor, the readings are 100 μA and 240 V. The ammeter has a resistance of 50 Ω, and the voltmeter sensitivity is 10 kΩ/V. If the voltmeter is on a 250-V range, calculate the value of measured resistance for each case and determine which of the two gives the most accurate result.

14-8 A series ohmmeter uses a 200-μA meter and a 20-kΩ precision resistor. The supply voltage is a battery with $E = 4$ V. Determine the value of resistance measured by the ohmmeter at pointer deflections of $\frac{1}{3}$ FSD, $\frac{1}{2}$ FSD, and $\frac{2}{3}$ FSD. Draw a sketch of the scale of the instrument, showing current levels and resistance values.

14-9 A series ohmmeter has a meter with 100-μA FSD and a resistance of $r_m = 30\ \Omega$. The supply battery has a terminal voltage of $E_B =$ 4 V, and the series resistor is 10 kΩ. An adjustable shunt resistor connected across the meter has a value of $R_2 = 30\ \Omega$. Determine the external resistance measured at $\frac{1}{4}$, $\frac{1}{2}$, and $\frac{3}{4}$ of FSD. Also determine the new value that R_2 must be adjusted to when E_B falls to 3 V.

14-10 A Wheatstone bridge has a 100-Ω precision resistor P connected in series with the unknown resistor R. Another precision resistor Q, having a value of 150 Ω, is connected in series with the variable resistor S. When the galvanometer indicates null, the value of S is found to be 119.25 Ω. Calculate the value of the unknown resistance.

14-11 A dc potentiometer has a 2-m slide wire with 2 V across it and uses a standard cell with a voltage of exactly 1.0186 V. Determine the point on the scale at which the moving contact should be set when calibrating the potentiometer. Also, calculate the voltage/cm for the instrument slide wire.

14-12 The dc potentiometer described in Problem 14-11 is used to calibrate an ammeter which is connected in series with 7.5-kΩ precision resistor, as in Figure 14-17(a).Galvanometer null is obtained when the sliding contact is exactly 75 cm from the end of the slide wire (point B in Figure 14-16). Determine the exact value of the current through the ammeter.

14-13 The dc potentiometer described in Problem 14-11 is used to calibrate a voltmeter. The precision resistors R_3 and R_4 in Figure 14-17(b) are 49 kΩ and 1 kΩ, respectively, and null is found at exactly 48.9 cm on the slide wire. Determine the exact value of the voltage across the meter.

15

INDUCTANCE

INTRODUCTION

Electromagnetic induction occurs when a magnetic flux in motion with respect to a single conductor or a coil *induces an emf* in the conductor or coil. Because the growth or decline of current through a coil generates a changing flux, an emf is induced in the coil by its own current change. The same effect can induce an emf in an adjacent coil. The amount of emf induced in each case depends upon the *self-inductance* of the coil, or upon the *mutual inductance* between the two coils. In all cases the polarity of the induced emf is such that it opposes the original change that induced the emf.

Components called *inductors* or *chokes* are constructed to have specified values of inductance. *Energy* is *stored* in an inductive coil when a current is flowing in the coil windings. Inductors can be operated in series or in parallel. Even the shortest of conductors has an inductance. This is usually an unwanted quantity, and is termed *stray inductance*.

15-1 ELECTROMAGNETIC INDUCTION

In Section 12-2 it was shown that when a magnetic flux cuts a conductor (or a coil), an emf is generated within the coil. This effect is termed *electromagnetic induction*. It is important to note that the emf is generated only when the flux is in motion with respect to the conductor or coil. When the flux and the conductor are stationary, no emf is generated.

Figure 15-1 illustrates electromagnetic induction in a single conductor in motion across a magnetic field. A center-zero galvanometer is connected in series with the conductor, and it is found that the galvanometer deflects to the left when the conductor is moved up through the field [Figure 15-1(a)]. This indicates that the current induced in the conductor is flowing through the galvanometer from right to left, as illustrated. It also shows that the polarity of the induced emf in the conductor is + on the right-hand side of the conductor and − on the left-hand side, as shown in the figure. When the direction of the current induced in the conductor is considered, it is seen that the current sets up its own flux in a clockwise direction around the conductor [see Figure 15-1(a)]. The effect of this (conductor) flux is to strengthen the magnetic field above the conductor and to weaken it below the conductor, as shown in Figure 15-1(b). The strengthening of the magnetic field above the conductor and the weakening of the field below the conductor make it more difficult for the conductor to move upward. Thus, it is seen that the magnetic flux set up by the current induced in the conductor actually opposes the direction of motion of the conductor.

A similar effect is produced when the conductor is moved down through the field, as illustrated in Figure 15-1(c) and (d). In this case, the induced emf is + on the conductor's left-hand side and − on its right-hand side. Galvanometer deflection is to the right, and the current direction is as shown in the figure. Figure 15-1(d) shows that the induced current in the conductor now generates a flux that strengthens the magnet's magnetic field below the conductor and weakens it above the conductor. Again, the field set up by the induced current opposes the conductor's direction of motion.

The fact that the flux set up by the induced current always opposes the direction of motion is stated in *Lenz's* law.*

Lenz's Law

The induced current always develops a flux which opposes the motion or change producing the current.

Lenz's law can be justified by considering the origin of the energy developed in the conductor. It is obviously generated as a result of the relative motion of the conductor and the magnetic field. Thus, work must be done in moving the conductor through the field, and for work to be done a force must oppose the motion of the conductor. This opposing force is the result of the flux set up by the induced current.

It should be noted that for electromagnetic induction to occur as illustrated in Figure 15-1, both the conductor and direction of motion must be at right angles to the magnetic field flux. If the conductor were

*Formulated by the Russian physicist Heinrich Lenz (1804–1865).

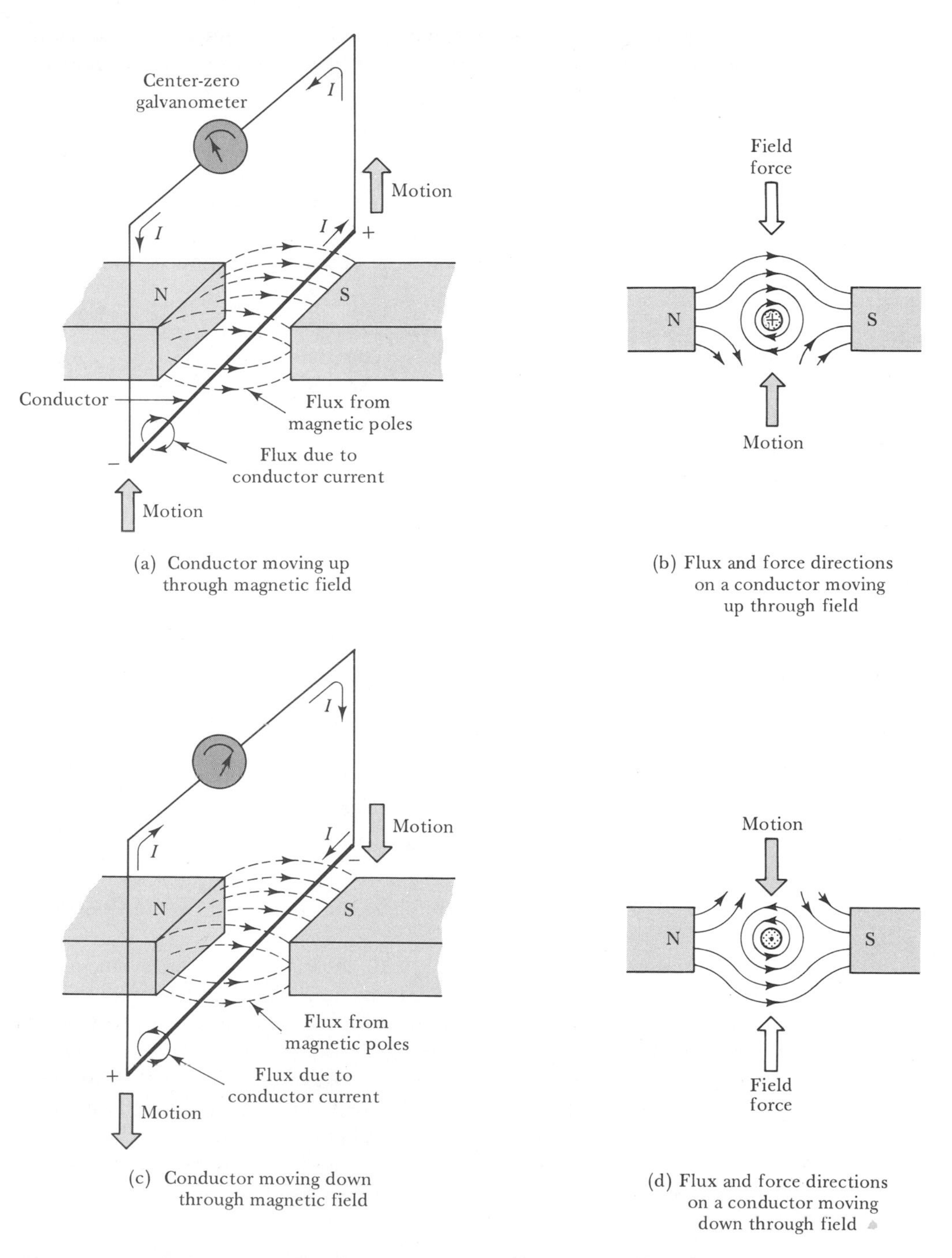

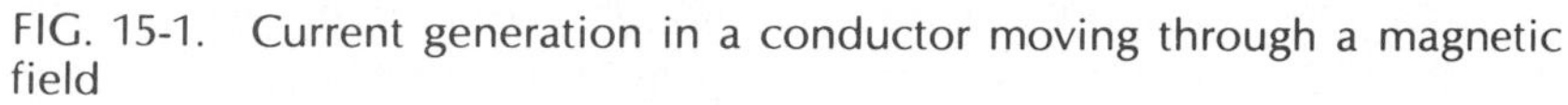

FIG. 15-1. Current generation in a conductor moving through a magnetic field

to be moved axially, for example, or if it was oriented in a horizontal direction from the N pole to the S pole no electromagnetic induction effect would occur.

15-2 INDUCED EMF AND CURRENT

In Section 12-4 the weber is defined as the magnetic flux which, linking a single-turn coil produces an emf of 1 V when the flux is reduced to zero at a uniform rate in 1 s. Therefore, the equation for induced emf can be written

$$\boxed{e_L = \frac{\Delta\Phi}{\Delta t}} \tag{15-1}$$

Here, e_L is in volts, $\Delta\Phi$ is in Wb, and Δt is in seconds. Equation (15-1) also originates from *Faraday's* law.*

Faraday's Law

The EMF induced in an electric circuit is proportional to the rate of change of flux linking the circuit.

If the total field flux in Figure 15-1 were 1 Wb, and if the conductor were moved through the field in exactly 1 s, the emf measured at the conductor terminals would be exactly 1 V. If the conductor were to be moved through the field in $\frac{1}{2}$ s, the voltage generated would be

$$e_L = \frac{1 \text{ Wb}}{0.5 \text{ s}}$$

$$= 2 \text{ V}$$

Similarly, if the flux were doubled, the generated voltage would be doubled.

Now consider Figure 15-2, which shows two coils on a ring-shaped iron core. Flux generated by a current flowing in the left-hand coil will pass through the iron core and link with the right-hand coil. Because the left-hand coil is the *input* coil and the right-hand coil is the *output*, the coils are identified as *primary* and *secondary*, respectively. While the current in the primary is constant, the core flux will not change and no emf will be induced in the secondary winding. When the primary current is increased or decreased (by adjustment of R), the core flux grows or declines, and in doing so it induces an emf in the secondary winding.

---

*Formulated by the English chemist and physicist Michael Faraday (1791–1867).

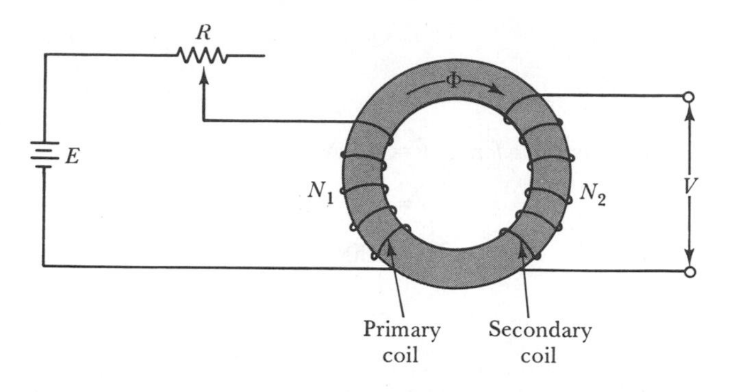

FIG. 15-2. Iron core used to channel the magnetic flux generated by one coil into another coil

If N is the number of turns on the secondary winding, the induced emf is

$$\boxed{e_L = \frac{\Delta\Phi N}{\Delta t}} \qquad (15\text{-}2)$$

EXAMPLE 15-1

In the circuit shown in Figure 15-2, the core flux is increased from zero to 0.05 Wb in a time of 4 s. Calculate the number of secondary turns required if the induced voltage is to be 1.5 V.

SOLUTION

Eq. (15-2):

$$e_L = \frac{\Delta\Phi N}{\Delta t}$$

Therefore,

$$N = \frac{e_L \Delta t}{\Delta\Phi}$$

$$= \frac{1.5 \text{ V} \times 4 \text{ s}}{0.05 \text{ Wb}}$$

$$\mathbf{N = 120 \text{ turns}}$$

EXAMPLE 15-2

The dimensions of the magnetic core shown in Figure 15-2 are: cross-sectional area $A = 3$ cm^2, magnetic path length $l = 10$ cm, and the relative permeability is 250. The primary coil has $N_p = 100$ turns and the secondary coil has $N_s = 75$ turns. If the current is increased from zero to 5 A in 0.1 s, determine the emf induced in the secondary.

SOLUTION

Eq. (12-2):

$$\textit{magnetomotive force},\ \mathcal{F} = I \times N_p$$
$$= 5\ \text{A} \times 100$$
$$= 500\ \text{A}$$

Eq. (12-3):

$$\textit{magnetic field strength},\ H = \frac{\mathcal{F}}{l}$$
$$= \frac{500\ \text{A}}{10 \times 10^{-2}\ \text{m}}$$
$$= 5000\ \text{A/m}$$

Eq. (13-5):

$$\textit{flux density},\ B = \mu_r \mu_o H$$
$$= 250 \times 4\pi \times 10^{-7} \times 5000\ \text{A/m}$$
$$= 1.57\ \text{T}$$

Eq. (12-1):

$$\textit{total flux},\ \Phi = B \times A$$
$$= 1.57 \times 3 \times 10^{-4}\ \text{m}^2$$
$$\Phi = 471\ \mu\text{Wb}$$

Eq. (15-2):

$$\textit{induced emf},\ e_L = \frac{\Delta\Phi N_s}{\Delta t}$$
$$= \frac{471\ \mu\text{Wb} \times 75}{0.1\ \text{s}}$$
$$\mathbf{e_L = 0.35\ V}$$

In Figure 15-3 an air-cored solenoid is shown, around which another (secondary) coil has been wound. When the switch is closed, the current through the coil causes the magnetic flux to grow from zero. All the flux from the solenoid cuts the secondary winding and induces an emf in the

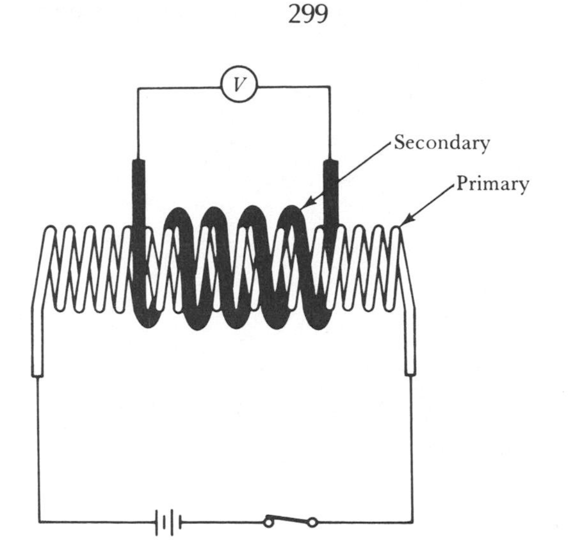

FIG. 15-3. Air-cored solenoid with a secondary winding

secondary coil. Once again, Eq. (15-2) can be used to calculate the induced emf.

EXAMPLE 15-3

The solenoid shown in Figure 15-3 has 1000 turns, is 30 cm long, and has a cross-sectional area of 5 cm^2. The secondary winding around it has 5 turns. Determine the emf induced in the secondary coil when the current through the solenoid is increased from zero to 20 A in a time period of 10 ms.

SOLUTION

Eq. (12-2):

$$\textit{magnetomotive force}, \mathcal{F} = N_p I_p$$
$$= 1000 \times 20 \text{ A}$$
$$= 20\ 000 \text{ A}$$

Eq. (12-3):

$$\textit{magnetic field strength}, H = \frac{\mathcal{F}}{l}$$
$$= \frac{20\ 000 \text{ A}}{30 \times 10^{-2} \text{ m}}$$
$$= 6.67 \times 10^4 \text{ A/m}$$

Eq. (13-4):

$$\textit{flux density}, B = H \times \mu_o$$
$$= 6.67 \times 10^4 \text{ A} \times 4\pi \times 10^{-7}$$
$$= 8.38 \times 10^{-2} \text{ T}$$

Eq. (12-1):

$$\textit{total flux}, \Phi = B \times A$$
$$= 8.38 \times 10^{-2} \text{ T} \times 5 \times 10^{-4} \text{ m}^2$$
$$= 41.9 \ \mu\text{Wb}$$

Eq. (15-2):

$$\textit{induced emf}, e_L = \frac{\Delta\Phi N_s}{\Delta t}$$
$$= \frac{41.9 \times 10^{-6} \text{ Wb} \times 5}{10 \times 10^{-3} \text{ s}}$$
$$\mathbf{e_L \simeq 21 \text{ mV}}$$

15-3 SELF-INDUCTANCE

It has been shown that a conductor moving through a magnetic field has an emf induced in it, and that the growth of current in a coil can induce an emf in another coil to which it is magnetically coupled. It is also possible for a coil to induce a voltage in itself as the current through it grows. This phenomenon is known as *self-inductance*, and the principle is illustrated in Figure 15-4.

A coil and its cross-sectional area are shown in Figure 15-4(a), with arrow tails and points indicating the current directions in each turn. Every turn of the coil has a flux around it due to the current flowing through the coil. However, for convenience the illustration shows the growth of flux around only one turn on the coil. It is seen that as the current grows, the flux expands outward and cuts the other turns. This causes currents to be induced in the other turns, and the direction of the induced currents are such that they set up a flux which opposes the flux inducing them. Remembering that the current through the coil causes the flux to grow around all turns at once, it is seen that the flux from every turn induces a current that opposes it in every other turn.

To set up opposing fluxes, the induced current in a coil must be in opposition to the current flowing through the coil from the external

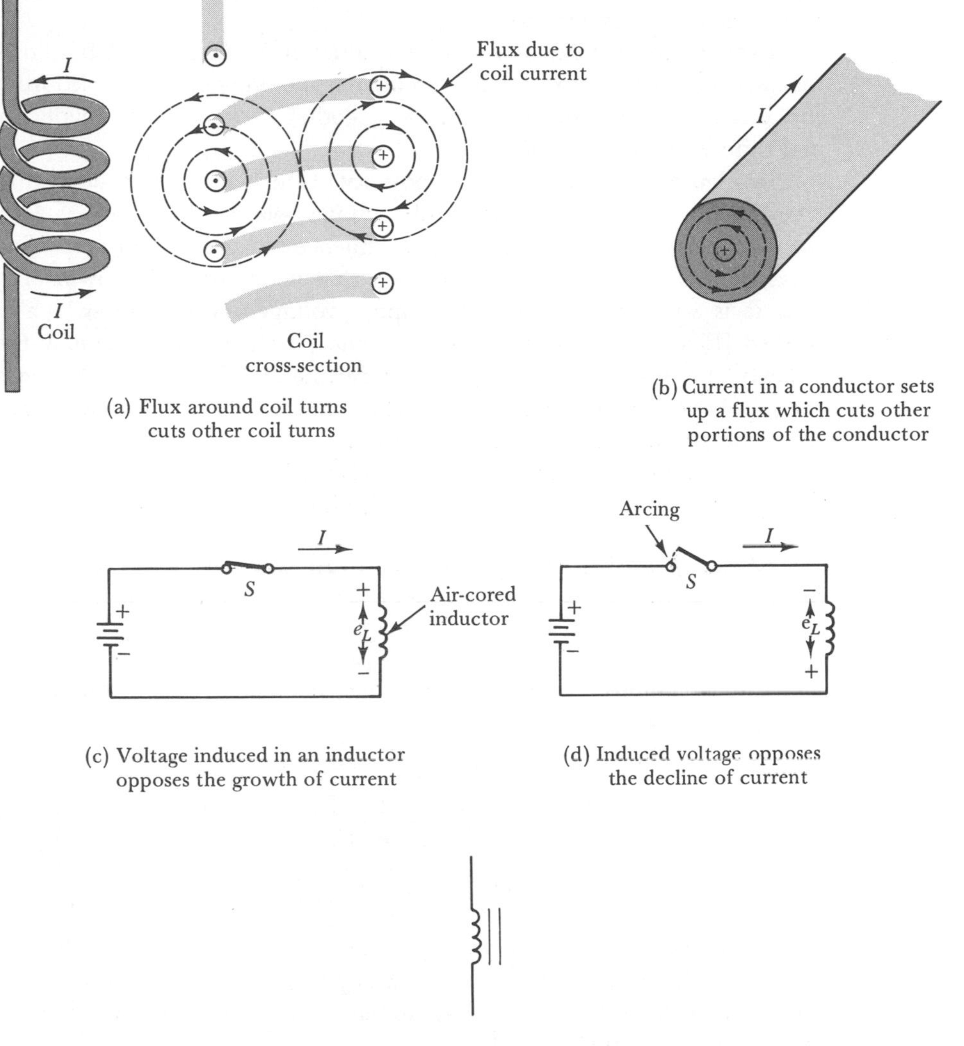

Flux due to coil current
I
Coil
Coil cross-section
(a) Flux around coil turns cuts other coil turns
(b) Current in a conductor sets up a flux which cuts other portions of the conductor
S
Air-cored inductor
e_L
Arcing
(c) Voltage induced in an inductor opposes the growth of current
(d) Induced voltage opposes the decline of current
(e) Iron-cored inductor symbol

FIG. 15-4. A counter-emf is induced in a coil (and in a conductor) when the current grows or declines

source of supply. The induced current is, of course, the result of an induced emf. Thus, it is seen that the self-inductance of a coil sets up an induced emf which opposes the external emf that is driving current through the coil. Because this induced emf is in opposition to the supply voltage, it is usually termed the *counter-emf*, or *back-emf*. The counter-emf occurs only when the coil current is growing or declining. When the current has reached a constant level, the flux is no longer changing and no counter-emf is generated.

Even a single conductor has self-inductance. Figure 15-4(b) shows that when current is growing in a conductor, flux may grow outward from the center of the conductor. This flux cuts other portions of the conductor and induces a counter-emf.

In Figure 15-4(c) and (d), the polarity of the counter-emf induced in a coil is illustrated for a given supply voltage polarity. In Figure 15-4(c), the switch is closed and current I commences to grow from zero. The polarity of the counter-emf (e_L) is such that it opposes the growth of I, thus it is *series-opposing* with the supply voltage. When the swhich is opened [Figure 15-4(d)], the current tends to fall to zero. But now the polarity of e_L is such that it opposes the decline of I. It is *series-aiding* with the supply voltage. In fact, e_L may cause arcing at the switch terminals as it attempts to maintain the flow of current.

The amplitude of the counter-emf induced in a coil by a given rate of change in current depends upon the coil's inductance.

The SI unit of inductance is the ***henry**** ***(H)****. The inductance of a circuit is* ***one henry*** *(1 H), when an emf of 1 V is induced by the current changing at the rate of 1 A/s.*

Thus, the relationship among inductance, induced voltage, and rate of change of current is

$$\boxed{L = \frac{e_L}{\Delta i / \Delta t}} \qquad \textbf{(15-3)}$$

where L is the inductance in henrys, e_L *is the induced counter-emf in volts, and* $(\Delta i / \Delta t)$ *is the rate of change of current in amperes/second.* A negative sign is sometimes included in front of e_L to show that the induced emf is in opposition to the applied emf.

A coil that is constructed to have a certain inductance is usually referred to as an *inductor* or a *choke*. Note the graphic symbols for an inductor shown in Figure 15-4(c), (d), and (e).

EXAMPLE 15-4 The current in a coil grows linearly from 0 to 10 A in a time of 0.25 s. If the coil has an inductance of 0.75 H, calculate the induced counter-emf.

---

*Named for the American physicist Joseph Henry (1797–1878).

SOLUTION

From Eq. (15-3):

$$e_L = L\left(\frac{\Delta i}{\Delta t}\right)$$

$$= 0.75\ \text{H} \times \frac{10\ \text{A} - 0\ \text{A}}{0.25\ \text{s}}$$

$$\mathbf{e_L = 30\ V}$$

An expression for inductance can also be derived involving the coil dimensions and the number of turns. From Eq. (15-2),

$$e_L = \frac{\Delta\Phi N}{\Delta t}$$

Substituting for e_L into Eq. (15-3) gives

$$L = \frac{\Delta\Phi N/\Delta t}{\Delta i/\Delta t}$$

or

$$\boxed{L = \frac{\Delta\Phi N}{\Delta i}} \qquad (15\text{-}4)$$

Also, $\Phi = B \times A$

and $B = \mu_r \mu_o H$

$$= \mu_r \mu_o \frac{IN}{l}$$

Therefore, $\Phi = \mu_r \mu_o IN\frac{A}{l}$

Since I is a maximum current level, it also represents the change in current (Δi) from zero to the maximum level. Therefore, change in flux is

$$\boxed{\Delta\Phi = \mu_r \mu_o \Delta i N \frac{A}{l}} \qquad (15\text{-}5)$$

Substituting for $\Delta\Phi$ in Eq. (15-4) gives

$$L = \frac{\left(\mu_r \mu_o \Delta i N \frac{A}{l}\right) \times N}{\Delta i}$$

or

$$\boxed{L = \mu_r \mu_o N^2 \frac{A}{l}} \qquad \mathbf{(15\text{-}6)}$$

Note that the inductance is proportional to the cross-sectional area of a coil and to the square of the number of turns. It is also inversely proportional to the coil length. Therefore, maximum inductance is obtained with a short coil with a large cross-sectional area and a large number of turns.

Equation (15-6) now affords a means of calculating the inductance of a coil of known dimensions. Alternatively, it can be used to determine the required dimensions for a coil that is to have a certain inductance. However, it is not so easily applied to iron-cored coils, because the permeability of ferromagnetic material changes when the flux density changes (see Figure 13-3). Consequently, the inductance of an iron-cored coil is constantly changing as the coil current increases and decreases.

EXAMPLE 15-5

A solenoid with 900 turns has a total flux of 1.33×10^{-7} Wb through its air core when the coil current is 100 mA. If the flux takes 75 ms to grow from zero to its maximum level, calculate the inductance of the coil. Also, determine the counter emf induced in the coil during the flux growth.

SOLUTION

$$\Delta\Phi = 1.33 \times 10^{-7}\ \text{Wb}$$

$$\Delta i = 100\ \text{mA}$$

$$\Delta t = 75\ \text{ms}$$

Eq. (15-4):

$$L = \frac{\Delta\Phi N}{\Delta i}$$

$$= \frac{1.33 \times 10^{-7}\ \text{Wb} \times 900}{100\ \text{mA}}$$

$$\mathbf{L \cong 1.2\ mH}$$

From Eq. (15-2):

$$e_L = \frac{\Delta\Phi N}{\Delta t}$$

$$= \frac{1.33 \times 10^{-7}\ \text{Wb} \times 900}{75\ \text{ms}}$$

$$\mathbf{e_L \cong 1.6\ mV}$$

EXAMPLE 15-6

The air-cored solenoid described in Example 15-5 has $l=15$ cm and inside diameter $D=1.5$ cm. Recalculate its inductance using the coil dimensions.

SOLUTION

$$\text{cross-sectional area, } A=\pi\left(\frac{D}{2}\right)^2$$

$$=\pi\times\left(\frac{1.5\times10^{-2}\,\text{m}}{2}\right)^2$$

$$=1.77\times10^{-4}\,\text{m}^2$$

Eq. (15-6):

$$L=\mu_r\mu_o N^2\frac{A}{l}$$

$$=1\times4\pi\times10^{-7}\times900^2\times\frac{1.77\times10^{-4}\,\text{m}^2}{15\times10^{-2}\,\text{m}}$$

$$\mathbf{L=1.2\ mH}$$

This checks with the answer to Example 15-5. *Note that the solenoid dimensions employed are those used in* Example 13-3.

EXAMPLE 15-7

An air-cored coil is to be 2.5 cm long and to have an average cross-sectional area of 2 cm^2. Determine the number of turns required if the coil is to have an inductance of 100 μH.

SOLUTION

From Eq. (15-6):

$$N^2=\frac{L\times l}{\mu_r\mu_o\times A}$$

Therefore,

$$N=\sqrt{\frac{L\times l}{\mu_r\mu_o\times A}}$$

$$=\sqrt{\frac{100\times10^{-6}\text{H}\times2.5\times10^{-2}\,\text{m}}{4\pi\times10^{-7}\times2\times10^{-4}\,\text{m}^2}}$$

$$\mathbf{N\simeq100\ turns}$$

EXAMPLE 15-8

A cast steel ring has a cross-sectional area of 4 cm^2 and a magnetic path length of 15 cm. A 300-turn coil is wound on the ring. Determine the coil inductance when the current is

a. $I=2$ A.

b. $I=0.5$ A.

SOLUTION

a. *For* $I=2$ A,

$$H=\frac{NI}{l}$$

$$=\frac{300\times 2\text{ A}}{15\times 10^{-2}\text{ m}}$$

$$H_1=4000\text{ A/m}$$

From Figure 13-3, *at* $H=4000$ A/m, $B=1.58$ T *for cast steel,*

$$\mu_r\mu_o=\frac{B}{H}$$

$$=\frac{1.58\text{ T}}{4000\text{ A/m}}$$

$$\mu_r\mu_o=3.95\times 10^{-4}$$

Eq. (15-6):

$$L=\mu_r\mu_o N^2\frac{A}{l}$$

$$=\frac{3.95\times 10^{-4}\times 300^2\times 4\times 10^{-4}\text{ m}^2}{15\times 10^{-2}\text{ m}}$$

$$\mathbf{L_1=94.8\ mH}$$

b. *For* $I=0.5$ A,

$$H=\frac{300\times 0.5\text{ A}}{15\times 10^{-2}\text{ m}}$$

$$H_2=1000\text{ A/m}$$

From Figure 13-3, *at* $H=1000$ A/m, $B=1.02$ T *for cast steel,*

$$\mu_r\mu_o=\frac{1.02\text{ T}}{1000\text{ A/m}}$$

$$=1.02\times 10^{-3}$$

Eq. (15-6):

$$L = \mu_r \mu_o N^2 \frac{A}{l}$$

$$= \frac{1.02 \times 10^{-3} \times 300^2 \times 4 \times 10^{-4}\ \text{m}^2}{15 \times 10^{-2}\ \text{m}}$$

$$L_2 = \mathbf{245\ mH}$$

The iron-cored coil in Example 15-7 actually has *less inductance* for a current change of 2 A than for a current change of 0.5 A. This is because the 2-A current tends to drive the core into magnetic saturation, while with 0.5 A the core is operating on the near-linear portion of the B/H curve. Therefore, it is seen that when inductance is specified for an iron-cored coil, the coil current must also be specified. Recall from Section 13-7 that the incremental permeability of ferromagnetic material varies along the length of the B/H curve. Thus, if the incremental permeability were used in Example 15-8, it would again be found that the greatest inductance is obtained when operating on the near-linear portion of the B/H curve.

In many cases it is desired to have a noninductive coil; for example, precision resistors are usually noninductive. To construct such a coil, the coil winding is made double, as illustrated in Figure 15-5. Thus, every coil turn has an adjacent turn which is carrying current in the opposite direction. The magnetic fields generated by adjacent turns cancel each other out. Therefore, no counter-emf is generated, and the coil is noninductive.

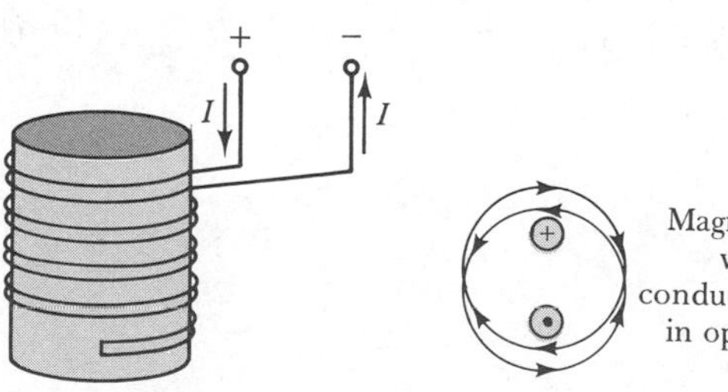
+
−
I
I
Magnetic fields cancel
when adjacent
conductors carry currents
in opposite directions
Noninductive coil with adjacent
conductors carrying currents
in opposite directions

FIG. 15-5. Noninductive coil construction

15-4 MUTUAL INDUCTANCE

When the flux from one coil cuts another adjacent (or magnetically coupled) coil, an emf is induced in the second coil. This was shown in Section 15-2. Following Lenz's law, the emf induced in the second coil sets up a flux that opposes the original flux from the first coil. Thus, the induced emf is again a counter-emf, and in this case the inductive effect is referred to as *mutual inductance*. Figure 15-6 shows the graphic symbols used for coils with mutual inductance, also termed *coupled coils*.

Like self-inductance, mutual inductance is measured in henrys (H):

Two coils have a mutual inductance of 1 H when an emf of 1 V is induced in one coil by current changing at the rate of 1 A/s in the other coil.

This definition gives rise to the equation relating mutual inductance to induced voltage and rate of change of current:

$$M = \frac{e_L}{\Delta i / \Delta t} \tag{15-7}$$

where M is the mutual inductance in henrys, e_L is the emf in volts induced in the secondary coil, and $(\Delta i/\Delta t)$ is the rate of change of current in the primary coil in amperes/second.

Recall from Section 15-2 that the coil through which a current is passed from an external source is termed the *primary*, and the coil that has an emf induced in it is referred to as the *secondary*.

An equation for the emf induced in the second coil can be written from Eq. (15-2):

$$e_L = \frac{\Delta \Phi N_s}{\Delta t} \tag{15-8}$$

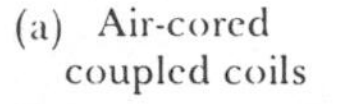

(a) Air-cored coupled coils

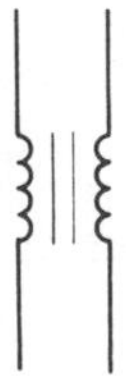

(b) Iron-cored coupled coils (or transformer)

FIG. 15-6. Circuit symbols for coupled coils

Here $\Delta\Phi$ is the total change in flux linking with the secondary winding, N_s is the number of turns on the secondary winding, and Δt is the time required for the flux change.

Substituting for e_L from Eq. (15-8) into Eq. (15-7) gives

$$M = \frac{\Delta\Phi N_s/\Delta t}{\Delta i/\Delta t}$$

Therefore,

$$\boxed{M = \frac{\Delta\Phi N_s}{\Delta i}} \qquad \textbf{(15-9)}$$

Figure 15-7(a) illustrates the fact that when the two coils are wound on a single ferromagnetic core, effectively all of the flux generated by the primary coil links with the secondary coil. However, when the coils are air-cored, only a portion of the flux from the primary may link with the secondary [see Figure 15-7(b)]. Depending upon how much of the primary flux cuts the secondary, the coils may be classified as *loosely coupled* or *tightly coupled*. One way to ensure tight coupling is shown in Figure 15-7(c), where each turn of the secondary winding is side by side with one turn of the primary winding. Coils wound in this fashion are said to be *bifilar*.

The amount of flux linkage from primary to secondary is also defined in terms of a *coefficient of coupling*, *k*. If all the primary flux links with the

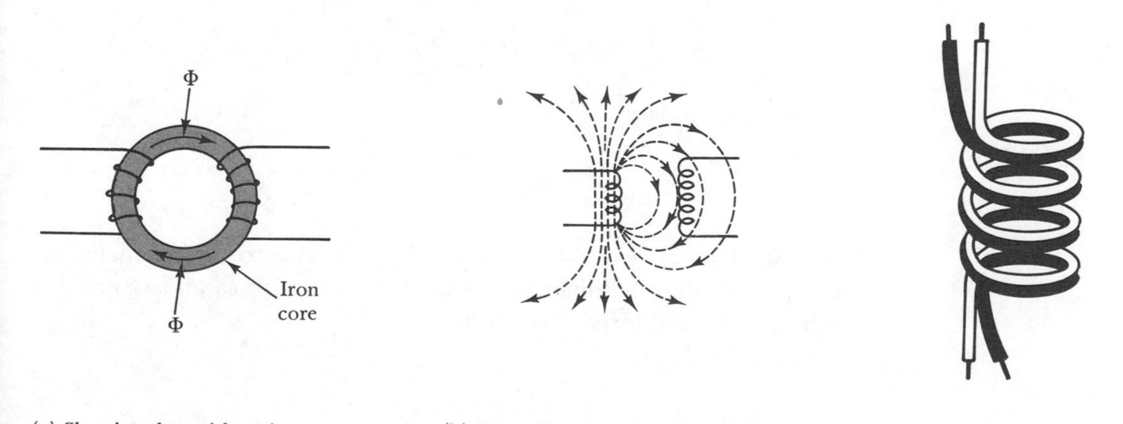

(a) Showing that with an iron core virtually all of the flux passes from one coil to the other

(b) Depending upon how closely air-cored coils are coupled, only a portion of the flux from the primary links with the secondary

(c) Bifilar winding for close coupling

FIG. 15-7. The amount of flux from a primary coil that links with the secondary depends upon how closely the coils are coupled

secondary, the coefficient of coupling is 1. When only 50% of the primary flux links with the secondary coil, the coefficient of coupling is 0.5. Thus,

$$k = \frac{\text{flux linkages between primary and secondary}}{\text{total flux produced by primary}}$$

Returning to Eq. (15-9), when $\Delta\Phi$ is the total flux change in the primary coil, the flux linking with the secondary is $k\Delta\Phi$. Therefore, the equation for M becomes

$$\boxed{M = k\frac{\Delta\Phi N_s}{\Delta i}} \qquad \textbf{(15-10)}$$

Also, substituting for $\Delta\Phi$ from Eq. (15-5) into Eq. (15-10) gives

$$M = \frac{kN_s}{\Delta i}\,\mu_r\mu_o\,\Delta i N_p\frac{A}{l}$$

or

$$\boxed{M = kN_pN_s\mu_r\mu_o\frac{A}{l}} \qquad \textbf{(15-11)}$$

Each winding considered alone has a self-inductance that can be calculated from Eq. (15-6). Thus, for the primary coil

$$L_1 = \mu_r\mu_o N_p^2\frac{A}{l}$$

and for the secondary

$$L_2 = \mu_r\mu_o N_s^2\frac{A}{l}$$

Assuming that the two windings share a common core (magnetic or nonmagnetic), the only difference in the expressions for L_1 and L_2 are the numbers of the coil turns. Therefore,

$$L_1 \times L_2 = N_p^2N_s^2\left(\mu_r\mu_o\frac{A}{l}\right)^2$$

or

$$\boxed{\sqrt{L_1L_2} = N_pN_s\mu_r\mu_o\frac{A}{l}} \qquad \textbf{(15-12)}$$

Comparing Eqs. (15-11) and (15-12), it is seen that

$$\boxed{M = k\sqrt{L_1 L_2}} \qquad \textbf{(15-13)}$$

EXAMPLE 15-9

Two identical coils are wound on a ring-shaped iron core that has a relative permeability of 500. Each coil has 100 turns, and the core dimensions are: cross-sectional area $A = 3\ \text{cm}^2$ and magnetic path length $l = 20$ cm. Calculate the inductance of each coil and the mutual inductance between the coils.

SOLUTION

From Eq. (15-6):

$$L_1 = L_2 = \mu_r \mu_o N^2 \frac{A}{l}$$

$$= 500 \times 4\pi \times 10^{-7} \times 100^2 \times \frac{3 \times 10^{-4}\ \text{m}^2}{20 \times 10^{-2}\ \text{m}}$$

$$\boldsymbol{L = 9.42\ \text{mH}}$$

Since the coils are wound on the same iron core, $k = 1$.

Eq. (15-13):

$$M = k\sqrt{L_1 L_2}$$

$$= \sqrt{9.42\ \text{mH} \times 9.42\ \text{mH}}$$

$$\boldsymbol{M = 9.42\ \text{mH}}$$

EXAMPLE 15-10

Two 100-turn end-to-end solenoids each have $l = 20$ cm and cross-sectional area $= 3\ \text{cm}^2$. Calculate their coefficient of coupling when the mutual inductance between them is measured as $0.62\ \mu\text{H}$.

SOLUTION

$$L_1 = L_2 = \mu_r \mu_o N^2 \frac{A}{l}$$

$$= 1 \times 4\pi \times 10^{-7} \times 100^2 \times \frac{3 \times 10^{-4}\ \text{m}^2}{20 \times 10^{-2}\ \text{m}}$$

$$= 18.8\ \mu\text{H}$$

$$M = k\sqrt{L_1 L_2}$$

Therefore,

$$k = \frac{M}{\sqrt{L_1 L_2}}$$

$$= \frac{0.62\ \mu\text{H}}{\sqrt{(18.8\ \mu\text{H})^2}}$$

$$\mathbf{k \cong 0.033}$$

The automobile ignition system relies on the mutual inductance between two coils. Consider Figure 15-8, which shows the basic circuit of an ignition coil, also known as a *Ruhmkorff coil*. The supply voltage from the automobile battery to the iron-cored primary winding is switched *on* and *off* rapidly by the make-and-break switch. This causes the flux generated by the primary winding to grow and collapse, and in doing so it cuts the turns on the secondary winding. Thus, an emf is induced in the secondary, first in one direction and then in the other; that is, it is an *alternating emf* (see Chapter 18). The secondary emf is proportional to the numbers of primary and secondary turns and the rate of change of primary current, as well as to the core dimensions and permeability. By selecting a large enough number of turns on each coil, a very large output voltage can be derived from a 12-V battery input.

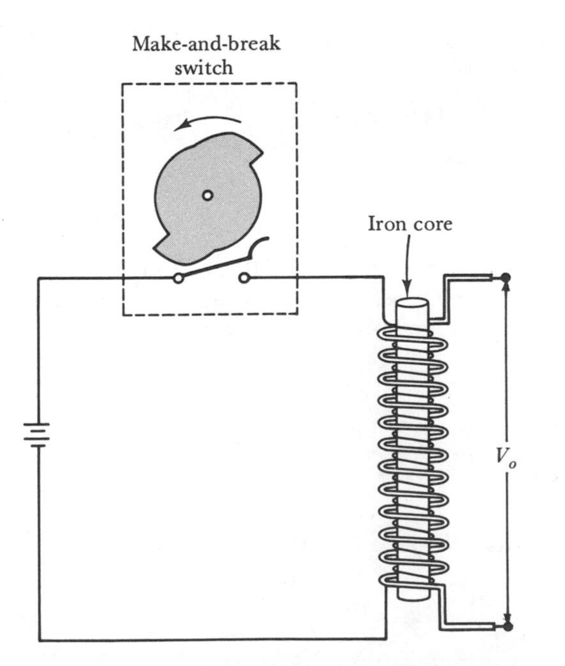

FIG. 15-8. Principle of the automobile ignition coil

15-5 TYPES OF INDUCTORS

There are a great many different types of inductors, ranging from large high-current iron-cored chokes to tiny resistor-style low-current coils.

Figure 15-9(a) shows an inductor used with a dc power supply. In this application, the inductor is usually required to pass a direct current which has a fluctuating level. Since the inductor opposes any change in the current level through its windings, it tends to smooth out the fluctuations (see the waveforms illustrated). This is exactly why the inductor is employed in this particular application. Because of the presence of the direct current through the windings, it is the *incremental inductance* of this component that is important. Therefore, the inductance value must be specified at a given level of direct current. Typical values for such inductors range from 50 mH to 20 H, with direct currents up to about 10 A and insulation voltage ratings up to 1000 V. Figure 15-9(b) shows the appearance of a typical power supply inductor.

In Figure 15-10, a low-current high-frequency type of inductor is illustrated. The core in this case is ferrite material (see Section 13-8) in two mating sections known as a *pot core*. As well as increasing the coil's inductance, the pot core screens the coil to protect adjacent components against flux leakage and to protect the coil from external magnetic fields. The coil is wound on a bobbin, so its number of turns is easily modified.

Three different types of low-current inductors are illustrated in Figure 15-11. Figure 15-11(a) shows a type that is available as an air-cored inductor or with a ferromagnetic core. With an air core, the inductance values range typically from 2.4 μH to 100 μH. With a

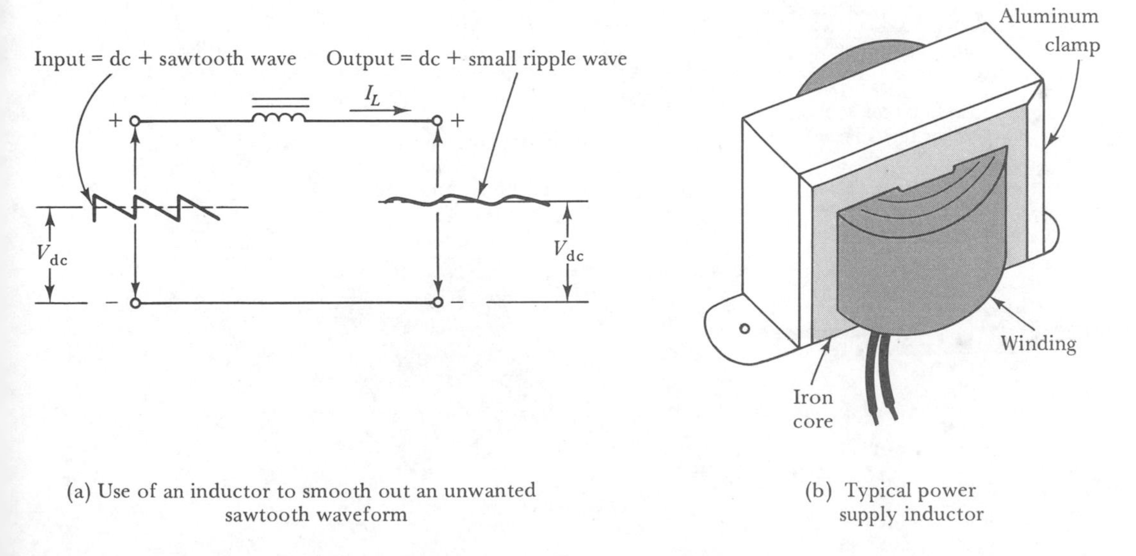

(a) Use of an inductor to smooth out an unwanted sawtooth waveform

(b) Typical power supply inductor

FIG. 15-9. Inductor used with a dc power supply

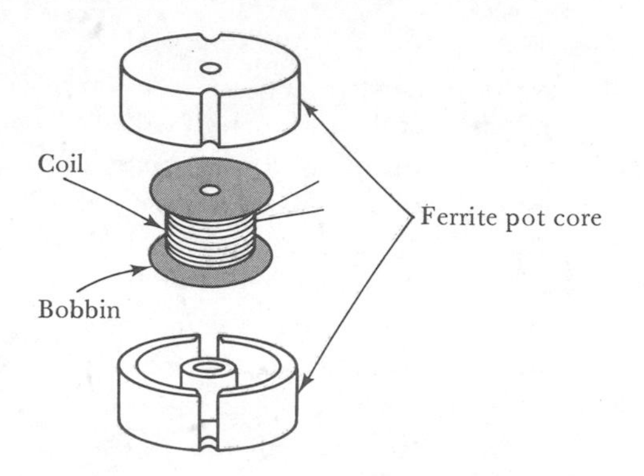

FIG. 15-10. Low-current high-frequency screened inductor

ferromagnetic core, inductance values up to about 10 mH can be obtained. Depending upon the thickness of wire used and the physical size of the inductor, the maximum current can be anything from about 50 mA to 1 A. The core in such an inductor may be made adjustable so that it can be screwed into or partially out of the coil. Thus, the coil inductance is variable. Note the graphic symbol for an inductor with an adjustable core [Figure 15-11(b)].

A small *molded inductor* is shown in Figure 15-11(c). Typical available values for this type range from 1.2 μH to 10 mH, with maximum currents

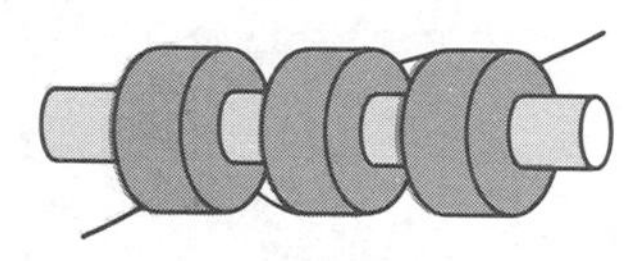

(a) Inductor with air core or ferromagnetic core

(b) Circuit symbol for an inductor with an adjustable ferromagnetic core

(c) Molded inductor

(d) Thin-film inductor

FIG. 15-11. Various types of low-current low-value inductors

of about 70 mA. The values of molded inductors are identified by a color code, similar to molded resistors. Figure 15-11(d) shows a tiny *thin-film* inductor used in certain types of electronic circuits. In this case the inductor is simply a thin metal film deposited in the form of a spiral on a ceramic base.

Laboratory-type variable inductors can be constructed in *decade box* format [see Figure 5-8(c)], in which precision inductors are switched into or out of a circuit by means of rotary switches. Alternatively, two coupled coils can be employed as a variable inductor. The coils may be connected in series or in parallel, and the total inductance is controlled by adjusting the position of one coil relative to the other.

15-6 ENERGY STORED IN AN INDUCTIVE CIRCUIT

From Eq. (15-3), the emf induced in an inductance is

$$e_L = L\left(\frac{\Delta i}{\Delta t}\right)$$

Therefore, if the current is controlled so that it grows linearly from zero to I in a time t, the average current is $\frac{1}{2}I$, and $\Delta i = I$. Thus the *constant* induced voltage is

$$e_L = L\left(\frac{I}{t}\right)$$

From Eq. (3-11), the electrical energy input is

$$W = EIt$$

Therefore, the energy supplied to the inductive circuit is

$$W = e_L \times \tfrac{1}{2}I \times t$$

$$= L\left(\frac{I}{t}\right) \times \tfrac{1}{2}I \times t$$

or energy stored is

$$\boxed{W = \tfrac{1}{2}LI^2} \qquad \textbf{(15-14)}$$

When L is in henrys and I is in amperes, W is given in joules.

From Eq. (15-6),

$$L = \mu_r\mu_o N^2 \frac{A}{l}$$

Substituting for L in Eq. (15-14) gives

$$W = \frac{1}{2}\mu_r\mu_o N^2 \frac{A}{l} I^2$$

This can be rewritten

$$W = \frac{1}{2}\mu_r\mu_o A\left(\frac{N^2 I^2}{l^2}\right)l$$

$$= \frac{1}{2}\mu_r\mu_o A H^2 l$$

and $$H = \frac{B}{\mu_r\mu_o}$$

Therefore, $$W = \frac{1}{2}\mu_r\mu_o A l\left(\frac{B}{\mu_r\mu_o}\right)^2$$

or

$$\boxed{W = \frac{B^2 A l}{2\mu_r\mu_o} \quad \text{joules}} \tag{15-15}$$

When μ_r is replaced by 1 for an air core, this equation is exactly the same as Eq. (13-6), which represents the energy stored in an air gap between two magnetic surfaces.

EXAMPLE 15-11 A 20-H choke with a resistance of 180 Ω has a 300-V supply. Calculate the energy stored.

SOLUTION

The steady-state current is

$$I = \frac{E}{R} = \frac{300\ \text{V}}{180\ \Omega}$$

$$\boldsymbol{I = 1.67\ \text{A}}$$

Eq. (15-14):

$$W = \frac{1}{2}LI^2$$

$$= \frac{1}{2} \times 20\ \text{H} \times (1.67\ \text{A})^2$$

$$\boldsymbol{W = 27.9\ \text{J}}$$

15-7 INDUCTORS IN SERIES AND PARALLEL

When inductors are connected in series or parallel and *there is no mutual induction between them*, they can be treated exactly like resistors. For series-connected inductors, the same current flows through each, and the counter-emf generated in each is proportional to the rate of change of current. Therefore, as illustrated in Figure 15-12(a), the total counter-emf is

$$e_{L(\text{total})} = e_{L1} + e_{L2} + e_{L3} + \cdots$$

and since $e_L = L(\Delta i/\Delta t)$,

$$e_{L(\text{total})} = L_1\left(\frac{\Delta i}{\Delta t}\right) + L_2\left(\frac{\Delta i}{\Delta t}\right) + L_3\left(\frac{\Delta i}{\Delta t}\right) + \cdots$$

$$= \left(\frac{\Delta i}{\Delta t}\right) L_s$$

Therefore total inductance is

$$\boxed{L_s = L_1 + L_2 + L_3 + \cdots} \qquad \textbf{(15-16)}$$

Note that Eq. (15-16) *is correct only when there is no mutual induction between the inductors.*

For parallel-connected inductors, the same voltage appears across each; consequently, the counter-emfs generated in each must be equal.

$$e_L = e_{L1} = e_{L2} = e_{L3} = \cdots$$

As illustrated in Figure 15-12(b), the total current change is

$$\Delta i_{\text{total}} = \Delta i_1 + \Delta i_2 + \Delta i_3 + \cdots$$

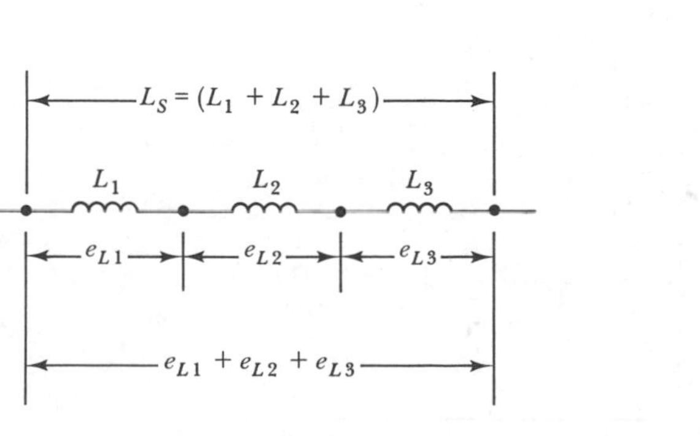

(a) Inductors in series

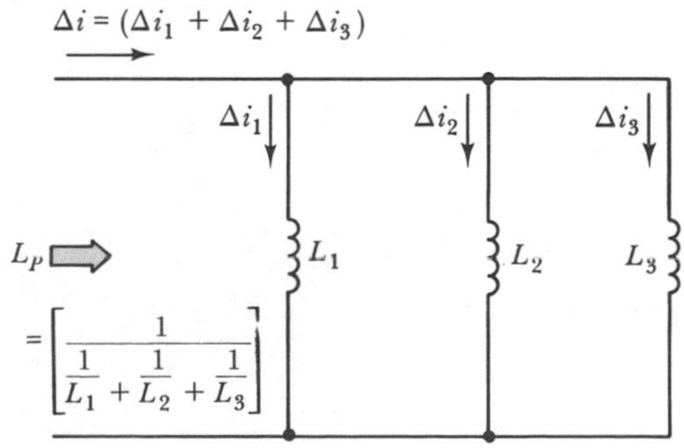

(b) Inductors in parallel

FIG. 15-12. Inductors in series and parallel

and
$$e_L = L\frac{\Delta i}{\Delta t}$$

Therefore,
$$\Delta i = \frac{e_L \Delta t}{L}$$

and
$$\Delta i_{\text{total}} = \left(\frac{e_L \Delta t}{L_1}\right) + \left(\frac{e_L \Delta t}{L_2}\right) + \left(\frac{e_L \Delta t}{L_3}\right) + \cdots$$

or
$$\frac{\Delta i_{\text{total}}}{e_L \Delta t} = \frac{1}{L_1} + \frac{1}{L_2} + \frac{1}{L_3} + \cdots$$

$$= \frac{1}{L_p}$$

Therefore, the total inductance is determined from

$$\boxed{\frac{1}{L_p} = \frac{1}{L_1} + \frac{1}{L_2} + \frac{1}{L_3} + \cdots} \qquad \textbf{(15-17)}$$

Note that Eq. (15-17) *is correct only when there is no mutual induction between the inductors.*

EXAMPLE 15-12

Three inductors have values of $L_1 = 10$ mH, $L_2 = 100$ μH, and $L_3 = 500$ μH. Determine the total inductance of the three:

a. When connected in series.

b. When connected in parallel.

Assume that there is no mutual induction between the inductors.

SOLUTION

a. *In series*:

$$L_s = L_1 + L_2 + L_3$$
$$= 10\text{ mH} + 100\ \mu\text{H} + 500\ \mu\text{H}$$
$$\mathbf{L_s = 10.6\ mH}$$

b. *In parallel*:

$$L_p = \frac{1}{\dfrac{1}{L_1} + \dfrac{1}{L_2} + \dfrac{1}{L_3}}$$

$$= \frac{1}{\dfrac{1}{10\text{ mH}} + \dfrac{1}{100\ \mu\text{H}} + \dfrac{1}{500\ \mu\text{H}}}$$

$$\mathbf{L_p = 82.6\ \mu H}$$

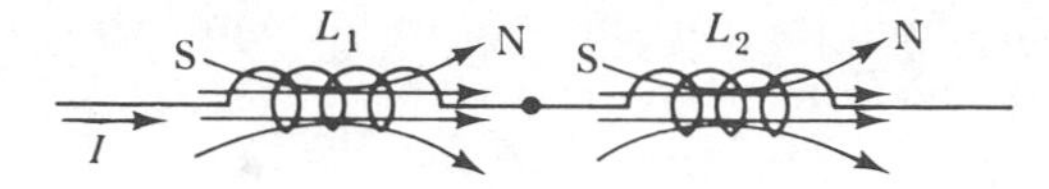

(a) Mutually coupled coils connected series-aiding

(b) Circuit symbols for mutually coupled coils connected series-aiding

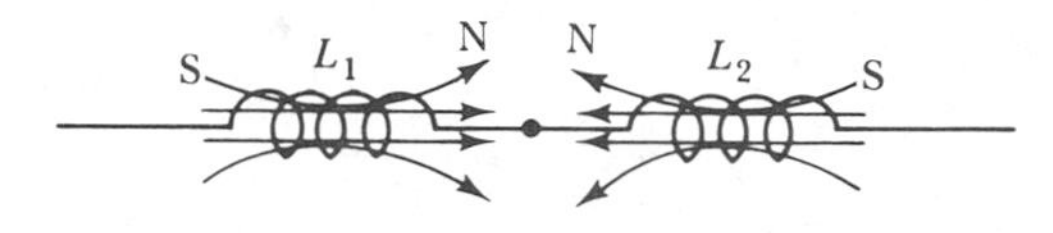

(c) Mutually coupled coils connected series-opposing

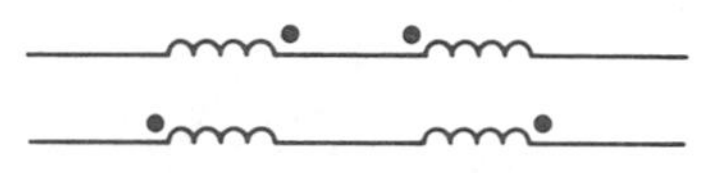
(d) Circuit symbols for mutually coupled coils connected series-opposing

FIG. 15-13. Mutually coupled coils connected series-aiding and series-opposing

Figure 15-13 illustrates the situation when mutual induction exists between two series-connected coils. In Figure 15-13(a) the coil connection is such that their fluxes assist each other, and the coils are said to be connected *series-aiding*. Figure 15-13(b) shows the graphic symbols used in circuit diagrams for mutually coupled coils which are connected series-aiding. The dots at the same ends of each coil indicate that the coils have fluxes in the same direction when a current flows.

The two coils shown in Figure 15-13(c) have their fluxes in opposite directions, and therefore they are said to be connected *series-opposing*. In Figure 15-13(d) the graphic symbols for series-opposing coils are illustrated. In this case the dots at opposite ends of each coil indicate that their fluxes are in opposition when a current flows.

For the series-aiding connections, the total emf induced in L_1 is the sum of the emf due to the coil's self-inductance and the emf induced from L_2. Thus,

$$e_{L1} = L_1 \frac{\Delta i}{\Delta t} + M \frac{\Delta i}{\Delta t}$$

$$= \frac{\Delta i}{\Delta t}(L_1 + M)$$

Similarly,

$$e_{L2} = \frac{\Delta i}{\Delta t}(L_2 + M)$$

and

$$e_{L(\text{total})} = \frac{\Delta i}{\Delta t}(L_1 + M + L_2 + M)$$

also

$$\frac{e_{L(\text{total})}}{\Delta i / \Delta t} = L_{\text{total}}$$

Therefore,

$$\boxed{L = L_1 + L_2 + 2M} \tag{15-18}$$

For the series-opposing connection, the emf induced in L_1 from L_2 is in opposition to the self-induced emf. Therefore,

$$e_{L1} = L_1 \frac{\Delta i}{\Delta t} - M \frac{\Delta i}{\Delta t}$$

$$= \frac{\Delta i}{\Delta t}(L_1 - M)$$

and

$$e_{L2} = \frac{\Delta i}{\Delta t}(L_2 - M)$$

Therefore,

$$e_{L(\text{total})} = \frac{\Delta i}{\Delta t}(L_1 - M + L_2 - M)$$

and

$$\boxed{L = L_1 + L_2 - 2M} \qquad \textbf{(15-19)}$$

EXAMPLE 15-13 Two 500-μH coils have a mutual inductance of 200 μH. Determine the total inductance of the two coils:

a. When they are connected series-aiding.

b. When they are connected series-opposing.

SOLUTION

a. *Series-aiding*:

Eq. 15-18

$$L = L_1 + L_2 + 2M$$

$$= 500\ \mu\text{H} + 500\ \mu\text{H} + 2(200\ \mu\text{H})$$

$$\mathbf{L = 1.4\ mH}$$

b. *Series-opposing*:

Eq. 15-19

$$L = L_1 + L_2 - 2M$$

$$= 500\ \mu\text{H} + 500\ \mu\text{H} - 2(200\ \mu\text{H})$$

$$\mathbf{L = 600\ \mu H}$$

If the mutual inductance between two adjacent coils is not known, it can be determined by measuring the total inductance of the coils in series-aiding and series-opposing connections. Then,

$$L_a = L_1 + L_2 + 2M \qquad \textit{for series-aiding}$$

and

$$L_b = L_1 + L_2 - 2M \qquad \textit{for series-opposing}$$

Subtracting,

$$L_a - L_b = 4M$$

Therefore,

$$\boxed{M = \frac{L_a - L_b}{4}} \tag{15-20}$$

Recall that the mutual inductance between two coils is given by Eq. (15-13) as

$$M = k\sqrt{L_1 L_2}$$

From these two equations, the coefficient of coupling of the two coils can be determined.

EXAMPLE 15-14 Use the results of Example 15-13 to determine the mutual inductance and coefficient of coupling for the two coils.

SOLUTION

$$L_a = 1.4 \text{ mH}$$

$$L_b = 600\ \mu\text{H}$$

Eq. (15-20):

$$M = \frac{L_a - L_b}{4}$$

$$= \frac{1.4 \text{ mH} - 600\ \mu\text{H}}{4}$$

$$\mathbf{M = 200\ \mu H}$$

From Eq. (15-13),

$$k = \frac{M}{\sqrt{L_1 L_2}}$$

$$= \frac{200\ \mu\text{H}}{\sqrt{500\ \mu\text{H} \times 500\ \mu\text{H}}}$$

$$\mathbf{k = 0.4}$$

15-8 STRAY INDUCTANCE

Since inductance is (change in flux linkages)/(change in current), every current-carrying conductor has some self-inductance, and every pair of conductors has inductance. These *stray inductances* are usually unwanted, although they are sometimes used as components in a circuit design. In dc applications, stray inductance is normally unimportant, but in radio frequency ac circuits, they can be a considerable nuisance. Stray inductance is normally minimized by keeping connecting wires as short as possible.

GLOSSARY OF FORMULAS

Induced emf:

$$e_L = \frac{\Delta\Phi}{\Delta t}$$

Induced emf:

$$e_L = \frac{\Delta\Phi N}{\Delta t}$$

Inductance:

$$L = \frac{e_L}{\Delta i/\Delta t}$$

Inductance:

$$L = \frac{\Delta\Phi N}{\Delta i}$$

Flux change:

$$\Delta\Phi = \mu_r \mu_o \Delta i N \frac{A}{l}$$

Self-inductance:

$$L = \mu_r \mu_o N^2 \frac{A}{l}$$

Mutual inductance:

$$M = \frac{e_L}{\Delta i/\Delta t}$$

Induced emf:

$$e_L = \frac{\Delta\Phi N_s}{\Delta t}$$

Mutual inductance:

$$M = \frac{\Delta\Phi N_s}{\Delta i}$$

Mutual inductance:

$$M = k\frac{\Delta\Phi N_s}{\Delta i}$$

Mutual inductance:

$$M = kN_p N_s \mu_r \mu_o \frac{A}{l}$$

Mutual inductance:

$$M = k\sqrt{L_1 L_2}$$

Energy stored:

$$W = \tfrac{1}{2}LI^2$$

Energy stored:

$$W = \frac{B^2 Al}{2\mu_r \mu_o}$$

Inductances in series:

$$L_s = L_1 + L_2 + L_3 + \cdots$$

Inductances in parallel:

$$\frac{1}{L_p} = \frac{1}{L_1} + \frac{1}{L_2} + \frac{1}{L_3} + \cdots$$

Total inductance:

$$L = L_1 + L_2 + 2M$$

Total inductance:

$$L = L_1 + L_2 - 2M$$

Mutual inductance:

$$M = \frac{L_a - L_b}{4}$$

REVIEW QUESTIONS

15-1 Draw sketches to show the direction of the induced current in a conductor in motion through a magnetic field. Also, show the field generated by the conductor. Briefly explain.

15-2 State Lenz's law and explain it in relation to the induced emf and current in a conductor moving through a magnetic field.

15-3 State Faraday's law, and define the weber. Use the definition of the weber to write an equation for induced emf.

15-4 Define: electromagnetic induction, primary, secondary, self-inductance, mutual inductance, counter-emf, back-emf, and coefficient of coupling.

15-5 Draw sketches to show how the current through a coil induces a counter-emf in the coil. Explain carefully.

15-6 Define the henry and use the definition to write an equation for the inductance of a coil.

15-7 Derive an equation for inductance L in terms of flux linkages and current change. Also, derive the equation that relates L to the dimensions of a core. Discuss the difficulties that arise with ferromagnetic cored inductors in relation to coil current.

15-8 Show how a coil can be wound to be noninductive, and draw the graphic symbol for a noninductive resistor.

15-9 Explain mutual inductance and define the relationship between two coils when they have a mutual inductance of 1 H. Using the definition, write an equation for the mutual inductance between two coils.

15-10 Derive an equation for mutual inductance in terms of flux linkages and current change. Also explain coefficient of coupling and show how it affects the equation for M.

15-11 Derive the equation relating M to the dimensions of a core. Also, derive the equation relating M to the individual coil inductances.

15-12 Sketch the basic circuit of an automobile ignition coil and briefly explain how it operates.

15-13 Describe the various types of inductors and discuss their characteristics and important parameters.

15-14 Derive an expression for the energy stored in an inductive circuit in terms of the inductance value and the current through the coil. Also, derive an expression for the energy stored in terms of the flux density and core dimensions.

15-15 Derive the equations for the inductance of several inductors:

a. Connected in series.

b. Connected in parallel.

Assume that there is no mutual induction between the inductors.

15-16 Using illustrations, explain what occurs when mutual inductance exists between two series-connected inductors:

a. When they are connected *series-aiding*.

b. When they are connected *series-opposing*.

Derive equations for the total inductance in each case.

15-17 Explain how the mutual inductance between two coils can be determined and derive the appropriate equation.

PROBLEMS

15-1 Write the equation for the emf induced in the secondary of two magnetically coupled coils. The flux linking the secondary of two coupled coils increases from zero to 0.12 Wb in a time of 0.3 s. Calculate the voltage induced in the secondary if it has 100 turns.

15-2 A cast iron core with two coils has a closed magnetic path 20 cm long and a cross-sectional area of 5 cm^2. The primary winding has $N_p = 400$ turns and the secondary winding has $N_s = 250$ turns. If the current through the primary increases from zero to 1 A in a time of 10 ms, determine the emf induced in the secondary.

15-3 Two coils are wound on a ring-shaped cast steel core. The secondary of the two has $N_s = 250$ turns, and the output voltage from it is to be 25 V when the primary current increases from zero to its maximum level in a time of 8 ms. If the primary winding has 200 turns, determine the maximum level of the current that must flow through the primary. The core cross-sectional area is 10 cm^2 and the magnetic path length is 30 cm.

15-4 A 5000-turn air-cored solenoid is 28 cm long and has an inside diameter of 2 cm. A secondary winding wound on top of the solenoid is to have 1 V induced in it when the solenoid current increases from zero to 15 A in a period of 2.5 ms. Determine the number of turns required on the secondary.

15-5 A coil current grows linearly from zero to a maximum of 30 A in a time period of 15 ms. If the coil has an inductance of 5 mH, determine the level of the counter-emf. Also, calculate the new maximum current level if the counter-emf is not to exceed 100 mV.

15-6 A solenoid is to be constructed with a length of 10 cm and inside diameter of 1 cm. If the inductance of the coil is to be 500 μH, calculate the number of turns required.

15-7 For the solenoid described in Problem 15-6, determine the total flux through the coil when the current is 25 mA. Also, calculate the counter-emf generated if the current increases linearly from zero in a time period of 50 ms.

15-8 Calculate the inductance of a 2000-turn coil that is 12 cm in length and has an average inside diameter of 1.2 cm.

15-9 A cast iron ring has a cross-sectional area of 1 cm^2 and a magnetic path length of 10 cm. A 100-turn coil is wound on the ring. Determine the coil inductance when the current through it is:

a. $I = 1$ A.

b. $I = 5$ A.

15-10 Two 75-turn coils are wound on an iron core that has a closed magnetic path. The core dimensions are $l=33$ cm and cross-sectional area $=9\ \text{cm}^2$. If the core has a relative permeability of 900, calculate the inductance of each coil and the mutual inductance between the coils.

15-11 Two 150-turn solenoids each have $l=22$ cm and cross-sectional area $=3.3\ \text{cm}^2$. The mutual inductance between the two is measured as 2 μH. Calculate their coefficient of coupling.

15-12 Calculate the energy stored in a 5-H choke that has a resistance of 105 Ω. The applied voltage is 250 V.

15-13 When 500 V is applied to a choke with a resistance of 120 Ω, it is found that the energy stored is 33 J. Calculate the inductance of the choke.

15-14 Four inductors have values of $L_1=100$ mH, $L_2=50$ mH, $L_3=750\ \mu$H, and $L_4=1$ H. Determine the total inductance of the four:

a. When connected in series.

b. When connected in parallel.

Assume that there is no mutual induction between the inductors.

15-15 Two inductors which have values of 500 μH and 800 μH are connected in parallel. Calculate the value of a third inductor which when connected in parallel with the other two will give a total inductance of 250 μH.

15-16 A 600-μH coil and a 400-μH coil have a mutual inductance between them of 100 μH. Determine the total inductance of the two:

a. When they are connected series-aiding.

b. When connected series-opposing.

15-17 Use the results of Problem 15-16 to determine the coefficient of coupling between the two coils.

15-18 Two identical coils when connected series-aiding give a total inductance of 850 μH. When connected series-opposing, their total inductance is 250 μH. Determine the mutual inductance between the two. Also, determine the values of individual inductances if the coefficient of coupling is 0.545.

16

CAPACITANCE

INTRODUCTION

A capacitor consists of a layer of insulating material sandwiched between two metal plates. The electric field in a capacitor has many parallels to a magnetic field, including a parameter known as the *permittivity*, which is analogous to magnetic permeability. The capacitance value of a capacitor is a measure of the amount of electric charge that can be stored in the device. The capacitance can be calculated from a knowledge of the capacitor dimensions and the permittivity of the insulating material. There are many different types of capacitors, each with its own particular characteristics and application. Parallel-connected and series-connected capacitors *cannot* be treated in the same way as parallel and series resistance circuits.

16-1 ELECTRIC CHARGE STORAGE

Figure 16-1(a) shows a device that consists of a layer of insulating material sandwiched between two metal plates. This device is known as a *capacitor* and, as will be shown, it has the ability to store an electric charge. The figure also shows a battery and switch S_1 for connecting the battery across the plates of the capacitor. Initially, there is no potential difference between the plates, but when the switch is closed, the plates assume the same potentials as the battery terminals. Therefore, there is now a potential difference between the capacitor plates [see Figure 16-1(b)].

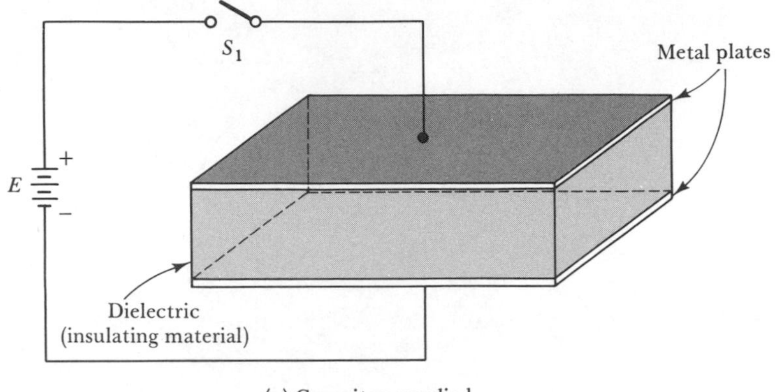

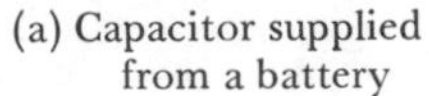
(a) Capacitor supplied from a battery

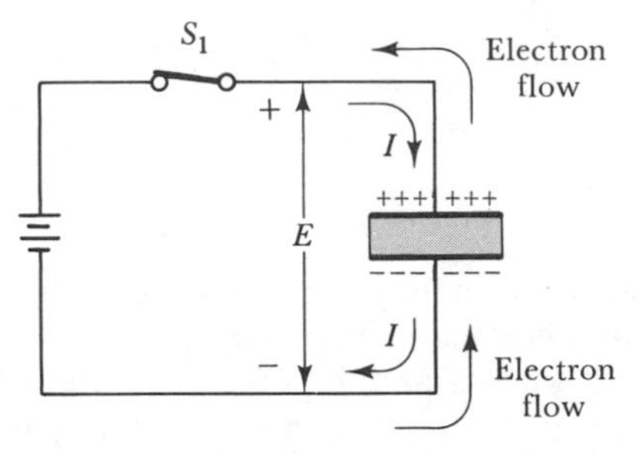

(b) Capacitor becomes charged to battery voltage when S_1 is closed

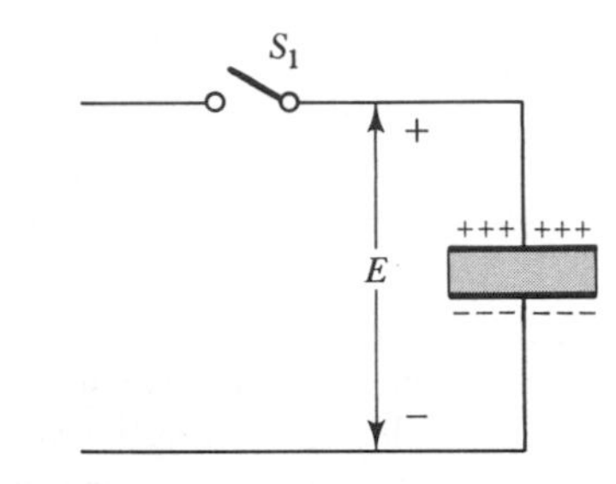

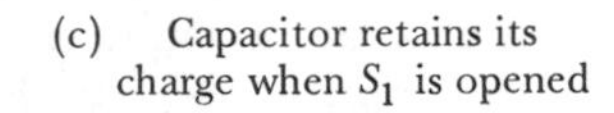
(c) Capacitor retains its charge when S_1 is opened

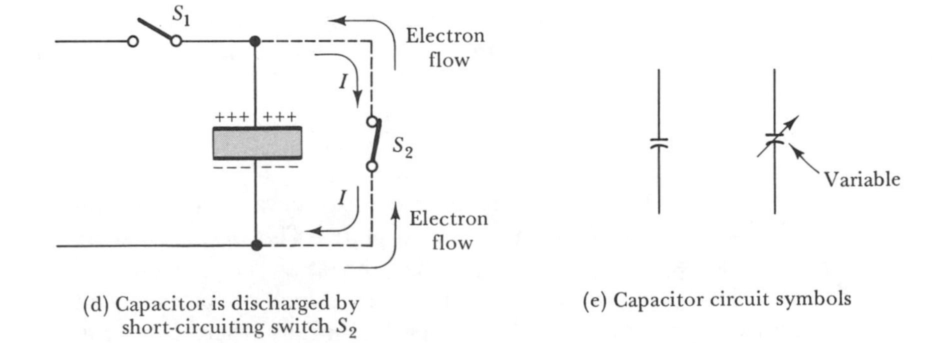

(d) Capacitor is discharged by short-circuiting switch S_2

(e) Capacitor circuit symbols

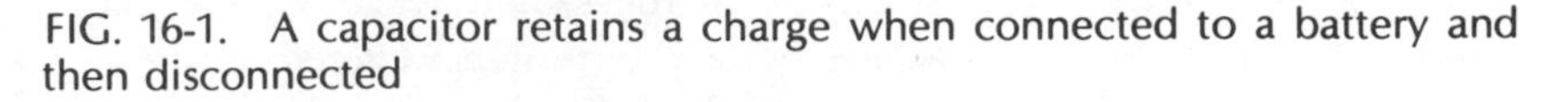
FIG. 16-1. A capacitor retains a charge when connected to a battery and then disconnected

Recall that the positive electrode of a battery is positive because many of its (*negative*) electrons have been removed. Also, the battery's negative electrode has an accumulation of electrons. For the capacitor plates in the illustration to have the same potential difference as the battery terminals, electrons must be removed from the top plate to make it positive, and the removed electrons must be delivered to the bottom plate to make it negative. This means that electrons flow out of the top plate and into the bottom plate, as illustrated in Figure 16-1(b). The flow of electrons constitutes a *conventional direction* current flow into the top plate and out of the bottom plate. The current flow lasts only a very brief time (less than milliseconds), and ceases just as soon as the plates are at the same potentials as the battery terminals.

If the switch is now opened, the capacitor plates are found to maintain their potential difference [see Figure 16-1(c)]. It can be shown that the capacitor has, in fact, stored an electric charge, and that the quantity of charge stored depends upon the dimension of the plates and insulating material as well as upon the battery voltage. The capacitor was *charged* by the brief pulse of current that flowed when switch S_1 was closed. If *short-circuiting switch* S_2 in Figure 16-1(d) is now closed, the surplus of electrons on the bottom plate returns to the top plate, as shown. Again, this is thought of as a conventional direction current flow from the positive plate to the negative plate. In any case, closing S_2 *discharges* the capacitor and returns the plates to the same potential.

The layer of insulating material between the capacitor plates is known as the *dielectric*. The dielectric has to be insulating material, because if it were conducting material, it would simply short-circuit the plates. Typical dielectric materials are rubber, mica, paper, and air. The capacitor plates must be conducting material in order to have large quantities of electrons which can be transferred between plates. The graphic symbols for fixed and variable capacitors are shown in Figure 16-1(e). The straight line identifies the terminal of the capacitor that should be always made positive.

16-2 ELECTRIC FIELD

In Chapter 1 it was explained that a force exists between electrically charged bodies—also, that bodies with like charges repel each other, and those with unlike charges experience a force of attraction. These forces are similar in many ways to the forces that occur between magnetic poles. As with magnetism, a *force field* exists around electrically charged bodies. *Electric lines of force* are spoken of, although their existence cannot be demonstrated as easily as in the case of magnetism. Also, the *electric field strength*, *electric flux*, and *electric flux density* can all be calculated.

In Figure 16-2(a), the bottom plate of a capacitor is shown grounded, and the top plate has a potential of $+E$ with respect to (w.r.t.) ground. In the space between the two plates, the potential at the halfway point becomes $+E/2$ w.r.t. ground. Similarly, at $\frac{1}{4}d$ and $\frac{3}{4}d$, the potentials are $+E/4$ and $(+3/4)E$, respectively, w.r.t. ground. At each of these points a line exists along which the potential is a constant quantity. Appropriately, the lines are termed *equipotential lines*.

In the space between the capacitor plates, the equipotential lines are found to be horizontal. However, at the edges of the plates, they curve outward, as illustrated in Figure 16-2(b). It is also found that the electric lines of force always cross the equipotential lines at right angles. Thus, the lines of force are vertical in the space between the plates and bulge outward at the plate edges, as shown in the figure.

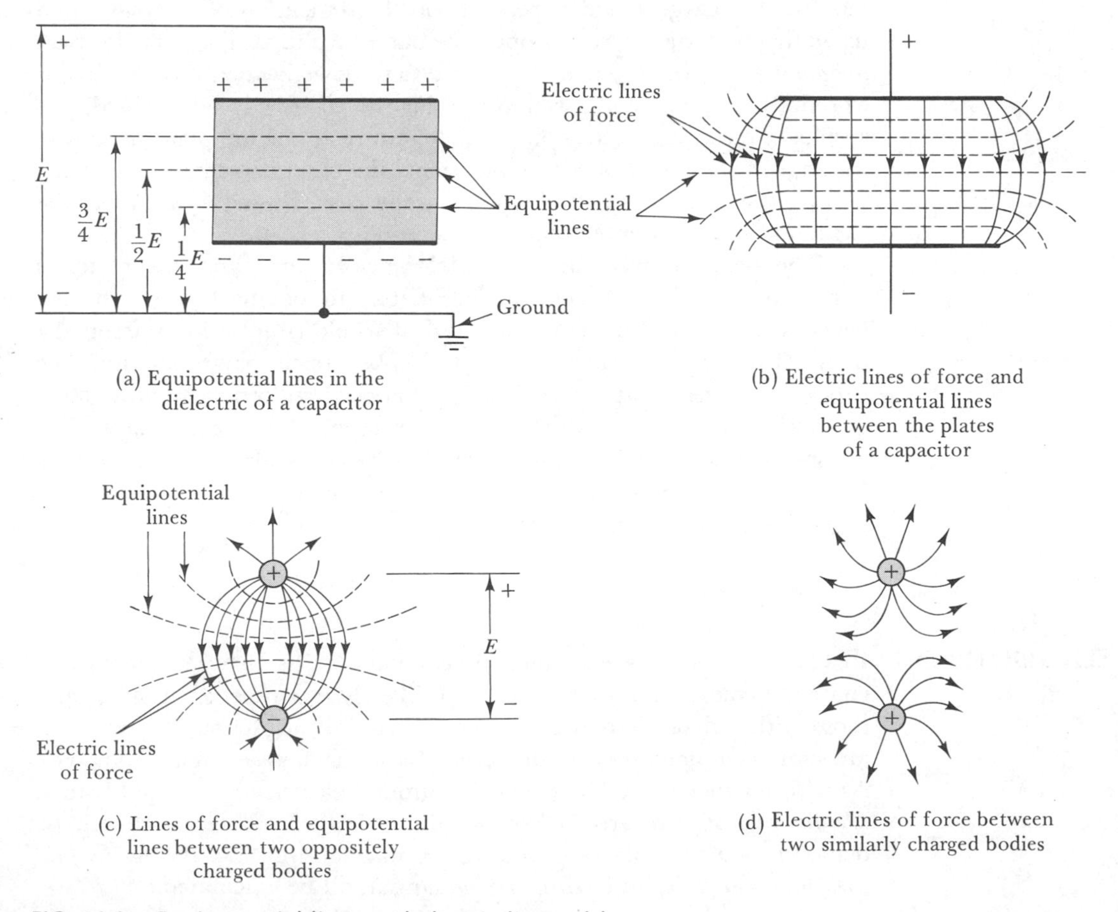

(a) Equipotential lines in the dielectric of a capacitor

(b) Electric lines of force and equipotential lines between the plates of a capacitor

(c) Lines of force and equipotential lines between two oppositely charged bodies

(d) Electric lines of force between two similarly charged bodies

FIG. 16-2. Equipotential lines and electric lines of force

The direction of the electric lines of force is assumed to be the direction in which a free positively charged particle would move if it were placed in the electric field. Such a particle would obviously be repelled from the positive plate of the capacitor and attracted to the negative plate. Therefore, the electric lines of force are said to flow from positive to negative. Figure 16-2(c) shows the electric field pattern that exists between two round bodies that are oppositely charged. The lines of force from each are seen to join together. In Figure 16-2(d), the lines of force from the two like-charged bodies are shown repelling each other. As with magnetic lines of force, it is found that electric lines of force repel each other, never intersect, and are always in a state of tension.

Electric charge is measured in coulombs (see Section 3-1), and the coulomb is also the unit of electric flux. Therefore, a body that is charged to Q coulombs emits a total electric flux of Q coulombs. Similarly, a capacitor that has Q coulombs of charge has a total electric flux of Q coulombs between its plates.

$$\text{electric flux, } \psi = Q \tag{16-1}$$

where ψ is the electric flux in coulombs, and Q is the charge in coulombs.

The electric flux density is simply the flux per unit area. Therefore,

$$\text{electric flux density, } D = \frac{Q}{A} \tag{16-2}$$

where D is electric flux density in C/m^2, Q is total charge in coulombs, and A is the area in m^2.

The electric field strength in the space between the plates of a capacitor depends upon the applied voltage and upon the distance between the plates:

$$\text{electric field strength, } \mathcal{E} = \frac{E}{d} \tag{16-3}$$

where E is the voltage difference between the plates in volts, d is the distance between the plates in meters, and $\mathcal{E}$ is the electric field strength in V/m.

The insulating dielectric material between the plates of a capacitor is subjected to electric stress; that is, the applied voltage (electric pressure) is pressing for a current flow. Where the voltage between the plates is low, the electric stress on the dielectric is low. When the voltage is high, the stress is greater, and if the stress becomes high enough, the material may break down. This was discussed in Section 5-2. The stress on the dielectric is measured in volts/meter, and it is, in fact, the electric field strength. The *dielectric strength* of a given material is the voltage per unit thickness at which the material would break down (see Table 5-1).

Since the dielectric is an insulating material, the electrons within the dielectric are never detached from their atoms (except in a breakdown situation). However, a certain amount of distortion of the atoms occurs, as illustrated in Figure 16-3. When the capacitor plates are at the same potential there is no electric stress on the dielectric, and in the dielectric atoms the electrons are orbiting normally [see Figure 16-3(a)]. When a potential difference exists between the plates, the orbiting electrons experience a force attracting them toward the positive plate, while the atomic nucleus is attracted toward the negative plate. The result is that the electrons tend to go into a distorted orbit around the nucleus, as illustrated in Figure 16-3(b). In this situation the atoms are said to be *polarized*.

When a capacitor is discharged by short-circuiting its plates, the polarized atoms may be expected to return to their normal state. With some dielectric materials, it is found that the polarized atoms do not return completely to their normal state. Consequently, when the short circuit is removed from the capacitor plates, a small voltage can again be measured across the plates. The plate voltage is, of course, the result of residual polarization within the dielectric. This phenomenon of energy being retained by the dielectric is referred to as *dielectric absorption*.

EXAMPLE 16-1 A capacitor has a plate area of 400 cm^2 and a dielectric thickness of 1 mm. If the capacitor has a charge of 10×10^{-3} C when the voltage

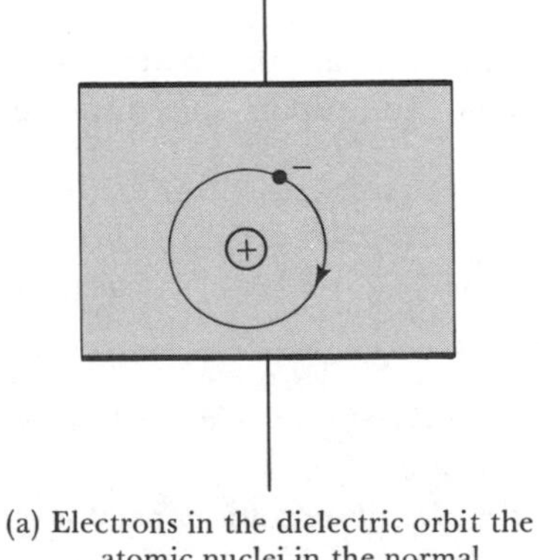

(a) Electrons in the dielectric orbit the atomic nuclei in the normal symmetrical way when the capacitor is uncharged

(b) Electron orbits are distorted when the capacitor is charged

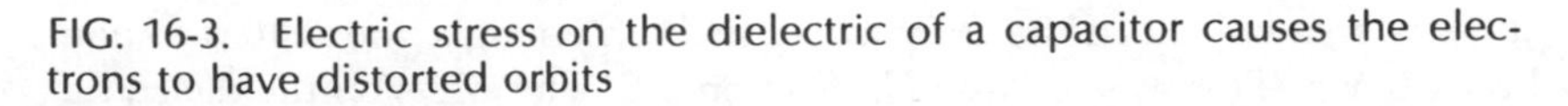

FIG. 16-3. Electric stress on the dielectric of a capacitor causes the electrons to have distorted orbits

between its plates is 25 V, determine the electric field strength and flux density within the dielectric.

SOLUTION

Eq. (16-3):

$$\textit{electric field strength}, \mathcal{E} = \frac{E}{d}$$

$$= \frac{25\ \text{V}}{1 \times 10^{-3}\ \text{m}}$$

$$\mathcal{E} = \mathbf{25\ 000\ V/m}$$

Eq. (16-2):

$$\textit{electric flux density}, D = \frac{Q}{A}$$

$$= \frac{10 \times 10^{-3}\ \text{C}}{400 \times 10^{-4}\ \text{m}^2}$$

$$D = \mathbf{0.25\ C/m^2}$$

16-3 CAPACITANCE AND CAPACITOR DIMENSIONS

The amount of charge that can be stored by a capacitor with a given terminal voltage is termed its *capacitance*. The capacitance of a capacitor has a definite relationship to the area of the plates and to the thickness of the dielectric.

Refer again to Figure 16-1(b) and recall that electrons are attracted to a positive potential. Therefore, the presence of the positive potential on the top plate causes electrons to be attracted into the adjacent bottom plate of the capacitor. The electrons cannot pass through the insulating dielectric to the positive top plate. Instead, they get as close to the top plate as possible by accumulating on the surface of the bottom plate close to the dielectric. Similarly, the negative potential on the bottom plate causes electrons to be driven out of the adjacent top plate. Consequently, a positive charge (i.e., a lack of electrons) is accumulated at the surface of the top plate close to the dielectric.

Now look at Figure 16-4(a), which shows a capacitor consisting of two metal plates with a very thick dielectric between them. With such a thick piece of insulating material between the two plates, the potential on each plate has very little influence on the electrons in the other plate. Therefore, only a relatively small charge can be stored. In Figure 16-4(b), a capacitor with a very thin dielectric is shown. In this case the potential on each plate has a strong influence on the very close adjacent plate. The

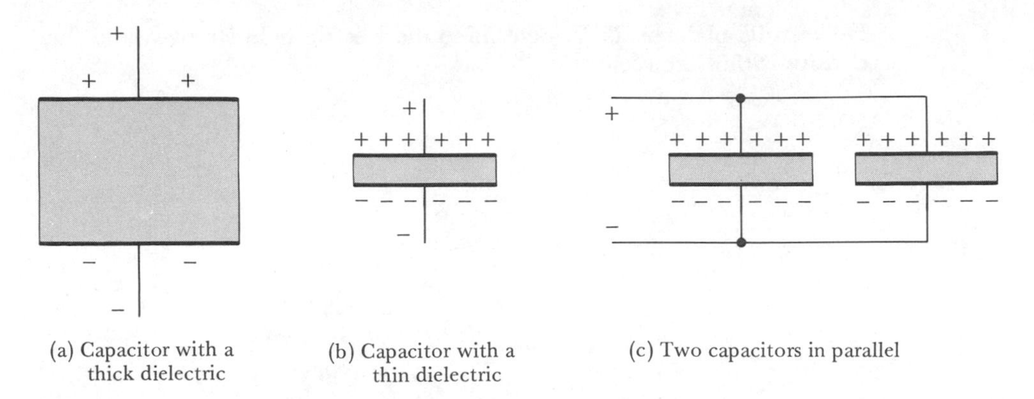

(a) Capacitor with a thick dielectric

(b) Capacitor with a thin dielectric

(c) Two capacitors in parallel

FIG. 16-4. Capacitance is directly proportional to plate area and inversely proportional to dielectric thickness

result is, of course, that a relatively large number of electrons is accumulated on the bottom plate, and an equally large number of electrons is driven out of the top plate, constituting a relatively large charge on the capacitor.

It can be experimentally demonstrated that the amount of charge stored in a capacitor is inversely proportional to the thickness of the dielectric.

Figure 16-4(c) shows two identical capacitors connected in parallel. If each capacitor has a charge of Q coulombs, the total charge stored by the two is $2Q$ coulombs. Paralleling the two similar capacitors means that the plate area for one capacitor has been doubled. Thus, a capacitor with twice the plate area would also store twice the charge. Now it can be written,

$$\text{capacitance} \propto \frac{\text{plate area}}{\text{distance between plates}}$$

or

$$C \propto \frac{A}{d}$$

Refer once more to Figure 16-4(b). It can be shown that for a given applied voltage, a definite number of electrons is attracted into the bottom plate, and the same number of electrons is driven out of the top plate of the capacitor. If the applied voltage is now doubled, twice as many electrons are displaced from the positive plate to the negative plate, and therefore the accumulated charge is doubled. Thus, the charge on a capacitor is directly proportional to the applied voltage as well as to the capacitance of the capacitor:

$$Q \propto CE$$

*The **farad*** *(F) is the SI unit of capacitance. The **farad** is the capacitance of a capacitor that contains a charge of 1 coulomb when the potential difference between its terminals is 1 volt.*

It is found that the farad is a very large unit; consequently, microfarads (μF), nanofarads (nF), and picofarads (pF) are most frequently used for expressing capacitance.

From the definition of the farad, the equation for charge becomes

$$\boxed{Q = CE} \tag{16-4}$$

where Q is charge in coulombs, C is capacitance in farads, and E is potential difference between the plates in volts.

TABLE 16-1

MATERIAL	TYPICAL DIELECTRIC CONSTANT (ϵ_r)
Vacuum	1
Air	1.0006
Ceramic (low loss)	6–20
Ceramic (high ϵ_r)	>1000
Glass	5–100
Mica	3–7
Mylar	3
Oxide film	5–25
Paper	4–6
Polystyrene	2.5
Teflon	2

Before proceeding further, another factor, known as the *permittivity*, must be introduced. Electric permittivity is analogous to magnetic permeability. It specifies the ease with which electric flux is *permitted* to pass through a given dielectric material. Like permeability, permittivity is subdivided into the *permittivity of free space* ϵ_o, and the *relative permittivity* ϵ_r, which is also known as the *dielectric constant*. Table 16-1 lists the relative permittivities of various dielectric materials. Note that the permittivity of air is 1.0006, and therefore it is usually taken as 1 (i.e., the same as for vacuum).

*Named for the English chemist and physicist Michael Faraday (1791–1867).

In SI units the permittivity of free space can be shown to be

$$\epsilon_o = \frac{1}{36\pi \times 10^9}$$

or

$$\epsilon_o \cong 8.84 \times 10^{-12}$$

It is interesting to note that

$$\frac{1}{\sqrt{\mu_o \epsilon_o}} = \frac{1}{\sqrt{4\pi \times 10^{-7} \times \dfrac{1}{36\pi \times 10^9}}}$$

$$= 3 \times 10^8$$

or

$$\frac{1}{\sqrt{\mu_o \epsilon_o}} = \left\{ \begin{array}{l} \text{velocity of light and other} \\ \text{electromagnetic waves, in meters per second*} \end{array} \right.$$

The dielectric permittivity has a similar relationship to field strength and flux density as does the magnetic permeability:

$$\boxed{\epsilon_r \epsilon_o = \frac{D}{\mathcal{E}}} \tag{16-5}$$

where $\mathcal{E}$ is electric field strength in V/m and D is electric flux density in C/m^2.

From Eq. (16-2),

$$D = \frac{Q}{A}$$

and from Eq. (16-3),

$$\mathcal{E} = \frac{E}{d}$$

Therefore,

$$\epsilon_r \epsilon_o = \frac{Q}{A} \times \frac{d}{E}$$

From Eq. (16-4),

$$Q = CE$$

---

*This relationship was discovered in 1865 by the Scottish physicist James C. Maxwell (1831–1879). From it he was able to predict the existence of electromagnetic waves many years before they were demonstrated experimentally.

Therefore,

$$\epsilon_r \epsilon_o = \frac{Cd}{A}$$

or

$$\boxed{C = \frac{\epsilon_r \epsilon_o A}{d}} \qquad \textbf{(16-6)}$$

When A is plate area in m^2, and d is dielectric thickness in meters, C is the capacitance in farads.

EXAMPLE 16-2

Calculate the capacitance of a capacitor with a plate area of 400 cm^2 and a dielectric thickness of 1 mm:

a. When the dielectric is air.

b. When the dielectric is mica with a relative permittivity of 5.

c. Determine the charge on the capacitor in each case when the applied voltage is 25 V.

SOLUTION

a. *From* Eq. (16-6),

$$C = \frac{\epsilon_r \epsilon_o A}{d}$$

$$= \frac{8.85 \times 10^{-12} \times 400 \times 10^{-4}\ \text{m}^2}{1 \times 10^{-3}\ \text{m}}$$

$$\boldsymbol{C = 354\ \text{pF}}$$

b.

$$C = \frac{5 \times 8.85 \times 10^{-12} \times 400 \times 10^{-4}\text{m}^2}{1 \times 10^{-3}\ \text{m}}$$

$$\boldsymbol{C = 1.77\ \text{nF}}$$

c. From Eq. (16-4),

$$Q = CE$$

For $C = 354$ pF,

$$Q = 354\ \text{pF} \times 25\ \text{V}$$

$$\boldsymbol{Q = 8.85 \times 10^{-9}\ \text{C}}$$

For $C = 1.77$ nF,

$$Q = 1.77 \text{ nF} \times 25 \text{ V}$$

$$\mathbf{Q = 4.425 \times 10^{-8} \text{ C}}$$

EXAMPLE 16-3

A 1-μF capacitor is to be constructed from rolled-up sheets of aluminum foil separated by a layer of paper 0.1 mm thick. Calculate the required area for each sheet of foil if the relative permittivity of the paper is 6.

SOLUTION

From Eq. (16-6),

$$A = \frac{Cd}{\epsilon_r \epsilon_o}$$

$$= \frac{1 \ \mu\text{F} \times 0.1 \times 10^{-3} \text{ m}}{6 \times 8.85 \times 10^{-12}}$$

$$\mathbf{A = 1.88 \text{ m}^2}$$

16-4 CAPACITOR TYPES AND CHARACTERISTICS

As well as the actual value of capacitance, there are many other important characteristics that must be considered in selecting a capacitor for a particular application.

WORKING VOLTAGE. The maximum voltage that may be safely applied to a capacitor is usually expressed in terms of its *dc working voltage*. A *maximum rms ac voltage* (see Chapter 18) may also be specified, and this is usually little more than half the maximum dc voltage. These limits should never be exceeded; otherwise, dielectric breakdown may occur.

CAPACITOR TOLERANCE. This is simply the accuracy with which the capacitance value is specified. Tolerances of $\pm 20\%$ are normal for most small-value capacitors, but more precise capacitors can be purchased at increased cost. In the case of larger-value capacitors, the capacitance has to be a certain minimum value in many applications. Consequently, manufacturers tend to specify the tolerance as $-10\% + 150\%$. This means, for example, that a 100-μF capacitor could have a value as low as 90 μF or as high as 250 μF.

TEMPERATURE EFFECTS. Every capacitor type has an operating temperature range specified by the manufacturer. Typical ranges are $-20°$C to +65°C, −40°C to +65°C, and −55°C to 125°C. Obviously, no capacitor should be employed in an environment where the temperatures may be beyond its range. The capacitance value is also likely to change over the temperature range. The maximum change that can occur is specified by the manufacturer either in *parts per million per degree celsius* (ppm/°C), or as a percentage change for the temperature extremes. Sometimes a graph is provided of capacitance percentage change versus temperature.

LEAKAGE CURRENT. Despite the fact that the dielectric is an insulator, small leakage currents flow between the plates of a capacitor. The actual level of leakage current depends upon the insulation resistance of the dielectric. *Plastic film* capacitors, for example, may have insulation resistances higher than 100 000 MΩ. At the other extreme, an *electrolytic* capacitor may have several milliamps of leakage current, with only 10 V applied to its terminals.

POLARIZATION. Electrolytic capacitors normally have one terminal identified as the most positive connection. Thus, they are said to be *polarized*. This usually limits their application to situations where the polarity of the applied voltage will not change.

CAPACITOR EQUIVALENT CIRCUIT. An *ideal capacitor* has a dielectric that has an infinite resistance, and plates that have zero resistance. However, an ideal capacitor does not exist, since all dielectrics have some leakage current and all capacitor plates have some resistance. The complete equivalent circuit for a capacitor shown in Figure 16-5(a) consists of an ideal capacitor C in series with a resistance R_D representing the resistance of the plates, and in parallel with a resistance R_L representing the leakage resistance of the dielectric. Usually, the plate resistance can be completely neglected, and the equivalent circuit becomes that shown in Figure 16-5(b). With capacitors that have a very high leakage resistance (e.g., mica and plastic film capacitors), the parallel resistor is frequently omitted in the equivalent circuit, and the capacitor is then treated as an ideal capacitor. This can normally *not* be done with

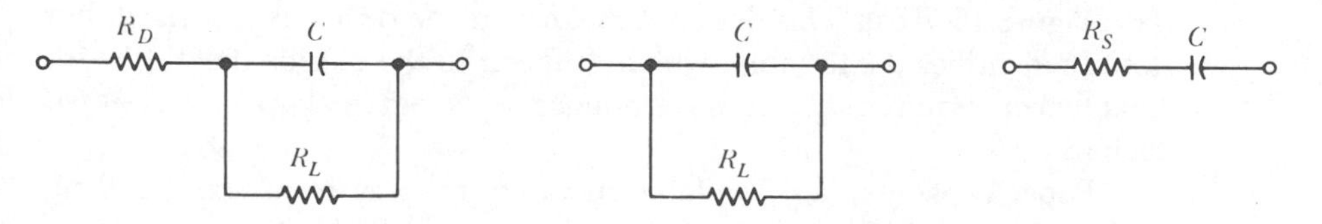

(a) Complete equivalent circuit (b) Parallel equivalent circuit (c) Series equivalent circuit

FIG. 16-5. Capacitor equivalent circuits

electrolytic capacitors, for example, which have relatively low leakage resistances. In Section 21-6 it is shown that the circuit of Figure 16-5(b) has an equivalent series circuit as shown in Figure 16-5(c).

Because a capacitor's dielectric is largely responsible for determining its most important characteristics, capacitors are usually identified by the type of dielectric used.

AIR CAPACITORS. A typical capacitor using air as a dielectric is illustrated in Figure 16-6. The capacitance is variable, as is the case with virtually all air capacitors. There are two sets of metal plates, one set fixed and one movable. The movable plates can be adjusted into or out of the spaces between the fixed plates by means of the rotatable shaft. Thus, the area of the plates opposite each other is increased or decreased, and the capacitance value is altered.

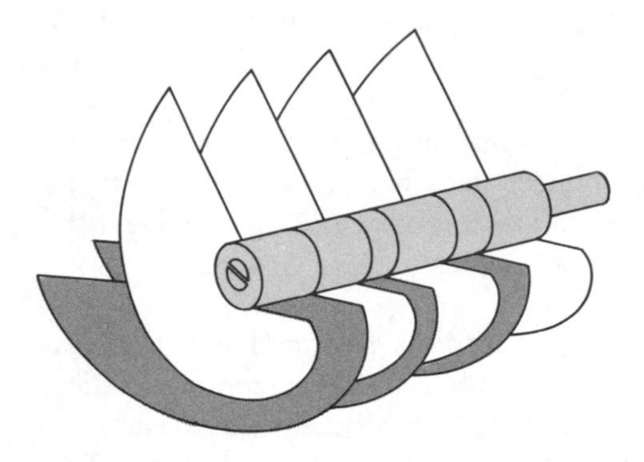

FIG. 16-6. Variable air capacitor

PAPER CAPACITORS. In its simplest form, a paper capacitor consists of a layer of paper between two layers of metal foil. The metal foil and paper are rolled up, as illustrated in Figure 16-7(a); external connections are brought out from the foil layers; and the complete assembly is dipped in wax or plastic. A variation of this is the *metalized paper* construction, in which the foil is replaced by thin films of metal deposited on the surface of the paper. One end of the capacitor sometimes has a band around it [see Figure 16-7(b)]. This does not mean that the device is polarized, but simply identifies the terminal which connects to the outside metal film, so that when required it can be grounded to avoid pickup of unwanted signals.

Paper capacitors are available in values ranging from about 500 pF to 50 μF, and in dc working voltages up to about 600 V. They are among the lowest-cost capacitors for a given capacitance value, but are physically larger than several other types having the same capacitance value.

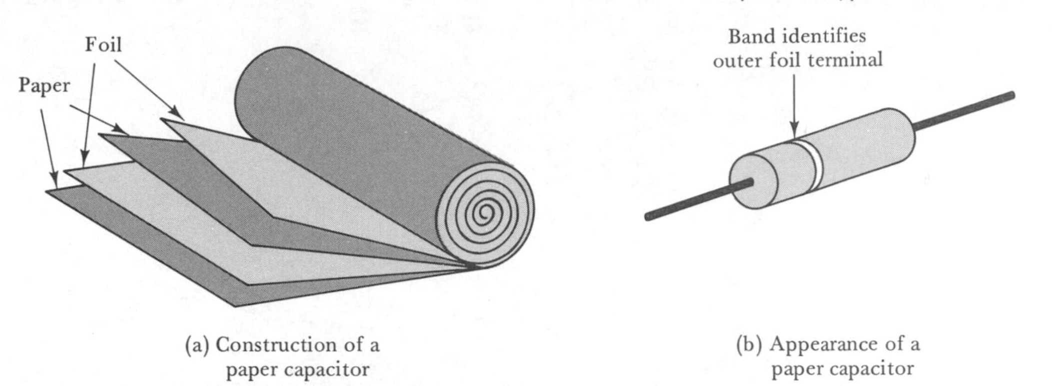

(a) Construction of a paper capacitor

(b) Appearance of a paper capacitor

FIG. 16-7. Construction and appearance of a paper capacitor

PLASTIC FILM CAPACITORS. The construction of plastic film capacitors is similar to that of paper capacitors, except that the paper is replaced by a thin film which is typically polystyrene or Mylar. This type of dielectric gives insulation resistances greater than 100 000 MΩ. Working voltages are as high as 600 V, with the capacitor surviving 1500-V surges for a brief period. Capacitance tolerances of ±2.5% are typical, as are temperature coefficients of 60 to 150 ppm/°C.

Plastic film capacitors are physically smaller but more expensive than paper capacitors. They are typically available in values ranging from 5 pF to 0.47 μF.

MICA CAPACITORS. As illustrated in Figure 16-8(a), mica capacitors consist of layers of mica alternated with layers of metal foil. Connections are made to the metal foil for capacitor leads, and the entire assembly is dipped in plastic or encapsulated in a molded plastic jacket. Typical capacitance values range from 1 pF to 0.1 μF, and voltage ratings as high as 35 000 V are possible. Precise capacitance values and wide operating temperatures are obtainable with mica capacitors. In a variation of the process, *silvered mica* capacitors use films of silver deposited on the mica layers instead of metal foil.

CERAMIC CAPACITORS. The construction of a typical ceramic capacitor is illustrated in Figure 16-8(b). Films of metal are deposited on each side of a thin ceramic disc, and copper wire terminals are connected to the metal. The entire unit is then encapsulated in a protective coating of plastic. Two different types of ceramic are used, one of which has extremely high relative permittivity. This gives capacitors which are much smaller than paper or mica capacitors having the same capacitance value. One disadvantage of this particular ceramic dielectric is that its leakage resistance is not as high as with other types. Another type of

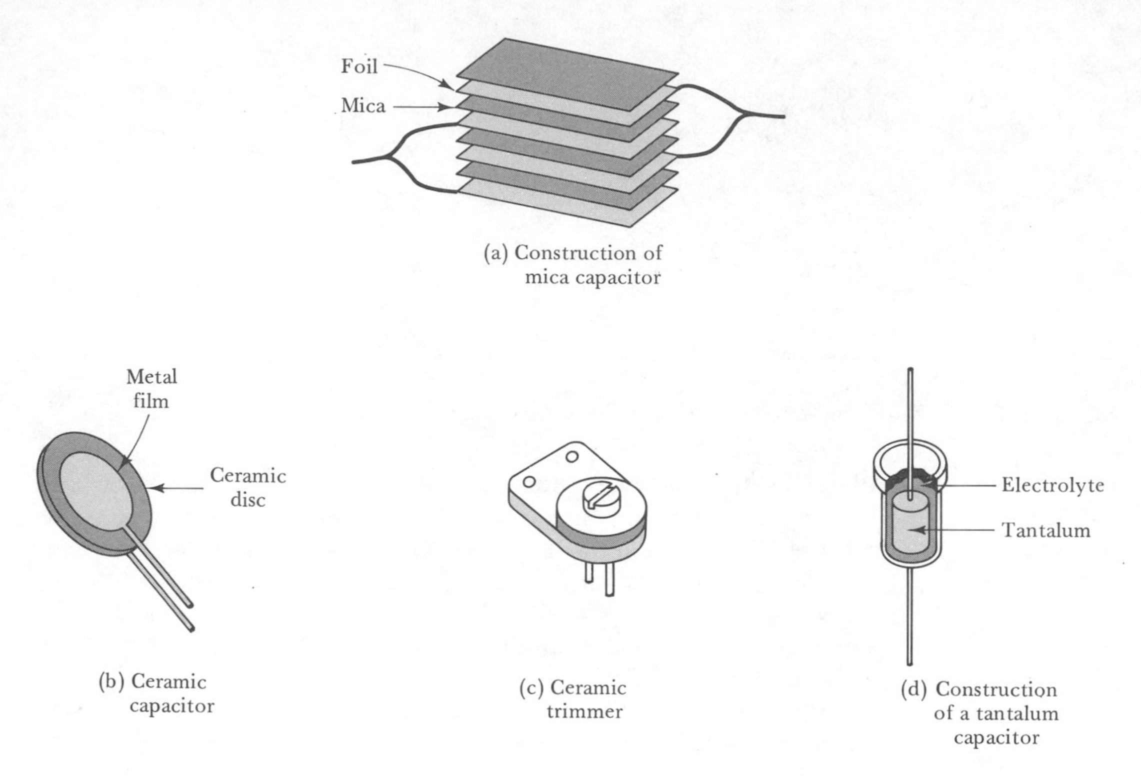

FIG. 16-8. Construction of various types of capacitors

ceramic is also used, and this gives leakage resistances on the order of 7500 MΩ. Because it has a lower permittivity, this type of ceramic produces capacitors that are relatively large for a given value of capacitance.

The range of capacitance values available with ceramic capacitors is typically 1 pF to 0.1 μF, with dc working voltages up to 1000 V.

Figure 16-8(c) shows a variable ceramic capacitor known as a *trimmer*. By means of a screwdriver, the area of plate on each side of a dielectric can be adjusted to alter the capacitance value. Typical ranges of adjustment available are 1.5 pF to 3 pF and 7 pF to 45 pF.

ELECTROLYTIC CAPACITORS. The most important feature of electrolytic capacitors is their very large capacitance value in a physically small container. For example, a capacitance of 5000 μF can be obtained in a cylindrical package approximately 5 cm long by 2 cm in diameter. In this case the dc working voltage is only 10 V. Similarly, a 1-F capacitor is available in a 22-cm by 7.5-cm cylinder, with a working voltage of only 3 V.

The construction of an electrolytic capacitor is basically the same as that of a paper capacitor (see Figure 16-7). Two sheets of aluminum foil separated by a fine gauze soaked in electrolyte are rolled up and encased in an aluminum cylinder for protection. When assembled, a direct voltage is applied to the capacitor terminals, and this causes a thin layer of aluminum oxide to form on the surface of the positive plate next to the electrolyte. The aluminum oxide is the dielectric, and the electrolyte and positive sheet of foil are the capacitor plates. The extremely thin oxide dielectric gives the very large value of capacitance.

It is very important that electrolytic capacitors be connected with the correct polarity. When incorrectly connected, gas forms within the electrolyte and the capacitor may explode. Nonpolarized electrolytic capacitors can be obtained. They consist essentially of two capacitors in one package connected *back to back*, so one of the oxide films is always correctly biased.

Electrolytic capacitors are available with dc working voltages greater than 400 V, but in this case capacitance values do not exceed 100 μF. In addition to their low working voltage and polarized operation, another disadvantage of electrolytic capacitors is their relatively high leakage current.

TANTALUM CAPACITORS. This is essentially another type of electrolytic capacitor. Powered tantalum is *sintered* (or baked), typically into a cylindrical shape. The resulting solid is quite porous, so that when immersed in a container of electrolyte, the electrolyte is absorbed into it, and the tantalum then has a large surface area in contact with the electrolyte [see Figure 16-8(d)]. When a dc *forming voltage* is applied, a thin oxide film is formed throughout the electrolyte–tantalum contact area. The result, again, is a large capacitance value in a small volume.

A typical tantalum capacitor in a cylindrical shape 2 cm by 1 cm might have a capacitance of 100 μF and a dc working voltage of 20 V. Other types are available with working voltage up to 630 V, but with capacitance values on the order of 3.5 μF. Like aluminum-foil electrolytic capacitors, tantalum capacitors must be connected with the correct polarity.

16-5 PARALLEL AND SERIES CAPACITORS

When capacitors are connected in *parallel*, the result is the same as increasing the total plate area. From Eq. (16-6), the capacitance of a capacitor is given by

$$C = \frac{\epsilon_r \epsilon_o A}{d}$$

Therefore, when three capacitors with plate areas of A_1, A_2, and A_3 are connected in parallel as shown in Figure 16-9(a), the total capacitance is

$$C = \frac{\epsilon_r \epsilon_o}{d}(A_1 + A_2 + A_3)$$

$$= \frac{\epsilon_r \epsilon_o A_1}{d} + \frac{\epsilon_r \epsilon_o A_2}{d} + \frac{\epsilon_r \epsilon_o A_3}{d}$$

or

$$\boxed{C = C_1 + C_2 + C_3 + \cdots} \quad (16\text{-}7)$$

Thus, the total capacitance of capacitors in parallel is the sum of the individual capacitance values.

For the parallel capacitors shown in Figure 16-9(a), the applied voltage is obviously common to all three. From Eq. (16-4), the charge on

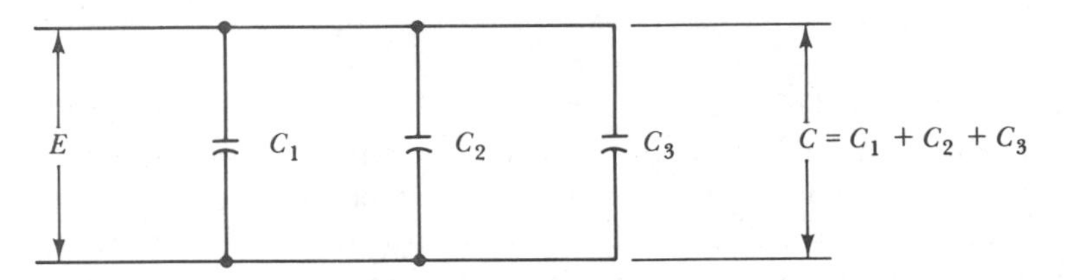

(a) Capacitors in parallel

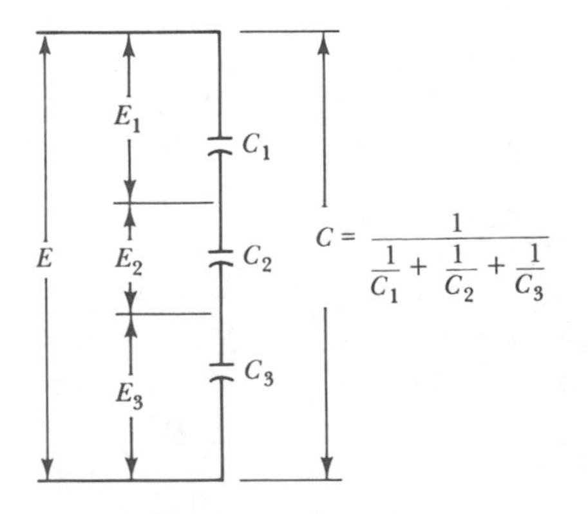

(b) Capacitors in series

FIG. 16-9. Capacitors connected in series and in parallel

a capacitor is

$$Q = CE$$

Therefore, the charge on each capacitor is

$$Q_1 = C_1E \qquad Q_2 = C_2E \qquad Q_3 = C_3E$$

and the total charge is

$$Q = Q_1 + Q_2 + Q_3$$
$$= C_1E + C_2E + C_3E$$

$$\boxed{Q = C_{\text{total}}E} \tag{16-8}$$

Three series-connected capacitors are shown in Figure 16-9(b). It is seen that connecting capacitors in series amounts to increasing the total thickness of the dielectric between the two outer plates. Equation (16-6) gives the capacitance as

$$C = \frac{\epsilon_r \epsilon_o A}{d}$$

Therefore, when three capacitors with similar plate areas and with dielectric thickness of d_1, d_2, and d_3 are connected in series, as shown in Figure 16-9(b), the total capacitance is

$$C = \frac{\epsilon_r \epsilon_o A}{d_1 + d_2 + d_3}$$

giving

$$\frac{1}{C} = \frac{d_1 + d_2 + d_3}{\epsilon_r \epsilon_o A}$$

$$= \frac{d_1}{\epsilon_r \epsilon_o A} + \frac{d_2}{\epsilon_r \epsilon_o A} + \frac{d_3}{\epsilon_r \epsilon_o A}$$

or

$$\boxed{\frac{1}{C} = \frac{1}{C_1} + \frac{1}{C_2} + \frac{1}{C_3} + \cdots} \tag{16-9}$$

Therefore, the total capacitance of capacitors connected in series is the reciprocal of the sum of the reciprocals of the individual capacitances.

Note that the formulas for total capacitance with parallel-connected and series-connected capacitors are the reverse of the formulas for similar resistance circuits.

For the three series-connected capacitors in Figure 16-9(b), the charging current must flow through all three capacitors. If a charging current I flows for a time t, then from Eq. (3-1), *the charge supplied to each capacitor is equal to the charge supplied to all three capacitors*:

$$Q = Q_1 = Q_2 = Q_3$$

$$Q = It$$

and, from Eq. (16-4),

$$E = \frac{Q}{C}$$

Therefore, the voltage across the individual capacitors can be calculated as

$$E_1 = \frac{Q}{C_1} \qquad E_2 = \frac{Q}{C_2} \qquad E_3 = \frac{Q}{C_3}$$

Note that the largest capacitor has the smallest terminal voltage. The total applied voltage is the sum of the individual capacitor voltages [see Figure 16-9(b)]:

$$E = E_1 + E_2 + E_3$$

EXAMPLE 16-4

Three capacitors have values $C_1 = 1\ \mu F$, $C_2 = 2\ \mu F$, and $C_3 = 3\ \mu F$. Determine the total capacitance and the charge on each capacitor when the three are connected in parallel across a 100-V supply.

SOLUTION

Eq. (16-7):

$$C = C_1 + C_2 + C_3$$

$$= 1\ \mu F + 2\ \mu F + 3\ \mu F$$

$$\mathbf{C = 6\ \mu F}$$

Eq. (16-4):

$$Q = CE$$

Therefore,

$$Q_1 = 1\ \mu F \times 100\ V$$

$$\mathbf{Q_1 = 100\ \mu C}$$

$$Q_2 = 2\ \mu F \times 100\ V$$

$$\mathbf{Q_2 = 200\ \mu C}$$

$$Q_3 = 3\ \mu F \times 100\ V$$

$$\mathbf{Q_3 = 300\ \mu C}$$

EXAMPLE 16-5

If the three capacitors in Example 16-4 are connected in series across the 100-V supply, determine the total capacitance and the voltage across each capacitor.

SOLUTION

Eq. (16-9):

$$\frac{1}{C}=\frac{1}{C_1}+\frac{1}{C_2}+\frac{1}{C_3}$$

$$=\frac{1}{1\ \mu\text{F}}+\frac{1}{2\ \mu\text{F}}+\frac{1}{3\ \mu\text{F}}$$

Therefore,

$$\boldsymbol{C \cong 0.545\ \mu\text{F}}$$

Eq. (16-4):

$$Q=CE=0.545\ \mu\text{F}\times 100\ \text{V}$$

$$=54.5\ \mu\text{C}$$

and

$$Q=Q_1=Q_2=Q_3$$

Therefore,

$$E_1=\frac{Q}{C_1}=\frac{54.5\ \mu\text{C}}{1\ \mu\text{F}}$$

$$\boldsymbol{E_1=54.5\ \text{V}}$$

$$E_2=\frac{Q}{C_2}=\frac{54.5\ \mu\text{C}}{2\ \mu\text{F}}$$

$$\boldsymbol{E_2=27.3\ \text{V}}$$

$$E_3=\frac{Q}{C_3}=\frac{54.5\ \mu\text{C}}{3\ \mu\text{F}}$$

$$\boldsymbol{E_3=18.2\ \text{V}}$$

$$E=E_1+E_2+E_3$$

$$=54.5\ \text{V}+27.3\ \text{V}+18.2\ \text{V}$$

$$\boldsymbol{E=100\ \text{V}}$$

16-6 ENERGY STORED IN CHARGED CAPACITORS

From Eq. (16-4):

$$C=\frac{Q}{E}$$

and

$$Q=It$$

Therefore, $C=\frac{It}{E}$

or

$$\boxed{I=\frac{CE}{t}} \tag{16-10}$$

When a capacitor is charged from a constant current source for a time t seconds, the voltage across it grows linearly from zero to E volts. The constant level of input current is given by Eq. (16-10), and the average input voltage is $\frac{1}{2}E$.

From Eq. (3-11), the electrical energy is

$$W=EIt$$

Therefore, the energy supplied to the capacitive circuit is

$$W=\tfrac{1}{2}E\times I\times t$$

$$W=\tfrac{1}{2}E\times\left(\frac{CE}{t}\right)\times t$$

or energy stored is:

$$\boxed{W=\tfrac{1}{2}CE^2} \tag{16-11}$$

When C is in farads and E is in volts, W is given in joules. Note the similarity between Eqs. (16-11) and (15-14).

EXAMPLE 16-6

Calculate the energy stored in each of the three series-connected capacitors referred to in Example 16-5. Also, calculate the total energy stored.

SOLUTION

For C_1:

$$W_1=\tfrac{1}{2}C_1E_1^2$$

$$=\tfrac{1}{2}\times 1\ \mu\text{F}\times(54.5\ \text{V})^2$$

$$\mathbf{W_1\simeq 1.49\times 10^{-3}\ J}$$

For C_2:

$$W_2 = \tfrac{1}{2} C_2 E_2^2$$
$$= \tfrac{1}{2} \times 2\ \mu\text{F} \times (27.3\ \text{V})^2$$
$$\boldsymbol{W_2 \simeq 0.745 \times 10^{-3}\ \text{J}}$$

For C_3:

$$W_3 = \tfrac{1}{2} C_3 E_3^2$$
$$= \tfrac{1}{2} \times 3\ \mu\text{F} \times (18.2\ \text{V})^2$$
$$\boldsymbol{W_3 \simeq 0.497 \times 10^{-3}\ \text{J}}$$

$$\textit{total energy stored, } W_T = \tfrac{1}{2} C_T E^2$$
$$= \tfrac{1}{2} \times 0.545 \mu\text{F} \times (100\ \text{V})^2$$
$$\boldsymbol{W \simeq 2.73 \times 10^{-3}\ \text{J}}$$

and

$$\textit{total energy stored, } W = W_1 + W_2 + W_3$$
$$= (1.49 + 0.745 + 0.497) \times 10^{-3}\ \text{J}$$
$$\boldsymbol{W \simeq 2.73 \times 10^{-3}\ \text{J}}$$

16-7 STRAY CAPACITANCE

Capacitance exists where any conductors are separated by an insulator. Thus, all pairs of adjacent conductors have capacitance between them, and every conductor has a capacitance to ground. This unwanted capacitance is termed *stray capacitance*, and because capacitance is inversely proportional to dielectric thickness, stray capacitance is most easily minimized by keeping conductors as far apart as possible. In dc applications, stray capacitance may be absolutely negligible, but in alternating current circuits, particularly at radio frequencies, the stray capacitance (like stray inductance) can be a severe problem.

GLOSSARY OF FORMULAS

Electric flux density:

$$D = \frac{Q}{A}$$

Electric field strength:

$$\mathcal{E} = \frac{E}{d}$$

Electric charge:

$$Q = CE$$

Permittivity:

$$\epsilon_o \epsilon_r = \frac{D}{E}$$

Capacitance:

$$C = \frac{\epsilon_r \epsilon_o A}{d}$$

Capacitors in parallel:
Total capacitance:

$$C = C_1 + C_2 + C_3 + \cdots$$

Total charge:

$$Q = C_{\text{total}} E$$

Capacitors in series:
Total capacitance:

$$\frac{1}{C} = \frac{1}{C_1} + \frac{1}{C_2} + \frac{1}{C_3} + \cdots$$

Total charge:

$$Q = Q_1 = Q_2 = Q_3 = \cdots$$

Energy stored:

$$W = \tfrac{1}{2} CE^2 \qquad \text{joules}$$

REVIEW QUESTIONS

16-1 Using illustrations, explain the process of electric charge storage in two metal plates separated by a thin layer of insulating material.

16-2 Discuss the electric field, comparing it to a magnetic field. Write the equations for electric field strength and electric flux density and explain each.

16-3 Explain electric lines of force, equipotential lines, dielectric strength, dielectric absorbtion, permittivity, and relative permittivity.

16-4 State the definition of the farad and from it write an equation for the charge stored in a capacitor.

16-5 Explain why the capacitance of a capacitor is proportional to the plate area and inversely proportional to dielectric thickness.

16-6 Derive an equation for the capacitance of a parallel-plate capacitor.

16-7 Discuss the following characteristics of capacitors: working voltage, tolerance, temperature effects, leakage current, and polarization.

16-8 Sketch the equivalent circuit of a capacitor and briefly explain.

16-9 Describe the construction of air capacitors, paper capacitors, plastic film capacitors, and mica capacitors. Also, explain the most important parameter for each type of capacitor.

16-10 Repeat Review Question 16-9 for ceramic, electrolytic, and tantalum capacitors.

16-11 Derive equations for total capacitance of capacitors connected in parallel, and for the total charge.

16-12 Derive equations for the total capacitance of capacitors connected in series, and for the terminal voltage on each capacitor.

16-13 Derive an equation for the energy stored in a charged capacitor.

16-14 Define stray capacitance and explain the problems associated with it.

PROBLEMS

16-1 A capacitor with a plate area of 50 cm^2 and a dielectric thickness of 0.5 mm has a charge of 10×10^{-9}C when the applied voltage is 20 V. Calculate the electric field strength and flux density.

16-2 If the dielectric strength for the capacitor described in Problem 16-1 is 100 000 V/m, calculate the applied voltage at which the dielectric is likely to break down.

16-3 Determine the capacitance of a capacitor with a plate area of 50 cm^2 and a dielectric thickness of 0.5 mm:

a. When the dielectric is air.

b. When the dielectric is glass with a relative permittivity of 75.

c. Determine the necessary applied voltage to the capacitor in each case to store a charge of 3.3×10^{-9} C.

16-4 A 20-μF capacitor is to be constructed using metal foil with a plastic film dielectric. If the dielectric is 0.05 mm thick and has a relative permittivity of 3, calculate the area of the metal foil plates.

16-5 A 50-μF electrolytic capacitor has aluminum-foil plates which are 10 cm $\times$ 75 cm. If the oxide dielectric has a relative permittivity of 20, calculate the thickness of the dielectric.

16-6 Four capacitors, having values of 100 μF, 50 μF, 25 μF, and 10 μF, are connected to a 25-V supply. Calculate the total capacitance, the voltage across each capacitor, and the charge on each capacitor:

a. When the capacitors are series-connected.

b. When the capacitors are parallel-connected.

16-7 A 100-μF capacitor is connected in parallel with a 50-μF capacitor, then the two are connected in series with a 25-μF capacitor.

a. Calculate the total capacitance.

b. If a 25-V supply is connected to the series combination, determine the voltage across each capacitor.

16-8 Calculate the energy stored in each of the capacitors described in Problem 16-7.

17

INDUCTANCE AND CAPACITANCE IN DC CIRCUITS

INTRODUCTION

When an inductive–resistive series circuit has a supply voltage switched to it, the inductance produces an initial maximum level of counter-emf which gradually falls to zero. The circuit current is zero initially, and grows gradually to its maximum level. The behavior of an *LR* circuit is most easily understood by plotting the graphs of instantaneous current versus time and instantaneous inductor voltage versus time. Care must be taken in open-circuiting an *LR* circuit, to avoid an extremely high induced voltage.

In the case of a capacitive–resistive series circuit, when the supply is first switched on, the charging current is initially at its maximum level and it gradually falls to zero. The capacitor voltage is zero at first and grows gradually to its maximum level. As with the *LR* circuit, the behavior of a *CR* circuit can be graphically represented by plotting instantaneous current and voltage versus time. Because energy is stored in a charged capacitor, a large current can flow when the capacitor terminals are short-circuited.

17-1 *LR* CIRCUIT OPERATION

When switch S_1 is closed in the simple battery and resistor circuit of Figure 17-1(a), the current tends to jump instantaneously to its maximum level of $I = E/R$. This is illustrated by the graph of i versus t in Figure 17-1(b), and it assumes that R is purely resistive. (Actually, there can be

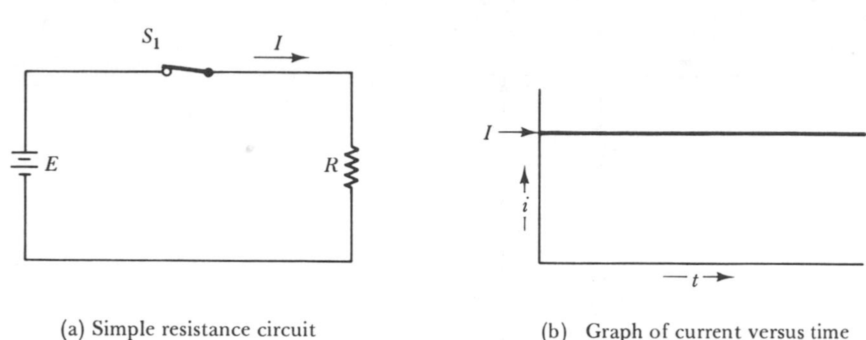

(a) Simple resistance circuit

(b) Graph of current versus time for the simple resistance circuit

FIG. 17-1. Simple resistance circuit and its current/time graph

no such thing as a *pure resistance*, and, as will be seen later, there is always some *rise time* involved in the current going from zero to its maximum level.)

Now consider the circuit shown in Figure 17-2(a), and assume that L is a pure inductance. Recall from Eq. (15-3) that when the current changes through an inductor, a counter-emf is generated which has the value

$$e_L = L\frac{\Delta i}{\Delta t} \qquad \text{volts}$$

The counter-emf opposes the current change that generates it; thus, its polarity is as shown in the circuit. Using Kirchhoff's law, the instantaneous level of current is

$$\boxed{i = \frac{E - e_L}{R}} \tag{17-1}$$

The initial rapid rate of change of current when S_1 is first closed causes the counter-emf e_L to be equal to E. Putting $e_L = E$ into Eq. (17-1) and using the voltage and component values shown in the circuit gives

$$i = \frac{10\text{ V} - 10\text{ V}}{1\text{ k}\Omega} = 0\text{ A}$$

Therefore, on the graph of i versus t in Figure 17-2(b), i is zero at $t=0$. At this instant the rate of change of current ($\Delta i_0/\Delta t_0$ on the graph) is at its maximum value and falling. Consequently, e_L also has its maximum level at $t=0$.

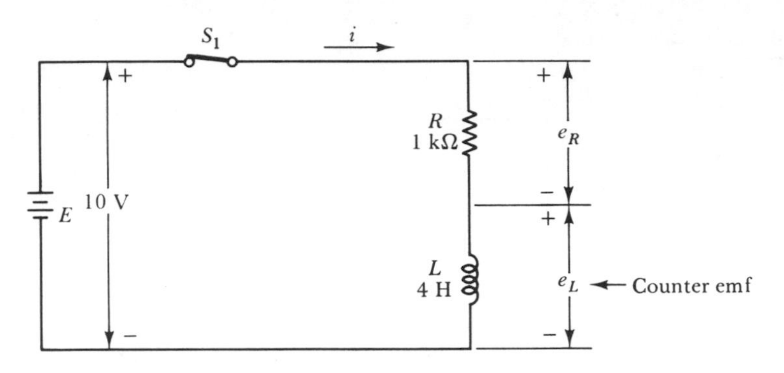

(a) Series LR circuit

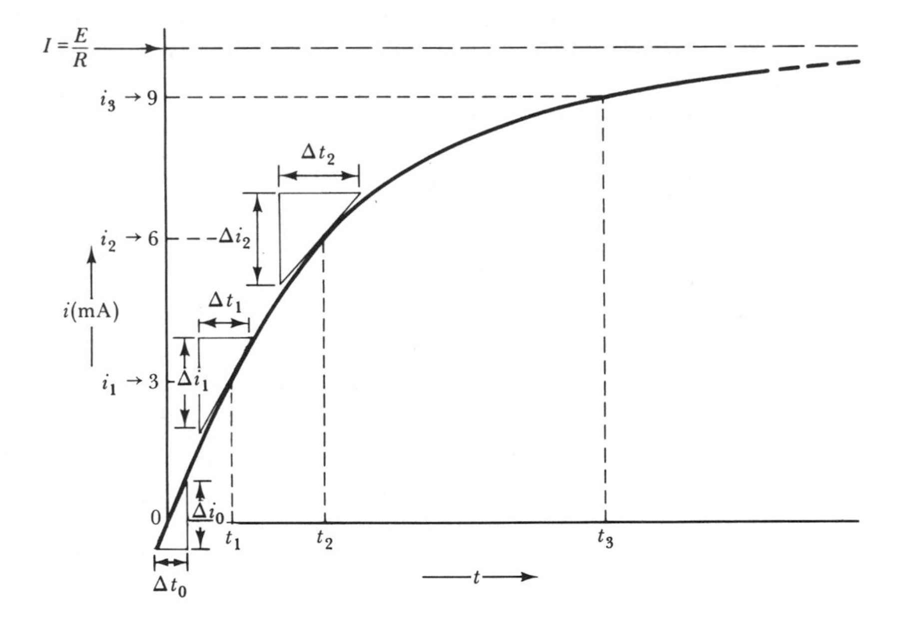

(b) Growth of current in LR circuit

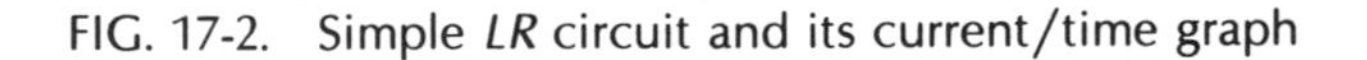

FIG. 17-2. Simple *LR* circuit and its current/time graph

Sometime t_1 after S_1 was closed, the rate of change of current has decreased to a value ($\Delta i_1/\Delta t_1$ on the graph) which gives $e_L = 7$ V. Therefore, using Eq. (17-1), at time t_1 the instantaneous current level is

$$i_1 = \frac{10\text{ V} - 7\text{ V}}{1\text{ k}\Omega}$$

$$= 3\text{ mA}$$

On the graph of i versus t, $i_1 = 3$ mA is plotted at time t_1.

Because e_L is falling, the current level is increasing. As the current level grows, it gets closer to its maximum level of $I = E/R$, and in doing so the rate of change of current $\Delta i/\Delta t$ continues to decrease. This decrease in $\Delta i/\Delta t$ is the cause of the decreasing level of counter-emf e_L.

After some time t_2, $\Delta i/\Delta t$ has fallen to a level that gives $e_L = 4$ V. Thus,

$$i_2 = \frac{10\text{ V} - 4\text{ V}}{1\text{ k}\Omega} = 6\text{ mA}$$

and at time t_3, $\Delta i/\Delta t$ is such that $e_L = 1$ V, giving

$$i_3 = \frac{10\text{ V} - 1\text{ V}}{1\text{ k}\Omega} = 9\text{ mA}$$

The values of i_2 and i_3 are plotted on the graph at times t_2 and t_3, respectively. The final current level when e_L becomes zero is, of course, $I = E/R$.

Because the counter-emf is initially large and falls off smoothly to zero, the instantaneous current grows continuously from zero to its maximum level of $I = E/R$. As the current approaches this maximum level, the rate of change of current keeps decreasing, and this produces the decreasing level of e_L.

17-2 INSTANTANEOUS CURRENT AND VOLTAGE IN *LR* CIRCUITS

By means of differential calculus it can be shown that the equation for instantaneous current in an inductive-resistive circuit is:

$$\boxed{i = \frac{E}{R}\left(1 - \epsilon^{-t(R/L)}\right)} \qquad (17\text{-}2)$$

where

i = *instantaneous current, in amperes, at time t*

E = *supply voltage*

R = *series resistance, in ohms (including the winding resistance of the inductor)*

ϵ = *exponential constant* = 2.718

t = *time, in seconds, from current commencement*

L = *inductance of the inductor, in henrys*

Using Eq. (17-2), the instantaneous current levels can be calculated for several different time intervals from $t=0$ for a given circuit. The corresponding values of i and t can be used to plot an accurate graph of i versus t for the circuit. The equation can also be manipulated to determine the required value of t, R, or L for a given current level:

$$i=\frac{E}{R}(1-\epsilon^{-t(R/L)})$$

$$\frac{iR}{E}=1-\epsilon^{-t(R/L)}$$

$$\epsilon^{-t(R/L)}=1-\frac{iR}{E}$$

$$=\frac{E-iR}{E}$$

$$\epsilon^{t(R/L)}=\frac{E}{E-iR}$$

$$\boxed{t\frac{R}{L}=\ln\left(\frac{E}{E-iR}\right)} \qquad \textbf{(17-3)}$$

EXAMPLE 17-1

Determine the instantaneous values of current at 2-ms intervals from $t=0$ for the circuit of Figure 17-2(a).

SOLUTION

Eq. (17-2):

$$i=\frac{E}{R}(1-\epsilon^{-t(R/L)})$$

At $t=0$, $\quad i=0$ *point* 1 *in* Figure 17-3

At $t=2$ ms, $\quad i=\frac{10\text{ V}}{1\text{ k}\Omega}(1-\epsilon^{-2\text{ ms}\times(1\text{ k}\Omega/4\text{ H})})\cong 3.93\text{ mA}$ *point* 2

At $t=4$ ms, $\quad i=\frac{10\text{ V}}{1\text{ k}\Omega}(1-\epsilon^{-4\text{ ms}\times(1\text{ k}\Omega/4\text{ H})})\cong 6.32\text{ mA}$ *point* 3

At $t=6$ ms, $\quad i=7.77$ mA *point* 4

At $t=8$ ms, $\quad i=8.65$ mA *point* 5

At $t=10$ ms, $\quad i=9.18$ mA *point* 6

At $t=12$ ms, $i=9.5$ mA *point* 7

At $t=14$ ms, $i=9.7$ mA *point* 8

At $t=16$ ms, $i=9.82$ mA *point* 9

At $t=\infty$, $i=10$ mA = *maximum current level*

The corresponding values of t and i are plotted in Figure 17-3.

Referring to the graph of i versus t plotted in Figure 17-3, it is seen that when $t=4$ ms, i is 6.32 mA. Also, 6.32 mA is 63.2% of the maximum current level and

$$t=\frac{L}{R}=\frac{4\text{ H}}{1\text{ k}\Omega}=4\times10^{-3}\text{ s}=4\text{ ms}$$

Therefore, when $t=L/R$, the instantaneous current level is always 63.2% of E/R. The quantity L/R is termed the *time constant* of an inductive–resistive circuit, and this is a very important quantity in determining the behavior of the circuit. Sometimes the greek letter τ is used as the symbol for the time constant. It can be shown that after a time of $t=5L/R$, the current is 99.3% of its maximum level (see the graph). By drawing a straight line from the origin at a tangent to the graph of i/t, it is seen that if the initial rate of change of current were maintained, the current would reach its maximum level in a time of $t=L/R$ (see Figure 17-3).

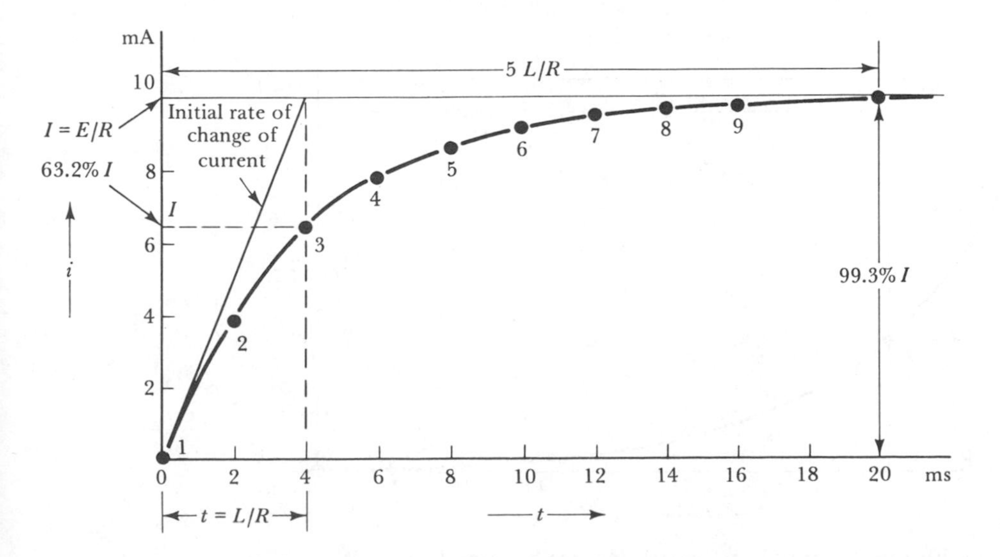

FIG. 17-3. Graph of current versus time for a series *LR* circuit

The graph of e_L versus t in Figure 17-4 shows how the counter-emf e_L changes with time. At $t=0$, e_L is equal to the supply voltage E. At $t=L/R$, e_L has fallen by 63.2% of E, while at $t=5L/R$, e_L has fallen through 99.3% of its initial level. Comparing the graphs of i versus t and e_L versus t, it is seen that inductor terminal voltage drops as the current increases.

EXAMPLE 17-2 Calculate the value of the counter-emf for the circuit in Figure 17-2(a) at $t=L/R$ and at $t=5L/R$.

SOLUTION

Eq. (17-2):

$$i=\frac{E}{R}(1-\epsilon^{-t(R/L)})$$

Therefore,

$$iR=E-E\epsilon^{-t(R/L)}$$

and from Eq. (17-1):

$$iR=E-e_L$$

Therefore,

$$E-e_L=E-E\epsilon^{-t(R/L)}$$

and

$$e_L=E\epsilon^{-t(R/L)}$$

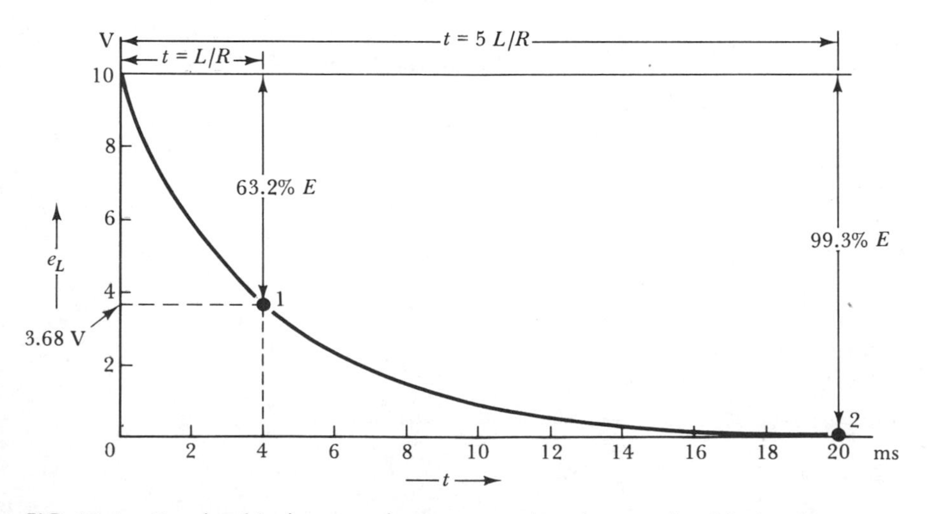

FIG. 17-4. Graph of inductor voltage versus time for a series *LR* circuit

At $t = L/R$,

$$e_L = E\epsilon^{-1}$$
$$= 10\ \text{V} \times \epsilon^{-1}$$
$$\mathbf{e_L = 3.68\ V} \qquad \textit{point } 1 \textit{ on } \text{Figure 17-4}$$

At $t = 5L/R$,

$$e_L = 10\ \text{V} \times \epsilon^{-5}$$
$$\mathbf{e_L = 0.067\ V} \qquad \textit{point } 2 \textit{ on } \text{Figure 17-4}$$

17-3 OPEN-CIRCUITING AN INDUCTIVE CIRCUIT

From Eq. (15-14), the energy stored in an inductive circuit is

$$W = \tfrac{1}{2}LI^2 \quad \text{joules}$$

The equation shows that when a steady-state current is flowing through an inductor, energy is stored in the inductor. Also, when the current goes to zero, there is obviously no energy stored.

Now consider what occurs when the switch is opened in an inductive circuit, such as that shown in Figure 17-5. Before the switch is opened, a constant current is flowing through the inductor, and energy is stored. When the switch is opened, the current must go to zero; consequently, the energy contained in the inductor must also go to zero. This means that the energy must somehow flow out of the inductor at the instant that the switch is opened. Now recall that the counter-emf generated in an

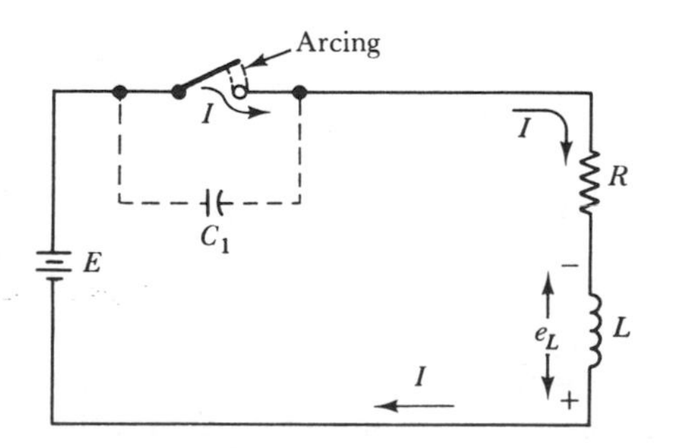

FIG. 17-5. Effect of open-circuiting an inductive circuit

inductor opposes the current change that produces it. Therefore, since the current is falling when the switch is opened, the counter-emf generated is such that it tends to oppose the fall in current; that is, its polarity is such that it tries to maintain the current flow. This is illustrated in Figure 17-5, where it is seen that the counter-emf is + at the bottom of the inductor and − at the top. This polarity is opposite to that of the counter-emf when the switch was first closed [see Figure 17-2(a)].

The effect of the counter-emf trying to maintain current flow when the switch is opened is that a spark jumps across the open switch contacts. The arcing across the contacts continues for the brief period necessary to discharge the energy contained in the inductor. To produce such an arc across open contacts requires a very high voltage. The high voltage is, of course, the counter-emf. It is not surprising that a very high voltage is generated, when it is remembered that its equation is

$$e_L = L\frac{\Delta i}{\Delta t}$$

Because the open circuit tries to make the current go to zero instantaneously, the rate of change of current $\Delta i/\Delta t$ is a very large quantity, and therefore, $L(\Delta i/\Delta t)$ is also very large.

To get some idea of the voltage generated when the switch is opened in an inductive circuit, consider again the circuit in Figure 17-2(a). The steady-state current level while the switch is closed is

$$I = \frac{E}{R} = \frac{10\text{ V}}{1\text{ k}\Omega} = 10\text{ mA}$$

At the instant that the switch is opened, the counter-emf is such that for a brief time period, this 10 mA continues to flow across the air gap between the switch contacts. Since the resistance of the air gap could very easily be 1 MΩ, the voltage across the air gap is at least

$$\begin{aligned} e &= IR \\ &= 10\text{ mA} \times 1\text{ M}\Omega \\ &= 10\ 000\text{ V} \end{aligned}$$

This 10 000 V is the generated counter-emf.

It is seen that when a switch is opened in an inductive circuit, the high level of counter-emf may damage the circuit insulation. Obviously, some means of discharging the energy in the inductor must be provided. A capacitor connected across the switch, as shown dashed in Figure 17-5 is sometimes employed for this purpose. The capacitor initially behaves as a short circuit when the switch is opened, and this allows a current flow to continue until the capacitor is charged to the level of the supply voltage E.

Another method of protecting an inductive circuit against high levels of counter-emf is shown in Figure 17-6. Resistor R_2, in parallel with R_1 and L, has no effect on the steady-state current through R_1 and L while the switch remains closed. When the switch is opened, the total resistance in series with L is (R_1+R_2). Therefore, the maximum level of counter-emf that must be generated to maintain the initial level of I is

$$e_L = I(R_1 + R_2)$$

For $R_1 = 5\ \text{k}\Omega$ connected in the circuit of Figure 17-2(a), the counter-emf is

$$\begin{aligned} e_L &= 10\ \text{mA}(5\ \text{k}\Omega + 1\ \text{k}\Omega) \\ &= 60\ \text{V} \end{aligned}$$

EXAMPLE 17-3

A circuit connected as shown in Figure 17-6 has $R_1 = 500\ \Omega$, $L = 500\ \text{mH}$, and $E = 100\ \text{V}$. Determine how long it takes the inductor current to get to 1.7 mA after the switch is closed. Also calculate the required value of R_2, if the counter-emf is not to exceed 300 V when the switch is opened.

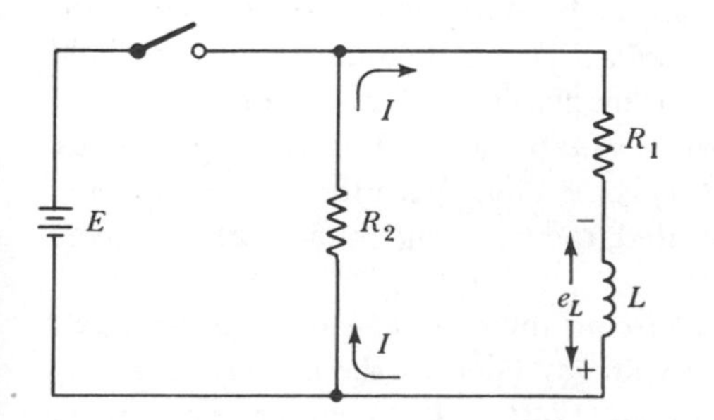

FIG. 17-6. Method of protecting an inductive circuit from high counter-emf levels when open circuited

SOLUTION

From Eq. (17-3),

$$t=\frac{L}{R_1}\ln\left(\frac{E}{E-iR_1}\right)$$

$$=\left(\frac{500\text{ mH}}{500\ \Omega}\right)\ln\left(\frac{100\text{ V}}{100\text{ V}-(1.7\text{ mA}\times 500\ \Omega)}\right)$$

$$\mathbf{t=8.5\ \mu s}$$

$$\textit{maximum current level, } I=\frac{E}{R_1}=\frac{100\text{ V}}{500\ \Omega}$$

$$\mathbf{I=200\ mA}$$

$$\textit{counter-emf, } e_L=I(R_1+R_2)$$

Therefore,

$$\frac{e_L}{I}=R_1+R_2$$

and

$$R_2=\frac{e_L}{I}-R_1$$

$$=\frac{300\text{ V}}{200\text{ mA}}-500\ \Omega$$

$$\mathbf{R_2=1\ k\Omega}$$

17-4 *LR* CIRCUIT WAVEFORMS

Consider the *LR* circuit once again, as illustrated in Figure 17-7(a). Suppose that switch S_1 is closed for an accurately measured time t_1, then opened again for an exactly equal time t_2, and the sequence repeated over and over. The resulting voltage applied to the circuit would look as shown in Figure 17-7(b). For obvious reasons the shape is termed a *square wave*, and the time t is referred to as the *pulse width* of the square wave. It would not really be practical to attempt to manually open and close a switch in order to generate a square wave; suitable electronic instruments are available for the purpose. However, going along with the assumption that a square wave is applied as illustrated, the resulting voltages across L and R can be investigated.

When E is switched *on* to the circuit, the counter-emf e_L immediately jumps to the level of the applied voltage. Then as the current grows, e_L slowly falls to zero, as shown in Figure 17-7(c). The counter-emf remains at zero until the switch is opened, and as already discussed, it now immediately jumps to $-I(R_1+R_2)$, from which it again slowly goes to zero, as illustrated. The positive and negative step and decline of e_L is

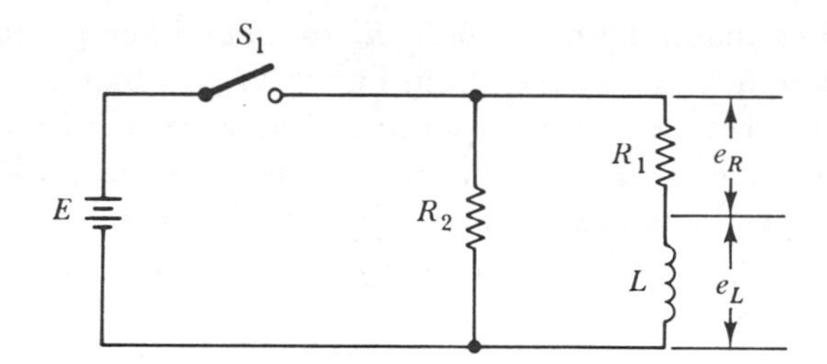

(a) Circuit for generating waveforms below

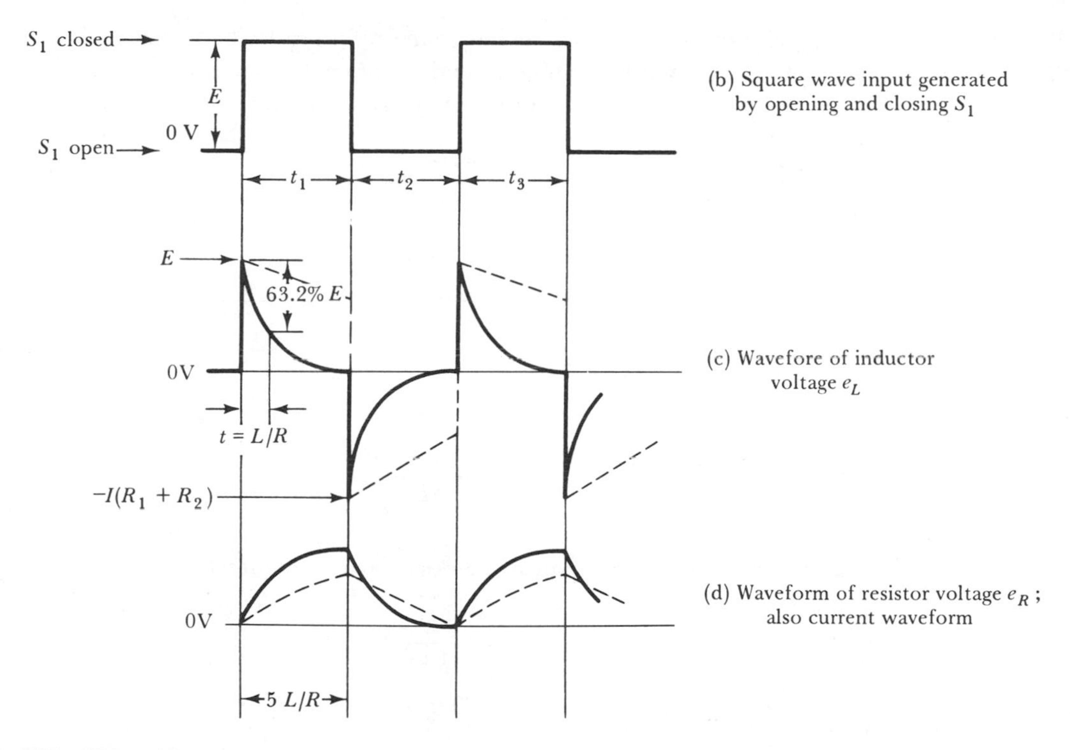

(b) Square wave input generated by opening and closing S_1

(c) Wavefore of inductor voltage e_L

(d) Waveform of resistor voltage e_R; also current waveform

FIG. 17-7. *LR* series circuit and waveforms generated by opening and closing S_1

repeated over and over as the switch is closed and opened, giving the *spike waveform* illustrated in Figure 17-7(c).

The current through R_1 and L is zero initially when S_1 is closed, and rises to its maximum level of $I = E/R_1$ when e_L has fallen to zero. When S_1 is opened, the current falls slowly as e_L moves from $-I(R_1 + R_2)$ back to zero. The resulting current waveform is shown in Figure 17-7(d). Since the voltage across R_1 is simply $I \times R_1$, the waveshape of e_R is exactly the same as the current waveshape.

Recalling that in a time $t=5(L/R)$ the counter-emf e_L falls by 99.3% of its maximum level, it is seen from Figure 17-7(c) that $t_1=t_2\cong 5(L/R)$. In this case, where the time constant L/R is short by comparison to the pulse width t, the circuit is said to have a *short time constant*. When L/R is greater than t, the typical shape of the waveforms are as shown by the dashed lines in Figure 17-7(c) and (d), and the circuit is said to have a *long time constant*.

17-5 *CR* CIRCUIT OPERATION

A capacitor and resistor are shown connected in series in Figure 17-8(a), together with a supply voltage E and a switch S_1. If the charge on the capacitor is zero at the instant the switch is closed, then $e_C=0$ and e_R is

$$e_R=E-e_C$$

The charging current is

$$i=\frac{e_R}{R}$$

or

$$\boxed{\boldsymbol{i=\frac{E-e_C}{R}}} \qquad \textbf{(17-4)}$$

For the voltage and component values shown in Figure 17-8(a),

$$i=\frac{10\text{ V}-0}{1\text{ k}\Omega}$$
$$=10\text{ mA}$$

The current flow causes the capacitor to charge with the polarity illustrated in the figure. After a time t_1, the capacitor voltage might be 3 V [see Figure 17-8(b)]. Then the charging current becomes

$$i_1=\frac{10\text{ V}-3\text{ V}}{1\text{ k}\Omega}$$
$$=7\text{ mA}$$

It is seen that because C has accumulated some charge, the voltage across R is reduced, and consequently the charging current is reduced

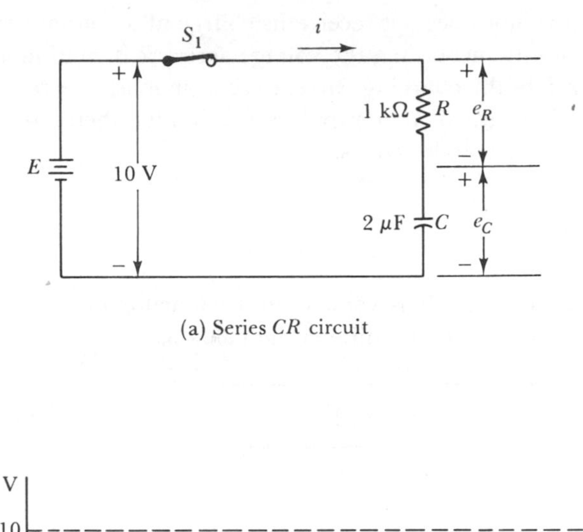

(a) Series CR circuit

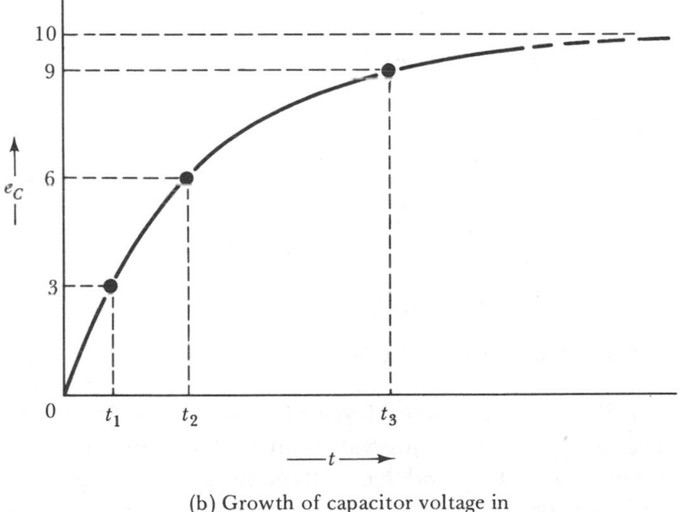

(b) Growth of capacitor voltage in series CR circuit

FIG. 17-8. Simple *CR* circuit and its voltage/time graph

from 10 mA to 7 mA. Because the charging current has been reduced, the capacitor voltage is now growing at a slower rate than before.

After time t_2, e_C has grown to 6 V, and the charging current becomes

$$i_2 = \frac{10\text{ V} - 6\text{ V}}{1\text{ k}\Omega} = 4\text{ mA}$$

The charging current has now been further reduced (from 7 mA to 4 mA), so the capacitor is charging at an even slower rate than before. An even longer time is now required for the capacitor voltage to grow by another 3 V. The capacitor does not receive its charge at a constant rate. Because e_C is continuously increasing, the voltage across R is continuously decreasing, and therefore the charging current is continuously decreasing. This means that C is charged at a rapid rate initially; then the rate decreases as the capacitor voltage grows.

17-6 INSTANTANEOUS CURRENT AND VOLTAGE IN *CR* CIRCUITS

The equation for the instantaneous voltage on a capacitor in a resistive capacitive circuit can be derived by differential calculus:

$$\boxed{e_C = E - (E - E_o)\epsilon^{-t/(CR)}} \tag{17-5}$$

where

e_C = *capacitor voltage at time t*

E = *supply voltage*

E_o = *initial level of capacitor voltage*

ϵ = *exponential constant* = 2.718

t = *time, in seconds, from commencement of charge*

C = *capacitance value, in farads*

R = *charging resistance, in ohms*

Using Eq. (17-5), the instantaneous levels of capacitor voltage can be calculated for several different time intervals from $t=0$ for a given circuit. The corresponding values of e_C and t can then be plotted to give an accurate graph of e_C versus t for the circuit.. The equation can also be manipulated to determine expressions for t, C, and R for a given capacitor voltage level.

When the capacitor is initially uncharged,

$$E_o = 0$$

and

$$e_C = E - E\epsilon^{-t/(CR)}$$

or

$$\boxed{e_C = E(1 - \epsilon^{-t/(CR)})} \tag{17-6}$$

Also,

$$E\epsilon^{-t/(CR)} = E - e_C$$

$$\epsilon^{-t/(CR)} = \frac{E - e_C}{E}$$

$$\epsilon^{t/(CR)} = \frac{E}{E - e_C}$$

giving

$$\boxed{\frac{t}{CR} = \ln\left(\frac{E}{E - e_C}\right)} \qquad \textbf{(17-7)}$$

EXAMPLE 17-4

Determine the instantaneous values of capacitor voltage at 1-ms intervals from $t=0$ for the circuit in Figure 17-8(a).

SOLUTION

Assuming that the initial level of capacitor voltage is zero, Eq. (17-6) *can be used*:

$$e_C = E(1 - \epsilon^{-t/(CR)})$$

At $t=0$, $e_C = 0\text{ V}$ — *point* 1 *in* Figure 17-9

At $t=1\text{ ms}$, $e_C = 10\text{ V}(1 - \epsilon^{-1\text{ ms}/(2\,\mu\text{F} \times 1\text{ k}\Omega)})$

$e_C = 3.93\text{ V}$ — *point* 2

At $t=2\text{ ms}$, $e_C = 6.32\text{ V}$ — *point* 3

At $t=3\text{ ms}$, $e_C = 7.77\text{ V}$ — *point* 4

At $t=4\text{ ms}$, $e_C = 8.65\text{ V}$ — *point* 5

At $t=5\text{ ms}$, $e_C = 9.18\text{ V}$ — *point* 6

At $t=6\text{ ms}$, $e_C = 9.5\text{ V}$ — *point* 7

At $t=7\text{ ms}$, $e_C = 9.7\text{ V}$ — *point* 8

At $t=8\text{ ms}$, $e_C = 9.82\text{ V}$ — *point* 9

At $t=\infty$, $e_C = 10\text{ V} =$ *maximum voltage level*

Referring to the graph of e_C versus t plotted in Figure 17-9, it is seen that when $t=2$ ms, e_C is 6.32 V. Also, 6.32 V is 63.2% of the maximum

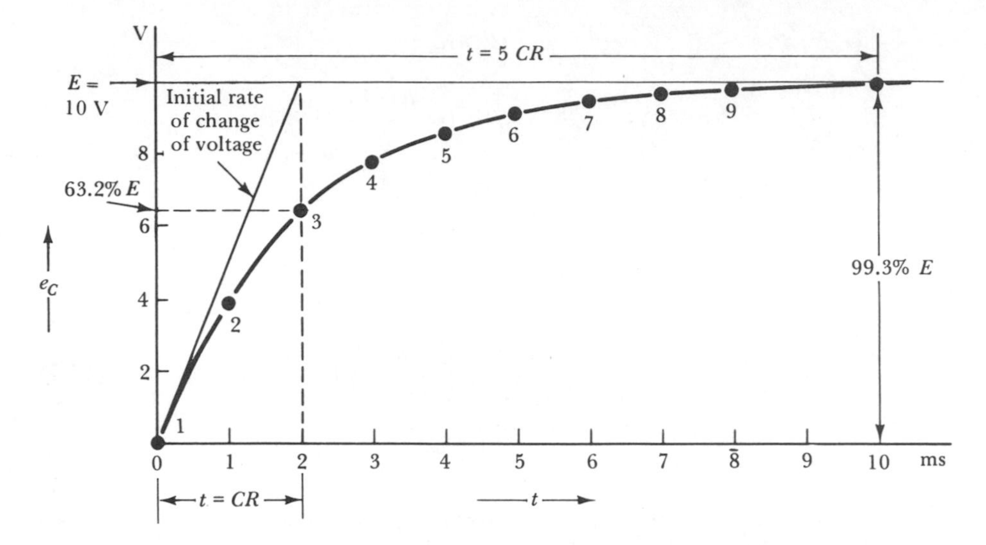

FIG. 17-9. Graph of capacitor voltage versus time for a series *CR* circuit

voltage level and

$$t = CR = 2\ \mu\text{F} \times 1\ \text{k}\Omega = 2 \times 10^{-3}\ \text{s} = 2\ \text{ms}$$

Therefore, when $t = CR$, the instantaneous capacitor voltage level is always 63.2% of E. The quantity CR is the *time constant* of a capacitive–resistive circuit, and, as in the case of an LR circuit, the time constant largely determines the behavior of the circuit. After a time period of $5CR$, the capacitor voltage is 99.3% of its maximum level. By drawing a straight line at a tangent to the graph of e_C/t, it can be shown that if the initial rate of charge were maintained, the capacitor voltage would reach its maximum level in a time of $t = CR$ (see Figure 17-9).

The graph of i versus t in Figure 17-10 shows how the charging current changes with time. At $t = 0$, $i = E/R$. At $t = CR$, i has fallen by 63.2% of E/R. And at $t = 5CR$, i has fallen by 99.3% of its initial level.

EXAMPLE 17-5 Calculate the value of capacitor charging current for the circuit in Figure 17-8 at $t = CR$ and $t = 5CR$.

SOLUTION

Eq. (17-6):

$$e_C = E(1 - \epsilon^{-t/(CR)})$$

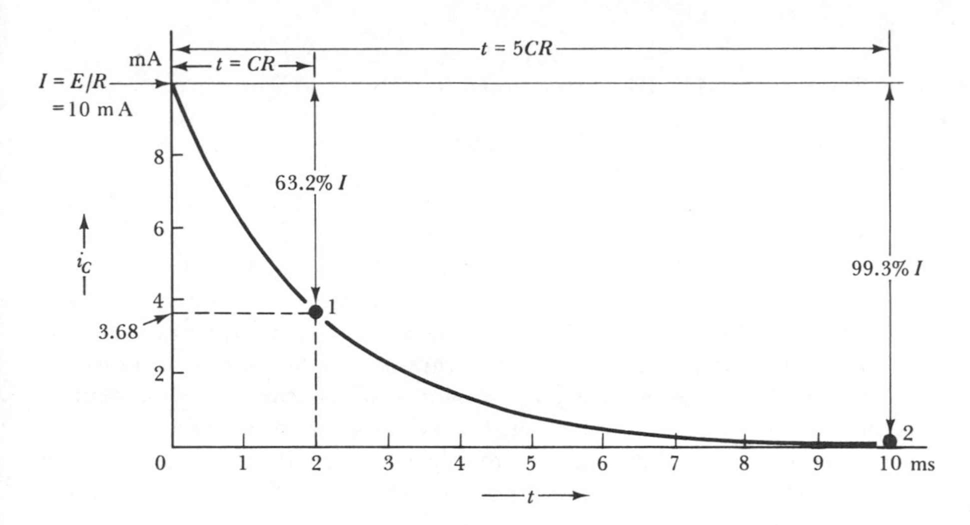

FIG. 17-10. Graph of capacitor charging current versus time for a series *CR* circuit

and from Eq. (17-4)

$$i = \frac{E - e_C}{R}$$

Therefore,

$$i = \frac{E - E(1 - \epsilon^{-t/(CR)})}{R}$$

$$= \frac{E\epsilon^{-t/(CR)}}{R}$$

At $t = CR$,

$$i = \frac{E\epsilon^{-1}}{R}$$

$$= \frac{10\text{ V} \times \epsilon^{-1}}{1\text{ k}\Omega}$$

$\mathbf{i = 3.68\text{ mA}}$ *point* 1 *on* Figure 17-10

At $t = 5CR$,

$$i = \frac{10\text{ V} \times \epsilon^{-5}}{1\text{ k}\Omega}$$

$\mathbf{i = 67\ \mu A}$ *point* 2 *on* Figure 17-10

17-7 DISCHARGING A CAPACITOR

From Eq. (16-11), the energy stored in a charged capacitor is

$$W=\tfrac{1}{2}CE^2 \qquad \text{joules}$$

From the equation, it is seen that when a steady-state voltage level exists across the capacitor terminals, there is a charge stored in the capacitor.

When the switch is opened in a series *CR* circuit such as that shown in Figure 17-8(a), there is no effect on the charge stored in the capacitor. The capacitor is able to hold its charge as a static quantity, and its terminal voltage is constant. This contrasts with the case of an inductor, where the charge is a dynamic quantity which cannot be maintained when the current is interrupted. The capacitor can hold its charge indefinitely, and if it is charged to a high voltage, it can be dangerous to someone working on a circuit believed to be safely disconnected from the supply.

Sometimes the switch employed in a *CR* circuit has two contacts, as illustrated in Figure 17-11(a). Then, when S_1 is switched to position 2, after being in position 1 for some time, C_1 is discharged via R_1. In this case the capacitor has already been charged up to E volts, with the polarity shown. When S_1 is in position 2, the capacitor acts like a voltage source and causes current to flow through R_1, as illustrated. This current discharges the capacitor after a time approximately equal to $5CR$. Note that the direction of the discharging current through R_1 is opposite to the charging current direction when S_1 was in position 1. Consequently, the voltage drop e_R across R_1 now also has reversed polarity.

During discharge, the capacitor voltage at any instant can be determined from Eq. (17-5). With S_1 in position 2 in the circuit of Figure 17-11(a), the supply voltage to the *CR* circuit is zero. Therefore, putting $E=0$ in Eq. (17-5) gives

$$\boxed{e_C = E_o\epsilon^{-t/(CR)}} \qquad \textbf{(17-8)}$$

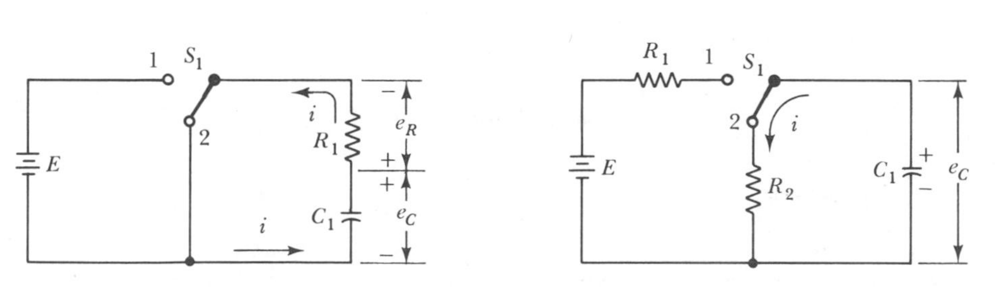

(a) Discharge of C_1 through R_1 (b) Discharge of C_1 through R_2

FIG. 17-11. Methods of discharging a charged capacitor

Figure 17-11(b) shows a slightly different discharge circuit arrangement. When S_1 is in position 1, C_1 is charged from the battery via R_1, as before. When S_1 goes to position 2, R_1 is taken out of the circuit and the capacitor is discharged via resistor R_2. By selecting R_2 larger than R_1, the discharge time for the capacitor may be made longer than the charge time. Alternatively, R_2 may be selected smaller than R_1, so the capacitor discharge time is shorter than the charging time. In this circumstance, the initial level of discharge current can be very large. For example, if e_C is 10 V and $R_2 = 0.1\ \Omega$, the initial level of discharge current is

$$i = \frac{10\ \text{V}}{0.1\ \Omega} = 100\ \text{A}$$

EXAMPLE 17-6

A CR circuit connected as shown in Figure 17-11(b) has $E = 100$ V and $C_1 = 1\ \mu F$. Determine the value of R_1 if C_1 is to become charged to 50 V in 20 ms. Also, calculate the value of R_2 that will limit the initial level of discharge current to 1 mA.

SOLUTION

From Eq. (17-7),

$$R_1 = \frac{t}{C \ln[E/(E - e_C)]}$$

$$= \frac{20\ \text{ms}}{1\ \mu\text{F} \ln[100\ \text{V}/(100\ \text{V} - 50\ \text{V})]}$$

$$\mathbf{R_1 = 28.9\ k\Omega}$$

The maximum level of e_C is

$$e_C = E = 100\ \text{V}$$

The initial level of discharge current is

$$i = \frac{e_C}{R_2}$$

Therefore,

$$R_2 = \frac{e_C}{i} = \frac{100\ \text{V}}{1\ \text{mA}}$$

$$\mathbf{R_2 = 100\ k\Omega}$$

17-8 *CR* CIRCUIT WAVEFORMS

As explained in Section 17-4, a *square wave* can theoretically be generated by opening and closing a switch connected between a circuit and the supply battery. Refer to the circuit of Figure 17-12(a), and assume that switch S_1 is held in position 1 for an accurately timed period t_1, and then moved to position 2 for an equal time t_2. During t_1, capacitor C_1 is charged from the battery via R_1, and during t_2, the capacitor discharges via R_1. The input waveform is as shown in Figure 17-12(b).

When the switch first goes to position 1, the capacitor voltage starts to grow slowly from 0 toward E, as illustrated in Figure 17-12(c). When e_C finally reaches E, it remains constant until S_1 is moved to position 2.

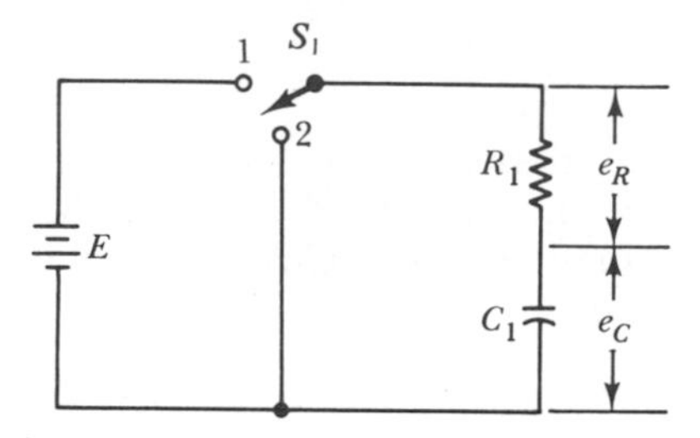

(a) Circuit for generating waveforms below

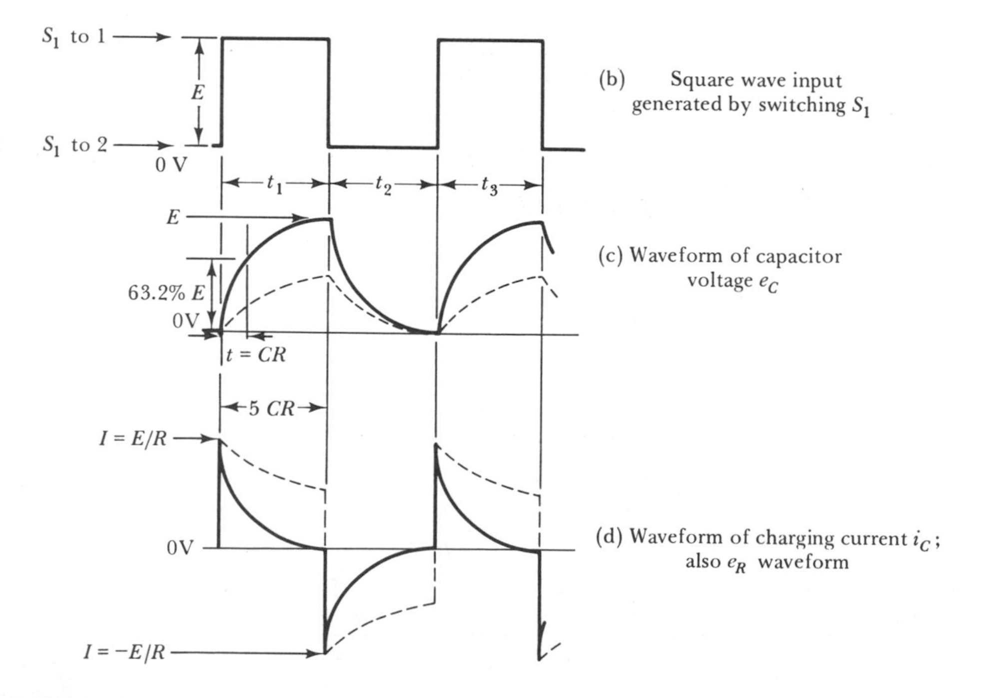

(b) Square wave input generated by switching S_1

(c) Waveform of capacitor voltage e_C

(d) Waveform of charging current i_C; also e_R waveform

FIG. 17-12. *CR* series circuit and waveforms generated by switching S_1

Then e_C immediately commences to fall slowly toward zero, at which level it remains until the switch is moved over again. This growth and decline of e_C is repeated over and over as the switch is moved from one position to the other, resulting in the waveform shown.

When S_1 is switched to position 1 and e_C is zero, the charging current is a maximum of $I=E/R$. Then as the capacitor voltage grows, the current falls off to zero. This is illustrated in Figure 17-12(d). Similarly, when S_1 is moved to position 2, the current is now a discharge current and it has an initial maximum level of $I=-E/R$. Thus, as shown in Figure 17-12(d), the current is a negative quantity falling off to zero during t_2. Once again, the sequence is repeated as S_1 is moved from one position to the other, and the *spike waveform* illustrated is generated. The resistor voltage e_R is, of course, $i \times R_1$; consequently, the waveform of e_R is identical in shape to the current waveform.

The waveforms shown in Figure 17-12 are those that are typically obtained when $t_1 = t_2 \cong 5CR$ [see Figure 17-12(c)]. CR is, therefore, thought of as a *short time constant* by comparison to the charge and discharge times of the capacitor. When CR is several times greater than the pulse width, the circuit is said to have a *long time constant*. In this case the typical waveforms would be as shown by the dashed lines in Figure 17-12(c) and (d).

GLOSSARY OF FORMULAS

Current in a series LR circuit:

$$i=\frac{E-e_L}{R}$$

Instantaneous current in an LR series circuit:

$$i=\frac{E}{R}(1-\epsilon^{-t(R/L)})$$

R, L, and t values in an LR series circuit:

$$t\frac{R}{L}=\ln\left(\frac{E}{E-iR}\right)$$

Current in a series CR circuit:

$$i=\frac{E-e_C}{R}$$

Instantaneous voltage in a series CR circuit:

$$e_C=E-(E-E_o)\epsilon^{-t/(CR)}$$

When $E_o=0$:

$$e_C=E(1-\epsilon^{-t/(CR)})$$

R, C, and t values in a series CR circuit:

$$\frac{t}{CR} = \ln\left(\frac{E}{E - e_C}\right)$$

Instantaneous discharging voltage in a series CR circuit:

$$e_C = E_o \epsilon^{-t/(CR)}$$

REVIEW QUESTIONS

17-1 Sketch an approximate representation of the current growth in a series inductive–resistive circuit. Carefully explain how the current and the counter-emf vary with time.

17-2 Define the *time constant* of an *LR* circuit and explain the special relationships between the time constant and the time at which the inductor's instantaneous current and voltage reach certain levels.

17-3 Explain what occurs when the supply switch in a series *LR* circuit is opened. Discuss the danger to the circuit that occurs and methods used to avoid damage.

17-4 Sketch the voltage and current waveforms that result when a square wave is applied to a series *LR* circuit:

a. For a long time constant.

b. For a short time constant.

Explain briefly.

17-5 Sketch an approximate representation of the voltage growth across a capacitor being charged via a series resistor. Carefully explain how the charging current and voltage vary with time.

17-6 Define the *time constant* of a *CR* circuit and explain the special relationship between the time constant and the time at which the capacitor's instantaneous voltage and charging current reach certain levels.

17-7 Discuss what occurs when the supply to a *CR* circuit is interrupted, and when a fully charged capacitor is short-circuited. Explain any danger that results, and the methods used to avoid damage.

17-8 Sketch the voltage and current waveforms that result when a square wave is applied to a series *CR* circuit:

a. For a long time constant.

b. For a short time constant.

Explain briefly.

PROBLEMS

17-1 A circuit consisting of a 100-mH inductor in series with a 500-Ω resistor has a 50-V supply. Calculate the instantaneous levels of circuit current at 0.2-ms intervals up to $t = 1$ ms, and plot the graph of i versus t.

17-2 For the circuit described in Problem 17-1, calculate the instantaneous levels of inductor voltage at 0.25-ms intervals, and plot the graph of e_L versus t.

17-3 A series *LR* circuit with a 35-V supply is to have an instantaneous current of 5 mA at a time of 22 ms from the instant the switch is closed. If $R = 4.7$ kΩ, determine the required value of L.

17-4 A 600-mH inductor with a winding resistance of 250 Ω is connected in series with a 330-Ω resistor and a 24-V supply. Determine the lengths of time that elapse from the instant the supply is switched *on* until the inductor voltage reaches 5 V and 17 V.

17-5 For the series circuit described in Problem 17-4, show how an additional resistor is connected to limit the inductor emf when the supply is switched *off*. Determine the value of the resistor required to limit the inductor emf to a maximum of 75 V.

17-6 A circuit consisting of a 10-μF capacitor in series with a 2.2-kΩ resistor has a 50-V supply. Calculate the instantaneous levels of capacitor voltage at 15-ms intervals up to $t = 105$ ms, and plot the graph of e_C versus t.

17-7 For the circuit described in Problem 17-6, calculate the instantaneous charging current levels at 20-ms intervals and plot the graph of i versus t.

17-8 A series *CR* circuit with a 30-V supply is to have an instantaneous capacitor voltage of 22 V at a time of 65 μs from $t = 0$. If $C = 0.1$ μF, determine the required value of R.

17-9 A 47-μF capacitor is connected in series with a 4.7-kΩ resistor and a 5-V supply. Determine how long it takes for the capacitor voltage to reach 2.2 V. Also calculate the length of time required for the voltage of the fully charged capacitor to fall to 3 V when the battery is replaced with a short circuit.

17-10 Sketch the series circuit described in Problem 17-9, and show how an additional resistor and switch should be included for safely discharging the capacitor. Determine the value of the additional resistor if the capacitor discharge current is not to exceed 1 mA. Also, calculate approximately how long it will take for the capacitor to be completely discharged.

18

ALTERNATING CURRENT AND VOLTAGE

INTRODUCTION

Alternating current (ac) is current that is continuously reversing direction, alternately flowing in one direction and then in the other. Alternating voltage can be similarly described. Usually ac voltages and currents are sinusoidal, and definite relationships exist among the *peak*, *average*, and *effective* values. The effective value, or *rms value*, of an ac quantity is the normally quoted value and is the value normally applied in Ohm's law and power calculations. The *frequency* and *phase angle* of an ac quantity can also be important, and the *wavelength* can be calculated from a knowledge of the frequency and the velocity of the wave.

18-1 GENERATION OF ALTERNATING VOLTAGE

Electromagnetic induction has been discussed in Sections 12-2 and 15-1. In Section 15-1 it was shown how an emf is induced in a conductor in motion through a magnetic field. Now consider two series-connected conductors rotating in a magnetic field, as shown in Figure 18-1. Conductors 1 and 2 form a loop situated within the field set up between two magnetic poles. The conductors are mechanically fastened to an axle equipped with a crank for hand turning. Each conductor is connected to a *slip ring*, and the slip rings have *brushes* for electrically connecting to external terminals. When the loop is rotated, the conductors cut the magnetic flux, and an emf is generated in each conductor.

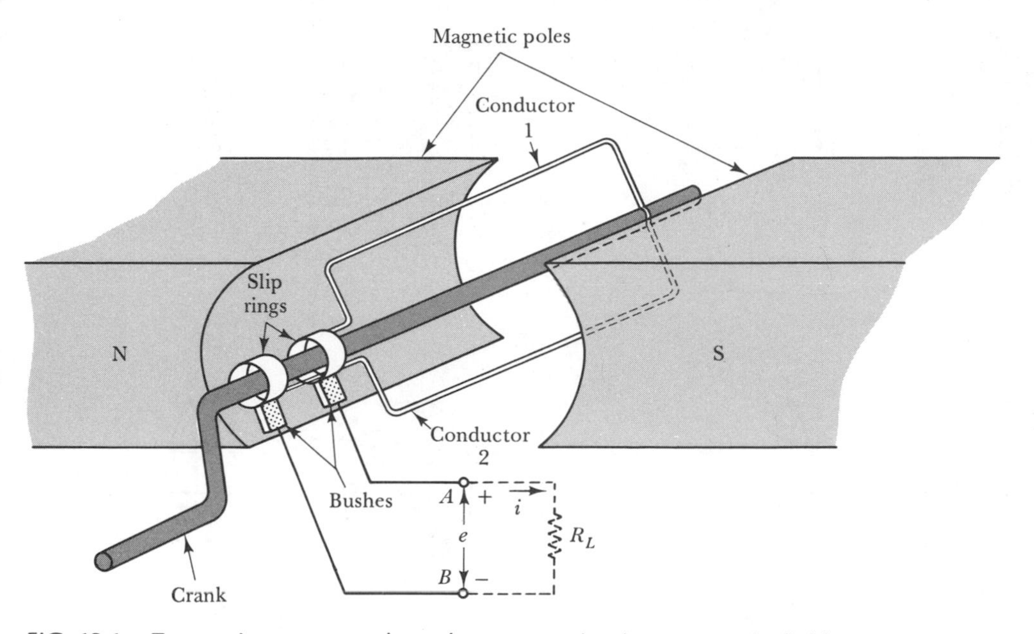

FIG. 18-1. Two series-connected conductors rotating in a magnetic field

Figure 18-2 illustrates the conditions that exist at each of several instants while the loop is being rotated in the magnetic field. These can be explained as follows:

- *Figure 18-2(a)*: Here the loop is vertical, and conductor 1 is moving to the left while conductor 2 is moving right. Because the motion of each conductor is parallel to the direction of the magnetic field at this instant, no flux is being cut by the conductors and so no emf is generated. Therefore, as illustrated by point 1 on the graph of e versus t, the voltage at output terminals A and B is zero, and no current flows through the external load resistor (R_L in Figure 18-1).
- *Figure 18-2(b)*: The conductors are now moving at an angle across the field, and so each is cutting some flux, and an emf is induced in each conductor. According to Lenz's law, the direction of the induced emf must be such that it opposes the motion that causes the emf. Thus, the emf induced in conductor 1 causes a current to flow *out of the paper*, as illustrated. This current produces a counterclockwise magnetic flux around the conductor, which strengthens the magnetic field in front of the conductor and weakens it behind (i.e., it opposes the conductor's motion). Similarly, the induced emf in conductor 2 causes a current in the direction shown, *into the page*. A clockwise flux is set up around conductor 2 and this again weakens the flux behind conductor 2 and strengthens it ahead of the conductor, again opposing the motion of the conductor. It is seen that the current directions in the two conductors are such that they assist each other (i.e., the

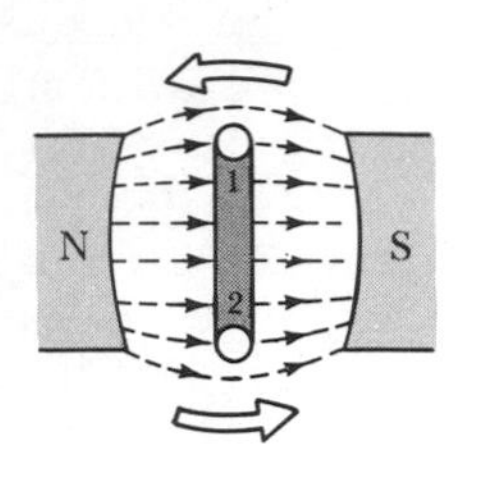

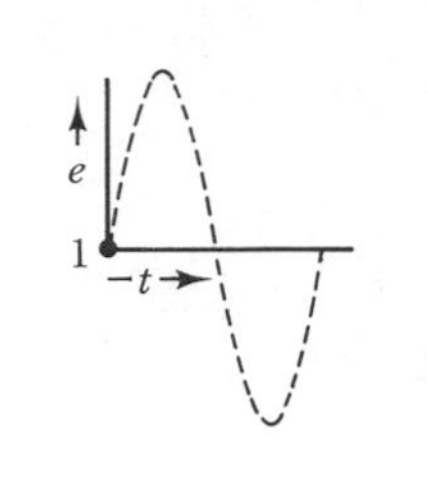

(a)

(b)

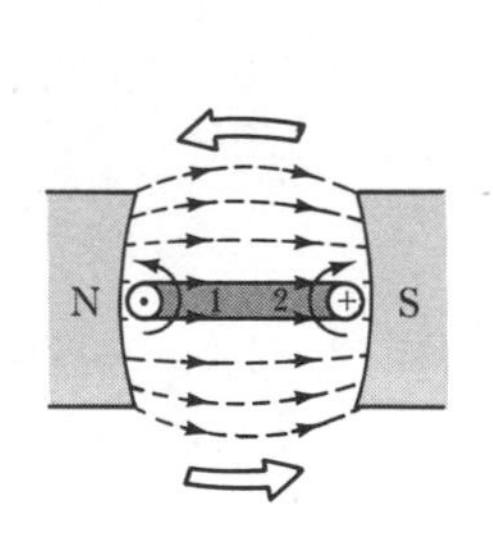

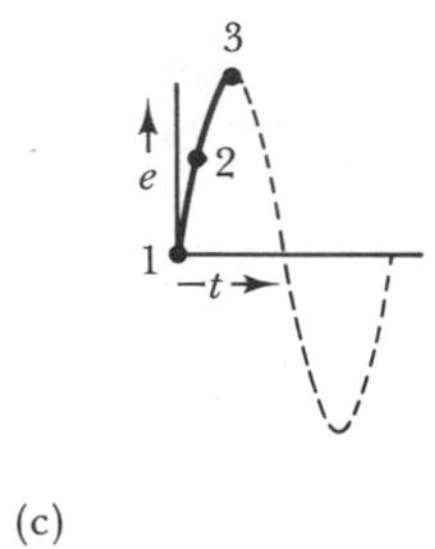

(c)

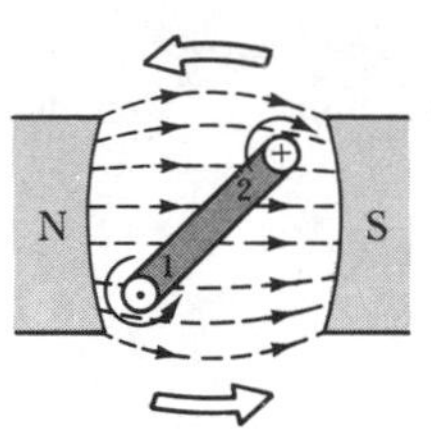

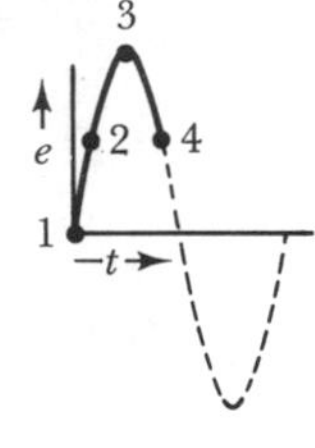

(d)

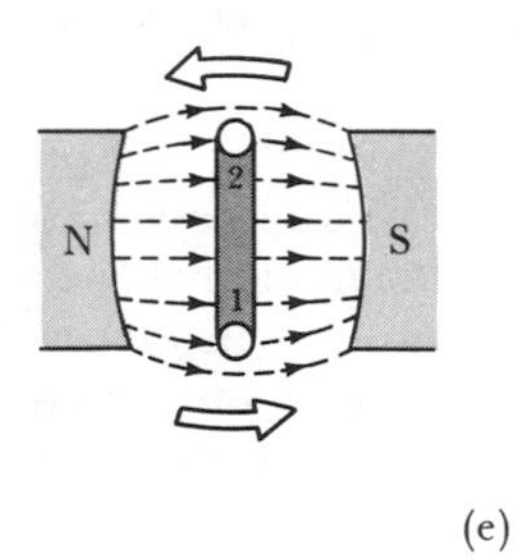

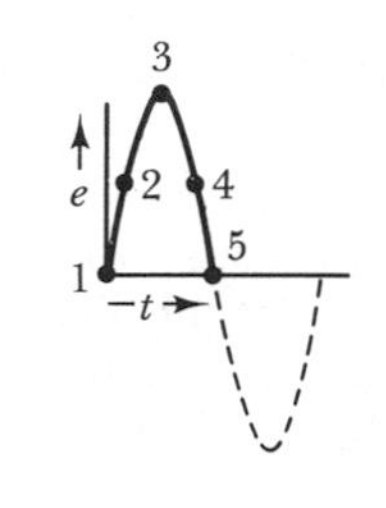

(e)

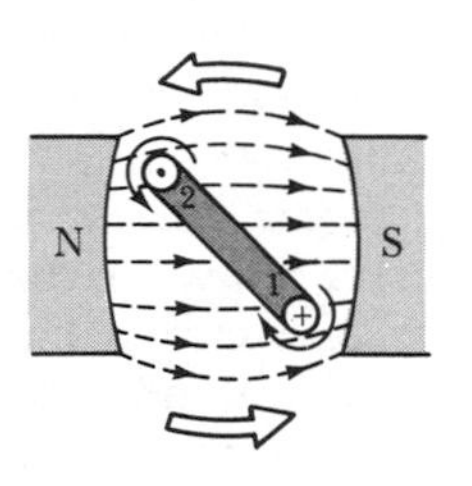

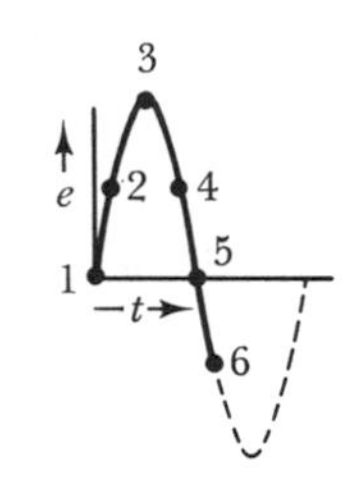

(f)

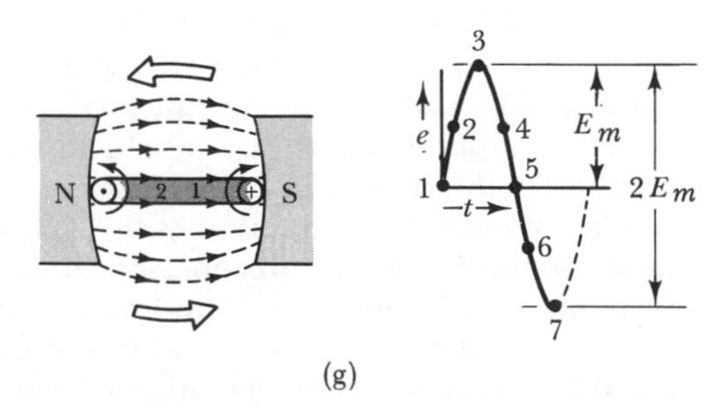

(g)

FIG. 18-2. Voltage generated by two conductors rotating in a magnetic field

current is flowing around the loop in a single direction). Consequently, the output voltage is positive at terminal A and negative at terminal B. (Conductor 1 is connected to A, and conductor 2 is connected to B.) A current flows in the direction shown (from A to B) through the external load R_L (see Figure 18-1). When the growth of the terminal voltage e is plotted versus time t, the result is as illustrated by the waveform in Figure 18-2(b). Point 1 represents the zero output voltage obtained for the conditions in Figure 18-2(a), and point 2 is the output voltage some time later when the conductor positions are as shown in Figure 18-2(b).

- *Figure 18-2(c)*: The conductors are now moving perpendicular to the magnetic field. Thus, they are cutting maximum flux and generating maximum emf. Again, the current directions are such that current flows around the loop, and terminal A remains positive while terminal B is negative. The output current through R_L is now greater than before, and as shown on the graph of e versus t, the output voltage has reached a peak value, point 3.
- *Figure 18-2(d)*: Now the conductors are again moving at an angle with respect to the direction of the magnetic field. Therefore, the amount of flux being cut is reduced, and consequently the generated emf is less than for the conditions in Figure 18-2(c). The current direction and output voltage polarity are the same as before. However, the output voltage has now dropped below the peak level, and is as shown at point 4 on the e versus t graph.
- *Figure 18-2(e)*: At this point the conductors are again moving parallel to the direction of the magnetic field. No flux is being cut, so no emf is generated. The output voltage is now zero, and is plotted at point 5 on the voltage versus time graph.
- *Figure 18-2(f)*: The conductors are once more moving at an angle with respect to the direction of the magnetic field. However, both conductors now have the induced current directions reversed. The current direction in conductor 1 is flowing *into the paper*, while that in conductor 2 is *out of the paper*. It is seen that the currents still assist each other, so that the current around the loop still has a single (but reversed) direction. The output voltage is now such that terminal B is positive, while terminal A is negative. The instantaneous output voltage e plotted versus time t gives point 6 on the graph.
- *Figure 18-2(g)*: The motion of the conductors is once again perpendicular to the magnetic field, and maximum flux is being cut. Thus, maximum emf is again generated. The output voltage is still a negative quantity and has reached its peak value, plotted at point 7 on the graph of e versus t. It takes only a little imagination to see that as the loop rotation continues, the output voltage goes to zero once more then reverses again, and the cycle is repeated over and over.

Since the voltage generated by the rotating loop is alternately positive and negative, it is referred to as an *alternating voltage*. The current produced in a load supplied by an alternating voltage also flows first in one direction and then in the other. Thus, it is an *alternating current* (*ac*). The designation *ac* is normally applied to both current and voltage. The voltage/time graph constructed in Figure 18-2 is spoken of as an *ac waveform*.

18-2 SINE WAVE

The waveform traced out on the graph of e versus t in Figure 18-2 has a shape that is termed *sinusoidal*, and therefore the wave is referred to as a *sine wave*. As will be shown, these terms arise because the instantaneous voltage is found to be dependent upon the trigonometrical *sine of an angle*.

In order to derive an equation for the instantaneous emf induced in the rotating loop, recall that the induced emf depends upon the amount of flux cut per second. From Eq. (15-1), the emf induced in a single conductor is

$$e = \frac{\Delta\Phi}{\Delta t}$$

where $\Delta\Phi$ is the total flux cut by the conductor during time Δt. Since $\Phi = B \times A$, the equation can be rewritten

$$e = B \times \frac{\Delta A}{\Delta t}$$

where B is the flux density between the magnetic poles and ΔA is the area swept by one conductor during time Δt.

Consider the situation illustrated in Figure 18-2(c), where each conductor is cutting maximum flux. If the total length of the two conductors within the magnetic field is l and if the conductors move through a distance Δd during time Δt, then the area swept by the conductors during Δt is

$$\Delta A = l \times \Delta d$$

and the equation for maximum emf E_m induced in the loop becomes

$$\boxed{E_m = B \times l \times \frac{\Delta d}{\Delta t}} \qquad \textbf{(18-1)}$$

Since $\Delta d/\Delta t$ is meters/second (i.e., a velocity v), the equation can be written

$$\boxed{E_m = Blv} \qquad \textbf{(18-2)}$$

where E_m is in volts, B is in teslas, l is in meters, and v is in m/s. In this case v

is the actual linear velocity of the conductors. It can also be described as the *peripheral velocity* with respect to the circle traced out by each conductor. It is important to note that Eq. (18-2) is an expression for the maximum or *peak value* E_m of the voltage generated [i.e., the voltage represented by point 3 on the e/t graph in Figure 18-2(g)]. The peak-to-peak value of the waveform is $2E_m$, as illustrated in Figure 18-2(g).

Now consider Figure 18-3, which illustrates the movement of conductor 1 from its position in Figure 18-2(a) toward that in Figure 18-2(b). It is seen that as the conductor moves through a distance Δd, the distance that it moves perpendicular to the flux is $\Delta d'$; and

$$\Delta d' = \Delta d \sin\alpha$$

where α is the angle moved through by the conductor in circular motion. Substituting in Eq (18-1), the instantaneous emf is

$$e = Bl\frac{\Delta d'}{\Delta t}$$

$$= Bl\frac{\Delta d}{\Delta t}\sin\alpha$$

$$\boxed{e = Blv\sin\alpha} \qquad \textbf{(18-3)}$$

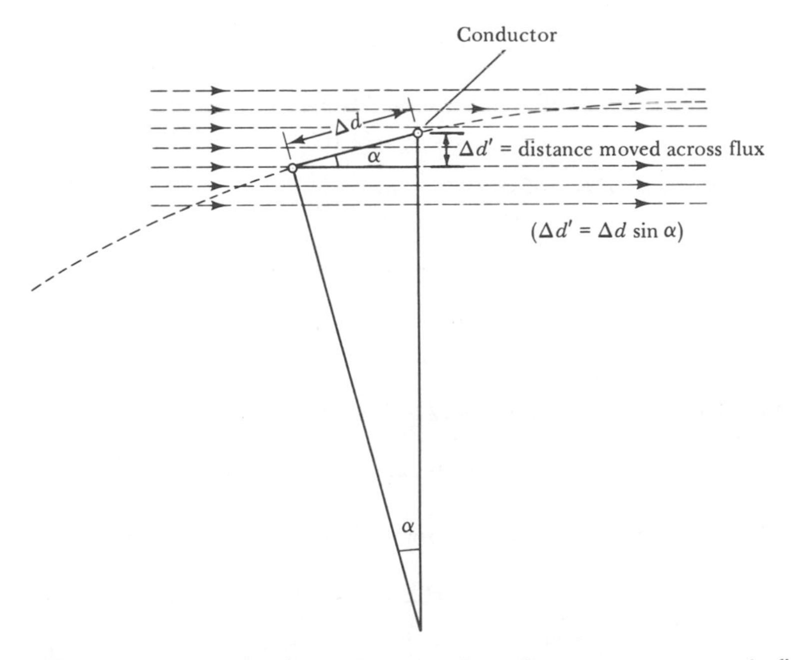

FIG. 18-3. Perpendicular movement of conductor across magnetic flux

Blv is the expression obtained in Eq. (18-2) for the peak value of the voltage generated. Therefore,

$$\boxed{e = E_m \sin\alpha} \qquad (18\text{-}4)$$

In Eq. (18-4), α is usually termed the *phase angle* of the instantaneous voltage e. E_m is, of course, the peak value, but it is also referred to as the *amplitude* of the waveform.

EXAMPLE 18-1

In the hand-cranked generator illustrated in Figure 18-1, the length of each conductor within the magnetic field is 25 cm. The distance from each conductor to the axis of rotation is $r = 5$ cm and the flux density of the magnetic field is $B = 0.1$ T. Calculate the maximum output voltage from the generator when the conductors are rotated at 100 revolutions/minute.

SOLUTION

Distance traveled by each conductor in one revolution is the circumference of the circle traced out by rotation. Therefore,

$$\text{distance, } d = 2\pi r$$

$$= 2\pi \times 5 \times 10^{-2}\ \text{m}$$

$$\mathbf{d = 0.1\pi\ m}$$

$$\text{distance traveled per minute} = d \times 100\ \text{rpm}$$

$$= 0.1\pi\ \text{m} \times 100$$

$$\textbf{\textit{distance traveled per minute}} = \mathbf{10\pi\ m}$$

Therefore, conductor peripheral velocity, $v = \dfrac{10\pi}{60}$ m/s

$$\mathbf{v = 0.524\ m/s}$$

From Eq. (18-2), *the peak emf generated in the loop is*

$$E_m = Blv$$

where $l = 2 \times$(length of each conductor). Therefore

$$E_m = 0.1\text{T} \times 2 \times 0.25\ \text{m} \times 0.524\ \text{m/s}$$

$$\mathbf{E_m = 26.2\ mV}$$

EXAMPLE 18-2

For the generator described in Example 18-1, calculate the instantaneous output voltage at points 2, 3, 4, and 6 on the graph of e/t illustrated in Figure 18-2. Assume that the phase angles of the conductors at these points with respect to the flux direction are 45°, 90°, 135°, and 225°, respectively.

SOLUTION

From Eq. (18-4), *the instantaneous emf generated is*

$$e = E_m \sin\alpha$$

At point 2,

$$e = 26.2 \text{ mV} \sin 45°$$

$$2e \cong \mathbf{18.5\ mV}$$

At point 3,

$$e = 26.2 \text{ mV} \sin 90°$$

$$e = \mathbf{26.2\ mV}$$

At point 4,

$$e = 26.2 \text{ mV} \sin 135°$$

$$e \cong \mathbf{18.5\ mV}$$

At point 6,

$$e = 26.2 \text{ mV} \sin 225°$$

$$e \cong \mathbf{-18.5\ mV}$$

18-3 FREQUENCY AND PHASE ANGLE

Consider Figure 18-4, which once again illustrates the sine-wave output from an ac generator. It has been shown that when the conductors in the simple generator are rotated through one complete revolution, the output voltage goes through one complete cycle of sinusoidal change. The time taken for *one cycle* of change (from zero volts through $+E_m$, zero, and $-E_m$ back to zero again) is referred to as the *time period* of the waveform, and is designated T (see Figure 18-4). If $T=1$ s, it is said that the waveform has a *frequency* (f) of 1 *cycle per second* or 1 *hertz* (Hz).

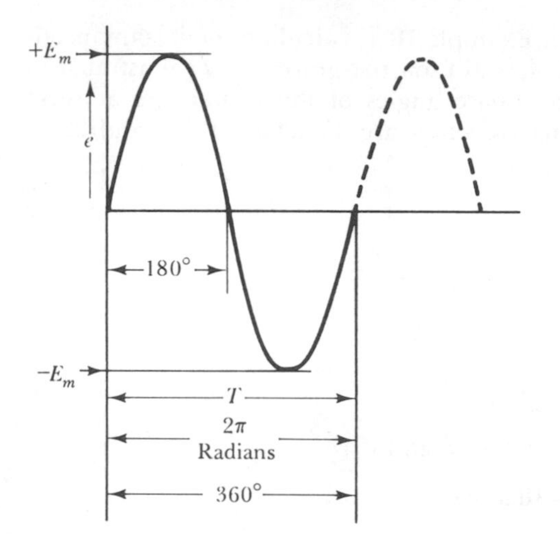

FIG. 18-4. One cycle of a sine wave

A waveform has a frequency of 1 ***hertz**** ***(Hz)*** *when it goes through one complete cycle of change in a period of 1 second.*

From this definition it can be written that

$$f = \frac{1}{T} \tag{18-5}$$

where f is in hertz and T is in seconds.

Obviously, a waveform that has a time period of 1 ms has a frequency of

$$f = \frac{1}{1 \text{ ms}}$$

$$= 1000 \text{ Hz}$$

or

$$f = 1 \text{ kHz}$$

The usual (North American) frequency for domestic and industrial ac supplies is 60 Hz. However, in electronics, the ac frequencies of waveforms can range from zero (i.e., dc) through kilohertz (kHz), megahertz (MHz), and gigahertz (GHz).

For one complete revolution, the conductors in the simple generator obviously rotate through 360°. If the angle of rotation of the conductors is measured in *radians* instead of degrees, one complete revolution represents

*Named for the German physicist Heinrich R. Hertz (1857–1894).

2π radians. Since T is the time period in seconds, the angle through which the conductors move in 1 s is given by

$$\text{angular velocity} = \frac{360}{T} \text{ degrees/second}$$

or

$$\text{angular velocity} = \frac{2\pi}{T} \text{ radians/second}$$

and the phase angle at any instant t_1, t_2, t_3, etc., measured from $t=0$ is

$$\text{angle} = \frac{2\pi}{T} \times t \text{ radians}$$

Since

$$f = \frac{1}{T}$$

the equation can be rewritten

$$\text{angle} = 2\pi f t \qquad \text{radians}$$

or

$$\text{angle} = \omega t \qquad \text{radians}$$

where ω is the angular velocity in rad/s.

Now Eq. (18-4) can be written in the form normally employed to represent ac waveforms:

$$\boxed{e = E_m \sin \omega t} \tag{18-6}$$

EXAMPLE 18-3

An ac waveform with a frequency of 1.5 kHz has a peak value of 3.3 V. Calculate the instantaneous levels of voltage at $t_1 = 0.65\ \mu s$ and $t_2 = 1.2$ ms.

SOLUTION

$$\begin{aligned}\omega &= 2\pi f \\ &= 2\pi \times 1.5 \text{ kHz} \\ &= 3\pi \times 10^3 \text{ rad/s}\end{aligned}$$

Eq. (18-6):

$$e = E_m \sin \omega t$$

At t_1,

$$e = 3.3\,\mathrm{V}\sin[(3\pi \times 10^3 \times 0.65\ \mu\mathrm{s})\mathrm{rad}]$$
$$\mathbf{e_1 \cong 20.2\ mV}$$

At t_2,

$$e = 3.3\,\mathrm{V}\sin[(3\pi \times 10^3 \times 1.2\ \mathrm{ms})\mathrm{rad}]$$
$$\mathbf{e_2 \cong -3.1V}$$

In Figure 18-5 four different waveforms are illustrated. Waveforms B, C, and D obviously have smaller amplitudes than waveform A. Because waveform B goes through its cycle of change exactly at the same time as waveform A, the two are said to be *in phase* with each other. Waveforms C and D, however, are *out of phase* with waveforms A and B. Because waveforms A and B commence their cycles ahead of C by an angle θ (degrees or radians), A and B are said to *lead* C by θ. Alternatively, waveform C may be said to *lag* waveforms A and B by θ. θ may also be referred to as the *phase difference* between the waveforms. Waveform D lags A and B by 180° and is said to be in *antiphase* to A and B because it is at its negative peak when A and B are at their positive peak levels.

The *wavelength* of an ac waveform depends upon its velocity. In the case of radio waves, the velocity is the speed of light, which is 3×10^8 m/s. For voltage waves moving along widely spaced conductors, the transmission speed is also 3×10^8 m/s. When the conductors are close together, the velocity depends upon the insulation employed. In Figure 18-6, part (a), e_1 represents the initial or zero level of a voltage waveform at the start of a long conductor. A fraction of a second later, part (b), e_1 has traveled

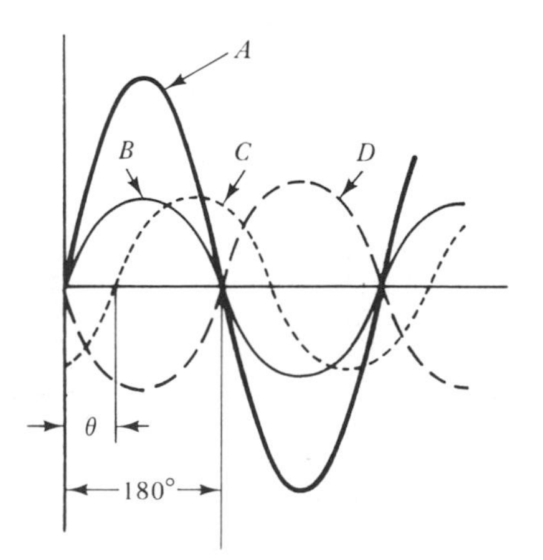

FIG. 18-5. In-phase and out-of-phase sine waves

along the conductor and the instantaneous level of the voltage at the generator terminals has changed to e_2. Now both e_1 and e_2 are traveling along the conductor at a velocity of 3×10^8 m/s, and in Figure 18-6(c) the voltage at the generator terminals has changed to e_3. By the time the generator terminal voltage has gone through one complete cycle [Figure 18-6(d)], e_1 has been moving along the conductor for the time period T. Therefore, e_1 has traveled a distance of

$$\text{distance} = (T \text{ seconds} \times 3 \times 10^8 \text{ m/s}) \qquad \text{meters}$$

and this distance represents the *wavelength* of the alternating voltage. The wavelength can be calculated from the equation

$$\lambda = c \times T$$

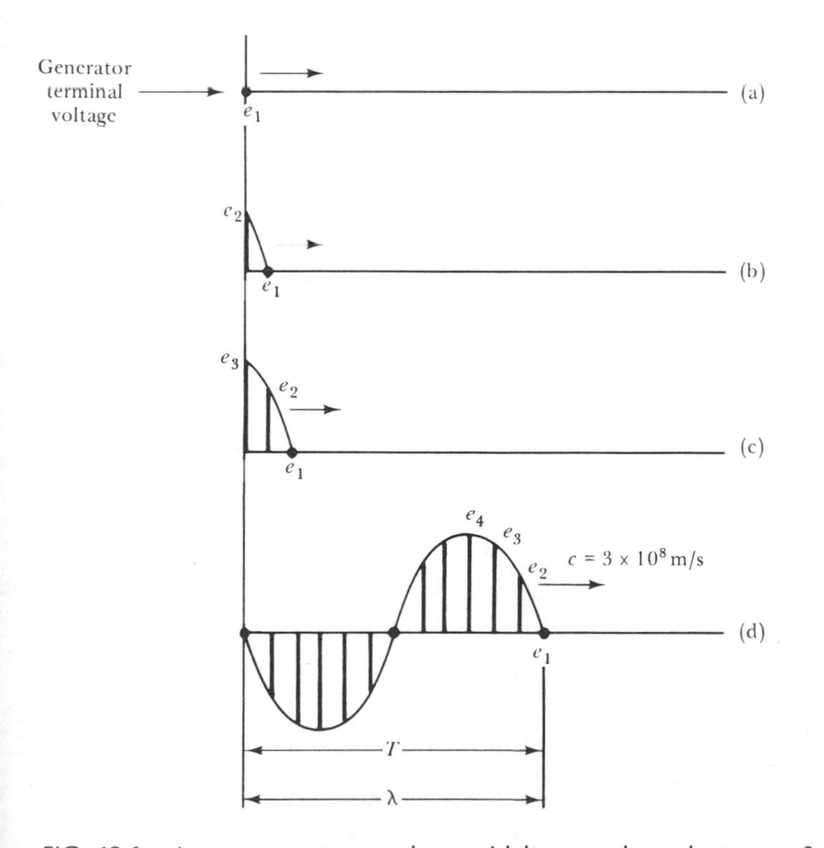

FIG. 18-6. An ac wave moves along widely spaced conductors at 3×10^8 m/s

or

$$\boxed{\lambda = \frac{c}{f}} \qquad (18\text{-}7)$$

where λ is wavelength in meters, c is velocity of light in m/s, and f is the frequency of the waveform in hertz.

EXAMPLE 18-4

Determine the wavelengths of 60-Hz and 1-MHz alternating waves. Also, calculate the frequency of a voltage that has a wavelength of 33 m.

SOLUTION

Eq. (18-7):

$$\lambda = \frac{c}{f}$$

For 60 Hz,

$$\lambda = \frac{3 \times 10^8 \text{ m/s}}{60 \text{ Hz}}$$

$$\boldsymbol{\lambda = 5 \times 10^6 \text{ m}}$$

For 1 MHz,

$$\lambda = \frac{3 \times 10^8 \text{ m/s}}{1 \text{ MHz}}$$

$$\boldsymbol{\lambda = 300 \text{ m}}$$

From Eq (18-7)

$$f = \frac{c}{\lambda} = \frac{3 \times 10^8 \text{ m/s}}{33 \text{ m}}$$

$$\boldsymbol{f = 9.09 \text{ MHz}}$$

18-4 RESISTIVE LOAD WITH AC SUPPLY

Figure 18-7(a) shows a resistor R connected to an ac voltage source. Note the graphic symbol for the ac voltage source. Unlike the case of a dc circuit, a continuous current direction cannot be shown, because the current through R will change direction each time the polarity of e reverses. However, it is convenient sometimes to show an instantaneous

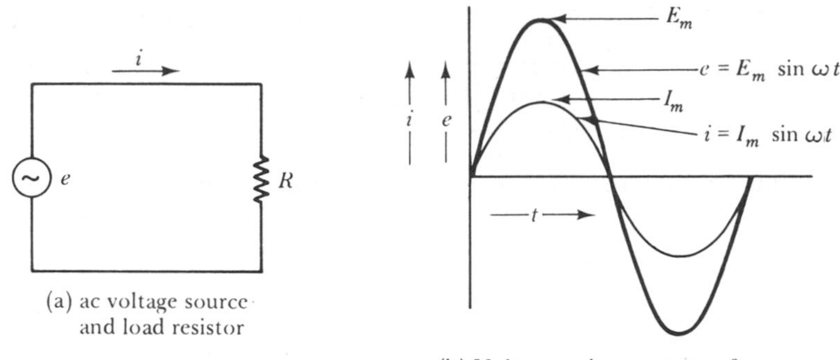

FIG. 18-7. The alternating current in a purely resistive circuit is in phase with the alternating voltage

current direction. The instantaneous current level can be found by Ohm's law:

$$i = \frac{e}{R}$$

$$\boxed{i = \frac{E_m}{R} \sin \omega t} \tag{18-8}$$

The peak value of the current occurs at the peak value of the instantaneous voltage (i.e., when $e = E_m$). Therefore,

$$\textit{maximum or peak current level, } I_m = \frac{E_m}{R}$$

and

$$\boxed{i = I_m \sin \omega t} \tag{18-9}$$

Since e is a sinusoidal quantity and i is directly proportional to e, the current in the circuit of Figure 18-7(a) is also a sinusoidal quantity. Also, when e is zero, i must be zero, and when e is a maximum, i is at its peak value. This means that i is in phase with e, and the two waveforms can be plotted as shown in Figure 18-7(b).

From Eq. (3-9), the power dissipated in a resistor is

$$P = I^2 R$$

and the instantaneous power is

$$\begin{aligned} p &= i^2 R \\ &= (I_m \sin \omega t)^2 R \end{aligned}$$

or

$$\boxed{p = I_m^2 R \sin^2 \omega t} \tag{18-10}$$

The maximum instantaneous power dissipation that occurs is

$$P_m = I_m^2 R$$

Therefore,

$$\boxed{p = P_m \sin^2 \omega t} \tag{18-11}$$

The instantaneous power dissipated in the load resistance at any instant can easily be calculated using the equations derived above. Note that when I_m is a negative quantity, Eq. (18-10) gives

$$p = (-I_m)^2 R \sin^2 \omega t$$

which becomes

$$p = I_m^2 R \sin^2 \omega t$$

which is a positive quantity. Obviously, power is dissipated by a current flowing in either direction through a resistor. The fact that p is always a positive quantity means that all the power supplied is dissipated in the resistor R. If the instantaneous power were to become a negative quantity at any time, it would imply that the resistor was supplying power to the signal source, which is something a resistor cannot do. The fact that the instantaneous power is always a positive quantity also explains the shape of the graph of instantaneous power plotted in Figure 18-8.

Examination of the waveform of instantaneous power dissipation in Figure 18-8 reveals that it is a perfect sine wave symmetrical about the level $\frac{1}{2}P_m$. Therefore, $\frac{1}{2}P_m$ is the average value of the power dissipated in the resistor; or, average power dissipated over one complete cycle is half the peak power:

$$\boxed{p = \tfrac{1}{2}P_m} \tag{18-12}$$

EXAMPLE 18-5

The circuit shown in Figure 18-7(a) has a 60-Hz supply voltage with a maximum value of 160 V. If $R = 10\ \Omega$, calculate the values of instantaneous current and power at phase angles of $\pi/4$, $\pi/2$, $5\pi/4$, and $3\pi/2$ radians.

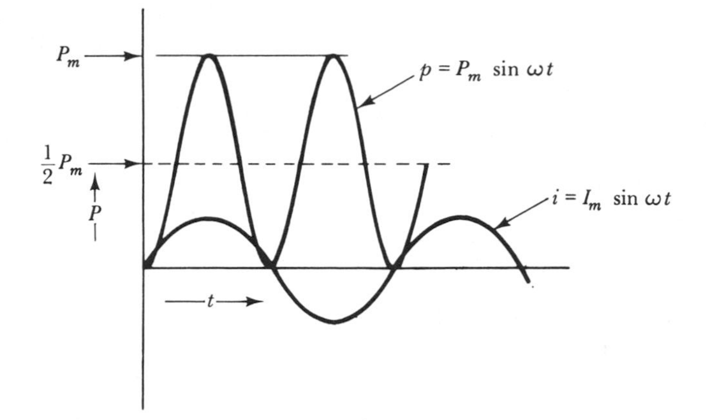

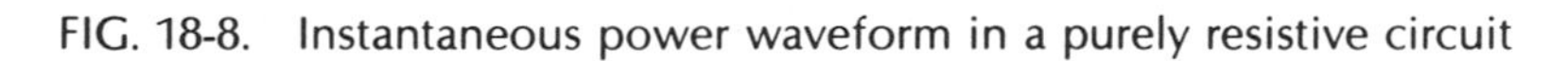
FIG. 18-8. Instantaneous power waveform in a purely resistive circuit

SOLUTION

$$I_m = \frac{E_m}{R} = \frac{160\ \text{V}}{10\ \Omega}$$

$$\boldsymbol{I_m = 16\ \text{A}}$$

$$P_m = I_m^2 R = (16\ \text{A})^2 \times 10\ \Omega$$

$$\boldsymbol{P_m = 2.56\ \text{kW}}$$

Current calculations:

Eq. (18-9):

$$i = I_m \sin \omega t$$

For $\omega t = \pi/4$,

$$i = 16\ \text{A} \sin \frac{\pi}{4}$$

$$\boldsymbol{i \cong 11.3\ \text{A}}$$

For $\omega t = \pi/2$,

$$i = 16\ \text{A} \sin \frac{\pi}{2}$$

$$\boldsymbol{i = 16\ \text{A}}$$

For $\omega t = 5\pi/4$,

$$i = 16\ \text{A} \sin \frac{5\pi}{4}$$

$$\boldsymbol{i \cong -11.3\ \text{A}}$$

For $\omega t = 3\pi/2$,

$$i = 16 \text{ A} \sin \frac{3\pi}{2}$$

$$\boldsymbol{i = -16 \text{ A}}$$

Power calculations:
Eq. (18-11):

$$p = P_m \sin^2 \omega t$$

For $\omega t = \pi/4$,

$$p = 2.56 \text{ kW} \sin^2 \frac{\pi}{4}$$

$$\boldsymbol{p = 1.28 \text{ kW}}$$

For $\omega t = \pi/2$,

$$p = 2.56 \text{ kW} \sin^2 \frac{\pi}{2}$$

$$\boldsymbol{p = 2.56 \text{ kW}}$$

For $\omega t = 5\pi/4$,

$$p = 2.56 \text{ kW} \sin^2 \frac{5\pi}{4}$$

$$\boldsymbol{p = 1.28 \text{ kW}}$$

For $\omega t = 3\pi/2$,

$$p = 2.56 \text{ kW} \sin^2 \frac{3\pi}{2}$$

$$\boldsymbol{p = 2.56 \text{ kW}}$$

18-5 AVERAGE AND RMS VALUES OF SINE WAVES

The average value of a waveform can be determined by differential calculus, or by the more laborious process of taking the average value of a number of equally spaced instantaneous levels. The process is illustrated in Figure 18-9, where each 180° half of the current waveform is divided into nine 20° sections. The instantaneous values of the waveform at the

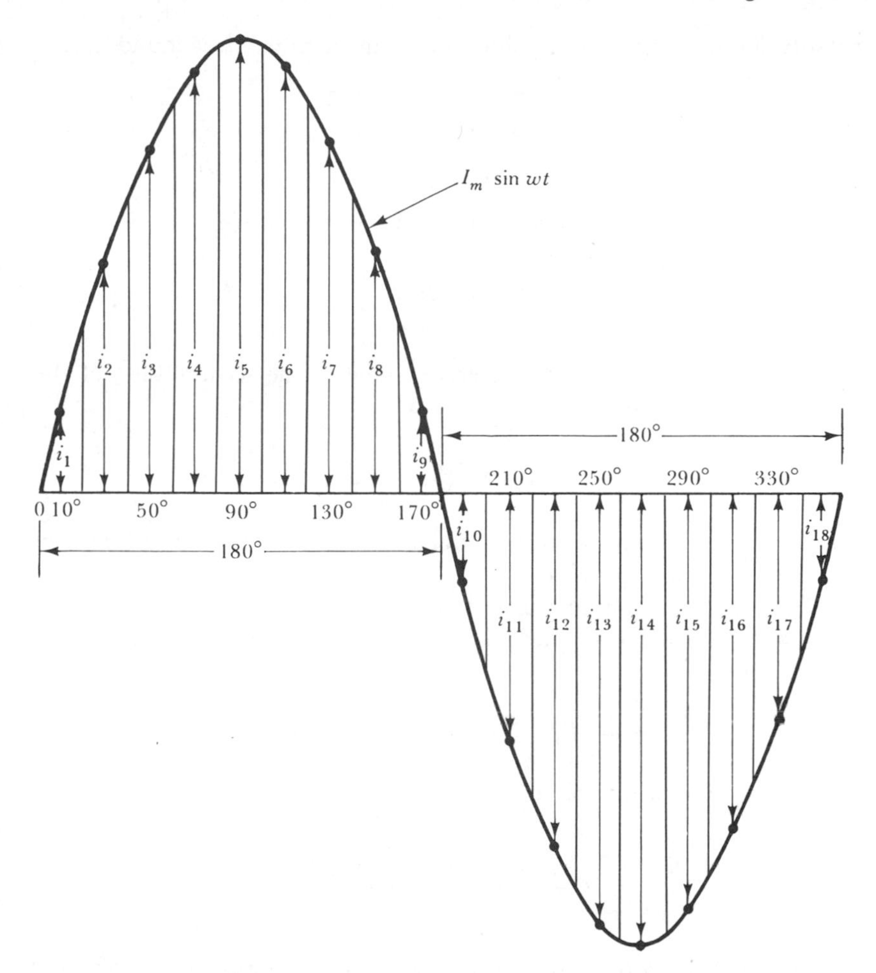

FIG. 18-9. Determination of average and rms values of a sine wave

center of each section is measured, and the average value calculated:

$$I_{av} = \frac{i_1 + i_2 + i_3 + \cdots + i_9}{9}$$

This process can be applied to waveforms of any shape; in fact, it may be the only way of determining the average value of some irregular waveforms. For the sine wave, however, it is not necessary to measure each instantaneous value, since they can be calculated from the maximum value and the sine of each angle. Thus, the average value is

$$I_{av} = \frac{I_m \sin 10° + I_m \sin 30° + I_m \sin 50° + \cdots + I_m \sin 170°}{9}$$

Substituting the appropriate values (or using calculus), it is found that

$$\boxed{I_{av} = \frac{2}{\pi} I_m \simeq 0.637 I_m} \tag{18-13}$$

Similarly, the average value of a sinusoidal voltage wave is

$$\boxed{E_{av} = 0.637 E_m} \tag{18-14}$$

Returning to Figure 18-9, the average value of the second 180° of the waveform is found to be

$$I_{av} = -0.637 I_m$$

Therefore, over the entire 360° sinusoidal waveform, the average value of the current (or voltage) is zero.

Taking each instantaneous current value in Figure 18-9 in turn, the instantaneous power dissipated in a resistor through which the current is flowing is obviously:

$$p_1 = i_1^2 R, \quad p_2 = i_2^2 R, \quad \cdots, \quad p_n = i_n^2 R$$

The average power dissipated in the resistor is then

$$P = \frac{i_1^2 R + i_2^2 R + \cdots + i_n^2 R}{n}$$

As explained in Section 18-4, the negative instantaneous current values also give a positive power dissipation in the load.

When a direct current is flowing through a resistor, the power dissipated is given by Eq. (3-9):

$$P = I^2 R$$

To find the equivalent alternating current that dissipates the same amount of power as a direct current I flowing through the resistor, use

$$I^2 R = \frac{i_1^2 R + i_2^2 R + \cdots + i_n^2 R}{n}$$

$$I^2 = \frac{i_1^2 + i_2^2 + \cdots + i_n^2}{n}$$

It is seen that I^2 is the average value of the instantaneous current squared values, or the *mean-squared value*. And

$$I = \sqrt{\frac{i_1^2 + i_2^2 + \cdots + i_n^2}{n}}$$

Here, I is the *root of the mean-squared value*, or *rms value* of the alternating current. Thus, the rms value of an alternating current is the effective value, or *dc equivalent value*, as far as power dissipation is concerned. For this reason, the rms value is frequently termed the *effective value*.

Like the average value, the relationship between the rms value and the peak value of an ac waveform can be determined by calculus. However, it can also be worked out quite simply. From Eq. (18-12)

$$p = \tfrac{1}{2} P_m$$

$$= \tfrac{1}{2} I_m^2 R$$

and from Eq. (3-9),

$$P = I^2 R$$

Therefore,

$$I^2 R = \tfrac{1}{2} I_m^2 R$$

$$I^2 = \tfrac{1}{2} I_m^2$$

$$I = \sqrt{\tfrac{1}{2} I_m^2}$$

$$\boxed{I = \frac{1}{\sqrt{2}} I_m = 0.707 I_m} \tag{18-15}$$

or

$$I = \frac{1}{1.414} I_m$$

Similarly, for an *alternating voltage*: the *effective value* is the rms value of the waveform, and where the waveform is sinusoidal,

$$\boxed{E = \frac{1}{\sqrt{2}} E_m = 0.707 E_m} \tag{18-16}$$

or

$$E = \frac{1}{1.414} E_m$$

With alternating current and voltage the rms values are the *normally quoted* values. A domestic alternating voltage supply of 115 V, for example, has a peak value of

$$E_m = 1.414 \times 115\ \text{V} \cong 163\ \text{V}$$

Rms values are also normally used in all Ohm's law calculations involving current, voltage, and resistance.

It is important to note that the relationships stated in Eqs. (18-13) through (18-16) apply only to pure sine waves. Where other waveforms are considered, the peak, average, and rms quantities are related by other (different) factors.

EXAMPLE 18-6 A 300-V sinusoidal ac supply is applied to a 50-Ω resistor. Determine the peak, rms, and average values of the current through the resistor. Also, calculate the power dissipated in the resistor.

SOLUTION

$$\textit{peak voltage}, E_m = 1.414 \times E_{\text{rms}}$$
$$= 1.414 \times 300\ \text{V}$$
$$\mathbf{E_m \cong 424\ V}$$

$$\textit{peak current}, I_m = \frac{E_m}{R}$$
$$= \frac{424\ \text{V}}{50\ \Omega}$$
$$\mathbf{I_m = 8.48\ A}$$

$$\textit{rms current}, I = \frac{E}{R}$$
$$= \frac{300\ \text{V}}{50\ \Omega}$$
$$\mathbf{I = 6\ A}$$

$$\textit{average current for half-cycle}, I_{av} = 0.637 I_m$$
$$= 0.637 \times 8.48\ \text{A}$$
$$\mathbf{I \cong 5.4\ A}$$

$$\textit{power dissipation}, P = I^2 \times R$$
$$= (6\text{A})^2 \times 50\ \Omega$$
$$\mathbf{P = 1.8\ kW}$$

18-6 CATHODE RAY OSCILLOSCOPE

The *cathode ray oscilloscope* is the basic instrument for the study of waveforms. It can be used for measuring voltage, frequency, time interval, and phase. For laboratory investigations involving alternating waveforms, it is essential that students become familiar with the use of the *oscilloscope*.

Figure 18-10 shows the front panel and controls of a typical *double-beam* oscilloscope. The waveforms under investigation are displayed on the *screen* located on the front panel (see the illustration). This is a circular flat end of a glass tube, with its inside surface coated with fluorescent chemicals that glow when struck by a *beam* of electrons. The screen is protected by a (vertically and horizontally calibrated) flat piece of hard plastic, called a *graticule*. A *double-beam* oscilloscope has two electron beams, so that it can simultaneously trace two separate waveforms for comparison with each other.

To facilitate the display of two waveforms, two separate sets of controls are provided, identified as *BEAM A* and *BEAM B* on the front panel of the instrument. The *INTENSITY* controls allow the brightness of each beam to be adjusted, and the *FOCUS* controls permit each to be focused to a fine point (or to a fine line when a waveform is being displayed).

The purpose of the *Y SHIFT* control is to move the displayed waveform (or beam) vertically up or down the screen, just to set it in a convenient position for viewing. The *V/DIV* selector switch determines the *sensitivity* of the display to input voltages. When this control is set to *1 V* (that is, 1 V per vertical division on the graticule), an input voltage with a peak-to-peak amplitude of 4 V, for example, would occupy four vertical divisions on the oscilloscope screen. Such a waveform is shown on the upper half of the screen in Figure 18-11. A small control knob at the center of the *V/DIV* knob provides continuous sensitivity adjustment, so that the display amplitude may be increased as desired. When the *V/DIV* selection is to be used for measuring the amplitude of a displayed waveform, the small center knob must be turned to its extreme clockwise (*calibrated*) position.

The typical sensitivity range available (as illustrated) is from 0.01 V/division (i.e., 10 mV/division) to 20 V/division. A *×1-×10* switch alongside the *V/DIV* selector knob allows the sensitivity to be multiplied 10 times. Sometimes, when a very low amplitude waveform is to be displayed, the greatest sensitivity (10 mV/division) does not give a large enough display. If the *×1-×10* switch is set to *×10*, the 10-mV/division sensitivity selection becomes 1 mV/division. Thus, the amplitude of the displayed waveform is increased 10 times, and this tenfold increase applies to all sensitivity selections.

To the left of the *V/DIV* control, the *AC-0-DC* switch permits the input signal to be directly connected (*DC* position), disconnected

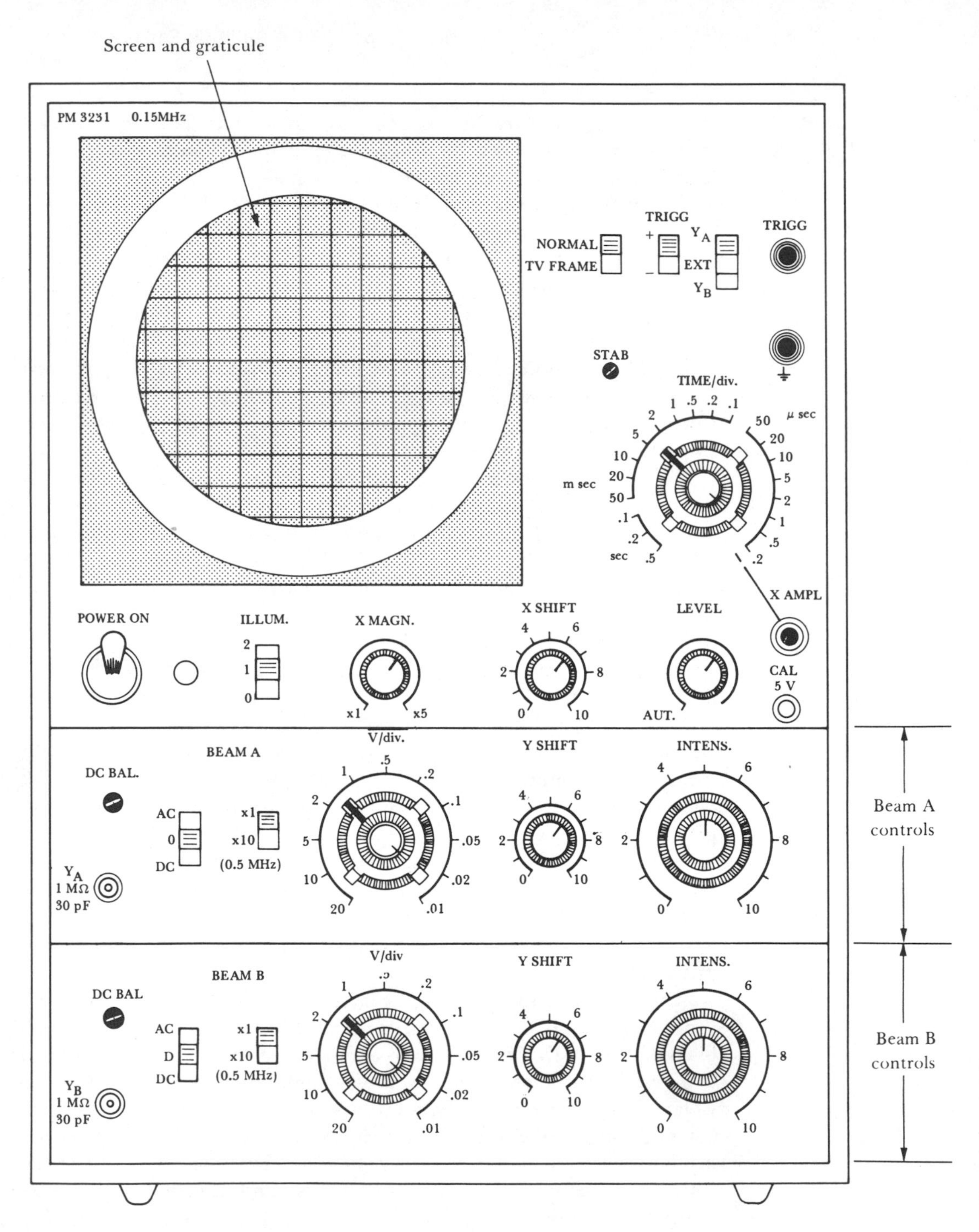

FIG. 18-10. Front panel and controls of a cathode ray oscilloscope (courtesy of Philips Electronics)

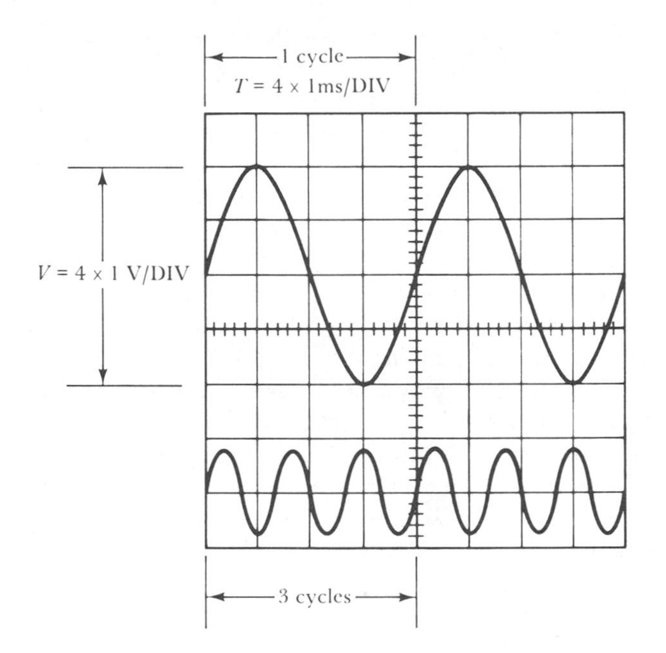

FIG. 18-11. Two sine waves displayed on an oscilloscope screen

(*O* position), or connected via a capacitor (*AC* position). Direct connection would be required, for example, to study dc voltage levels, or to investigate ac–dc voltage combinations. Sometimes it is necessary to display an ac voltage but not the dc level that it may be superimposed upon. This is achieved by setting the *AC-O-DC* switch to *AC*, where the coupling capacitor passes the ac quantity and blocks the dc.

The small (screwdriver adjustable) *DC BAL* (balance) control performs essentially the same function as the *Y SHIFT* knob. However, it is much more sensitive than the *Y SHIFT* control, and is normally adjusted only to calibrate certain internal circuits.

The *coaxial*-type input connections for each beam (or each *channel*) are identified on the front panel as Y_A and Y_B. Additional labeling shows their *input resistance* and *input capacitance* as 1 MΩ and 30 pF, respectively.

As already explained, the *V/DIV* switch adjusts the vertical amplitude of the display, by selecting the voltage sensitivity. In a similar way, the *TIME/DIV* switch (or *time base*) determines the horizontal amplitude of each cycle of input waveform displayed. This control applies to both beams. Refer to the two waveforms displayed on the screen in Figure 18-11 and assume that the *TIME/DIV* switch is set to 1 ms (i.e., 1 ms/horizontal division on the graticule). One cycle of the upper waveform occupies four horizontal divisions; therefore, its time period is 4×1 ms, or 4 ms. For the lower waveform, three complete cycles occupy 4 ms; consequently, the time period of the lower waveform is $\frac{4}{3}$ ms, or 1.3 ms.

The typical time-base range available (as illustrated) is from 0.5 s/division to 0.2 μs/division. At the extreme clockwise position of the *TIME/DIV* switch, the internal circuits which produce horizontal motion of the electron beams are disconnected. This allows horizontal deflection to be produced by an external voltage connected to the terminal identified as *X AMPL (X amplifier)*. A small control knob at the center of the *TIME/DIV* knob provides continuous time-base adjustment, so that the displayed waveform may be increased horizontally (widened) as desired. When the *TIME/DIV* selection is to be used for measuring the time period of a displayed waveform, the small center knob must be turned to its extreme clockwise (calibrated) position.

In order to provide a useful display of waveforms, it is necessary to ensure that the display commences (at the lefthand side of the screen) exactly when one of the input waveforms is at its zero position. Thus, the time-base circuits are said to be *triggered* at this instant. The triggering waveform is selected by the Y_A–*EXT*–Y_B switch above the TIME/DIV control. Thus, at Y_A, the waveform on *beam A* (or *channel A*) is used to trigger the time base. At the Y_B position, the *channel B* waveform triggers the time base. At *EXT*, triggering is provided by an external voltage connected across the *TRIGG* terminal and the ground terminal just below it. Triggering may also be + or −, selected by the *TRIGG* switch alongside the Y_A-*EXT*-Y_B switch. This simply means that the displayed waveform may be made to commence either when it is *going positive* or when it is *going negative* (see Figure 18-12). The *NORMAL-TV FRAME* switch applies only for certain tests on television sets and for most purposes is left at the *NORMAL* position. Also, the (screwdriver adjustable) *STAB* control is used for calibrating internal time-base circuitry and does not usually require adjustment.

Just below the *TIME/DIV* switch, the *LEVEL* knob is a *triggering level* control for the time-base circuits. This determines the voltage level of the

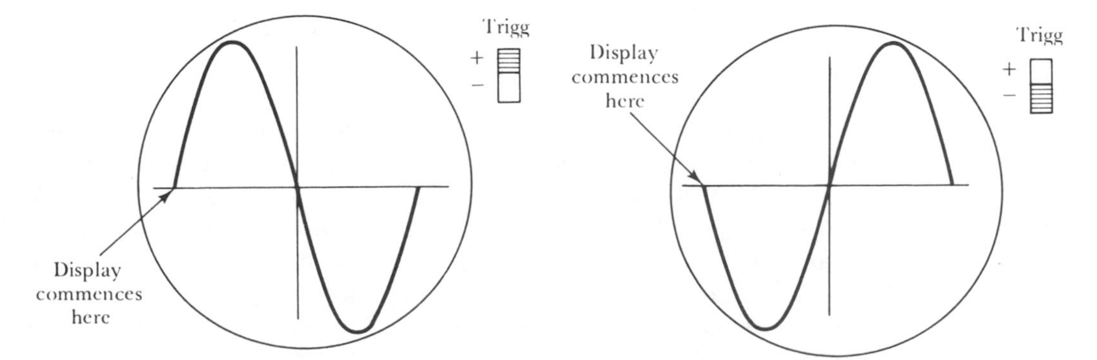

FIG. 18-12. A displayed waveform may be made to commence positive-going or negative-going

trigger waveform at which the display commences. Usually, this control is left at its extreme *counterclockwise AUT* (*automatic*) position, but sometimes it must be adjusted clockwise to obtain a display. To the right of the *LEVEL* control is a terminal labeled *CAL* 5 *V*. This provides an accurate internally generated square wave used for calibrating the oscilloscope.

The *X SHIFT* control has a similar function to the *Y SHIFT* knobs. It shifts the displayed waveforms horizontally about the screen, as desired for the best viewing. The *X MAGN* (*magnitude*) knob varies the sensitivity of the *TIME/DIV* selection. When the *X MAGN* is extremely *counterclockwise* (at ×*1*), all the *TIME/DIV* selections are exactly as identified. When the *X MAGN* control is at its extreme clockwise position (×*5*), the *TIME/DIV* selections all have their sensitivities increased five times. Thus, the 0.2 μs/division position becomes 0.04 μs/division. Between the ×*1* and ×*5* positions the time per division is indeterminant; therefore, the *X MAGN* control is usually kept at its ×*1* position.

To the left of the *X MAGN* knob, the *ILLUM* switch selects the level of illumination on the graticule's horizontal and vertical lines. Further left are a pilot light and power ON/OFF switch.

Figure 18-13 illustrates the method that should be used to achieve the most accurate measurement of voltage and time period. By means of the appropriate Y SHIFT control, the unused beam is shifted off the screen. The V/div and TIME/div controls are then adjusted to give the largest possible display of one cycle of input waveform. Note that the center knobs on the V/div and TIME/div controls must be turned fully clockwise, while the X MAGN knob should be at its counterclockwise extreme. From Figure 18-13(a), the time period of the waveform is (6.6 divisions)×(time/division).

Figure 18-13(b) and (c) shows how the displayed waveform should be moved horizontally by the X SHIFT control, in order to accurately measure the peak-to-peak amplitude of the waveform. From the illustration, the peak-to-peak amplitude is (3.3+3.5 divisions)×(volts/division).

The illustration in Figure 18-14 shows how the phase difference between two waveforms can be measured. The lower waveform is seen to be commencing its cycle 1.3 horizontal divisions after the upper waveform. Also, one cycle of the upper waveform occupies four horizontal divisions. Since

$$1 \text{ cycle} = 360°$$

$$4 \text{ div} = 360°$$

and

$$1 \text{ div} = \frac{360°}{4}$$

$$= 90° \text{ (i.e., degrees/div)}$$

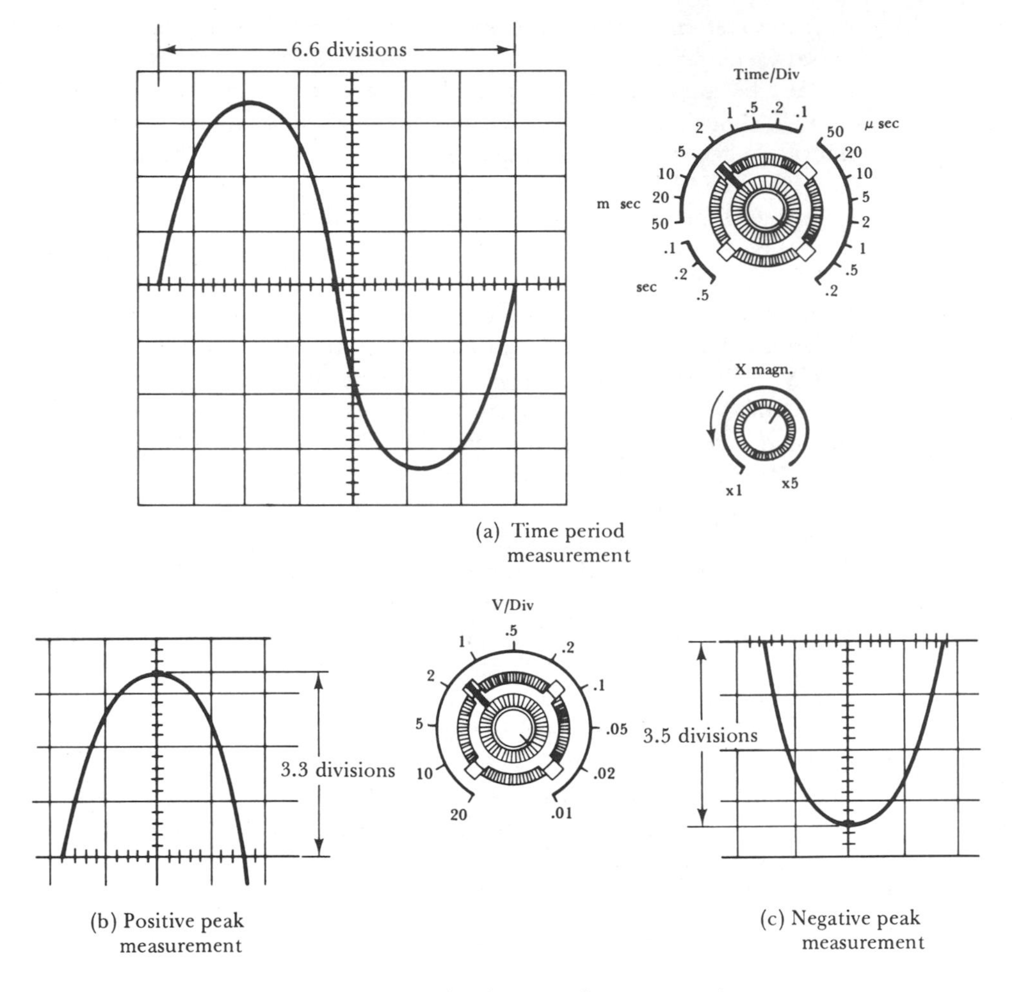

(a) Time period measurement

(b) Positive peak measurement

(c) Negative peak measurement

FIG. 18-13. Accurate measurement of voltage and time period

Therefore, the phase difference is

$$\phi = 1.3 \text{ div} \times (\text{degrees/div})$$
$$= 1.3 \times 90^\circ$$
$$= 117^\circ$$

The input signals are normally connected to the oscilloscope via coaxial cables with *probes* on their ends (see Figure 18-15). These are ordinarily just convenient-to-use insulated connecting clips. Each probe has two connections, as illustrated, one *input* and one *ground*. It is important to connect both ground terminals to the same (grounded) point in

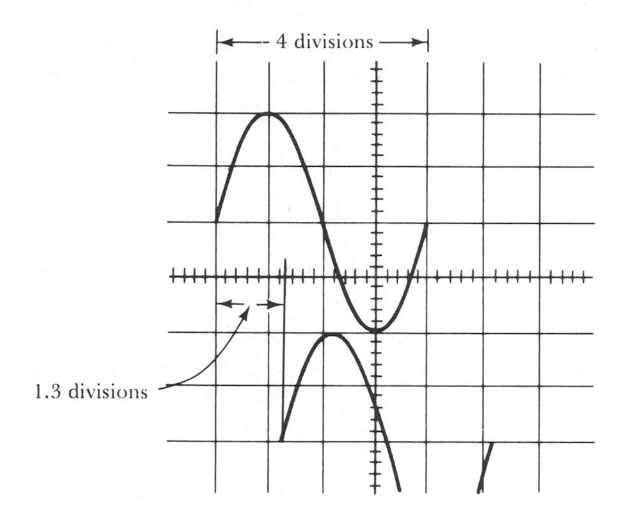

FIG. 18-14. Measurement of phase difference between two waveforms

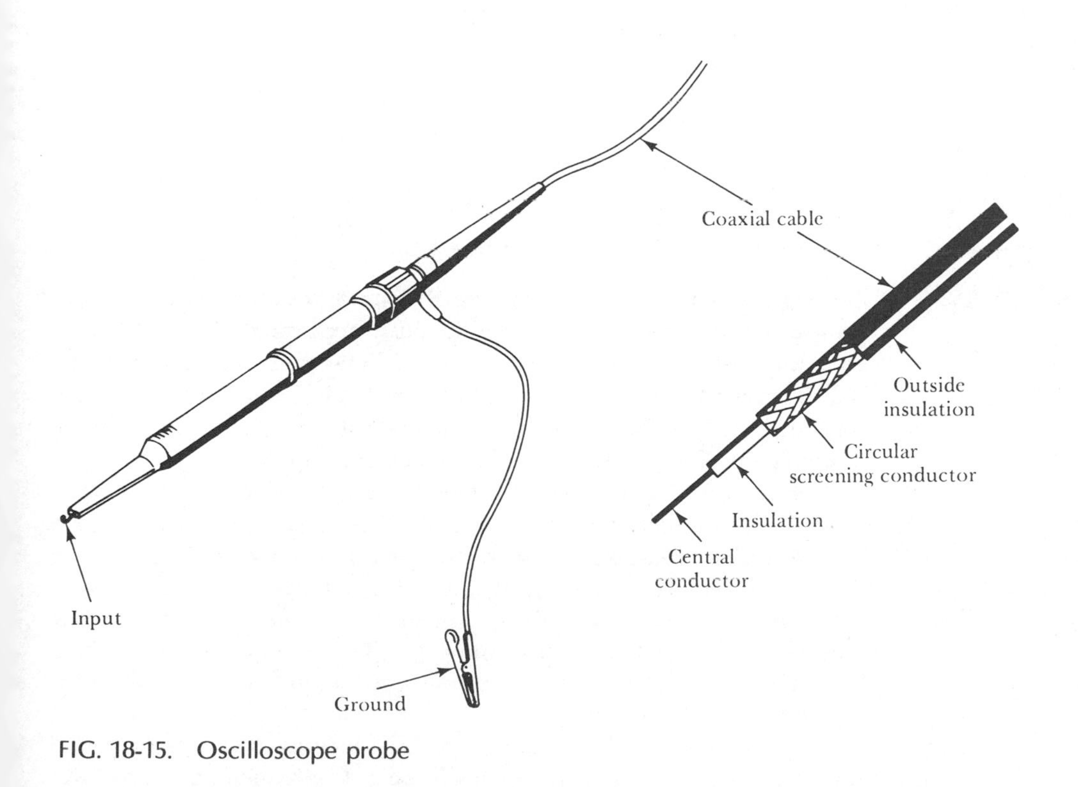

FIG. 18-15. Oscilloscope probe

the circuit under investigation; otherwise, there will be more than one grounded point in the circuit and it may not function correctly. Some probes, termed *attenuator probes*, have resistors inside them to increase the input impedance from the normal 1-MΩ input impedance of the oscilloscope to 10 MΩ. These probes also have the effect of reducing the voltage applied to the oscilloscope by a factor of 10. Thus, they are usually referred to as *10:1 probes*, and the ordinary (nonattenuator) probes are termed *1:1 probes.* As illustrated in Figure 18-15, a coaxial cable consists of a central conductor with its insulation surrounded by another (plaited circular) conductor. The circular conductor is grounded, and it acts as a *screen* which helps to prevent unwanted radio frequency signals being picked up by the oscilloscope input.

The type of oscilloscope described above has a typical specification as follows:

		MEASUREMENT ERROR
Frequency range	0 Hz to 15 MHz	$<5\%$
Time measurement	0.2 μs to 4 s	$<5\%$
Voltage range	1 mV to 160 V	$<4\%$
Input impedance	1 MΩ‖30 pF	

18-7 THREE-PHASE AC

The ac generator described in Section 18-1 produces an output waveform which is simply one sine wave continuously repeating. Figure 18-16(a) shows three similar rotating loops situated between two magnetic poles. Each loop generates a sinusoidal alternating voltage as it rotates in the magnetic field. So three separate sine waves are produced at the three separate pairs of output terminals [see Figure 18-16(b)].

Returning to Figure 18-16(a), note that the two sides (or conductors) of each loop are identified as: A and a, B and b, C and c. Assuming a counterclockwise rotation, it is seen that for conductor A moving past the N pole and conductor a moving past the S pole (as shown), the instantaneous emf generated in A and a is a maximum, with output terminal A being positive with respect to terminal a. (This is explained in Section 18-1.) Therefore, at this instant the sine wave output from A and a is at its positive peak level.

At this same instant, conductor B is moving away from the S magnetic pole, and b is leaving the N pole. Consequently, the voltage output from terminals B and b is past its negative peak value (terminal B being negative with respect to b) and moving toward zero.

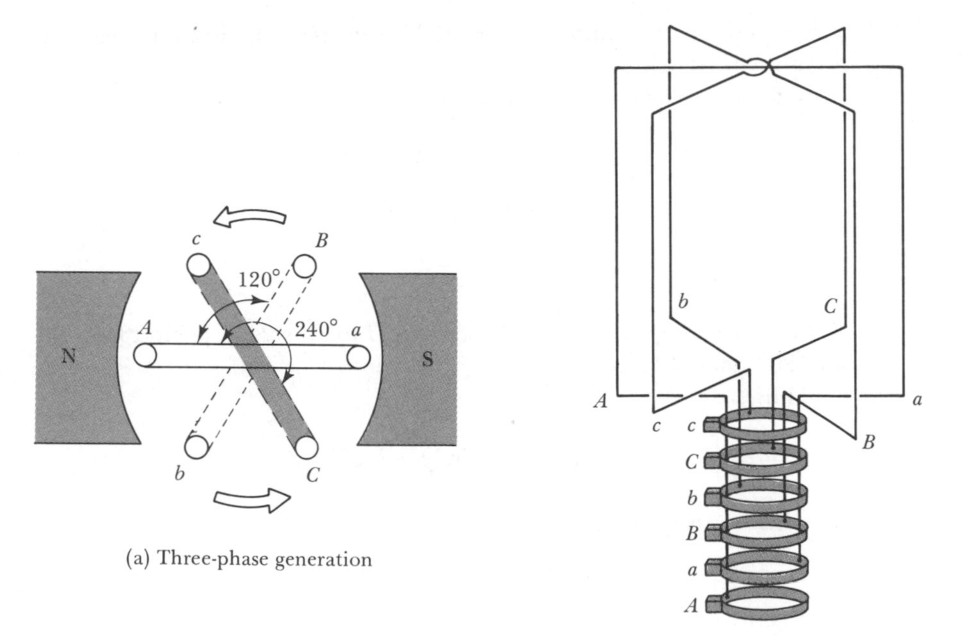

(a) Three-phase generation

(b) Output connections of generator

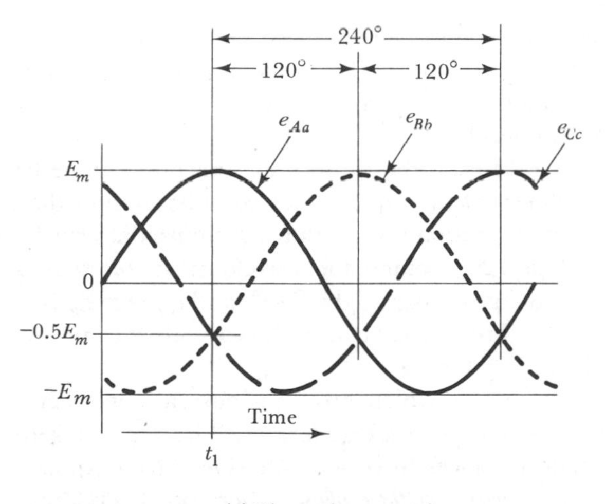

(c) Three-phase waveform

FIG. 18-16. Generation of three-phase alternating voltage

Loop *Bb* (reading from end *B* to *b*) is 120° behind loop *Aa* (reading from end *A* to *a*). Therefore, the output voltage e_{Bb} can easily be calculated with respect to e_{Aa}.

$$e_{Aa} = E_m \sin\phi$$

and

$$e_{Bb} = E_m \sin(\phi - 120°)$$

e_{Aa} is a positive maximum when $\phi = 90°$; that is at the instant illustrated in Figure 18-16(a).

Therefore, when e_{Aa} is at its positive peak,

$$e_{Bb} = E_m \sin(90° - 120°)$$

$$= E_m \sin(-30°)$$

or

$$e_{Bb} = -0.5\ E_m$$

Similarly, loop *Cc* (reading from end *C* to *c*) is 240° behind loop *Aa*, and 120° behind loop *Bb*.

Thus,

$$e_{Cc} = E_m \sin(\phi - 240°)$$

and when e_{Aa} is at its positive peak,

$$e_{Cc} = E_m \sin(90° - 240°)$$

$$e_{Cc} = -0.5\ E_m$$

Conductors *C* and *c* are seen to be approaching the S and N poles respectively [Fig. 18-16(a)]. Thus, the output from terminals *C* and *c* is a negative voltage (*C* negative with respect to *c*), growing toward its negative peak level.

In Figure 18-6(c), the three instantaneous voltage levels determined above are shown at time t_1. Further consideration of the three loops and the direction of rotation reveals that the sinusoidal waveform generated at terminal *Bb* is 120° behind the waveform at *Aa*; that is e_{Bb} *lags* e_{Aa} by 120°. Also, voltage e_{Cc} lags e_{Aa} by 240°, and e_{Cc} lags e_{Bb} by 120°. One cycle of each of the three waveforms is shown with the correct phase relationships in Figure 18-6(c).

An *ac* generator which produces only a single sine wave output is referred to as a *single-phase* generator. The type of generator discussed above, which produces three separate sine wave outputs with 120° phase differences, is termed a *three-phase* generator. Therefore, *ac* supplies may be classified as single-phase or three-phase.

Three-phase systems have distinct advantages over single-phase systems, especially for supplying heavy industrial loads. For a given load, the conductor currents are lower in a three-phase system than with a single-phase supply. Consequently, thinner conductors can be employed, and less copper is used than in a single-phase installation. The three-phase alternator also experiences a more constant load than an equivalent single-phase alternator.

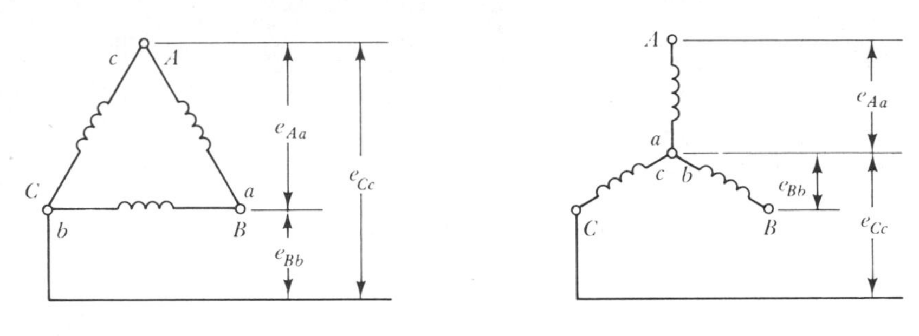

(a) Delta connection (b) Wye connection

FIG. 18-17. Delta-connected and wye-connected three-phase generators

Two methods of connecting the terminals of a three-phase generator are shown in Figure 18-17. The *delta* (Δ) connection [Figure 18-17(a)] gives a three-terminal output. In this case the output voltage, or *line voltage*, is equal to the *phase voltage* produced at each pair of terminals: *Aa*, *Bb*, or *Cc*. A four-terminal output is possible with the *wye* (Y) arrangement shown in Figure 18-17(b). The common terminal is referred to as a *neutral*, and the output voltage between neutral and any one of the other three terminals is equal to the generator phase voltage.

GLOSSARY OF FORMULAS

Peak voltage generated in a conductor rotating in a magnetic field:

$$E_m = Blv$$

Frequency:

$$f = \frac{1}{T}$$

Wavelength:

$$\lambda = \frac{c}{f}$$

Instantaneous ac voltage:

$$e = E_m \sin \omega t$$

Instantaneous ac current:

$$i = I_m \sin \omega t$$

Instantaneous ac power:

$$p = p_m \sin^2 \omega t$$

Average ac power:

$$P = \tfrac{1}{2} P_m$$

Average ac current:

$$I_{\text{av}} = \frac{2}{\pi} I_m = 0.637 I_m$$

Average ac voltage:

$$E_{\text{av}} = \frac{2}{\pi} E_m = 0.637 E_m.$$

Rms ac current:

$$I = \frac{1}{\sqrt{2}} I_m = 0.707 I_m$$

Rms ac voltage:

$$E = \frac{1}{\sqrt{2}} E_m = 0.707 E_m$$

REVIEW QUESTIONS

18-1 Draw sketches to illustrate, and explain how an alternating current is generated in a conducting loop rotated in a magnetic field.

18-2 Derive an expression for the maximum voltage induced in a conducting loop rotated in a magnetic field. Also, derive an expression for the instantaneous value of the voltage.

18-3 Define: ac current, ac voltage, sinusoidal waveform, sine wave, peak value, phase angle, and amplitude.

18-4 Sketch two sine waves which:

a. Are in-phase.

b. Are in antiphase.

c. Have a phase difference of $\pi/2$ radians.

18-5 Derive equations for the instantaneous level of current through a resistor and for the instantaneous power dissipated when a sinusoidal ac voltage is applied to the resistor.

18-6 Draw a graph of instantaneous power versus time for the power dissipated in a resistor when a sinusoidal ac voltage is applied. Also, show that the average power dissipated is half the maximum instantaneous power.

18-7 Explain how the average value of a sinusoidal voltage may be calculated and write an expression for the average value in terms of the maximum value:

a. For a half-cycle of the waveform.

b. For a full cycle of the waveform.

18-8 Define the rms value of an alternating current and write an expression for the rms value in terms of the maximum value. Also, show that the rms value is the equivalent dc value, or effective value.

18-9 Referring to Figure 18-10, identify each control of the cathode ray oscilloscope, and explain the function of each.

18-10 Using illustrations, explain how an oscilloscope is used to measure voltage, frequency, and phase difference.

18-11 Draw illustrations to show how three-phase ac is generated. Show the two methods of connecting the output from a three-phase generator. Briefly explain.

PROBLEMS

18-1 A conducting loop rotated in a magnetic field has an axial length of $l=30$ cm and a distance between sides of $D=8$ cm. The flux density of the magnetic field is $B=0.25$ T and the loop is rotated at 140 revolutions/minute.

a. Calculate the maximum output voltage from the loop.

b. If the loop has 10 turns instead of one, determine the maximum output voltage.

18-2 For the generator described in Problem 18-1, calculate the instantaneous levels of output voltage e, at $\pi/4$, $3\pi/4$, and $5\pi/4$ radians from $e=0$.

18-3 For the generator described in Problem 18-1, calculate the instantaneous levels of output voltage e at 5 ms, 10 ms, 25 ms, and 30 ms from $e=0$.

18-4 An ac voltage has a maximum value of 9 V and a frequency of 150 kHz. Determine the instantaneous voltage levels at $t_1=1.1$ μs, $t_2=5$ μs, and $t_3=29$ μs.

18-5 Calculate the frequency of an ac voltage that has a wavelength of 300 km. Also, calculate the wavelength of 120-kHz and 12-MHz ac voltages.

18-6 A 120-Hz ac supply with a peak value of 100 V is applied to a resistor of $R=27$ Ω. Calculate the instantaneous power dissipations at phase angles of $\pi/3$, $2\pi/3$, $5\pi/3$ and $3\pi/2$ radians.

18-7 For Problem 18-6 calculate the rms and average values of the current through R. Also, calculate the average power dissipation.

19

PHASORS AND COMPLEX NUMBERS

INTRODUCTION

Sinusoidal alternating currents and voltages can be graphically represented by a line of fixed length rotating about one of its ends. Such a line is termed a *phasor*, and the instantaneous value of a sinusoidal quantity can be determined from the length of the phasor and its angle with respect to the horizontal at the particular instant.

A phasor can be mathematically represented by its length and angle, or by its vertical and horizontal components. The vertical and horizontal components of a phasor can be written in the form of a *complex number*. Provided that certain rules are followed, phasors can be added, substracted, multiplied, or divided. The same is true for complex numbers.

19-1 PHASOR REPRESENTATION OF ALTERNATING VOLTAGE

Quantities that can be completely represented by a number of units are termed *scalar* quantities. Some examples of scalar quantities are temperature, volume, and resistance. Other quantities, like force and velocity, must have a direction indicated as well as the number of units stated if they are to be completely specified. Such quantities can be graphically represented by a line which has a length proportional to the number of units and which is drawn at an angle with respect to some reference direction. This line is termed a *vector*, and the quantity it represents is classified as a *vector quantity*.

The instantaneous levels of alternating current and voltage are vector quantities, but since the instantaneous levels are continuously changing, an ac waveform must be represented by a *rotating vector*, or *phasor*. A phasor is a vector that is rotating at a constant angular velocity.

The sinusoidal output voltage from the simple generator discussed in Section 18-1 can be represented by the *phasor diagram* shown in Figure 19-1. Here OA is a rotating vector (or phasor) with a constant angular velocity and a length that represents the peak output voltage E_m. The arrowhead identifies the end of the phasor which moves, and the other end is the axis of rotation. By convention, the direction of phasor rotation is taken as counterclockwise.

The instantaneous value of the generated voltage represented by the phasor obviously depends upon the angle of the phasor with respect to the zero level. At time $t=0$, the angle is $\theta=0$ and the instantaneous voltage level is zero. At t_1, the angle is θ_1 and the instantaneous level is e_1, as illustrated in the figure, where

$$e_1 = (OA)\sin\theta_1$$
$$= E_m \sin\theta_1$$

At t_2,

$$e_2 = E_m \sin\theta_2$$

and at t_3,

$$e_3 = E_m \sin\theta_3$$

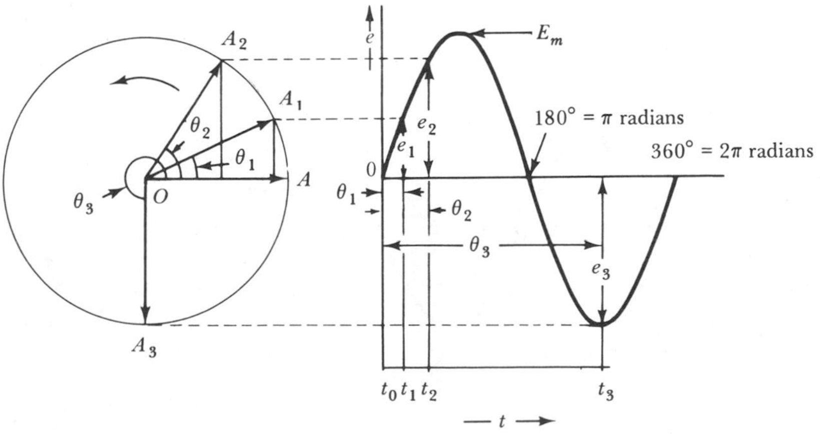

FIG. 19-1. Phasor representation of a sine wave

At each of these instants, the voltage level is represented by the *stopped phasor* (or vector) OA_1 at an angle θ_1, OA_2 at angle θ_2, and OA_3 at angle θ_3. Any sinusoidal ac voltage or current can be represented by a phasor, and instead of degrees the angles may be expressed in radians, as explained in Section 18-3.

19-2 ADDITION AND SUBTRACTION OF PHASORS

In Figure 19-2 two sinusoidal waveforms A and B are shown together with the phasor representing each. It is seen that waveform B has a larger amplitude than waveform A. Therefore, phasor OB is longer than phasor OA. Also, waveform B starts to grow positively from its zero level $\phi°$ after the beginning of waveform A. Consequently, waveform B *lags* waveform A by $\phi°$, and the phasor OB is shown $\phi°$ behind OA. It can also be said that waveform A *leads* waveform B by $\phi°$.

The instantaneous level of waveform A can be written

$$e_A = OA \sin\theta$$

and the instantaneous level of waveform B is

$$e_B = OB \sin(\theta - \phi)$$

Now consider Figure 19-3(a), in which the two waveforms are shown to be outputs from each of two series-connected generators. The resultant output of the two in series is the *phasor sum* of the two waveforms. Figure 19-3(b) shows how the instantaneous levels of A and B may be added to give the resultant waveform C. At t_1, $a_1 + b_1$ gives point 1 on waveform C. At t_2, $a_2 + b_2$ gives point 2 on waveform C, and so on. It is seen that

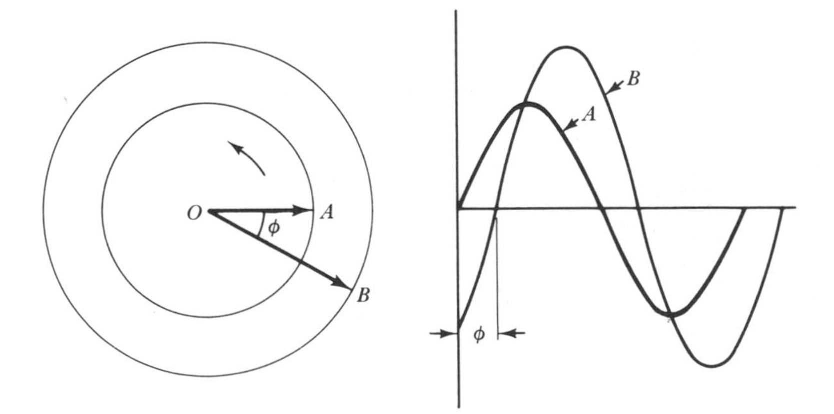

FIG. 19-2. Two waveforms and their phasors

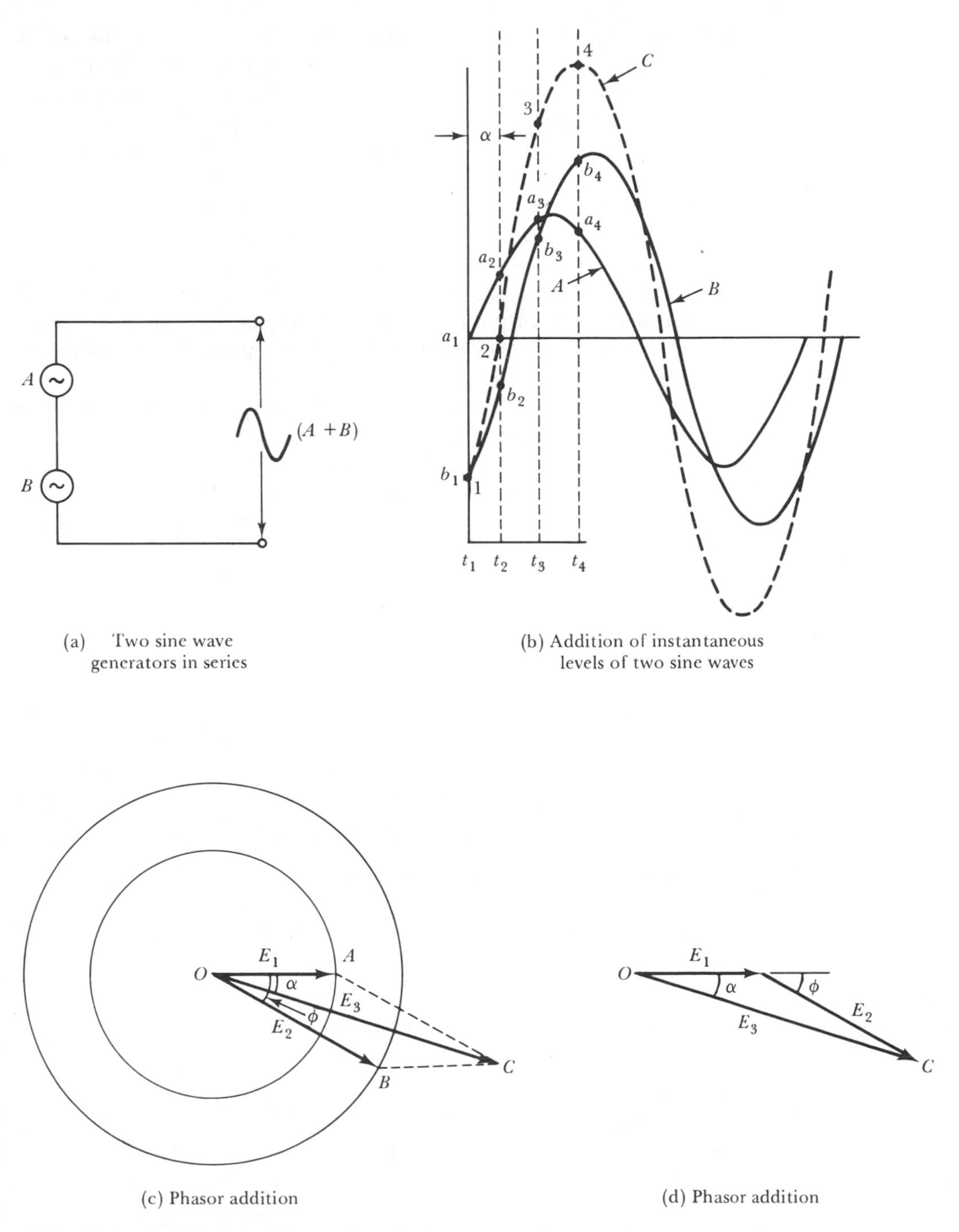

(a) Two sine wave generators in series

(b) Addition of instantaneous levels of two sine waves

(c) Phasor addition

(d) Phasor addition

FIG. 19-3. Combined effect of two sine waves from series-connected generators

waveform C lags waveform A by an angle $\alpha°$. Therefore, waveform C is represented by phasor OC lagging $\alpha°$ behind OA [see Figure 19-3(c)].

Referring to Figure 19-3(c), another method of obtaining the resultant of waveforms A and B is illustrated. A parallelogram is constructed, with AC being drawn equal to and in parallel with OB, and BC equal to and in parallel with OA. The resultant diagonal of the parallelogram OC represents the amplitude of waveform C, and the angle $\alpha°$ is the angle with which waveform C lags waveform A. Another approach to obtain the same result is shown in Figure 19-3(d). OA, identified as E_1, is first drawn horizontally. Then E_2 (which is OB) is drawn from the end of E_1 and at angle $-\phi$ with respect to E_1. The resultant E_3 (or OC) is found by drawing a line from O to the end of E_2. The angle of E_3 with respect to E_1 is again seen to be $-\alpha°$.

Instead of graphically adding the two phasors, the resultant may be obtained mathematically. Referring to Figure 19-4, each phasor is resolved into horizontal and vertical components (by trigonometry). The arithmetical sum of the horizontal components is then the horizontal component of the resultant. Similarly, the arithmetical sum of the vertical components is the vertical component of the resultant.

$$\text{horizontal component of } E_1 = OA$$

$$\text{horizontal component of } E_2 = OD = E_2 \cos\phi$$

$$\text{total of horizontal components} = OE = OA + OD$$

$$= E_1 + E_2 \cos\phi$$

$$\text{vertical component of } E_1 = 0$$

$$\text{vertical component of } E_2 = OF = -E_2 \cos(90° - \phi)$$

$$\text{total of vertical components} = 0 + OF = -E_2 \cos(90° - \phi)$$

The negative sign here indicates that the vertical component is measured

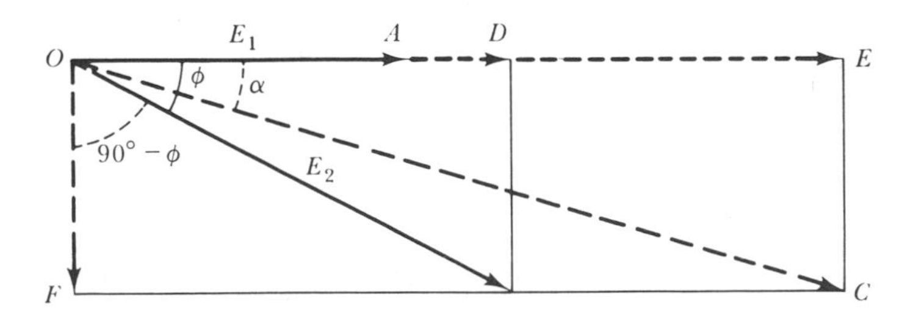

FIG. 19-4. Addition of two phasors by resolving each into horizontal and vertical components

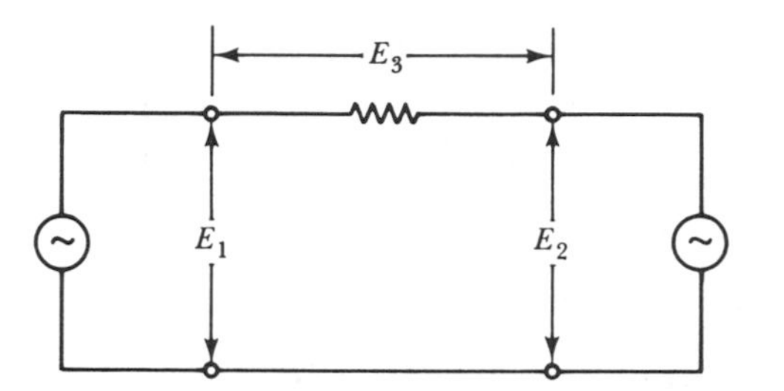

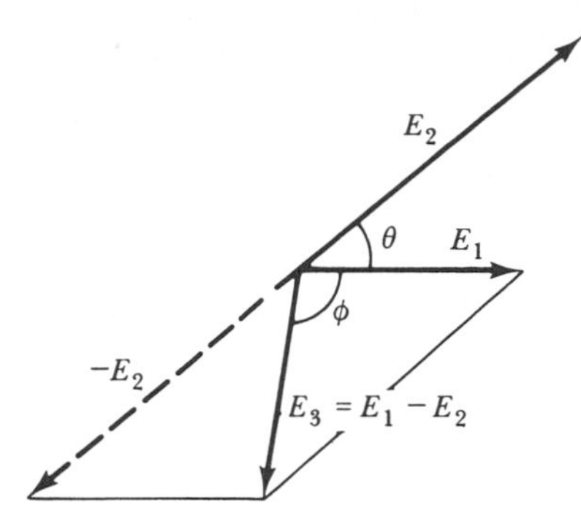

(a) E_3 is the phasor difference of E_1 and E_2

(b) Determination of E_3

FIG. 19-5. An alternating voltage which is the phasor difference of two voltages is obtained by summing E_1 and $-E_2$

down from the horizontal, rather than up.

$$\text{resultant } E_3 = OC = \sqrt{(OE)^2 + (OF)^2}$$

$$\text{angle } \alpha = \arctan\left(\frac{OF}{OE}\right)$$

Sometimes instead of phasor addition, *phasor subtraction* is required. An example of this occurs with the circuit shown in Figure 19-5(a), where

$$E_3 = E_1 - E_2$$

Figure 19-5(b) shows the process of phasor subtraction. E_2 is shown as $\theta°$ ahead of E_1. For subtraction, E_2 is reversed to become $-E_2$, as illustrated. The resultant E_3 is then the phasor sum of E_1 and $-E_2$. Again, the resultant may be found mathematically, or by either of the two graphical methods described above.

Phasor addition and subtraction are not limited to two phasor quantities. The resultant of any number of quantities can be found, and some of these may require phasor addition while others may have to be subtracted.

EXAMPLE 19-1 Two sinusoidal quantities are given as $v_1 = 120 \sin\theta$ and $v_2 = 75 \sin(\theta - 30°)$. Determine the resultant obtained when:

a. The quantities are added.

b. When v_2 is subtracted from v_1.

SOLUTION

a. $v_1 + v_2$ (*refer to* Figure 19-6):

$$\textit{horizontal component of } v_1 = 120$$
$$\textit{horizontal component of } v_2 = 75\cos 30° \cong 65$$
$$\textit{horizontal component of } v_3 = 120 + 65 = 185$$

$$\textit{vertical component of } v_1 = 0$$
$$\textit{vertical component of } v_2 = -75\sin 30° = -37.5$$
$$\textit{vertical component of } v_3 = 0 - 37.5 = -37.5$$

$$v_3 = \sqrt{185^2 + 37.5^2}$$
$$v_3 = 188.8$$
$$\textit{angle } \alpha = \textit{arctan } \frac{37.5}{185}$$
$$\boldsymbol{\alpha = 11.5°}$$

$$\boldsymbol{v_3 = 188.8 \sin(\theta - 11.5°)}$$

b. $v_1 - v_2$ (*refer to* Figure 19-6):

$$\textit{horizontal component of } v_1 = 120$$
$$\textit{horizontal component of } -v_2 = 75\cos(180° - 30°) \cong -65$$
$$\textit{horizontal component of } v_4 = 120 - 65 = 55$$

$$\textit{vertical component of } v_1 = 0$$
$$\textit{vertical component of } -v_2 = 75\sin(180° - 30°) = 37.5$$
$$\textit{vertical component of } v_4 = 0 + 37.5 = 37.5$$
$$v_4 = \sqrt{55^2 + 37.5^2}$$
$$\boldsymbol{v_4 = 66.6}$$
$$\textit{angle } \phi = \textit{arctan } \frac{37.5}{55}$$
$$\boldsymbol{\phi = 34.3°}$$

$$\boldsymbol{v_4 = 66.6 \sin(\theta + 34.3°)}$$

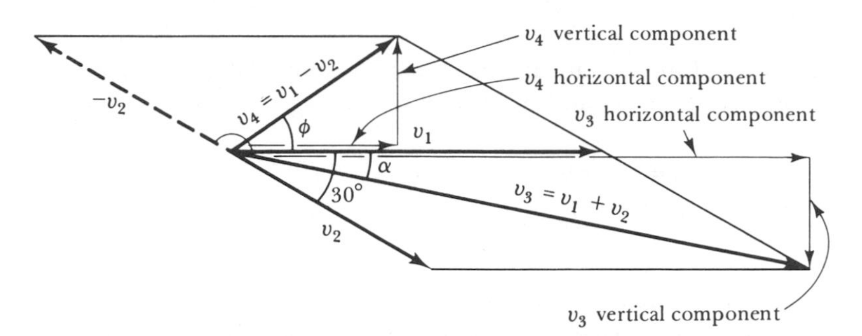

FIG. 19-6. Sum and difference of two phasors

EXAMPLE 19-2 Graphically find the resultant of $v_1 + v_2 - v_3$, where $v_1 = 50 \sin\theta$, $v_2 = 30 \sin(\theta + 25°)$, and $v_3 = 25 \sin(\theta - 90°)$.

SOLUTION

1. *The phasor diagram for v_1, v_2, and v_3 is drawn in* Figure 19-7(a).
2. *In* Figure 19-7(b), *v_1 is drawn to scale.*
3. *v_2 is drawn to scale from the end of v_1, and at an angle of +25° with respect to v_1.*
4. *v_3 is drawn to scale from the end of v_2, and at an angle of −90° with respect to v_1.*
5. *$-v_3$ is drawn, as illustrated.*
6. *v_4 is drawn by joining the origin at* 0 *to the end of $-v_3$.*
7. *v_4 and ϕ are measured as*: 86 and 26°:

$$v_4 = 86 \sin(\theta + 26°)$$

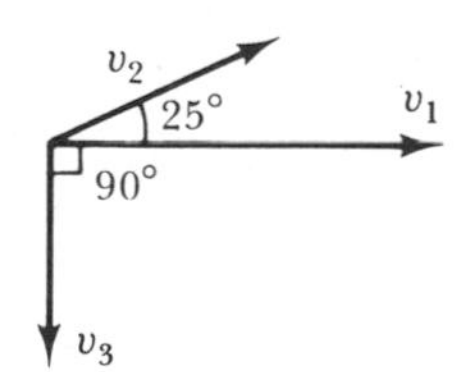

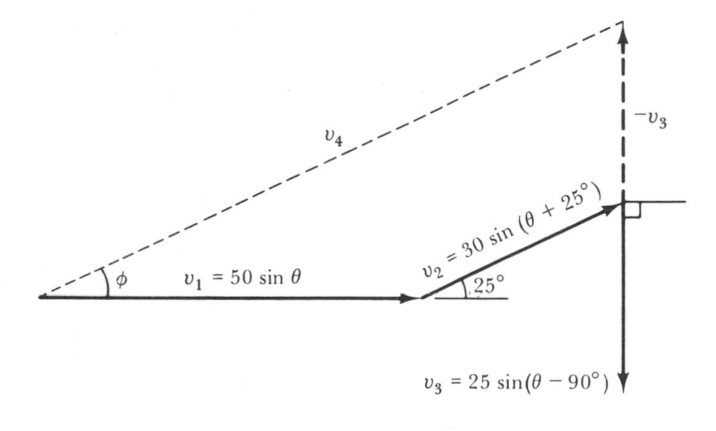

(a) Phasors v_1, v_2, and v_3

(b) Graphical solution of $v_1 + v_2 - v_3$

FIG. 19-7. Graphical method of phasor addition

19-3 POLAR AND RECTANGULAR FORM

Another way of mathematically expressing a sinusoidal quantity, termed *polar form*, leaves out *sin* and substitutes ∠ for angle. Thus

$$e_1 = E_m \sin\theta \qquad \text{becomes } e_1 = E_m \underline{/\theta}$$

$$\text{and} \qquad i_1 = I_m \sin(\theta + \alpha) \quad \text{becomes } i_1 = I_m \underline{/\theta + \alpha}$$

Polar form is found to be very convenient for phasor quantities. The terms *modulus* and *argument* are sometimes used for the magnitude and angle, respectively, of a polar quantity. For e_1 above, the modulus is E_m, while θ is the argument.

Since there is a direct (0.707) relationship between the rms value and the peak value of an ac waveform, rms values are normally employed when using polar form. In fact, *with alternating voltages and currents, all quantities are assumed to be rms quantities unless otherwise indicated.* Phasor addition and subtraction using rms quantities follows exactly the same graphical or mathematical procedure as when using peak quantities. The resultant obtained is, of course, an rms quantity.

As already explained, phasors must be resolved into horizontal and vertical components for phasor addition and subtraction. Thus, in Figure 19-8(a) the polar form, $e_1 = E\underline{/\theta}$, becomes

$$\text{horizontal component, } a = E\cos\theta$$

$$\text{vertical component, } b = E\sin\theta$$

For convenience in writing vertical and horizontal components the *j operator* is employed, as illustrated in Figure 19-8(b). All positive vertical components are given the prefix *j*, while all negative vertical components

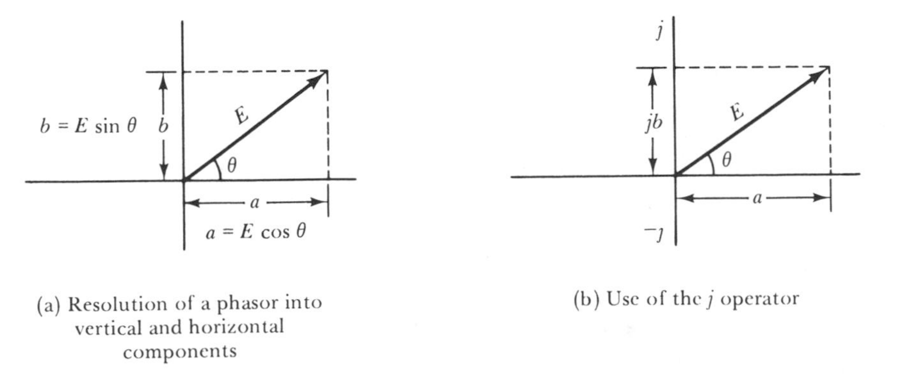

(a) Resolution of a phasor into vertical and horizontal components

(b) Use of the *j* operator

FIG. 19-8. Use of the *j* operator to identify vertical components of a phasor

are prefixed with $-j$. $E\angle\theta$ can now be written

$$E\angle\theta = a + jb$$

or

$$\boxed{E\angle\theta = E\cos\theta + jE\sin\theta} \tag{19-1}$$

A phasor stated in this way is said to be expressed in *rectangular form*. Conversion from polar to rectangular form is simply a matter of resolving the phasor into its horizontal and vertical components. Conversion from rectangular to polar form is then the reverse of that process:

$$E = \sqrt{a^2 + b^2}$$

and

$$\theta = \arctan\left(\frac{b}{a}\right)$$

or,

$$\boxed{E\angle\theta = \sqrt{a^2 + b^2}\ \angle\arctan\left(\frac{b}{a}\right)} \tag{19-2}$$

The j operator is not merely a prefix employed for convenience in identifying the vertical component of a phasor. Expressions such as Eq. (19-1), using the j operator, are termed *complex numbers*, and the mathematics of complex numbers require that certain rules be followed. Consider phasor $E\angle 0°$ in Figure 19-9. When multiplied by j, the phasor is

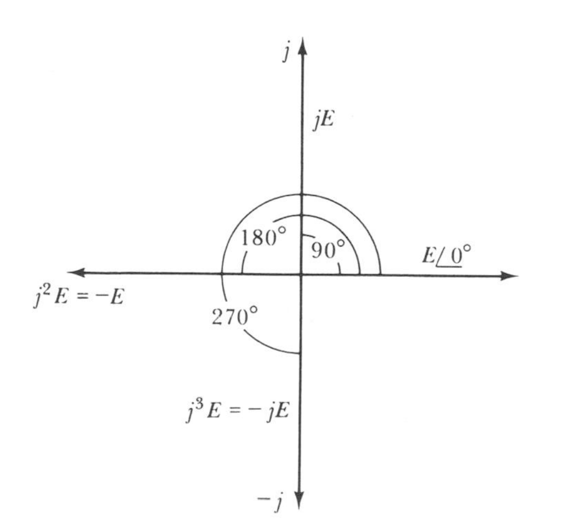

FIG. 19-9. Effect of multiplying $E\angle 0°$ by j, j^2, and j^3

rotated counterclockwise by 90° or $\pi/2$ radians. Thus, it becomes

$$E\angle 90° = jE$$

When jE is multiplied by j, the phasor is

$$E\angle 180° = j^2E = -E$$

Therefore, to convert E to $-E$,

$$j^2 = -1$$

or

$$j = \sqrt{-1}$$

The actual mathematical square root of $+1$ is 1, but the square root of -1 cannot be determined. For this reason j is frequently referred to as an *imaginary number*. For the complex number $a+jb$, a is sometimes termed the *real part*, while jb is called the *imaginary part*.

Returning to Figure 19-9, when $-E$ is once more multiplied by j, the phasor is rotated counterclockwise by a further 90°, giving

$$E\angle 270° = -jE$$

and when multiplied by a fourth j, the quantity is

$$\begin{aligned} E\angle 0° &= -j^2E \\ &= -(-1)E \\ &= E \end{aligned}$$

EXAMPLE 19-3

Convert the following quantities into rectangular form:

a. $12\angle 30°$.

b. $270\angle 1.7\pi$.

c. $40\angle 105°$.

SOLUTION

a.

Eq. (19-1):

$$\begin{aligned} E\angle\theta &= E\cos\theta + jE\sin\theta \\ 12\angle 30° &= 12\cos 30° + j12\sin 30° \\ 12\angle 30° &= 10.4 + j6 \qquad \textit{see Figure 19-10(a)} \end{aligned}$$

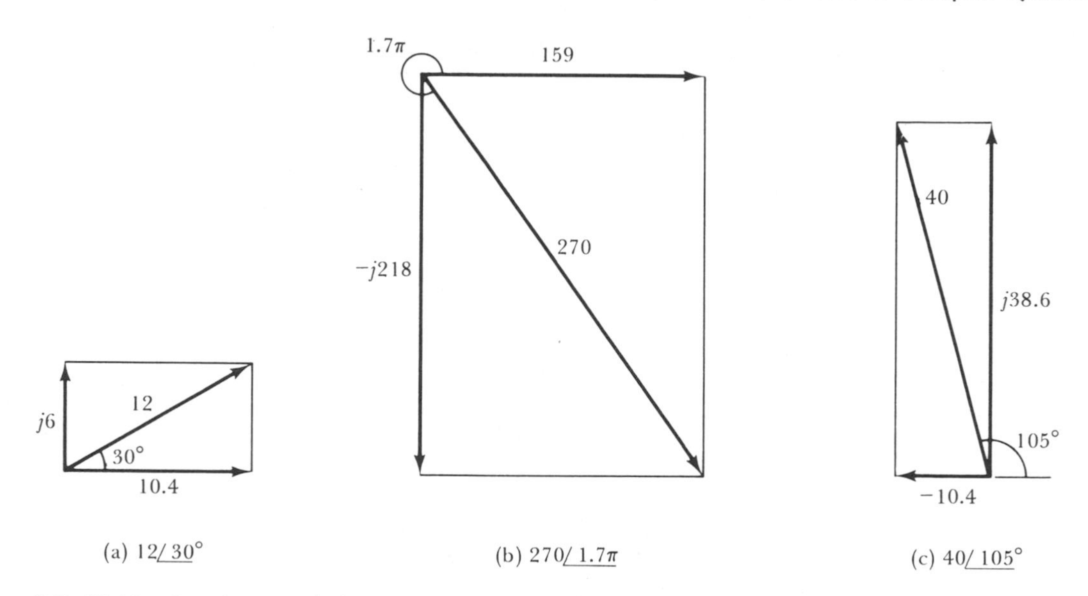

FIG. 19-10. Resolution of phasors into horizontal and vertical components

b. $270\angle 1.7\pi = 270\cos(1.7\pi) + j270\sin(1.7\pi)$

$270\angle 1.7\pi = 159 - j218$ *see* Figure 19-10(b)

c. $40\angle 105° = 40\cos 105° + j40\sin 105°$

$40\angle 105° = -10.4 + j38.6$ *see* Figure 19-10(c)

19-4 MATHEMATICS OF COMPLEX QUANTITIES

When mathematically processing complex quantities in rectangular form, the usual rules of algebra apply as long as it is remembered that $j^2 = -1$.

For addition and subtraction, the real parts of the number are added or subtracted, and the imaginary parts are added or subtracted. This is, of course, simply adding or subtracting the horizontal and vertical components of a phasor.

$$\boxed{(a+jb)+(c+jd)=(a+c)+j(b+d)} \qquad (19\text{-}3)$$

$$\boxed{(a+jb)-(c+jd)=(a-c)+j(b-d)} \qquad (19\text{-}4)$$

Multiplication of complex quantities in rectangular form follows the normal procedure for multiplication of algebraic expressions:

$(a+jb)\times(c+jd)$,

$$\begin{array}{r} a+jb \\ \times(c+jd) \\ \hline \end{array}$$

multiplying the top line by c, $\qquad ac+jbc$

multiplying the top line by jd, $\qquad jad+j^2bd$

$$\overline{ac+jbc+jad-bd}$$

Therefore,

$$\boxed{(a+jb)\times(c+jd)=(ac-bd)+j(bc+ad)} \tag{19-5}$$

Division of complex quantities in rectangular form requires the use of the *conjugate* of the denominator. The conjugate of a complex number is the same number with the sign changed on the j component. Thus, the conjugate of $(a+jb)$ is $(a-jb)$.

$$\frac{a+jb}{c+jd}=\frac{a+jb}{c+jd}\times\frac{c-jd}{c-jd}$$

Multiplying by $(c-jd)/(c-jd)$ *is the same as multiplying by* 1. *Therefore, nothing has been changed.*

$$\frac{a+jb}{c+jd}\times\frac{c-jd}{c-jd}=\frac{ac+jbc-jad-j^2bd}{c^2+jcd-jcd-j^2d^2}$$

$$=\frac{(ac+bd)+j(bc-ad)}{c^2+d^2}$$

Therefore,

$$\boxed{\frac{a+jb}{c+jd}=\frac{ac+bd}{c^2+d^2}+j\frac{bc-ad}{c^2+d^2}} \tag{19-6}$$

Another important consideration in processing j quantities is the reciprocal of a j quantity:

$$\frac{1}{jx}=\frac{1}{jx}\times\left(\frac{-j}{-j}\right)$$

$$=\frac{-j}{-j^2x}=\frac{-j}{-(-1)x}=\frac{-j}{x}$$

Therefore, $$\frac{1}{jx} = -j\frac{1}{x}$$

or $$\frac{1}{j} = -j$$

EXAMPLE 19-4 Determine:

a. $(75 - j50) \times (25 + j5)$.

b. $(75 - j50) \div (25 + j5)$.

SOLUTION

a.

$$\begin{array}{l} 75 - j50 \\ \times(25 + j5) \\ \hline 1875 - j1250 \\ \quad + j375 - j^2 250 \\ \hline 1875 - j1250 + j375 + 250 \\ = \mathbf{2125 - j875} \end{array}$$

b.

$$\frac{75 - j50}{25 + j5} \times \frac{25 - j5}{25 - j5} = \frac{1875 - j1250 - j375 + j^2 250}{625 + j125 - j125 - j^2 25}$$

$$= \frac{1625 - j1625}{650}$$

$$= \mathbf{2.5 - j2.5}$$

Multiplication and division in polar form are much simpler than in rectangular form:

$$\boxed{E_1 \angle\theta_1 \times E_2 \angle\theta_2 = E_1 E_2 \angle\theta_1 + \theta_2} \qquad (19\text{-}7)$$

and

$$\boxed{\frac{E_1 \angle\theta_1}{E_2 \angle\theta_2} = \frac{E_1}{E_2} \angle\theta_1 - \theta_2} \qquad (19\text{-}8)$$

For addition and subtraction, quantities stated in polar form must first be converted to rectangular form.

EXAMPLE 19-5

Determine:

a. $90\angle -33.7^\circ \times 25.5\angle 11.3^\circ$.

b. $90\angle -33.7^\circ \div 25.5\angle 11.3^\circ$.

SOLUTION

a.

$$90\angle -33.7^\circ \times 25.5\angle 11.3^\circ = (90\times 25.5)\angle -33.7^\circ + 11.3^\circ$$
$$= \mathbf{2295\angle -22.4^\circ}$$

b.

$$90\angle -33.7^\circ \div 25.5\angle 11.3^\circ = \frac{90}{25.5}\angle -33.7^\circ - 11.3^\circ$$
$$= \mathbf{3.5\angle -45^\circ}$$

Note that $90\angle -33.7^\circ$ *and* $25.5\angle 11.3^\circ$ *are the polar form of the quantities in* Example 19-4, *and that when the answers above are checked, they are found to be the polar form of the answers in* Example 19-4.

GLOSSARY OF FORMULAS

Polar to rectangular conversion:

$$E\angle\theta = E\cos\theta + jE\sin\theta$$

Rectangular to polar conversion:

$$a + jb = \sqrt{a^2 + b^2}\ \angle \arctan\left(\frac{b}{a}\right)$$

Addition:

$$(a+jb)+(c+jd)=(a+c)+j(b+d)$$

Subtraction:

$$(a+jb)-(c+jd)=(a-c)+j(b-d)$$

Multiplication:

$$(a+jb)\times(c+jd)=(ac-bd)+j(bc+ad)$$

Division:

$$\text{for } \frac{a+jb}{c+jd} \text{ multiply by } \frac{c-jd}{c-jd}$$

Multiplication:

$$E_1\angle\theta_1 \times E_2\angle\theta_2 = E_1E_2\angle\theta_1+\theta_2$$

Division:

$$\frac{E_1\angle\theta_1}{E_2\angle\theta_2} = \frac{E_1}{E_2}\angle\theta_1-\theta_2$$

REVIEW QUESTIONS

19-1 Sketch a sinusoidal waveform and show how its instantaneous value may be represented by a phasor.

19-2 Show how two sinusoidal waveforms that have a phase difference may be represented by phasors. Briefly explain.

19-3 Sketch the waveform from two series connected ac generators which have ac voltages that differ in amplitude and phase. Show how the resultant of the two waveforms may be obtained by adding the instantaneous voltage values. Also, show how the resultant may be obtained by phasor addition.

19-4 Sketch two phasors with different amplitudes and phase angles. Show how the resultant of the two can be obtained by resolving each into horizontal and vertical components.

19-5 Draw the diagram of a circuit that involves phasor subtraction. Sketch two phasors differing in amplitude and phase and show how one should be subtracted from the other.

19-6 Sketch a phasor diagram to show the relationship between polar and rectangular forms of phasor representation. Also, show the effect of multiplying $E\angle 0$ by j, j^2, and j^3. Explain bricfly.

PROBLEMS

19-1 Find the resultant of:

a. v_1+v_2.

b. v_1-v_2, where $v_1=47\sin\phi$ and $v_2=33\sin(\phi+20°)$.

19-2 Graphically determine the resultant of $v_1+v_2+v_3-v_4$, where $v_1=10\sin\theta$, $v_2=15\sin(\theta-15°)$, $v_3=20\sin(\theta+10°)$, and $v_4=18\sin(\theta+25°)$.

19-3 Convert the quantities stated in Example 19-4 into polar form.

19-4 Convert the following quantities into rectangular form:

a. $150\angle 22°$.

b. $85\angle 2.5\pi$.

c. $64\angle 72°$.

19-5 Find $v_1+v_2-v_3$ and $v_1-v_2-v_3$, where, $v_1=45\angle 30°$, $v_2=27\angle 21°$. and $v_3=30\angle 42°$.

19-6 For the quantities stated in Problem 19-5, determine $(v_1\times v_2)/v_3$.

19-7 Solve:

$$\frac{(16+j12)\times(22-j18)}{14+j25}$$

19-8 Graphically solve Problem 19-5.

19-9 For the quantities given in Problem 19-4, find $a+b-c$: (a) graphically, (b) algebraically.

19-10 For the quantities given in Problem 19-4, determine:

$$\frac{a-2b}{b+c}.$$

19-11 Solve:

$$\frac{\left[(25+j15)+(45-j50)\right]\times(33-j29)}{(62+j70)-(32+j100)}$$

19-12 For the quantities given in Problem 19-1, find:

$$\frac{v_1\times v_2}{2(v_1+v_2)}$$

20

INDUCTANCE AND CAPACITANCE IN AN AC CIRCUIT

INTRODUCTION

Alternating current flow in an inductor depends upon the applied voltage and upon the *inductive reactance* of the inductor. The inductive reactance is proportional to the inductance value and the frequency of the alternating supply voltage. Similarly, the current flow in a capacitor is dependent upon the supply voltage and the *capacitive reactance*. The capacitance value and the supply frequency determine the capacitive reactance.

When an alternating voltage is applied to a pure inductance, the current that flows through the inductance lags its terminal voltage by 90°. Conversely, the alternating current through a capacitor leads the capacitor terminal voltage by 90°. For *RL* and *RC* series circuits, the phase angle of current with respect to supply voltage is less than 90°.

A series *RLC* circuit may behave as a resistive–inductive circuit or as a resistive–capacitive circuit, depending upon the component values and supply frequency. The same can be said of a parallel *RLC* circuit. In series circuits the current flows through all components; therefore, in drawing a phasor diagram, the phasors for each of the component voltages are drawn with respect to current as a reference phasor. In parallel circuits, the applied voltage appears across the terminals of each component; consequently, in the phasor diagram all current phasors are referred to the supply voltage phasor.

20-1 ALTERNATING CURRENT AND VOLTAGE IN AN INDUCTIVE CIRCUIT

In Section 18-4 it was explained that when an alternating voltage is applied to a purely resistive circuit, the resultant current through the resistor is in phase with the applied voltage. This is illustrated once again in Figure 20-1. Because of the presence of a *counter-emf* in an inductive circuit, the current and voltage phase relationships are considerably different from the case of the resistive circuit.

Figure 20-2 shows an inductance L with an ac voltage applied to it. For the waveforms illustrated, it is assumed that the resistance of the coil is very much smaller than the inductance. Thus, the circuit is assumed to be purely inductive. Since the counter-emf is proportional to the rate of change of current through the inductance, it is appropriate to start with the current waveform.

As shown in the illustration, the current waveform is sinusoidal with a zero level at $t=0$, positive peak at t_1, zero again at t_2, and so on. At t_1, the current has stopped increasing positively, and has not yet started to decrease. Therefore, the rate of change of current at t_1 is zero. This gives point 1 on the waveform representing *rate of change of current*. At t_2, the current is decreasing at its maximum rate, which means it has a maximum negative rate of change. Thus, point 2 on the $\Delta i/\Delta t$ waveform is a peak negative value. At t_3, the current has a zero rate of change once again, and the $\Delta i/\Delta t$ value is therefore plotted at point 3. Finally, at t_4, the current has a maximum positive rate of change, giving point 4 on the waveform of current rate of change.

Now recall from Eq. (15-3) that the counter-emf in an inductor is

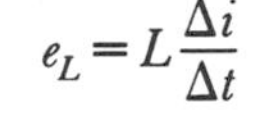

$$e_L = L\frac{\Delta i}{\Delta t}$$

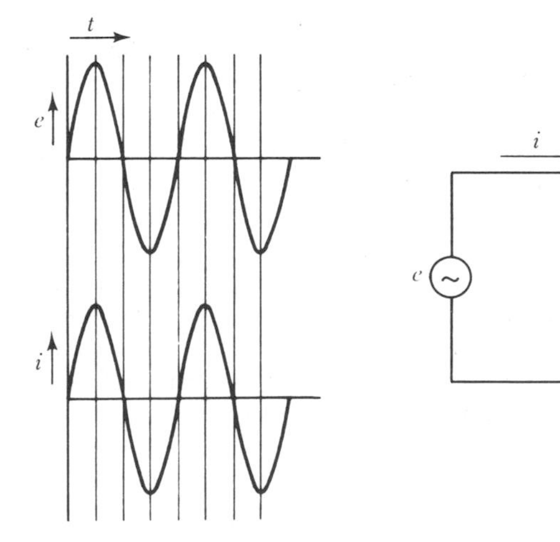

FIG. 20-1. Alternating voltage and current in a purely resistive circuit

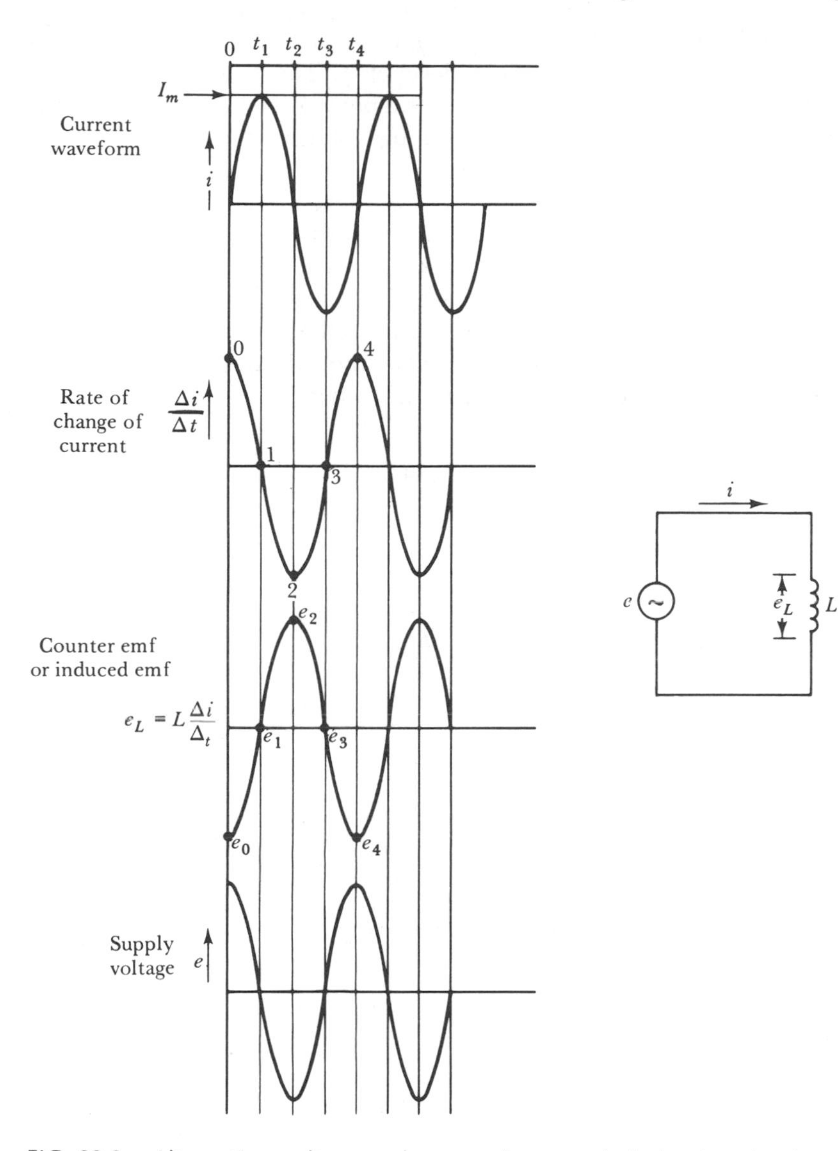

FIG. 20-2. Alternating voltage and current in a purely inductive circuit

Therefore, the counter-emf is directly proportional to the rate of change of current. However, because the counter-emf opposes the change of current, it has a maximum positive value when the rate of change of current is at its maximum negative value, and vice versa. The outcome of this is that the waveform of counter-emf is in *antiphase* with the waveform of rate of change of current. In Figure 20-2, $e_L = e_1 = 0$ when $\Delta i/\Delta t$ is zero at t_1. At t_2, $\Delta i/\Delta t$ is a peak negative quantity, and $e_L = e_2$, which is a peak positive quantity. Similarly, at t_4, $\Delta i/\Delta t$ has a peak positive value, and $e_L = e_4$, which is the peak negative level of e_L. The counter-emf is always in antiphase to the applied emf, or ac supply voltage. Therefore, the final waveform in Figure 20-2 shows the supply voltage waveform in antiphase to that of the counter-emf.

Putting the waveforms of supply voltage and inductor current together, as in Figure 20-3(a), it is seen that *the current in a purely inductive circuit* ***lags*** *the supply voltage by* 90°, *or* $\pi/2$ *radians*. Alternatively, it may be stated that the supply voltage leads the current by 90°. The phasor diagram of voltage and current in an inductive circuit [Figure 20-3(b)] uses the voltage as a reference and shows the current phasor 90° behind the voltage. Where the inductor terminal voltage is

$$e = E_m \sin \omega t$$

the inductor current is represented as

$$\boxed{i = I_m \sin\left(\omega t - \frac{\pi}{2}\right)} \qquad \textbf{(20-1)}$$

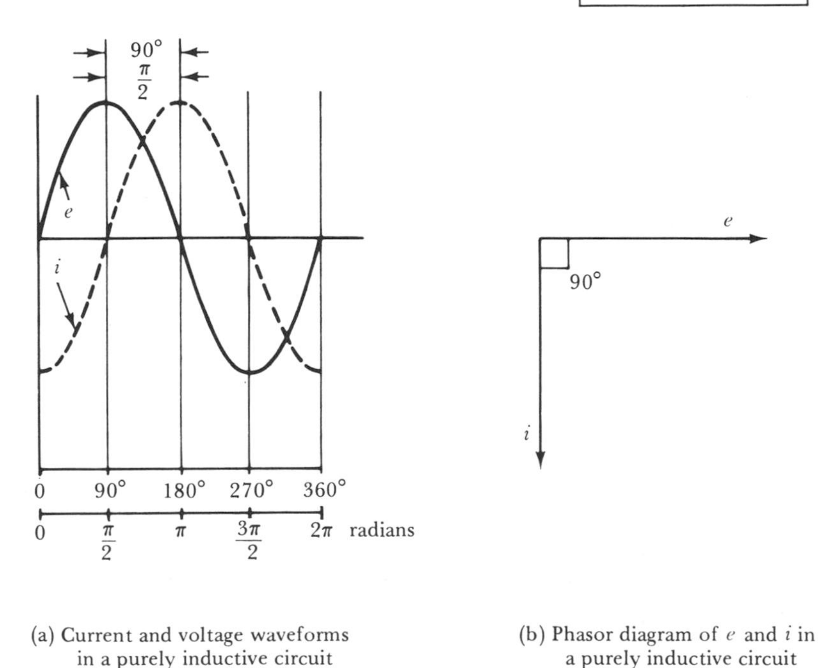

(a) Current and voltage waveforms in a purely inductive circuit

(b) Phasor diagram of e and i in a purely inductive circuit

FIG. 20-3. In a purely inductive circuit, current lags voltage by 90° or $\pi/2$ radians

20-2 INDUCTIVE REACTANCE

Referring once again to the current waveform in Figure 20-2, it is seen that the current grows from zero to its maximum value I_m in one fourth of a cycle. The time period of one cycle is

$$T=\frac{1}{f}$$

Therefore, the time taken for the current to grow from zero to I_m is

$$t=\frac{1}{4f}$$

The *average* rate of change of current during time t is

$$\frac{\Delta i}{\Delta t}=\frac{I_m}{1/(4f)}=4fI_m$$

From Eq. (15-3), $e_L=L(\Delta i/\Delta t)$. Substituting for $\Delta i/\Delta t$, the average counter-emf induced is

$$e_L=4fLI_m$$

And since the counter-emf is equal to the supply voltage in a purely inductive circuit, the average supply voltage is

$$E_{av}=4fLI_m$$

Actually there is always a slight difference between counter-emf and supply voltage; otherwise, no current would flow and no counter-emf would be generated. From Eq. (18-14),

$$E_{av}=\frac{2}{\pi}E_m$$

Therefore,

$$\frac{2}{\pi}E_m=4fLI_m$$

or

$$\frac{E_m}{I_m}=\frac{4\pi fL}{2}$$

$$\frac{E_m}{I_m}=2\pi fL$$

Substituting rms quantities for maximum quantities,

$$\frac{1.414E}{1.414I}=2\pi fL$$

which gives

$$\boxed{\frac{E}{I}=2\pi fL} \qquad (20\text{-}2)$$

where E and I are rms quantities.

From Ohm's law, (E volts)/(I amps) give resistance in ohms. However, in a pure inductance there is zero resistance, but $2\pi fL$ does represent an *opposition* to current flow. The name given to this quantity is *inductive reactance*, and like resistance, inductive reactance is measured in ohms. The symbol X_L is employed for inductive reactance.

Since $2\pi f=\omega$, the angular velocity in radians per second, Eq. (20-2) can be rewritten as

$$\frac{E}{I}=X_L=\omega L$$

It is seen that the inductive reactance of any inductor is

$$\boxed{X_L=\omega L=2\pi fL} \qquad (20\text{-}3)$$

When f is in hertz and L is in henrys, X_L is given in ohms. Note that X_L is directly proportional to frequency and inductance.

Substituting for I_m in Eq. (20-1), the expression for the instantaneous current in an inductive circuit becomes

$$\boxed{i=\frac{E_m}{\omega L}\sin\left(\omega t-\frac{\pi}{2}\right)} \qquad (20\text{-}4)$$

The reciprocal of resistance R is conductance G, which is the measure of a resistive circuit's ability to pass current. Similarly, inductive reactance X_L has its reciprocal in *inductive susceptance*, for which the symbol is B_L. Inductive susceptance is a measure of a purely inductive circuit's ability to pass current, and like conductance its unit is the siemens, S.

$$\boxed{\textbf{inductive susceptance, } B_L=\frac{1}{X_L}} \qquad (20\text{-}5)$$

When X_L is in ohms, B_L is in siemens.

EXAMPLE 20-1

A 500-mH inductor is supplied from a 115-V source with a frequency of 60 Hz. Calculate the inductive reactance and the current that flows in the circuit.

SOLUTION

Eq. (20-3):

$$\textit{inductive reactance}, X_L = 2\pi f L$$

$$= 2\pi \times 60 \text{ Hz} \times 500 \text{ mH}$$

$$\mathbf{X_L \cong 188.5\ \Omega}$$

$$\textit{current}, I = \frac{E}{X_L} = \frac{115 \text{ V}}{188.5\ \Omega}$$

$$\mathbf{I = 610\ mA}$$

EXAMPLE 20-2

An inductor supplied with 50 V ac with a frequency of 10 kHz passes a current of 7.96 mA. Determine the inductance of the inductor.

SOLUTION

$$X_L = \frac{E}{I} = \frac{50 \text{ V}}{7.96 \text{ mA}}$$

$$\mathbf{= 6.28\ k\Omega}$$

From Eq. (20-3),

$$L = \frac{X_L}{2\pi f} = \frac{6.28 \text{ k}\Omega}{2\pi \times 10 \text{ kHz}}$$

$$\mathbf{L \cong 100\ mH}$$

By substituting Equation (15-7) instead of Equation (15-3) in the derivation of Equation (20-2), an expression is obtained for the alternating voltage induced in the secondary of two mutually coupled coils (see Section 15-4). When an alternating current flows in the primary,

$$\frac{E_2}{I_1} = 2\pi f M$$

or

$$\boxed{E_2 = \omega M I_1} \qquad (20\text{-}6)$$

where E_2 is the rms voltage induced in the secondary winding, M is the mutual inductance, and I_1 is the rms current in the primary.

EXAMPLE 20-3

When a 100-mA alternating current with a frequency of 1 kHz flows in the primary of two coupled coils, the secondary voltage is 1 V. Calculate the mutual inductance between the two coils.

SOLUTION

From Eq. (20-6),

$$M = \frac{E_2}{\omega I_1}$$

$$= \frac{1\ \text{V}}{2\pi \times 1\ \text{kHz} \times 100\ \text{mA}}$$

$$\mathbf{M = 1.59\ mH}$$

20-3 ALTERNATING CURRENT AND VOLTAGE IN A CAPACITIVE CIRCUIT

In the circuit shown in Figure 20-4, it is assumed that C is a pure capacitance with a dielectric having infinite resistance, and that the connecting wires have zero resistance. The first waveform illustrated is that of the instantaneous level of supply voltage e. This is also the voltage that appears at the capacitor terminals. The waveform representing rate of change of capacitor voltage is drawn by considering the supply voltage waveform.

At t_1, the capacitor voltage has stopped increasing positively and has not yet started to decrease. Consequently, the rate of change of voltage at t_1 is zero. This gives point 1 on the waveform of *rate of change* of voltage. At t_2, the voltage is decreasing at its maximum rate; thus, it has a maximum negative rate of change, and this is plotted at point 2 on the $\Delta v/\Delta t$ waveform. At t_3, the voltage has zero rate of change again, giving point 3 on the $\Delta v/\Delta t$ wave. The voltage is increasing positively at its maximum rate of change at t_4; therefore, point 4 is plotted as the positive peak of the $\Delta v/\Delta t$ waveform.

From Eq. (16-4),

$$It = CE$$

giving

$$I = \frac{CE}{t}$$

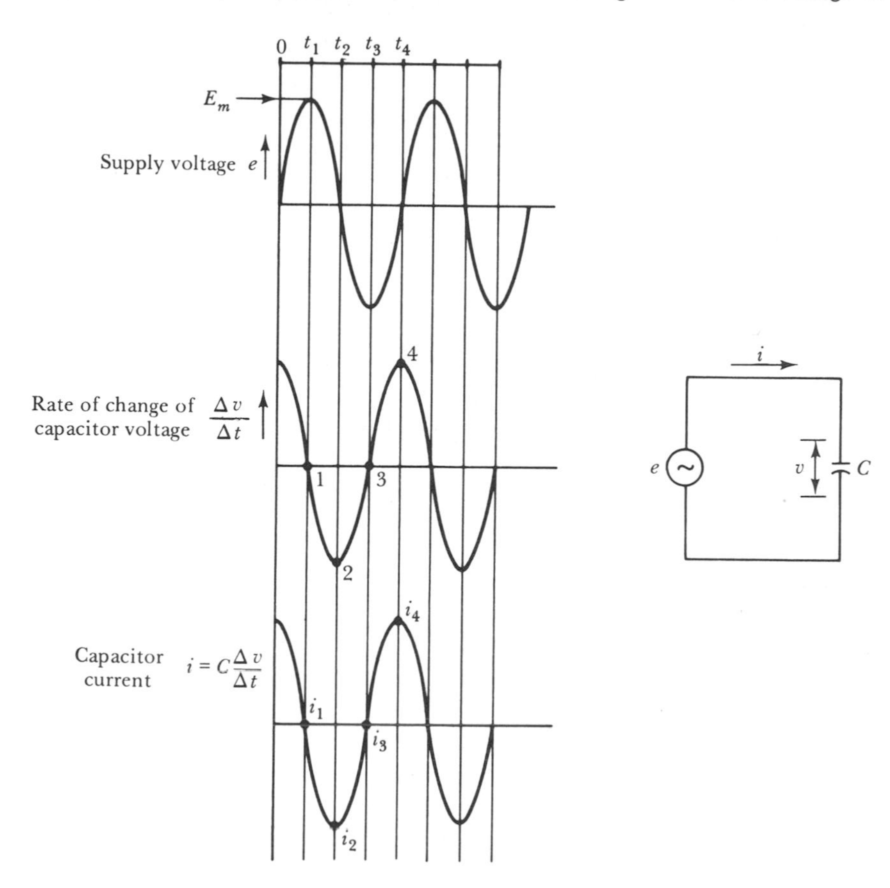

FIG. 20-4. Alternating voltage and current in a purely capacitive circuit

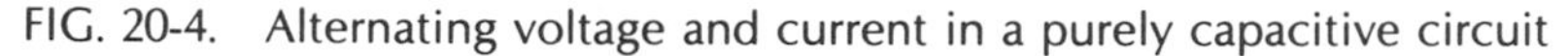

While a capacitor is being charged through Δv volts for a time Δt seconds, the instantaneous current is

$$i = C\frac{\Delta v}{\Delta t}$$

It is seen that the instantaneous current in a purely capacitive circuit is directly proportional to the rate of change of capacitor voltage. Now consider the $\Delta v/\Delta t$ waveform once again in Figure 20-4. The value of $\Delta v/\Delta t$ is zero at time t_1. Therefore, the instantaneous current i_1 is plotted as zero on the current waveform. At time t_2, the instantaneous value of $\Delta v/\Delta t$ is maximum negative rate of change; consequently, the instantaneous current at t_2 is a maximum negative quantity i_2, as illustrated. At t_3, $\Delta v/\Delta t$ is zero once again, giving i_3 equal zero. When the rate of change of voltage is a maximum positive quantity at t_4, the instantaneous current is also a positive maximum, plotted as i_4 on the current waveform.

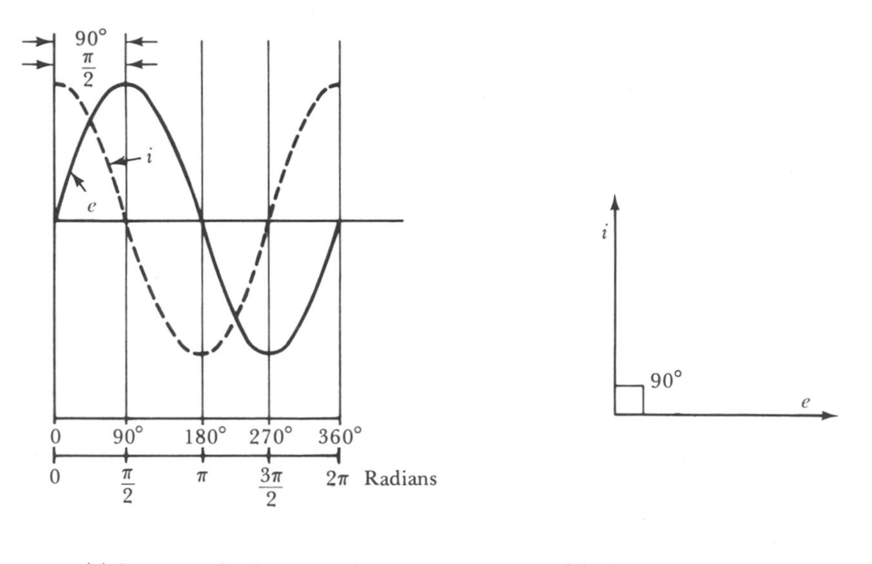

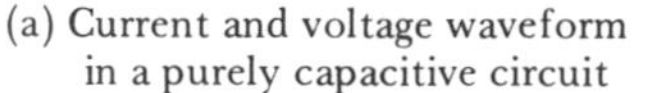
(a) Current and voltage waveform in a purely capacitive circuit

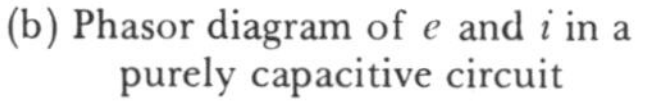
(b) Phasor diagram of e and i in a purely capacitive circuit

FIG. 20-5. In a purely capacitive circuit, current leads voltage by 90° or $\pi/2$ radians

In Figure 20-5(a) the waveforms of supply voltage and capacitor current are shown together. The waveforms, and the phasor diagram in Figure 20-5(b), show that *in a purely capacitive circuit the current* ***leads*** *the voltage by* 90°, *or* $\pi/2$ *radians*. When the capacitor terminal voltage is represented as

$$e = E_m \sin \omega t$$

the capacitor current is

$$\boxed{i = I_m \sin\left(\omega t + \frac{\pi}{2}\right)} \tag{20-7}$$

20-4 CAPACITIVE REACTANCE

As already explained for the inductive circuit, the capacitor current grows from zero to its maximum level I_m in a time of $t = 1/4f$. From Eq. (18-13),

$$I_{av} = \frac{2}{\pi} I_m$$

and

$$Q = CE_m = I_{av} \times t$$

Therefore, $$CE_m = \frac{2}{\pi} I_m \times \frac{1}{4f}$$

$$= I_m \times \frac{1}{2\pi f}$$

from which $$\frac{E_m}{I_m} = \frac{1}{2\pi fC}$$

Substituting rms quantities for maximum voltage and current,

$$\frac{1.414E}{1.414I} = \frac{1}{2\pi fC}$$

or

$$\boxed{\frac{E}{I} = \frac{1}{2\pi fC}} \qquad \textbf{(20-8)}$$

where E and I are rms quantities.

As in the case of the pure inductor, a pure capacitor has no resistive component, but $1/2\pi fC$ represents an opposition to current flow. Termed *capacitive reactance*, the quantity is measured in ohms and is given the symbol X_C:

$$\boxed{X_C = \frac{1}{\omega C} = \frac{1}{2\pi fC}} \qquad \textbf{(20-9)}$$

When f is in hertz and C in farads, X_C is given in ohms. Note that X_C is inversely proportional to frequency and capacitance.

Substituting for I_m in Eq. (20-7), the instantaneous current level in a capacitive circuit is given by

$$\boxed{i = E_m(\omega C)\sin\left(\omega t + \frac{\pi}{2}\right)} \qquad \textbf{(20-10)}$$

The reciprocal of capacitive reactance X_C is *capacitive susceptance* B_C, which is a measure of a purely capacitive circuit's ability to pass current:

$$\textbf{\textit{capacitive susceptance,}}\ \boxed{B_C = \frac{1}{X_C}} \qquad \textbf{(20-11)}$$

where X_C is in ohms and B_C is in siemens.

EXAMPLE 20-4

A 50-μF capacitor is supplied from a 115-V source with a frequency of 60 Hz. Determine the capacitive reactance and calculate the current that flows in the circuit.

SOLUTION

Eq. (20-9):

$$\textit{capacitive reactance},\ X_C = \frac{1}{2\pi f C}$$

$$= \frac{1}{2\pi \times 60\ \text{Hz} \times 50\ \mu\text{F}}$$

$$\mathbf{X_C \cong 53.1\ \Omega}$$

$$\textit{current},\ I = \frac{E}{X_C} = \frac{115\ \text{V}}{53.1\ \Omega}$$

$$\mathbf{I \cong 2.2\ A}$$

EXAMPLE 20-5

A capacitor passes a current of 12.6 mA when supplied with 20 V ac with a frequency of 1 kHz. Determine the capacitance of the capacitor.

SOLUTION

$$X_C = \frac{E}{I} = \frac{20\ \text{V}}{12.6\ \text{mA}}$$

$$\mathbf{X_C \cong 1.59\ k\Omega}$$

From Eq. (20-9),

$$C = \frac{1}{2\pi f X_C}$$

$$= \frac{1}{2\pi \times 1\ \text{kHz} \times 1.59\ \text{k}\Omega}$$

$$\mathbf{C = 0.1\ \mu F}$$

20-5 SERIES *RL* CIRCUIT

A series circuit consisting of inductance L and resistance R is shown in Figure 20-6(a), while the waveforms and phasor diagram for the circuit are shown in Figure 20-6(b) and (c), respectively. Referring to the circuit diagram, it is seen that (as for all series circuits) the current i is common to both R and L. The circuit waveforms are therefore drawn starting with the waveform of current.

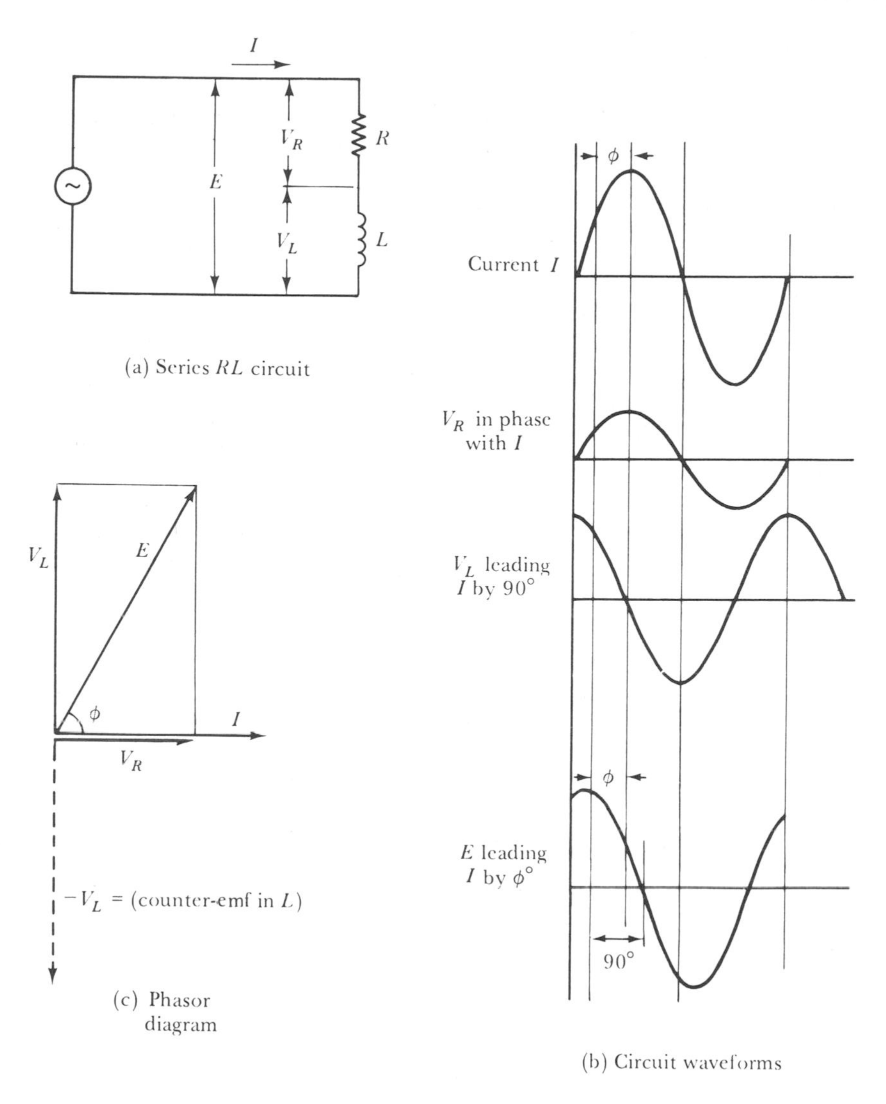

FIG. 20-6. Series *RL* circuit, waveforms, and phasor diagram

The voltage V_R across the resistance R is always in phase with the current through the resistance. Thus, the waveform of V_R in Figure 20-6(b) is drawn in phase with the current waveform. In the case of the inductor, the voltage V_L leads the current through the inductor by 90° (see Section 20-1). The waveform of V_L is therefore drawn in Figure 20-6(b) 90° ahead of the waveform of I. The applied voltage E is obviously the resultant of the two component voltages V_R and V_L, and its waveform is obtained simply by summing the instantaneous levels of V_R and V_L. It is seen that E leads I by an angle $\phi°$ which is less than 90°.

The phasor diagram for the circuit is drawn by once again starting with current, since the current is the common quantity in a series circuit. A horizontal line is drawn to scale representing current I [see Figure 20-6(c)]. Since V_R is in phase with I, another horizontal line is drawn alongside I to represent V_R. V_L is 90° ahead of I; therefore, the phasor for V_L is drawn vertically at an angle of 90° with respect to I. The phasor addition of V_L and V_R gives a resultant that represents the applied voltage E. Once again, it is seen that the applied voltage leads the circuit current by an angle of $\phi°$, which is less than the 90° angle that would exist between E and I in a pure inductive circuit. Note that the phasor representing the counter-emf in the inductor is $-V_L$, as shown by the dashed line in the diagram.

As explained in Chapter 19, E can be expressed in rectangular or polar form:

$$\boxed{E = V_R + jV_L} \tag{20-12}$$

or from Eq. (19-2),

$$\boxed{E = \sqrt{V_R^2 + V_L^2}\ \big/ \arctan\left(\frac{V_L}{V_R}\right)} \tag{20-13}$$

Dividing Eq. (20-12) by I gives

$$\boxed{\frac{E}{I} = \frac{V_R}{I} + j\frac{V_L}{I}} \tag{20-14}$$

V_R/I is, of course, the voltage across the resistance R, divided by the current through the resistance. Therefore,

$$\frac{V_R}{I} = R$$

Also, V_L/I is the inductor voltage divided by the current through the inductor. Thus,

$$\frac{V_L}{I} = X_L$$

In Eq. (20-14) E/I obviously represents opposition to current flow; however, it is neither resistance nor inductive reactance, since it has both as component parts. In this case E/I is termed *impedance* and is given the symbol Z. Thus, Eq. (20-14) can be restated:

$$\textit{impedance,}\ \boxed{Z = R + jX_L} \tag{20-15}$$

When R and X_L are in ohms, the units of Z are also ohms.

The numerical value or modulus of Z (i.e., not including the angle) is written $|Z|$, and

$$\boxed{|Z| = \sqrt{R^2 + X_L^2}} \tag{20-16}$$

Also, the angle of Z is

$$\boxed{\phi = \arctan \frac{X_L}{R}} \tag{20-17}$$

Therefore,

$$\boxed{Z = \sqrt{R^2 + X_L^2}\ \angle \arctan \frac{X_L}{R}} \tag{20-18}$$

Z, R, and X_L can be represented on a vector diagram. They are not phasor quantities because they have fixed values; that is, unlike ac voltage and current, Z, R, and X_L do *not* have continuously changing instantaneous values. To distinguish it from a phasor diagram, the vector diagram of impedance, reactance, and resistance is usually drawn in triangular form, and is referred to as an *impedance diagram*. Figure 20-7 shows the impedance diagram for the series *RL* circuit of Figure 20-6(a). A horizontal line is first drawn to represent the resistive component R.

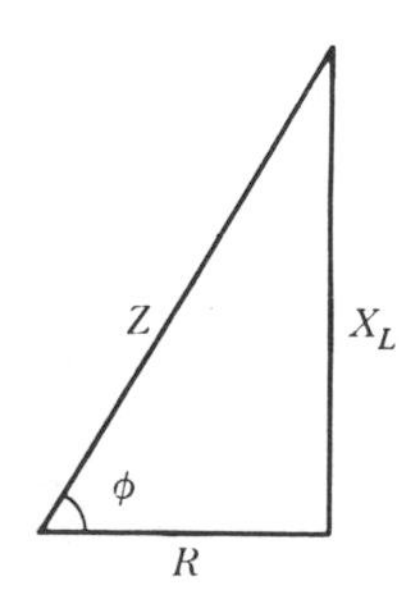

FIG. 20-7. Impedance diagram for a series *RL* circuit

The j component (i.e., X_L) is then drawn at 90° with respect to R, as illustrated. The Z component is the hypotenuse of the triangle, and the angle ϕ is the phase angle of the impedance Z with respect to resistance R.

As with resistance and reactance, the reciprocal of impedance is sometimes employed. The *admittance*, symbol Y, is the reciprocal of impedance Z, and its unit is the siemens, S.

$$\textbf{admittance,} \quad \boxed{Y = \frac{1}{Z}} \qquad \textbf{(20-19)}$$

When Z is in ohms, Y is in siemens.

All inductors are coils that have some winding resistance, and, as already explained, they can be represented by a pure inductance in series with the winding resistance. Therefore, the diagrams in Figures 20-6 and 20-7 could apply to the case of an inductor connected directly to an ac supply. Where an external resistance is connected in series with the inductor, the coil resistance should also be shown as a series component in the equivalent circuit.

EXAMPLE 20-6

A series circuit consists of $R = 20\ \Omega$, $L = 20$ mH, and an ac supply of 60 V with $f = 100$ Hz. Calculate the current, the voltage across R, the voltage across L, and the phase angle of current with respect to the supply voltage.

SOLUTION

$$X_L = 2\pi f L$$
$$= 2\pi \times 100\ \text{Hz} \times 20\ \text{mH}$$
$$\mathbf{X_L \simeq 12.57\ \Omega}$$

Eq. (20-16):

$$|Z| = \sqrt{R^2 + X_L^2}$$
$$= \sqrt{20^2 + 12.57^2}$$
$$\mathbf{|Z| \simeq 23.6\ \Omega}$$

$$|I| = \frac{E}{|Z|} = \frac{60\ \text{V}}{23.6\ \Omega}$$
$$\mathbf{I \simeq 2.54\ A}$$

Eq. (20-17):

$$\phi = \arctan \frac{X_L}{R}$$

$$= \arctan \frac{12.57\ \Omega}{20\ \Omega}$$

$$\boldsymbol{\phi = 32.1°}$$

$$V_R = E \cos \phi$$

$$= 60\ \text{V} \cos 32.1°$$

$$\boldsymbol{V_V = 50.8\ \text{V}}$$

$$V_L = E \sin \phi$$

$$= 60\ \text{V} \sin 32.1°$$

$$\boldsymbol{V_L = 31.9\ \text{V}}$$

EXAMPLE 20-7 A 64-mH inductor with a winding resistance of $R_W = 700\ \Omega$ is connected in series with a resistor $R_1 = 3.3\ \text{k}\Omega$. A 10-V ac supply with a frequency of 5 kHz is connected to the series circuit. Calculate the circuit current and the terminal voltage of the inductor.

SOLUTION

$$X_L = 2\pi f L$$

$$= 2\pi \times 5\ \text{kHz} \times 64\ \text{mH}$$

$$\boldsymbol{X_L \cong 2\ \text{k}\Omega}$$

$$\textit{total resistance} = R_1 + \textit{winding resistance of } L$$

$$= 3.3\ \text{k}\Omega + 700\ \Omega$$

$$\boldsymbol{R = 4\ \text{k}\Omega}$$

$$Z = \sqrt{R^2 + X_L^2}$$

$$= \sqrt{(4\ \text{k}\Omega)^2 + (2\ \text{k}\Omega)^2}$$

$$\boldsymbol{Z = 4.47\ \text{k}\Omega}$$

$$I = \frac{E}{Z} = \frac{10\ \text{V}}{4.47\ \text{k}\Omega}$$

$$\boldsymbol{I \cong 2.24\ \text{mA}}$$

$$V_R = I \times R$$

$$= 2.24\ \text{mA} \times 4\ \text{k}\Omega$$

$$\boldsymbol{V_R = 8.96\ \text{V}}$$

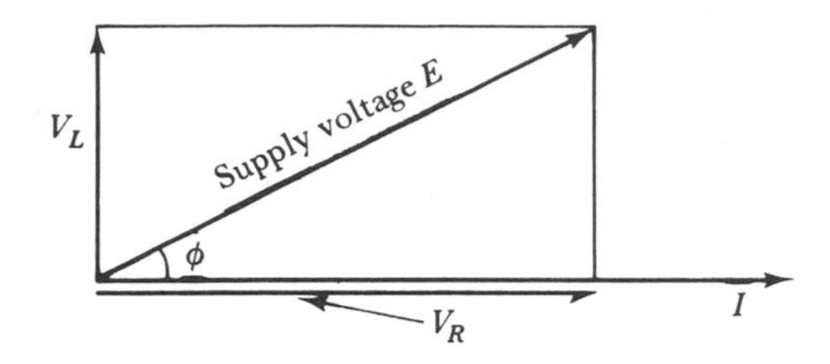

(a) Phasor diagram showing relationship between supply voltage, current, and the inductive and resistive voltage components

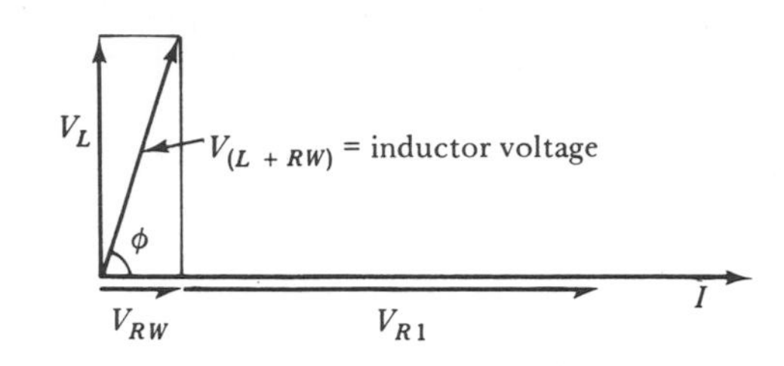

(b) Phasor components of inductor voltage

FIG. 20-8. Phasor diagrams for a series *LR* circuit

The phasor diagram in Figure 20-8(a) *shows the total resistive component of the input voltage* (V_R) *in phase with the current. The inductive component* V_L *leads* V_R *by* 90°.

$$V_L = I \times X_L$$
$$= 2.24 \text{ mA} \times 2 \text{ k}\Omega$$
$$\mathbf{V_L = 4.48 \text{ V}}$$

The voltage drop across the resistive component of the inductor is

$$V_{RW} = I \times R_W$$
$$= 2.24 \text{ mA} \times 700 \ \Omega$$
$$\mathbf{V_{RW} = 1.57 \text{ V}}$$

In Figure 20-8(b) *it is seen that* V_R *consists of two components* $V_{RW} + V_{R1}$, *and that the voltage across the inductor is the resultant of* V_L *and* V_{RW}. *Therefore,*

$$V_{(L+RW)} = \sqrt{V_{RW}^2 + V_L^2}$$
$$= \sqrt{1.57^2 + 4.48^2}$$
$$\textit{inductor voltage, } V_{(L+RW)} = 4.75 \text{ V}$$

The inductor voltage leads the current by an angle θ [*see* Figure 20-8(b)], *where*

$$\theta = \arctan \frac{V_L}{V_{RW}}$$
$$= \arctan \frac{4.48 \text{ V}}{1.57 \text{ V}}$$
$$\mathbf{\theta = 70.7^\circ}$$

20-6 SERIES *RC* CIRCUIT

Figure 20-9(a) shows a capacitor C connected in series with a resistor R, with an ac supply voltage. The circuit waveforms are shown in Figure 20-9(b), and the phasor diagram for the circuit is drawn in Figure 20-9(c).

Starting with the waveform of current I which is common to both R and C, the waveform of voltage V_R across R is drawn in phase with I. The current through the capacitor leads the capacitor terminal voltage by 90° (see Section 20-3); therefore, the V_C waveform is drawn 90° behind the

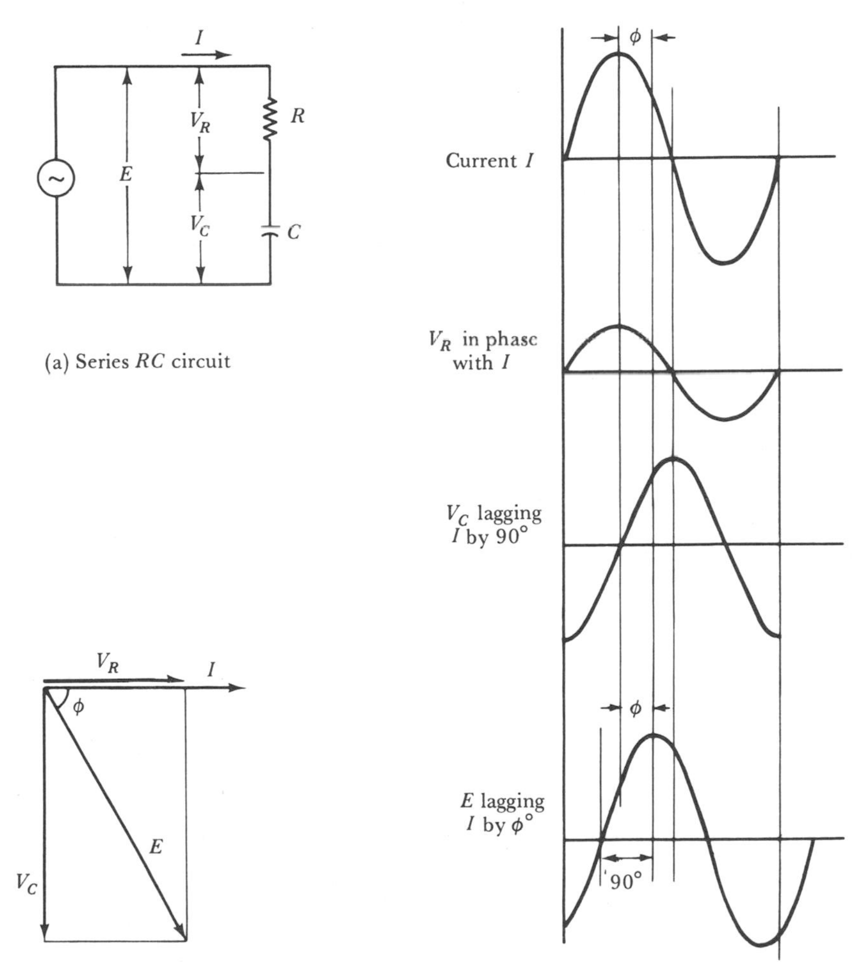

(c) Phasor diagram

(b) Circuit waveforms

FIG. 20-9. Series *RC* circuit, waveforms, and phasor diagram

current wave. The applied voltage E is the vector sum of V_R and V_C, and it is seen that I leads E by an angle ϕ which is less than 90°.

For the phasor diagram in Figure 20-9(c), a horizontal line is drawn to represent I, and the phasor for the voltage V_R is drawn in phase with I. The capacitor voltage phasor V_C is drawn 90° behind the current phasor. The supply voltage E is the resultant of V_R and V_C, and it lags I by the angle ϕ, as illustrated.

From Chapter 19 and Figure 20-9(c), the rectangular form expression for E in the series resistance capacitance circuit is

$$\boxed{E = V_R - jV_C} \tag{20-20}$$

and from Eq. (19-2),

$$\boxed{E = \sqrt{V_R^2 + V_C^2} \;\angle \arctan \frac{V_C}{V_R}} \tag{20-21}$$

Dividing Eq. (20-20) through by I:

$$\frac{E}{I} = \frac{V_R}{I} - j\frac{V_C}{I}$$

or

$$\textit{impedance,}\quad \boxed{Z = R - jX_C} \tag{20-22}$$

where R is resistance in ohms, X_C is capacitive reactance in ohms and Z is impedance in ohms. Also,

$$\boxed{|Z| = \sqrt{R^2 + X_C^2}} \tag{20-23}$$

and

$$\boxed{\phi = \arctan \frac{X_C}{R}} \tag{20-24}$$

giving

$$\boxed{Z = \sqrt{R^2 + X_C^2} \;\angle \arctan \frac{X_C}{R}} \tag{20-25}$$

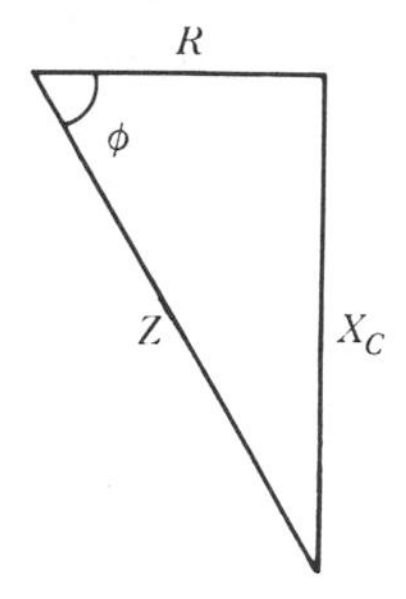

FIG. 20-10. Impedance diagram for a series *RC* circuit

The impedance diagram for a series resistance–capacitance circuit is shown in Figure 20-10. Note that the $-j$ component X_C is drawn 90° lagging R, and the phase angle of Z with respect to R is $-\phi$.

EXAMPLE 20-8

A series circuit consists of $R=47\ \Omega$, $C=10\ \mu F$, and an ac supply of 100 V with $f=300$ Hz. Calculate the current, the voltage across R, the voltage across C, and the phase angle of the current with respect to the supply voltage.

SOLUTION

$$X_C=\frac{1}{2\pi fC}$$

$$=\frac{1}{2\pi\times 300\ \text{Hz}\times 10\ \mu\text{F}}$$

$$\mathbf{X_C \cong 53.1\ \Omega}$$

Eq. (20-23):

$$|Z|=\sqrt{R^2+X_C^2}$$

$$=\sqrt{47^2+53.1^2}$$

$$\mathbf{|Z|=70.9\ \Omega}$$

$$|I|=\frac{E}{|Z|}=\frac{100\ \text{V}}{70.9\ \Omega}$$

$$\mathbf{I=1.41\ A}$$

Eq. (20-24):

$$\phi = \arctan \frac{X_C}{R}$$

$$= \arctan \frac{53.1\ \Omega}{47\ \Omega}$$

$$\boldsymbol{\phi = 48.5°}$$

$$V_R = E \cos\phi$$

$$= 100\ \text{V} \cos 48.5°$$

$$\boldsymbol{V_R \cong 66.3\ \text{V}}$$

$$V_C = E \sin\phi$$

$$= 100\ \text{V} \sin 48.5°$$

$$\boldsymbol{V_C \cong 74.9\ \text{V}}$$

20-7 SERIES *R, L,* AND *C*

When resistance, inductance, and capacitance are connected in series, the circuit behaves either as a series *RL* circuit or as a series *RC* circuit, depending upon which of the two reactances (X_L and X_C) is the larger. The special case where X_L and X_C are equal is considered in Chapter 24.

In the *RLC* series circuit shown in Figure 20-11(a), the current is once again common to all series components. The phasor diagram is drawn as before, by starting with the horizontal current phasor [see Figure 20-11(b)]. The resistor voltage V_R is in phase with the current; therefore, the phasor of V_R is drawn in phase with the I phasor, as shown. Since the current lags the voltage V_L across the inductor, the inductor voltage phasor is drawn 90° ahead of the current phasor. Similarly, the capacitor voltage phasor V_C is drawn 90° behind the current phasor, because the current leads the voltage across the capacitor by 90°.

Figure 20-11(b) shows a phasor diagram for a circuit where the inductive reactance is greater than the capacitive reactance. This means that V_L is greater than V_C, and the circuit behaves as an *RL* series circuit. The two reactive phasors (V_L and V_C) must be added vectorially to find the resultant reactive phasor V_X. Then the resultant of V_X and V_R is the supply voltage E, and ϕ is its phase angle with respect to the current. In Figure 20-11(c), the phasor diagram illustrates the case where X_C is greater than X_L, giving a capacitor voltage V_C which is larger than the

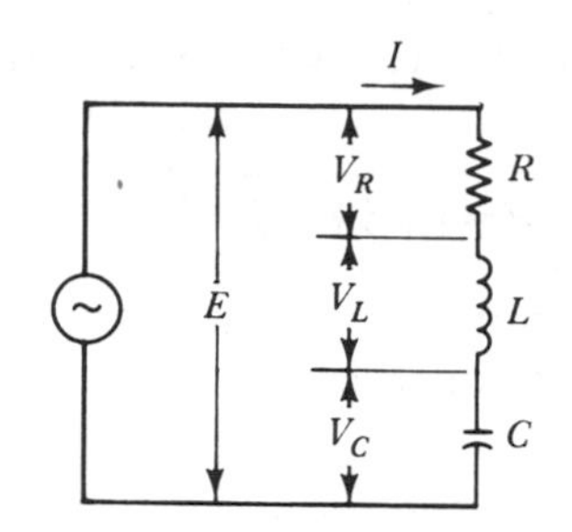

(a) Series RLC circuit

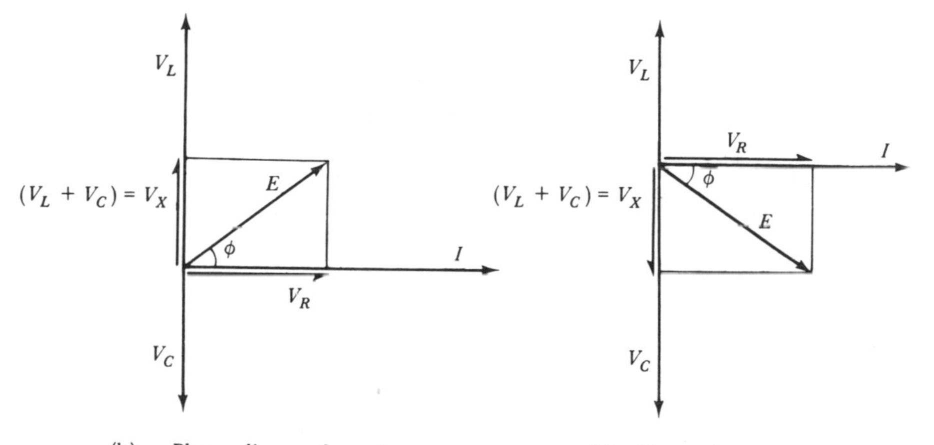

(b) Phasor diagram for series RLC circuit in which $X_L > X_C$

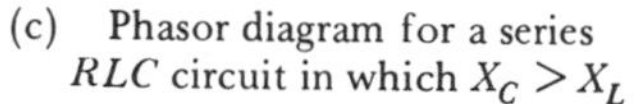

(c) Phasor diagram for a series RLC circuit in which $X_C > X_L$

FIG. 20-11. Series *RLC* circuit and phasor diagrams

inductor voltage V_L. In this case V_X is capacitive, E lags I by the phase angle ϕ, and the circuit behaves as a RC series circuit.

The rectangular expression for the voltage in a series RLC circuit is

$$\boxed{E = V_R + j(V_L - V_C)} \tag{20-26}$$

or

$$\boxed{E = V_R + jV_X} \tag{20-27}$$

where

$$V_X = V_L - V_C$$

The expression for the voltage in polar form is

$$\mathbf{E = \sqrt{V_R^2 + V_X^2} \;\angle\; \arctan \frac{V_X}{V_R}} \qquad \text{(20-28)}$$

Dividing Eq. (20-26) through by I,

$$\frac{E}{I} = \frac{V_R}{I} + j\left(\frac{V_L}{I} - \frac{V_C}{I}\right)$$

which gives

$$\textit{impedance,} \quad \mathbf{Z = R + j(X_L - X_C)} \qquad \text{(20-29)}$$

Also,

$$\mathbf{|Z| = \sqrt{R^2 + X_{eq}^2}} \qquad \text{(20-30)}$$

where

$$X_{eq} = X_L - X_C$$

and

$$\mathbf{\phi = \arctan \frac{X_{eq}}{R}} \qquad \text{(20-31)}$$

giving

$$\mathbf{Z = \sqrt{R^2 + X_{eq}^2} \;\angle\; \arctan \frac{X_{eq}}{R}} \qquad \text{(20-32)}$$

The impedance diagrams shown in Figure 20-12(a) and (b) relate to the phasor diagrams in Figure 20-11(b) and (c) respectively, and are self-explanatory.

EXAMPLE 20-9

A series RLC circuit has $R = 33\ \Omega$, $L = 50$ mH, and $C = 10\ \mu$F. The supply voltage is 75 V with a frequency of 200 Hz. Calculate I, V_R, V_L, V_C, and the phase angle of the current with respect to the supply voltage.

SOLUTION

$$X_L = 2\pi f L$$

$$= 2\pi \times 200 \text{ Hz} \times 50 \text{ mH}$$

$$\mathbf{X_L = 62.8\ \Omega}$$

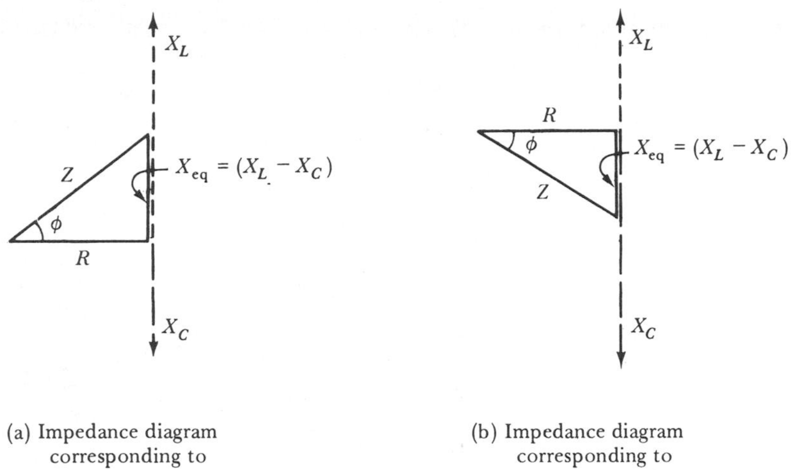

(a) Impedance diagram corresponding to Fig. 20-11(b)

(b) Impedance diagram corresponding to Fig. 20-11(c)

FIG. 20-12. Impedance diagrams for series *RLC* circuits

$$X_C = \frac{1}{2\pi fC}$$

$$= \frac{1}{2\pi \times 200\ \text{Hz} \times 10\ \mu\text{F}}$$

$$\mathbf{X_C = 79.6\ \Omega}$$

Eq. (20-29):

$$Z = R + j(X_L - X_C)$$

$$= 33\ \Omega + j(62.8 - 79.6)\ \Omega$$

$$= 33 - j16.8\ \Omega$$

$$X_{eq} = -16.8\ \Omega \qquad (\textit{because } X_C > X_L,\ X_{eq} \textit{ is capacitive})$$

$$Z = \sqrt{R^2 + X_{eq}^2}\ \angle \arctan \frac{X_{eq}}{R}$$

$$= \sqrt{33^2 + 16.8^2}\ \angle \arctan \frac{-16.8\ \Omega}{33\ \Omega}$$

$$\mathbf{|Z| \cong 37\ \Omega\ \angle -27^\circ}$$

$$|I| = \frac{E}{|Z|}$$

$$= \frac{75\ \text{V}}{37\ \Omega\ \angle -27^\circ}$$

$$\mathbf{I = 2.03\ A\ \angle 27^\circ}$$

Since X_{eq} is a capacitive reactance, the circuit current I leads the supply voltage E.

$$V_R = I \times R$$
$$= 2.03\ \text{A} \times 33\ \Omega$$
$$\mathbf{V_R \cong 67\ V}$$

$$V_L = I \times X_L$$
$$= 2.03\ \text{A} \times 62.8\ \Omega$$
$$\mathbf{V_L = 127\ V}$$

$$V_C = I \times X_C$$
$$= 2.03\ \text{A} \times 79.6\ \Omega$$
$$\mathbf{V_C = 162\ V}$$

20-8 PARALLEL *R, L,* AND *C*

When resistance, inductance and capacitance are connected in parallel as in Figure 20-13(a), the supply voltage E is common to all components. The phasor diagram is commenced by first drawing the phasor for E horizontally [see Figure 20-13(b)]. The rest of the phasor diagram consists of current phasors, with i_R being drawn in phase with E. The inductive current i_L lags the applied voltage; therefore, the i_L phasor is drawn 90° behind E. Similarly, the capacitive current i_C leads the applied voltage, so its phasor is drawn 90° ahead of E. The phasors of i_L and i_C are in antiphase, so, as illustrated in Figure 20-13(b), their resultant is simply

$$i_X = i_C - i_L$$

The supply current, I, is the resultant of the phasor sum of i_X and i_R, and the phase angle of I with respect to E is ϕ, as shown.

In the phasor diagram illustrated, i_L is greater than i_C, and the complete circuit behaves as a resistive–inductive parallel combination. Obviously, it is also possible for i_C to be greater than i_L, giving a circuit that performs as a resistor and capacitor in parallel. The special case where i_L and i_C are equal is considered in Chapter 24.

Consideration of the circuit and phasor diagram in Figure 20-13 reveals that the equation for the total current is written

$$\boxed{\mathbf{I = i_R + j(i_C - i_L)}} \qquad \textbf{(20-33)}$$

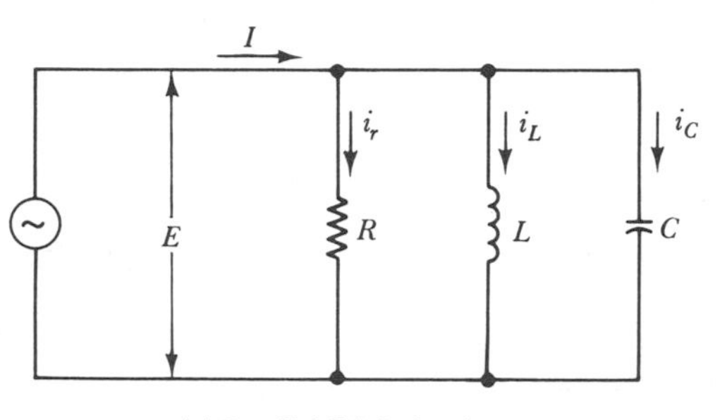

(a) Parallel RLC circuit

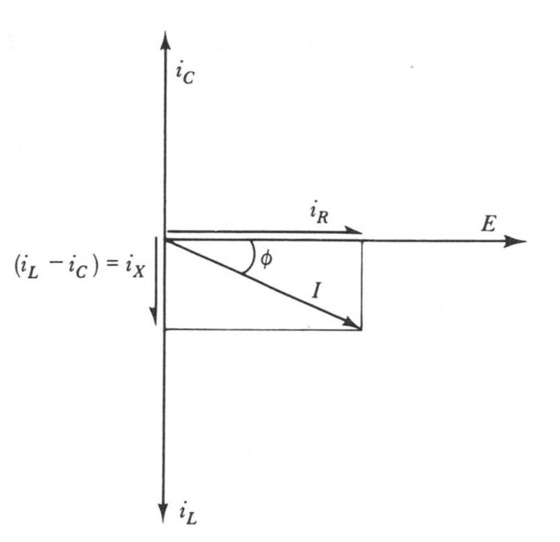

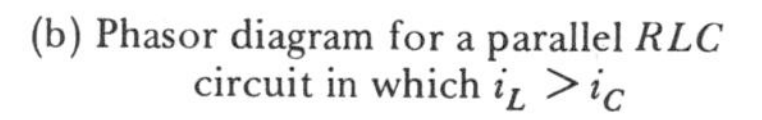

(b) Phasor diagram for a parallel RLC circuit in which $i_L > i_C$

FIG. 20-13. Parallel RLC circuit and phasor diagram

or

$$\boxed{\boldsymbol{I = i_R + j i_X}} \qquad \textbf{(20-34)}$$

where $\qquad i_X = i_C - i_L$

The expression for the current in polar form is

$$\boxed{\boldsymbol{I = \sqrt{i_R^2 + i_X^2}\ \Big/ \arctan \frac{i_X}{i_R}}} \qquad \textbf{(20-35)}$$

Dividing Eq. (20-33) through by E:

$$\frac{I}{E} = \frac{i_R}{E} + j\left(\frac{i_C}{E} - \frac{i_L}{E}\right)$$

which gives

$$\boxed{\boldsymbol{\frac{1}{Z} = \frac{1}{R} + j\left(\frac{1}{X_C} - \frac{1}{X_L}\right)}} \qquad \textbf{(20-36)}$$

or

$$\boxed{\textbf{\textit{admittance}}, \boldsymbol{Y = G + j(B_C - B_L)}} \qquad \textbf{(20-37)}$$

where Y is admittance, G is conductance, B_L is inductive susceptance, and B_C is *capacitive susceptance, all of which are measured in siemens, S. Also,*

$$Y = G + jB_{eq}$$

where $$B_{eq} = B_C - B_L$$

and

$$\boxed{|Y| = \sqrt{G^2 + B_{eq}^2}} \quad (20\text{-}38)$$

and

$$\boxed{\phi = \arctan \frac{B_{eq}}{G}} \quad (20\text{-}39)$$

giving

$$\boxed{Y = \sqrt{G^2 + B_{eq}^2} \angle \arctan \frac{B_{eq}}{G}} \quad (20\text{-}40)$$

It is impossible to draw an impedance diagram for a parallel *RLC* circuit; instead, an *admittance diagram* is prepared, as shown in Figure 20-14. A horizontal line is first drawn to scale to represent the conductance *G*. The capacitive susceptance vector B_C is next drawn vertically at an angle of 90° with respect to the conductance vector. Then the inductive susceptance B_L vector is drawn at an angle of −90° with respect to *G*. The resultant susceptance vector B_{eq} is found by vectorially summing B_C and B_L. Finally, the resultant of B_{eq} and *G* is found, giving the admittance *Y*.

EXAMPLE 20-10 A parallel *RLC* circuit has $R = 100\ \Omega$, $L = 20$ mH, and $C = 10\ \mu$F. The supply voltage is 35 V with a frequency of 500 Hz. Calculate i_R, i_L, i_C, and the supply current *I*. Also, determine the phase angle of *I* with respect to the supply voltage. Draw a phasor diagram and an admittance diagram for the circuit, and determine the impedance of the total circuit.

SOLUTION

$$X_L = 2\pi f L$$

$$= 2\pi \times 500 \text{ Hz} \times 20 \text{ mH}$$

$$\mathbf{X_L = 62.8\ \Omega}$$

$$X_C = \frac{1}{2\pi f C}$$

$$= \frac{1}{2\pi \times 500 \text{ Hz} \times 10\ \mu\text{F}}$$

$$\mathbf{X_C = 31.8\ \Omega}$$

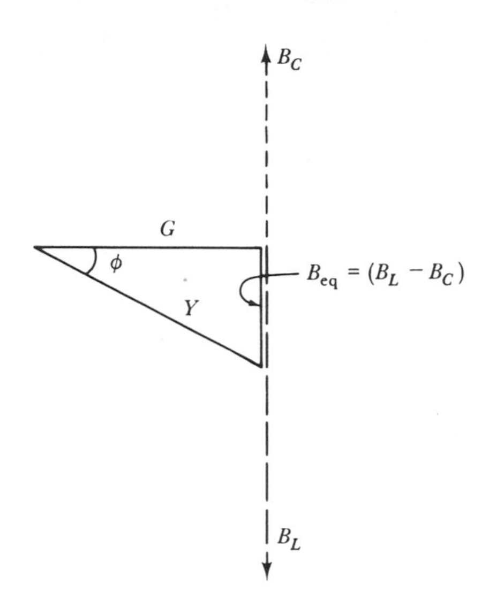

FIG. 20-14. Admittance diagram for a parallel *RLC* circuit

$$i_R = \frac{E}{R} = \frac{35 \text{ V}}{100\ \Omega}$$

$$\mathbf{i_R = 350 \text{ mA}}$$

$$i_L = \frac{E}{X_L} = \frac{35 \text{ V}}{62.8\ \Omega}$$

$$\mathbf{i_L = 557 \text{ mA}}$$

$$i_C = \frac{E}{X_C} = \frac{35 \text{ V}}{31.8\ \Omega}$$

$$\mathbf{i_C = 1.1 \text{ A}}$$

$$i_X = i_C - i_L$$

$$= 1.1 \text{ A} - 557 \text{ mA}$$

$$\mathbf{i_X = 543 \text{ mA}}$$

Eq. (20-34):

$$I = i_R + j i_X$$

$$I = 350 \text{ mA} + j543 \text{ mA}$$

Eq. (20–35):

$$I = \sqrt{i_R^2 + i_X^2}\ \angle \arctan \frac{i_X}{i_R}$$

$$= \sqrt{0.35^2 + 0.543^2}\ \angle \arctan \frac{0.543 \text{ A}}{0.35 \text{ A}}$$

$$\mathbf{I = 646 \text{ mA}\ \angle 57.2^\circ}$$

Phasor diagram:

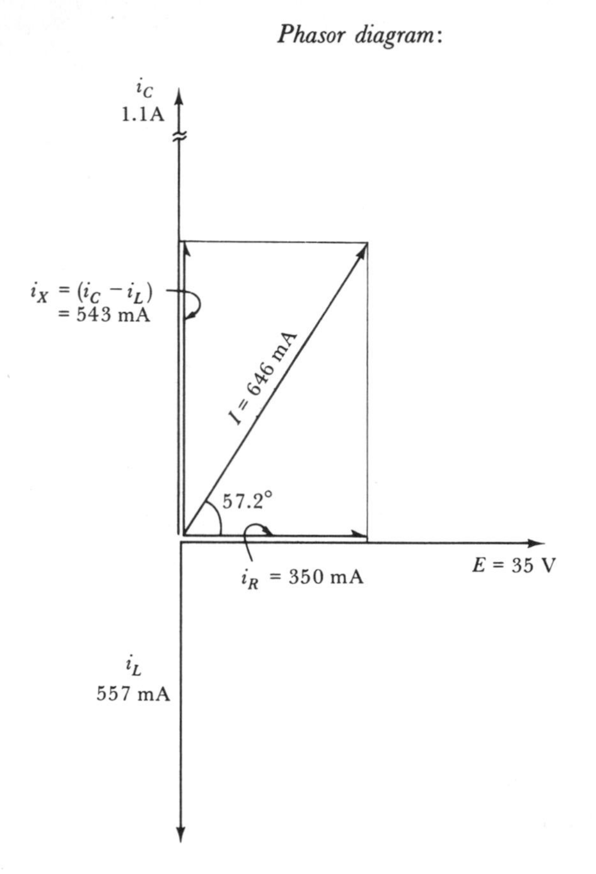

FIG. 20-15. Example 20-10

$$G = \frac{1}{R} = \frac{1}{100\ \Omega}$$

$$\boldsymbol{G = 10 \times 10^{-3}\ \mathrm{S}}$$

$$B_L = \frac{1}{X_L} = \frac{1}{62.8\ \Omega}$$

$$\boldsymbol{B_L = 15.9 \times 10^{-3}\ \mathrm{S}}$$

$$B_C = \frac{1}{X_C} = \frac{1}{31.8\ \Omega}$$

$$\boldsymbol{B_C = 31.4 \times 10^{-3}\ \mathrm{S}}$$

$$B_{\mathrm{eq}} = B_C - B_L$$

$$= (31.4 \times 10^{-3}) - (15.9 \times 10^{-3})$$

$$\boldsymbol{B_{\mathrm{eq}} = 15.5 \times 10^{-3}\ \mathrm{S}}$$

Eq. (20-40):

$$Y = \sqrt{G^2 + B_{eq}^2} \;\angle \arctan \frac{B_{eq}}{G}$$

$$= \sqrt{(10 \times 10^{-3})^2 + (15.5 \times 10^{-3})^2} \;\angle \arctan \frac{15.5 \times 10^{-3}}{10 \times 10^{-3}}$$

$$Y = 18.4 \times 10^{-3} \text{S} \angle 57.2°$$

Eq. (20-19):

$$Z = \frac{1}{Y} = \frac{1}{18.4 \times 10^{-3} \angle 57.2°}$$

$$\mathbf{Z = 54.3\ \Omega \;\angle -57.2°}$$

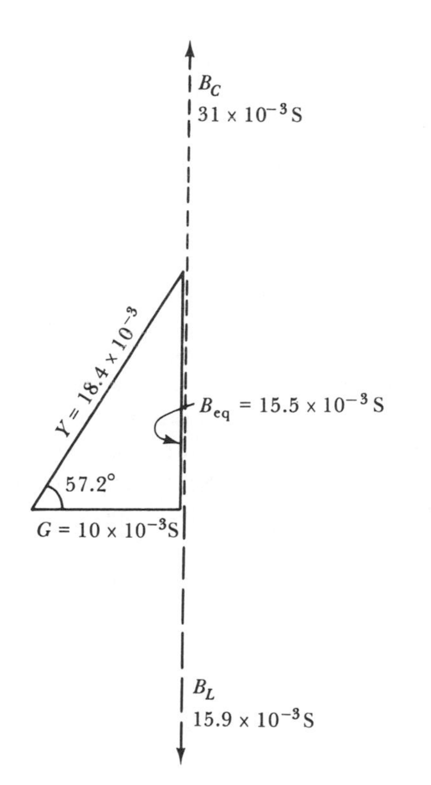

FIG. 20-16. Admittance diagram

Example 20-9 shows that the parallel *RLC* circuit with the component values and frequency as stated has an impedance of

$$Z = 54.3\ \Omega \;\angle -57.2°$$

The impedance can be resolved from the polar form into rectangular form:

$$Z = 54.3\cos(-57.2°) + j54.3\sin(-57.2°)$$

$$Z = 29.4\ \Omega - j45.6\ \Omega$$

This value of Z would also be obtained with a resistance of $R = 29.4\ \Omega$, connected *in series* with a capacitive reactance of $X_C = 45.6\ \Omega$. Thus, it is seen that the parallel RLC circuit has an equivalent series RC circuit. Depending upon the component values, the equivalent series circuit could alternatively be an RL circuit.

20-9 LOW-PASS FILTERS

Sometimes a direct voltage has an unwanted alternating voltage component, which must be reduced or removed. The circuit employed to do this is known as a *low-pass filter*. Similarly, an unwanted high frequency that is superimposed upon a low-frequency alternating voltage can be attenuated by the use of a low-pass filter. Therefore, the function of a low-pass filter is to pass low frequencies, and reduce or *attenuate* high frequencies.

The circuit of Figure 20-17(a) shows a simple CR low-pass filter. The input is a direct voltage with an ac waveform superimposed, and the output is a direct voltage and an attenuated alternating quantity. Apart from a voltage drop across R due to any direct-current flow, the presence of C and R have no significant effect upon the dc input voltage. However, the capacitor offers a definite impedance ($X_C = 1/\omega C$) to alternating input voltages, and this impedance decreases as the input frequency increases. Thus, as illustrated in Figure 20-17(b), the ac input is potentially divided, or attenuated, by the resistance and the capacitive impedance. The higher the ac frequency, the lower the capacitive impedance, and consequently the greater the attenuation.

The filter attenuation is easily calculated for a given input frequency: input current

$$|I| = \frac{V_i}{\sqrt{R^2 + X_C^2}}$$

and ac output voltage is

$$|V_o| = I \times X_C$$

$$V_o = \frac{V_i \times X_C}{\sqrt{R^2 + X_C^2}}$$

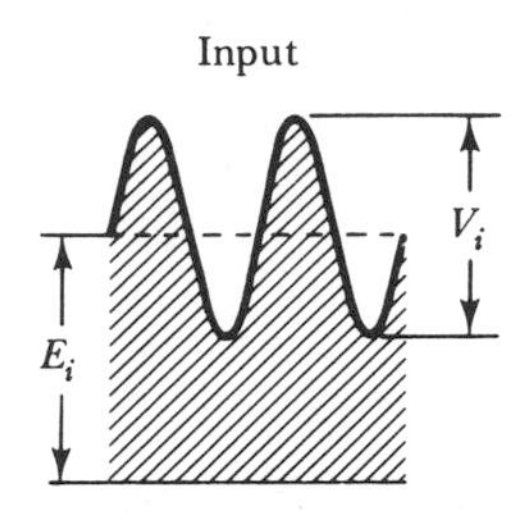

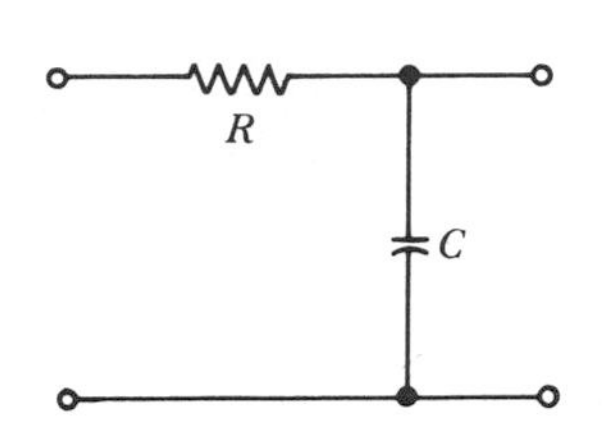

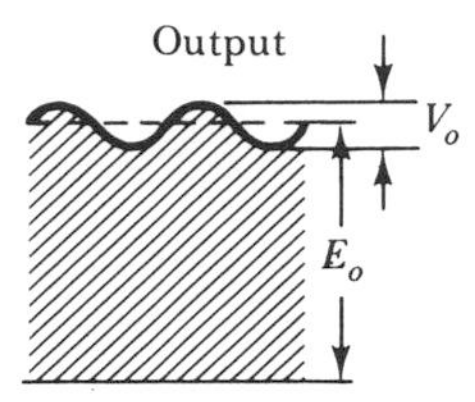

(a) *CR* filter low-pass filter circuit

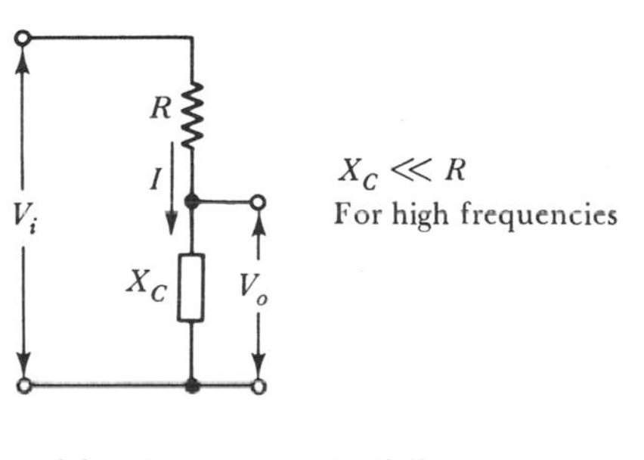

(b) ac inputs are potentially divided across R and X_C

FIG. 20-17. *CR* low-pass filter circuit, waveforms, and equivalent circuit

The filter attenuation is

$$\boxed{\frac{V_o}{V_i} = \frac{X_C}{\sqrt{R^2 + X_C^2}}} \tag{20-41}$$

EXAMPLE 20-11 The low-pass filter in Figure 20-17(a) has $R = 50\ \Omega$ and $C = 200\ \mu\text{F}$. The ac input voltage is $V_i = 20$ V, and the input frequency is 100 Hz. Calculate the level of the alternating output voltage.

SOLUTION

$$X_C = \frac{1}{2\pi fC} = \frac{1}{2\pi \times 100\ \text{Hz} \times 200\ \mu\text{F}}$$

$$X_C \cong 7.96\ \Omega$$

From Eq. (20-41),

$$V_o = \frac{20\ \text{V} \times 7.96\ \Omega}{\sqrt{50^2 + 7.96^2}}$$

$$\mathbf{V_o = 3.14\ V}$$

Figure 20-18(a) shows the circuit of a low-pass filter which uses a series-connected inductance L in place of the resistance R. The input is shown as a low-frequency sine wave with a low-amplitude high-frequency ac component superimposed. The inductive reactance of L is much higher for the high-frequency input than for the low-frequency input, and the capacitive reactance is much lower for the high frequency than for the low frequency. The combined effect is that both ac quantities are attenuated by X_L and X_C [see Figure 20-18(b)], with the high-frequency component suffering the greatest attenuation. For the kind of input waveform shown in Figure 20-18(a), the unwanted high-frequency component is sometimes referred to as *noise*, while the wanted voltage is

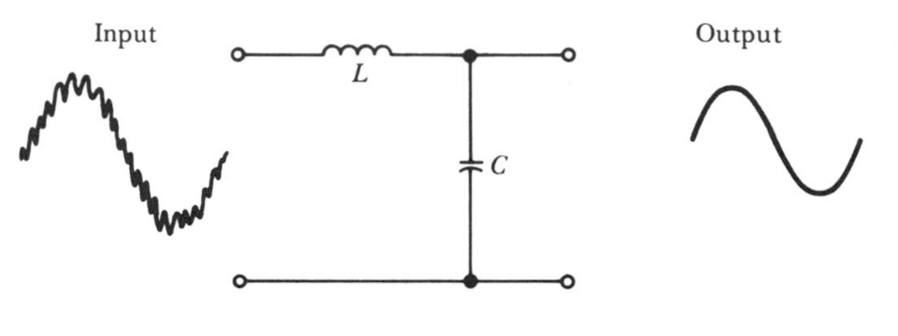

(a) LC low-pass filter circuit

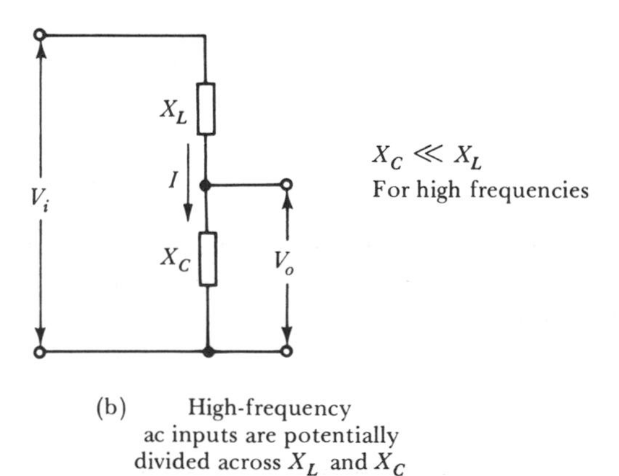

(b) High-frequency ac inputs are potentially divided across X_L and X_C

FIG. 20-18. *LC* low-pass filter circuit, waveforms, and equivalent circuit

termed the *signal*. The *signal-to-noise ratio* is then an important quantity. At the output of a filter, the signal-to-noise ratio should be much greater than that at the input. The type of waveform shown in Figure 20-17(a) could also be filtered by the circuit in Figure 20-18(a).

Assuming that the resistive component of the inductance is much smaller than the inductive reactance, the input current to the LC filter is

$$|I| = \frac{V_i}{X_L - X_C}$$

and the output voltage is

$$|V_o| = I \times X_C$$

$$V_o = \frac{V_i \times X_C}{X_L - X_C}$$

The filter attenuation is

$$\boxed{\frac{V_o}{V_i} = \frac{X_C}{X_L - X_C}} \qquad (20\text{-}42)$$

EXAMPLE 20-12

A low-pass filter has $L = 20$ mH and $C = 0.12\ \mu$F. The input signal amplitude is 2 V peak to peak, and its frequency is 5 kHz. An unwanted noise input is also present with an amplitude of 0.2 V peak to peak and a frequency of 50 kHz. Calculate the signal-to-noise ratio at the input and at the output.

SOLUTION

The input signal-to-noise ratio is

$$\frac{S_i}{N_i} = \frac{2\ \text{V}}{0.2\ \text{V}} = 10$$

At 5 kHz,

$$X_L = 2\pi fL = 2\pi \times 5\ \text{kHz} \times 20\ \text{mH}$$

$$= 628\ \Omega$$

$$X_C = \frac{1}{2\pi fC} = \frac{1}{2\pi \times 5\ \text{kHz} \times 0.12\ \mu\text{F}} = 265\ \Omega$$

From Eq. (20-42) *the signal output is*

$$S_o = \frac{S_i \times X_C}{X_L - X_C} = \frac{2\ \text{V} \times 265\ \Omega}{628\ \Omega - 265\ \Omega}$$

$$\mathbf{S_o = 1.46\ V\ peak\ to\ peak}$$

At 50 kHz,

$$X_L = 2\pi fL = 2\pi \times 50\ \text{kHz} \times 20\ \text{mH}$$

$$= 6.28\ \text{k}\Omega$$

$$X_C = \frac{1}{2\pi fC} = \frac{1}{2\pi \times 50\ \text{kHz} \times 0.12\ \mu\text{F}}$$

$$= 26.5\ \Omega$$

Using Eq. (20-42) *the noise output is*

$$N_o = \frac{N_i \times X_C}{X_L - X_C}$$

$$= \frac{0.2\ \text{V} \times 26.5\ \Omega}{6.28\ \text{k}\Omega - 26.5\ \Omega}$$

$$\mathbf{N_o \cong 0.85\ mV\ peak\ to\ peak}$$

The output signal-to-noise ratio is

$$\mathbf{\frac{S_o}{N_o} = \frac{1.46\ V}{0.85\ mV} \cong 1700}$$

In Example 20-12, the input signal amplitude is just 10 times the unwanted noise voltage amplitude. The output signal is seen to be 1700 times the output noise.

The filter circuits shown in Figure 20-19(a) and (b) are slightly more complicated than the simple circuit considered in Example 20-12. Since additional reactive components are employed, these circuits provide greater attenuation of the unwanted frequency. The *T filter* shown in Figure 20-19(a) is most suited to signal sources that have a low source impedance Z_S. The presence of L_1 increases the effective source impedance, so the high-frequency inputs are attenuated by the combined effect of X_{L1} and X_C. Further attenuation is then provided by X_{L2} and R_L. For signals that have a high source impedance, the π*-filter* circuit shown in

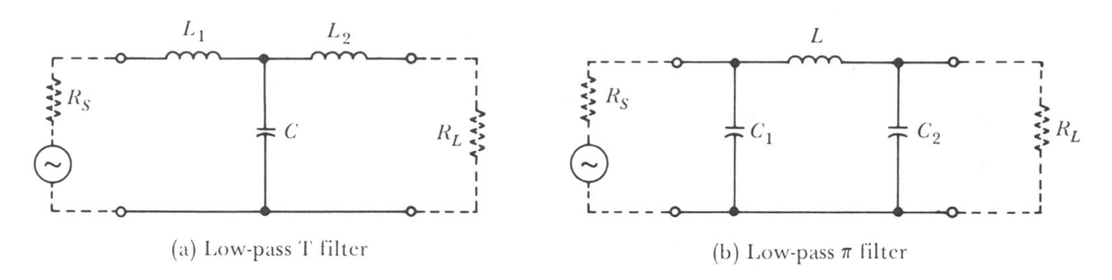

FIG. 20-19. Low-pass T and π filter circuits

Figure 20-19(b) gives the best performance. A first stage of attenuation is provided by R_S and X_{C1} and then further attenuation occurs across X_L and X_{C2}.

20-10 HIGH-PASS FILTERS

The function of a *high-pass filter* is the converse of that of a low-pass filter. A high-pass filter passes high-frequency input voltages, and blocks low frequencies and dc quantities. Usually, direct voltages are completely blocked, while low-frequency ac components are attenuated. Figure 20-20(a) shows the circuit of a *CR* high-pass filter, with a combined dc and ac input. The presence of the series-connected capacitor prevents the direct voltage input from passing to the output. The ac input is potentially divided across X_C and R [see Figure 20-20(b)], so the ac output is an attenuated version of the input. Where a large enough capacitor is employed, the ac attenuation may be negligible.

In the circuit of Figure 20-20(a), capacitor C is sometimes termed a *coupling capacitor*. The size of the capacitor can be determined by deciding upon a capacitive impedance that will give an acceptable attenuation at the lowest frequency to be passed. For example, if an output amplitude of 90% of V_i is acceptable, the relative values of R and X_C should allow 10% of V_i to be developed across C, and 90% to appear across R. Then,

$$X_C \cong \frac{R}{9}$$

or

$$\frac{1}{2\pi fC} \cong \frac{R}{9}$$

giving

$$\boxed{C \cong \frac{9}{2\pi fR}} \qquad \textbf{(20-43)}$$

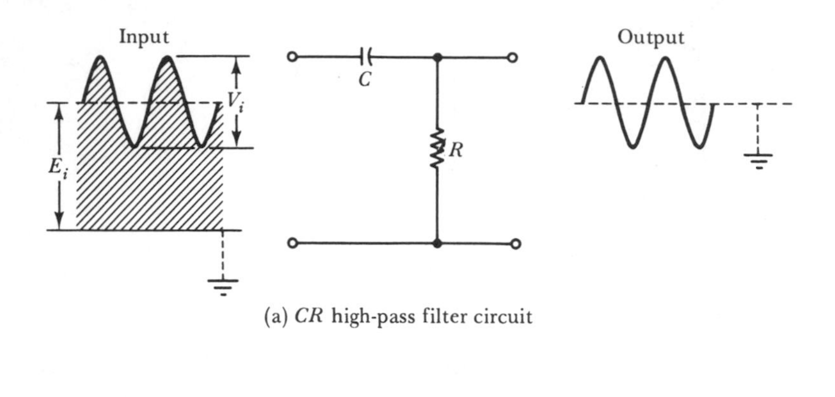

(a) CR high-pass filter circuit

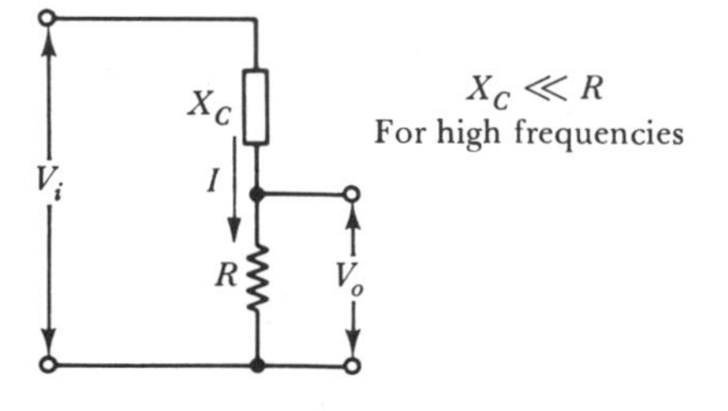

(b) ac inputs are only slightly affected by potential division across X_C and R

FIG. 20-20. *CR* high-pass filter circuit, waveforms, and equivalent circuit

The actual attenuation of the input signal can be calculated in a similar way as for the low-pass filter;

$$|I| = \frac{V_i}{\sqrt{R^2 + X_C^2}}$$

$$V_o = IR$$

$$= \frac{V_i R}{\sqrt{R^2 + X_C^2}}$$

Therefore,

$$\boxed{\frac{V_o}{V_i} = \frac{R}{\sqrt{R^2 + X_C^2}}} \tag{20-44}$$

EXAMPLE 20-13

A high-pass CR filter has a resistance of $R = 2\ \text{k}\Omega$. The lowest input frequency to be passed is 7.5 kHz. Calculate the value of a suitable coupling capacitor. Also determine the attenuation of the filter for 60-Hz frequencies.

SOLUTION

Allowing a 10% *attenuation on the* 7.5 kHz *signal:*
Eq. (20-43):

$$C \cong \frac{9}{2\pi f R}$$

$$\cong \frac{9}{2\pi \times 7.5\ \text{kHz} \times 2\ \text{k}\Omega}$$

$$\mathbf{C \cong 0.1\ \mu F}$$

For $f = 60$ Hz,

$$X_C = \frac{1}{2\pi \times 60\ \text{Hz} \times 0.1\ \mu\text{F}}$$

$$= 26.5\ \text{k}\Omega$$

From Eq. (20-44), *the attenuation at* 60 Hz *is*

$$\frac{V_o}{V_i} = \frac{2\ \text{k}\Omega}{\sqrt{(2\ \text{k}\Omega)^2 + (26.5\ \text{k}\Omega)^2}}$$

$$\mathbf{\frac{V_o}{V_i} = 7.5 \times 10^{-2}}$$

The LC high-pass filter circuit in Figure 20-21(a) substitutes an inductance L in place of the resistance in the CR circuit. Since the inductance offers a low impedance to low frequencies, and the capacitance offers a high impedance, X_L is much smaller than X_C [see Figure 20-21(b)]. Therefore, low-frequency inputs are attenuated. With high frequency, however, the inductance has a high impedance and the capacitor impedance is low. Consequently, high-frequency input voltages experience very little attenuation. The input waveform shown in Figure 20-21(a) is a high-frequency wanted signal combined with an unwanted low-frequency component. If the filter does its job, the low frequency is severely attenuated, while the high-frequency signal is passed to the output.

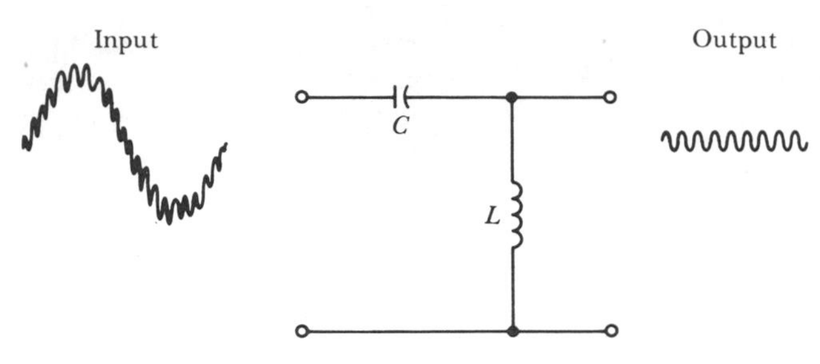

(a) LC high-pass filter circuit

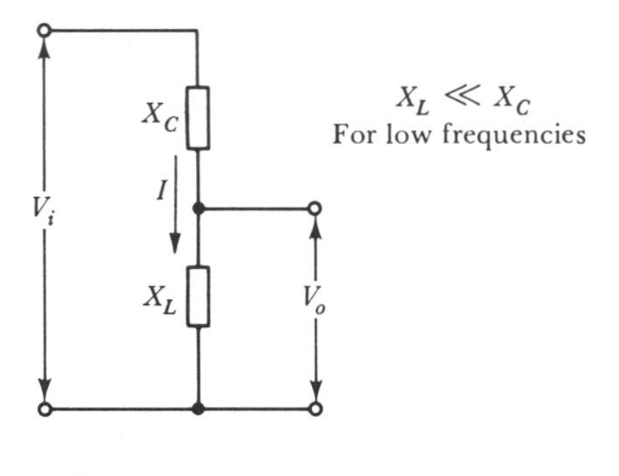

$X_L \ll X_C$
For low frequencies

(b) Low-frequency ac inputs are potentially divided across X_L and X_C

FIG. 20-21. *LC* high-pass filter circuit, waveforms, and equivalent circuit

As before, the attenuation of the filter is easily calculated:

$$\text{input current, } |I| = \frac{V_i}{X_C - X_L}$$

$$V_o = IX_L$$

$$= \frac{V_i X_L}{X_C - X_L}$$

$$\boxed{\textit{attenuation, } \frac{V_o}{V_i} = \frac{X_L}{X_C - X_L}} \qquad \textbf{(20-45)}$$

EXAMPLE 20-14 The high-pass filter circuit shown in Figure 20-21(a) has $C = 1\ \mu\text{F}$ and $L = 0.47$ mH. A 5-kHz input signal is applied, with a 60-Hz unwanted noise frequency combined with the signal. If the signal amplitude is 2 V peak to peak, and the 60-Hz frequency is 10 V peak to peak at the input, calculate the output amplitudes of each.

SOLUTION

At 5 kHz,

$$X_L = 2\pi fL = 2\pi \times 5\ \text{kHz} \times 0.47\ \text{mH}$$

$$X_L = 14.8\ \Omega$$

$$X_C = \frac{1}{2\pi fC} = \frac{1}{2\pi \times 5\ \text{kHz} \times 1\ \mu\text{F}}$$

$$\mathbf{X_C = 31.8\ \Omega}$$

From Eq. (20-45):
5-kHz *signal output*:

$$S_o = \frac{S_i X_L}{X_C - X_L}$$

$$= \frac{2\ \text{V} \times 14.8\ \Omega}{31.8\ \Omega - 14.8\ \Omega}$$

$$\mathbf{S_o = 1.74\ V}$$

At 60 Hz,

$$X_L = 2\pi fL = 2\pi \times 60\ \text{Hz} \times 0.47\ \text{mH}$$

$$\mathbf{X_L = 0.177\ \Omega}$$

$$X_C = \frac{1}{2\pi fC} = \frac{1}{2\pi \times 60\ \text{Hz} \times 1\ \mu\text{F}}$$

$$\mathbf{X_C = 2.65\ k\Omega}$$

From Eq. (20-44):
60-Hz *noise output*:

$$N_o = \frac{N_i X_L}{X_C - X_L}$$

$$= \frac{10\ \text{V} \times 0.177\ \Omega}{2.65\ \text{k}\Omega - 0.177\ \Omega}$$

$$\mathbf{N_o = 668\ \mu V}$$

High-pass T- and π-filter circuits are shown in Figure 20-22. As in the case of the low-pass T and π filters, the circuit that should be used

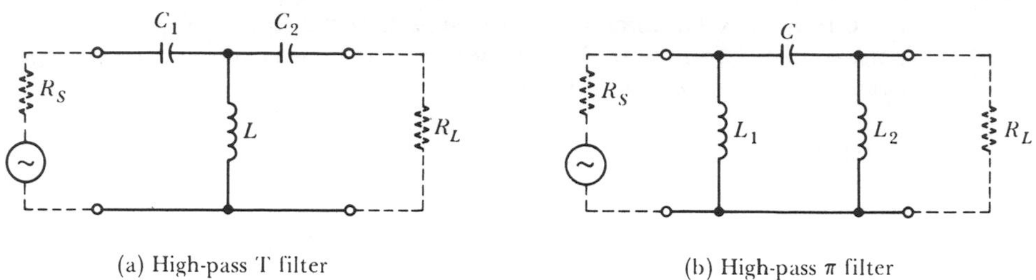

FIG. 20-22. High-pass T and π filter circuits

depends upon whether the signal source has a high impedance or a low impedance. With low-impedance signal sources, the circuit of Figure 20-22(a) is most suitable. Unwanted low-frequency inputs are initially attenuated by X_{C1} and X_L, and then further attenuated by X_{C2} and R_L. For high-impedance sources, Figure 20-22(b) gives the greatest attenuation. X_{L1} combines with R_S to attenuate the unwanted inputs, then X_C and X_{L2} produce further attenuation.

GLOSSARY OF FORMULAS

Inductive reactance:

$$X_L = \omega L = 2\pi f L$$

Inductive susceptance:

$$B_L = \frac{1}{X_L}$$

Capacitive reactance:

$$X_C = \frac{1}{\omega C} = \frac{1}{2\pi f C}$$

Capacitive susceptance:

$$B_C = \frac{1}{X_C}$$

For LR series circuit:

$$E = \sqrt{V_R^2 + V_L^2} \;\angle \arctan \frac{V_L}{V_R}$$

$$Z = \sqrt{R^2 + X_L^2} \;\angle \arctan \frac{X_L}{R}$$

For CR series circuit:

$$E=\sqrt{V_R^2+V_C^2}\ \angle \arctan \frac{V_C}{V_R}$$

$$Z=\sqrt{R^2+X_C^2}\ \angle \arctan \frac{X_C}{R}$$

For RLC series circuit:

$$E=\sqrt{V_R^2+V_X^2}\ \angle \arctan \frac{V_X}{V_R}$$

$$Z=\sqrt{R^2+X_{eq}^2}\ \angle \arctan \frac{X_{eq}}{R}$$

For RLC parallel circuit:

$$I=\sqrt{i_R^2+i_X^2}\ \angle \arctan \frac{i_X}{i_R}$$

Admittance:

$$Y=\frac{1}{Z}$$

$$Y=G+j(B_C-B_L)$$

$$Y=\sqrt{G^2+B_{eq}^2}\ \angle \arctan \frac{B_{eq}}{G}$$

Attenuation of low-pass CR filter:

$$\frac{V_o}{V_i}=\frac{X_C}{\sqrt{R^2+X_C^2}}$$

Attenuation of low-pass LC filter:

$$\frac{V_o}{V_i}=\frac{X_C}{X_L-X_C}$$

Coupling capacitor:

$$C=\frac{9}{2\pi fR}$$

Attenuation of high-pass CR filters:

$$\frac{V_o}{V_i}=\frac{R}{\sqrt{R^2+X_C^2}}$$

Attenuation of high-pass LC filters:

$$\frac{V_o}{V_i}=\frac{X_L}{X_C-X_L}$$

REVIEW QUESTIONS

20-1 Define:inductive reactance, inductive susceptance, capacitive reactance, capacitive susceptance, impedance, admittance, impedance diagram, and admittance diagram.

20-2 Define low-pass filter, high-pass filter, noise, signal-to-noise ratio, T filter, π filter, and coupling capacitor.

20-3 For an inductance with an alternating voltage supply, sketch the waveforms of current, rate of change of current, induced voltage, and supply voltage. Show the phase relationships between each of the waveforms, and carefully explain the origin of the relationships.

20-4 Sketch the phasor diagram and the waveforms of terminal voltage and current for an inductance with an ac supply. Also write equations for instantaneous current. Explain briefly.

20-5 Using the waveforms sketched for Review Question 20-3, develop equations for $\Delta i/\Delta t$, e_L, E_{av}, and X_L. Define X_L and B_L.

20-6 For a capacitor with an alternating voltage supply, sketch the waveforms of supply voltage, rate of change of capacitor voltage, and instantaneous capacitor current. Show the phase relationships between each waveform and carefully explain the origin of the relationships.

20-7 Sketch the phasor diagram and the waveforms of terminal voltage and current, for a capacitor with an ac supply. Also write equations for instantaneous supply voltage and instantaneous current. Explain briefly.

20-8 Using the waveforms sketched for Review Question 20-6, develop an equation for X_C. Define X_C and B_C.

20-9 For a series *RL* circuit with an ac supply, sketch typical waveforms of current, resistor voltage, inductor voltage, and supply voltage. Explain the phase relationships between the waveforms.

20-10 Sketch a typical phasor diagram for an *RL* series circuit, and briefly explain. Also, in rectangular and polar form, write the equations for supply voltage and for the circuit impedance.

20-11 For a series *RC* circuit with an ac supply, sketch typical waveforms of current, resistor voltage, capacitor voltage, and supply voltage.

20-12 Sketch a typical phasor diagram for an *RC* circuit and explain briefly. Also, in rectangular and polar form, write the equations for supply voltage and for circuit impedance.

20-13 For a series *RLC* circuit, sketch typical phasor diagrams for the case of:

a. $X_C > X_L$.
b. $X_L > X_C$.

Explain briefly.

20-14 For a series *RLC* circuit sketch typical impedance diagrams for:

a. $X_C > X_L$.
b. $X_L > X_C$.

Explain briefly. Also, in polar and rectangular form, write equations for supply voltage and for circuit impedance.

20-15 For a parallel *RLC* circuit, derive rectangular form expressions for the circuit admittance. Also, write an equation for the circuit impedance.

20-16 Sketch typical phasor and admittance diagrams for a parallel *RLC* circuit. Explain briefly.

20-17 Sketch the circuits of low-pass filters:

a. Using a capacitor and resistor.
b. Using a capacitor and inductor.

Explain briefly how each circuit operates.

20-18 Sketch the circuits of high-pass filters:

a. Using a capacitor and resistor.
b. Using a capacitor and inductor.

Explain briefly how each circuit operates.

20-19 Derive equations for the attenuation of a *CR* high-pass filter and for an *LC* high-pass filter. Also derive the equations for *CR* and *LC* low-pass filters.

PROBLEMS

20-1 A 100-V 250-Hz supply is applied to a 300-mH inductor. Calculate the inductive reactance and the current that flows. Sketch the voltage and current waveforms and draw a phasor diagram for the circuit.

20-2 An inductor with a value of 100 μH passes a current of 10 mA when its terminal voltage is 6.3 V. Calculate the frequency of the ac supply.

20-3 A 100-V 250-Hz supply is applied to a 30-μF capacitor. Calculate the capacitive reactance and the current that flows. Sketch the voltage and current waveforms and draw a phasor diagram for the circuit.

20-4 A 100-μF capacitor passes a current of 0.25 A when its terminal voltage is 4 V. Determine the frequency of the ac supply.

20-5 A series *RL* circuit has $R=68\ \Omega$, $L=10$ mH. The supply voltage is 33 V with a frequency of 1 kHz. Determine the circuit current, the resistor voltage, the inductor voltage, and the phase angle of the current with respect to the supply voltage. Sketch a phasor diagram and an impedance diagram for the circuit.

20-6 A 25-V 10-kHz supply is applied to a 10-mH inductor in series with a 1.2-kΩ resistor. The inductor has a winding resistance of 220 Ω. Calculate the circuit current and the inductor terminal voltage. Also, draw a phasor diagram and an impedance diagram for the circuit.

20-7 A series *RC* circuit has $R=120\ \Omega$, $C=3.3\ \mu$F, and an ac supply of 12 V with $f=1$ kHz. Calculate the current, the voltage across R, the voltage across C, and the phase angle of the current with respect to the supply voltage. Sketch a phasor diagram and an impedance diagram for the circuit.

20-8 A 33-V 500-Hz supply is applied to a capacitor in series with 220-Ω resistor. If the circuit current is 66 mA, calculate the value of the capacitor.

20-9 A 20-mH inductor is connected in series with a 2-μF capacitor and a 200-Ω resistor. The supply voltage is 15 V with a frequency of 600 Hz. Calculate I, V_R, V_L, V_C, and the phase angle of the current with respect to the voltage.

20-10 For the circuit described in Problem 20-9, accurately sketch phasor and impedance diagrams.

20-11 A 220-mH inductor is connected in parallel with an 8-μF capacitor and a 330-Ω resistor. The supply is 10 V with a frequency of 100 Hz. Calculate i_R, i_L, i_C, and the current taken from the supply. Also, determine the phase angle between supply voltage and supply current.

20-12 For the circuit described in Problem 20-11, accurately sketch phasor and admittance diagrams.

20-13 A low-pass *CR* filter has $C=100\ \mu$F and $R=100\ \Omega$. Determine the attenuation of the circuit:

a. For a 10-Hz input frequency.

b. For a 250-Hz frequency.

20-14 A low-pass filter has $L=50$ mH and $C=0.2\ \mu$F. A 1-V peak to peak input signal is provided with a frequency of 3 kHz and an unwanted *noise* is also present at the input. The noise has an amplitude of 0.3 V and a frequency of 25 kHz. Determine the output signal and noise amplitudes and the output signal-to-noise ratio.

20-15 A capacitor couples a 500-Hz signal to a resistive load of $R=3$ kΩ. Calculate a suitable capacitor value if the voltage developed across R is to be not less than $0.95 \times V_i$.

20-16 A high-pass filter has a 12-kHz input signal with an unwanted 100-Hz noise voltage. The signal amplitude is 5 V, and the noise has a 100-mV amplitude. If $L=0.1$ mH and $C=0.1$ μF, calculate the signal-to-noise ratio at the filter output.

21

SERIES AND PARALLEL AC CIRCUITS

INTRODUCTION

AC series impedance circuits can be analyzed in a similar way to dc series resistance circuits, as long as the phase angle of each impedance is taken into consideration. For addition, impedances must be in rectangular form, and for division or multiplication, it is convenient to convert to polar form. The ac voltage divider can also be treated in a similar way to the dc voltage divider as long as the phase angles are correctly considered. Parallel impedances should be converted into admittances and added together; then the result may be inverted to determine the equivalent total impedance. In dealing with series–parallel impedance combinations, each parallel group of impedances should be resolved into a single equivalent impedance, so that the entire configuration reduces to a simple series impedance circuit. For every simple series impedance circuit there is an equivalent parallel impedance circuit, and vice versa. The formula for conversion from one to the other is sometimes very useful.

21-1 SERIES IMPEDANCES

Impedances in series in an ac circuit can be treated similarly to resistances in series in a dc circuit. The equivalent total impedance is determined, and using the applied voltage, the current through the circuit can be calculated. However, the series impedances cannot be added directly, but must be resolved into rectangular form; then the real and imaginary components are added separately. Figure 21-1 illustrates the process. The three series impedances in Figure 21-1(a) are each converted to their rectangular form, Figure 21-1(b). The total values of R, L, and C are then calculated and converted back to polar form, to give the equivalent circuit of Figure 21-1(c).

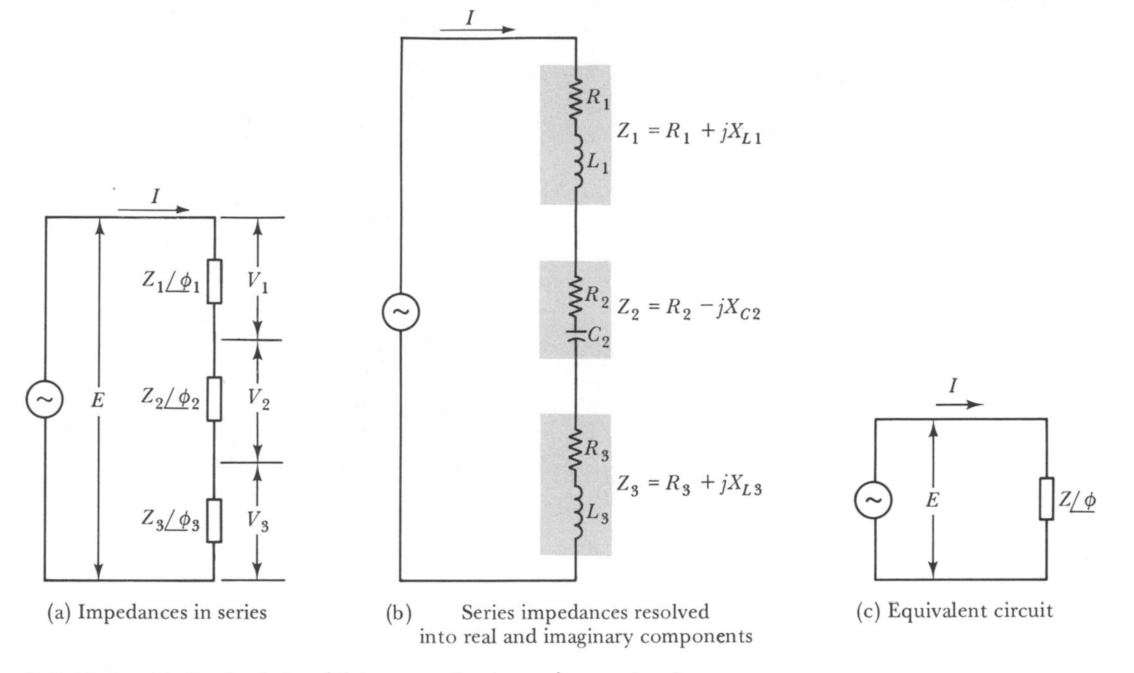

(a) Impedances in series (b) Series impedances resolved into real and imaginary components (c) Equivalent circuit

FIG. 21-1. Method of simplifying a series impedance circuit

EXAMPLE 21-1 The three impedances shown in Figure 21-1 are $Z_1 = 70.7\ \Omega\ \angle 45°$, $Z_2 = 92.4\ \Omega\ \angle 330°$, $Z_3 = 67\ \Omega\ \angle 60°$, and the supply voltage is 100 V. Analyze the circuit to determine the current taken from the supply. Also, draw a phasor diagram of supply voltage and current.

SOLUTION

Converting to rectangular form:
From Eq. (19-1):

$$Z_1 = 70.7 \cos 45° + j70.7 \sin 45°$$
$$= (50 + j50)\ \Omega$$
$$Z_2 = 92.4 \cos 330° + j92.4 \sin 330°$$
$$= (80 - j46.2)\ \Omega$$
$$Z_3 = 67 \cos 60° + j67 \sin 60°$$
$$= (33.5 + j58)\ \Omega$$

Adding impedances in series:

$$Z = Z_1 + Z_2 + Z_3$$
$$= (50 + 80 + 33.5) + j(50 - 46.2 + 58)$$
$$= (163.5 + j61.8)\ \Omega$$

Converting to polar form:
From Eq. (20-18):

$$Z = \sqrt{163.5^2 + 61.8^2} \;\angle \arctan \frac{61.8}{163.5}$$

$$= 174.8\ \Omega \;\angle 20.7^\circ$$

Determining the current:

$$I = \frac{E}{Z}$$

$$= \frac{100\ \text{V}}{174.8\ \Omega \;\angle 20.7^\circ}$$

$$\boldsymbol{I = 572\ \text{mA} \;\angle -20.7^\circ}$$

The current is 572 mA, *lagging the supply voltage by* 20.7° (*see* Figure 21-2).

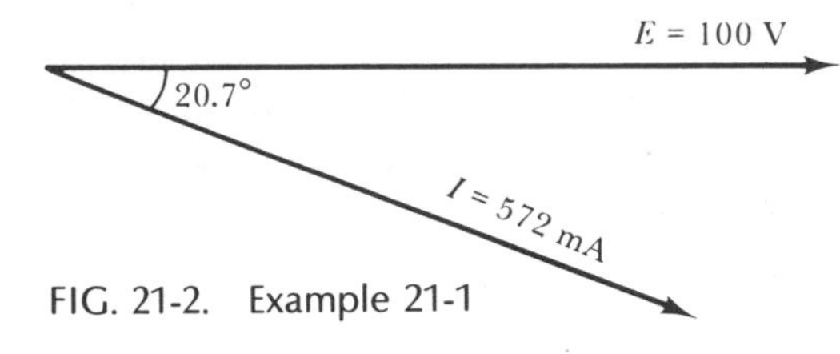

FIG. 21-2. Example 21-1

21-2 AC VOLTAGE DIVIDER

The voltage divider rule as applied to series resistors in a dc circuit can also be applied to series impedances in an ac circuit. As always in an ac circuit, care must be taken to ensure that phase angles are taken into consideration. Referring again to Figure 21-1(a), it is seen that the voltage across Z_3 is

$$V_3 = IZ_3$$

Also, as already shown, the current through Z_1, Z_2, and Z_3 is

$$I = \frac{E}{Z_1 + Z_2 + Z_3}$$

Therefore,

$$\boxed{\boldsymbol{V_3 = \frac{EZ_3}{Z_1 + Z_2 + Z_3}}} \tag{21-1}$$

Similarly,

$$V_2 = \frac{EZ_2}{Z_1 + Z_2 + Z_3}$$

and

$$V_1 = \frac{EZ_1}{Z_1 + Z_2 + Z_3}$$

EXAMPLE 21-2

For the circuit described in Example 21-1, determine the values of V_1, V_2, and V_3, and draw a complete phasor diagram for the circuit voltages.

SOLUTION

From Example 21-1,

$$Z_1 + Z_2 + Z_3 = 174.8\ \Omega\ \angle 20.7^\circ$$

From Eq. (21-1),

$$V_3 = \frac{100\text{ V }(67\ \Omega\angle 60^\circ)}{174.8\ \Omega\ \angle 20.7^\circ}$$

$$= \frac{100\text{ V} \times 67\ \Omega}{174.8\ \Omega}\ \angle(60^\circ - 20.7^\circ)$$

$$\mathbf{V_3 = 38.3\ V\ \angle 39.3^\circ}$$

$$V_2 = \frac{100\text{ V }(92.4\ \Omega\angle 330^\circ)}{174.8\ \Omega\ \angle 20.7^\circ}$$

$$\mathbf{V_2 = 52.9\ V\ \angle 309.3^\circ}$$

$$V_1 = \frac{100\text{ V }(70.7\ \Omega\ \angle 45^\circ)}{174.8\ \Omega\ \angle 20.7^\circ}$$

$$\mathbf{V_1 = 40.4\ V\ \angle 24.3^\circ}$$

The voltage phasor diagram is as shown in Figure 21-3. *Note that the phasor sum of V_1, V_2, and V_3 gives the supply voltage E.*

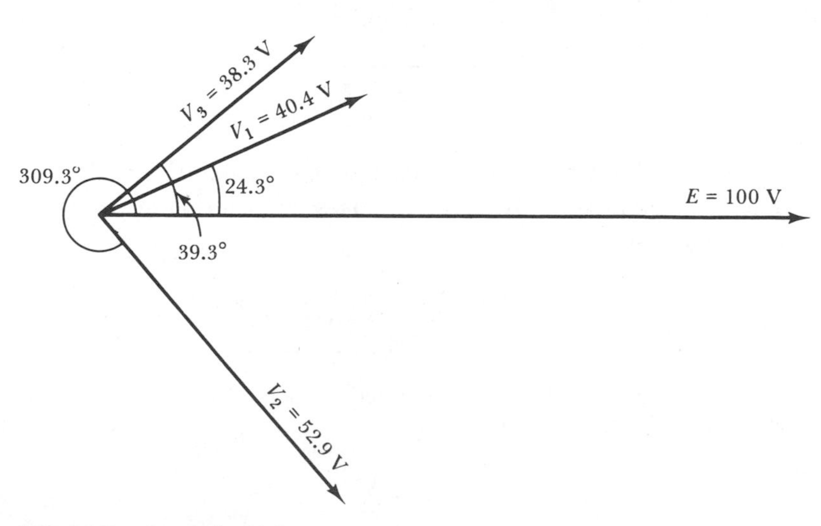

FIG. 21-3. Example 21-2

21-3 IMPEDANCES IN PARALLEL

As discussed in Section 20-8, impedances in parallel must be converted into admittances in order to determine the total circuit admittance. This can then be inverted to give the total impedance of the circuit. Once the total impedance is known, the current taken from the supply can be determined simply by dividing the impedance into the supply voltage. Alternatively, the supply current can be found by first calculating the individual branch currents in rectangular form, and then adding the real and imaginary components.

As illustrated in Figure 21-4, the total circuit impedance is determined in a series of steps:

1. Operating in polar form each individual impedance is converted into an admittance [see Figure 21-4(b)], simply by using Eq. (20-19):

$$Y=\frac{1}{Z}$$

2. The individual admittances are resolved into rectangular form [see Figure 21-4(c)] using Eq. (20-37):

$$Y=G+jB_{eq}$$

or

$$\boxed{\boldsymbol{Y=Y\cos\theta+jY\sin\theta}} \qquad \textbf{(21-2)}$$

3. The total parallel admittance is found as

$$\boxed{\boldsymbol{Y=Y_1+Y_2+Y_3}} \qquad \textbf{(21-3)}$$

where the admittances are in rectangular form [see Figure 21-4(d)].

4. The total admittance is converted back to polar form, using Eq. (20-40) [Figure 21-4(e)]:

$$Y=\sqrt{G^2+B_{eq}^2}\ \Big/\ \arctan\frac{B_{eq}}{G}$$

5. The polar-form admittance is inverted, to determine the total circuit impedance [Figure 21-4(f)]:

$$Z=\frac{1}{Y}$$

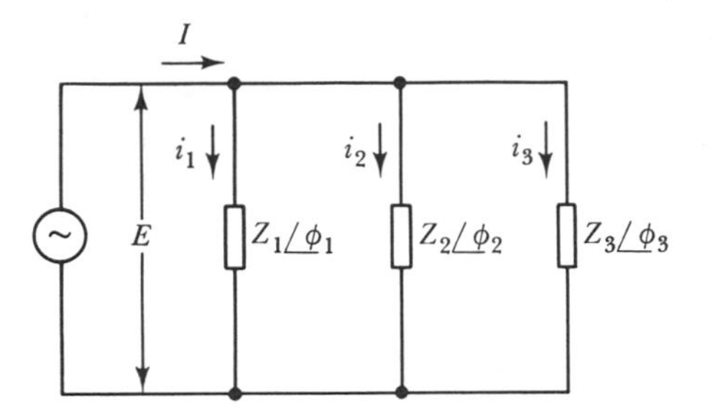

(a) Parallel impedance circuit

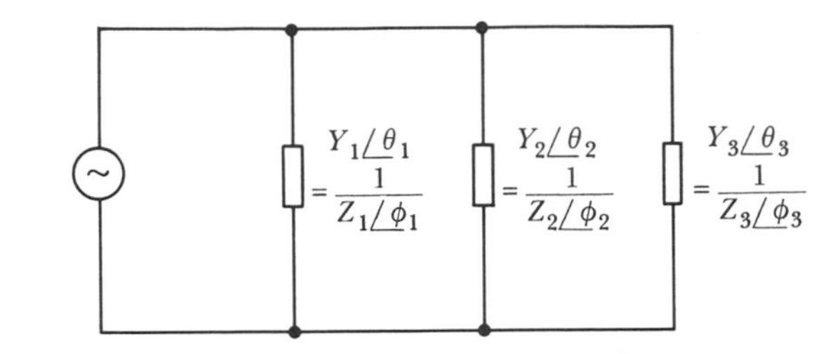

(b) Impedances converted to admittances

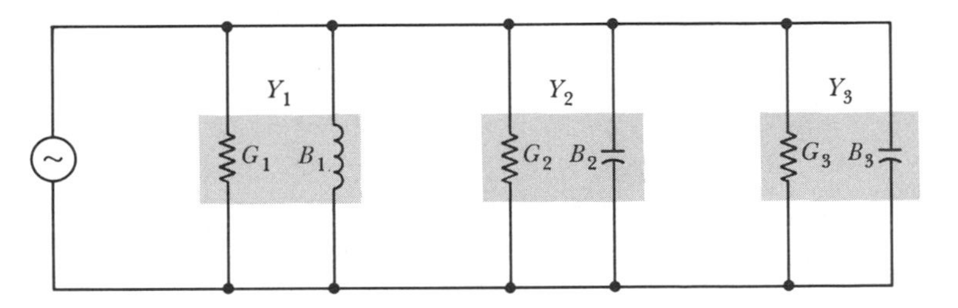

(c) Admittances resolved into conductances and susceptances

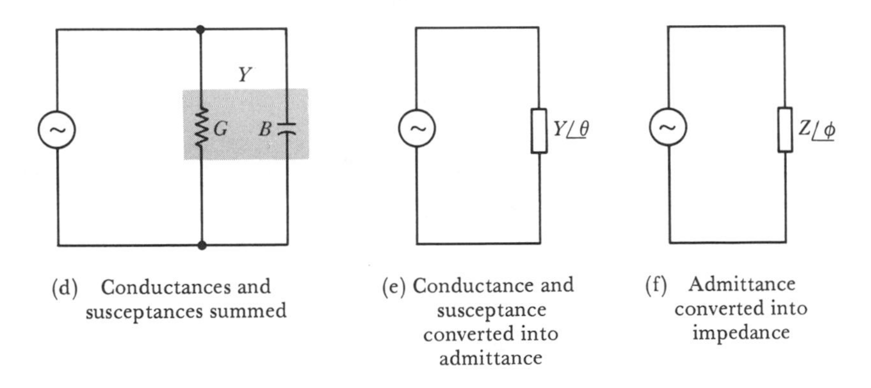

(d) Conductances and susceptances summed

(e) Conductance and susceptance converted into admittance

(f) Admittance converted into impedance

FIG. 21-4. Simplifying a parallel impedance circuit

EXAMPLE 21-3 Three impedances in parallel consist of $Z_1 = 1606\ \Omega\ \angle 51^\circ$, $Z_2 = 977\ \Omega\ \angle -33^\circ$, and $Z_3 = 953\ \Omega\ \angle -19^\circ$. If the supply voltage is 33 V, calculate the total circuit impedance and the current taken from the supply.

SOLUTION

Refer to Figure 21-4.

Determining individual admittances:
From Eq. (20-19):

$$Y_1 = \frac{1}{Z_1}$$

$$= \frac{1}{1606\ \Omega \angle 51^\circ}$$

$$\mathbf{Y_1 = 622.7\ \mu S \angle -51^\circ}$$

$$Y_2 = \frac{1}{977\ \Omega \angle -33^\circ}$$

$$\mathbf{Y_2 = 1.02\ mS \angle 33^\circ}$$

$$Y_3 = \frac{1}{953\ \Omega \angle -19^\circ}$$

$$\mathbf{Y_3 = 1.05\ mS \angle 19^\circ}$$

converting to rectangular form:
From Eq. (21-2)

$$Y_1 = Y_1 \cos \phi_1 + jY_1 \sin \phi_1$$

$$Y_1 = 622.7\ \mu S[\cos(-51^\circ) + j\sin(-51^\circ)]$$

$$\mathbf{Y_1 = 392\ \mu S - j484\ \mu S}$$

$$Y_2 = 1.02\ mS(\cos 33^\circ + j\sin 33^\circ)$$

$$\mathbf{Y_2 = 855\ \mu S + j556\ \mu S}$$

$$Y_3 = 1.05\ mS(\cos 19^\circ + j\sin 19^\circ)$$

$$\mathbf{Y_3 = 993\ \mu S + j342\ \mu S}$$

Adding admittances:

$$Y = Y_1 + Y_2 + Y_3$$

$$= (392 + 855 + 993)\ \mu S + j(-484 + 556 + 342)\ \mu S$$

$$\mathbf{Y = 2.24\ mS + j0.414\ mS}$$

Converting back to polar form:
From Eq. (20-40):

$$Y = \sqrt{G^2 + B_{eq}^2} \angle \arctan \frac{B_{eq}}{G}$$

$$= \sqrt{(2.24\ mS)^2 + (0.414\ mS)^2} \angle \arctan \frac{0.414\ mS}{2.24\ mS}$$

$$\mathbf{Y = 2.28\ mS \angle 10.5^\circ}$$

Total impedance:

$$Z=\frac{1}{Y}$$

$$=\frac{1}{2.28\text{ mS}\angle 10.5^\circ}$$

$$\mathbf{Z=439\ \Omega\angle -10.5^\circ}$$

Supply current:

$$I=\frac{33\text{ V}}{439\ \Omega\angle -10.5^\circ}$$

$$\mathbf{I=75.2\text{ mA}\angle 10.5^\circ}$$

EXAMPLE 21-4 For the parallel impedance circuit in Example 21-3, determine the individual branch currents, and the total supply current.

SOLUTION

Refer to Figure 21-4.
Branch currents:

$$i_1=\frac{E}{Z_1}=\frac{33\text{ V}}{1606\ \Omega\angle 51^\circ}$$

$$\mathbf{i_1=20.5\text{ mA}\angle -51^\circ}$$

$$i_2=\frac{33\text{ V}}{977\ \Omega\angle -33^\circ}$$

$$\mathbf{i_2=33.8\text{ mA}\angle 33^\circ}$$

$$i_3=\frac{33\text{ V}}{953\ \Omega\angle -19^\circ}$$

$$\mathbf{i_3=34.6\text{ mA}\angle 19^\circ}$$

Converting to rectangular form:

$$i_1=i_1\cos\phi+ji_1\sin\phi$$

$$=20.5\text{ mA}[\cos(-51^\circ)+j\sin(-51^\circ)]$$

$$\mathbf{i_1=12.9\text{ mA}-j15.9\text{ mA}}$$

$$i_2=33.8\text{ mA}(\cos 33^\circ+j\sin 33^\circ)$$

$$\mathbf{i_2=28.3\text{ mA}+j18.4\text{ mA}}$$

$$i_3=34.6\text{ mA}(\cos 19^\circ+j\sin 19^\circ)$$

$$\mathbf{i_3=32.7\text{ mA}+j11.3\text{ mA}}$$

Adding the currents:

$$I = i_1 + i_2 + i_3$$
$$= (12.9 + 28.3 + 32.7)\ \text{mA} + j(-15.9 + 18.4 + 11.3)\ \text{mA}$$
$$I = 73.9\ \text{mA} + j13.8\ \text{mA}$$

Converting back to polar form:

$$I = \sqrt{(73.9\ \text{mA})^2 + (13.8\ \text{mA})^2}\ \angle\ \arctan\frac{13.8\ \text{mA}}{73.9\ \text{mA}}$$
$$\mathbf{I = 75.2\ mA\ \angle\ 10.6°}$$

21-4 AC CURRENT DIVIDER

The current divider rule can be applied to ac impedances in parallel as well as to parallel resistors in a dc circuit. Referring to Figure 21-5, the equation for branch currents developed in Section 8-4 is rewritten:

$$\boxed{i_1 = I\frac{Z_2}{Z_1 + Z_2}} \quad \textbf{(21-4)}$$

and

$$\boxed{i_2 = I\frac{Z_1}{Z_1 + Z_2}} \quad \textbf{(21-5)}$$

In applying these equations, it is necessary to convert the denominator impedances into rectangular form, to add them together. Once Z_1 and Z_2

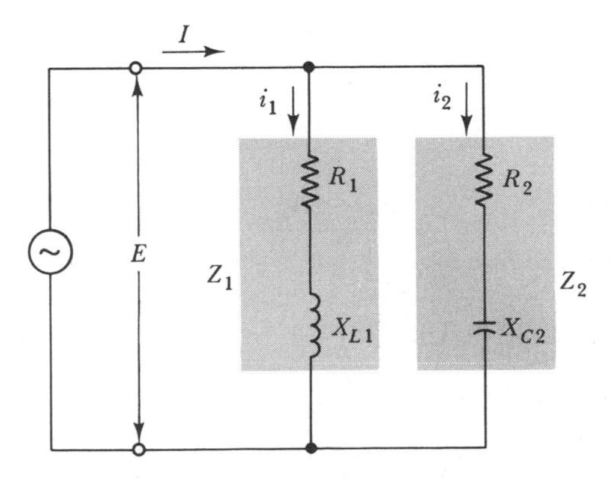

FIG. 21-5. ac current division between two parallel impedances

are added, the denominator should be returned to polar form so that it may be easily divided into the numerator.

EXAMPLE 21-5

The two impedances in the circuit shown in Figure 21-5 are $Z_1 = (1\ \text{k}\Omega + j2.7\ \text{k}\Omega)$ and $Z_2 = (790\ \Omega - j1.6\ \text{k}\Omega)$. The total current taken from the supply is 15 mA. Using the current divider rule, calculate the two branch currents.

SOLUTION

$$Z_1 = \sqrt{(1\ \text{k}\Omega)^2 + (2.7\ \text{k}\Omega)^2}\ \angle \arctan \frac{2.7\ \text{k}\Omega}{1\ \text{k}\Omega}$$

$$\mathbf{Z_1 = 2.88\ k\Omega\ \angle 69.7^\circ}$$

$$Z_2 = \sqrt{(790\ \Omega)^2 + (1.6\ \text{k}\Omega)^2}\ \angle \arctan \frac{-1.6\ \text{k}\Omega}{790\ \Omega}$$

$$\mathbf{Z_2 = 1.78\ k\Omega\ \angle -63.7^\circ}$$

$$\begin{aligned} Z_1 + Z_2 &= (1\ \text{k}\Omega + j2.7\ \text{k}\Omega) + (790\ \Omega - j1.6\ \text{k}\Omega) \\ &= 1.79\ \text{k}\Omega + j1.1\ \text{k}\Omega \\ &= \sqrt{(1.79\ \text{k}\Omega)^2 + (1.1\ \text{k}\Omega)^2}\ \angle \arctan \frac{1.1\ \text{k}\Omega}{1.79\ \text{k}\Omega} \end{aligned}$$

$$\mathbf{Z_1 + Z_2 = 2.1\ k\Omega\ \angle 31.6^\circ}$$

Eq. (21-4):

$$\begin{aligned} i_1 &= I \frac{Z_2}{Z_1 + Z_2} \\ &= (15\ \text{mA}) \frac{1.78\ \text{k}\Omega\ \angle -63.7^\circ}{2.1\ \text{k}\Omega\ \angle 31.6^\circ} \end{aligned}$$

$$\mathbf{i_1 = 12.7\ mA\ \angle -95.3^\circ}$$

Eq. (21-5):

$$\begin{aligned} i_2 &= I \frac{Z_1}{Z_1 + Z_2} \\ &= (15\ \text{mA}) \frac{2.88\ \text{k}\Omega\ \angle 69.7^\circ}{2.1\ \text{k}\Omega\ \angle 31.6^\circ} \end{aligned}$$

$$\mathbf{i_2 = 20.6\ mA\ \angle 38.1^\circ}$$

When a phasor diagram is drawn for i_1 and i_2, it is seen that their resultant is $I = 15$ mA (i.e., the supply current).

21-5 SERIES–PARALLEL IMPEDANCES

In analyzing a series–parallel impedance circuit, the procedure is as follows:

1. Add together (vectorially) all impedances that are directly connected in series.
2. Resolve parallel connected impedances into a single equivalent impedance, so that the circuit becomes a group of series-connected impedances.
3. Determine the single equivalent impedance for the whole circuit.
4. Calculate the total current taken from the supply.
5. Using the current divider rule, determine the individual branch current.

Figure 21-6(a) shows a series–parallel impedance circuit, and Figure 21-6(b) and (c) shows the stages in analyzing the circuit. Impedances Z_3 and Z_4 are added together (in rectangular form) to give a single equivalent impedance. Then converting Z_2 and (Z_3+Z_4) into admittances, the

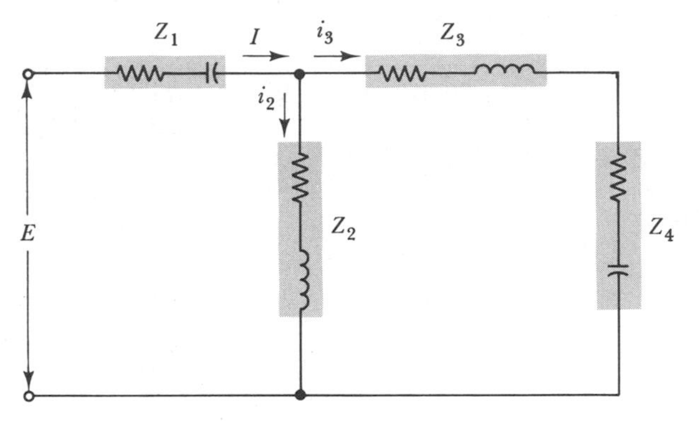

(a) Series-parallel impedance circuit

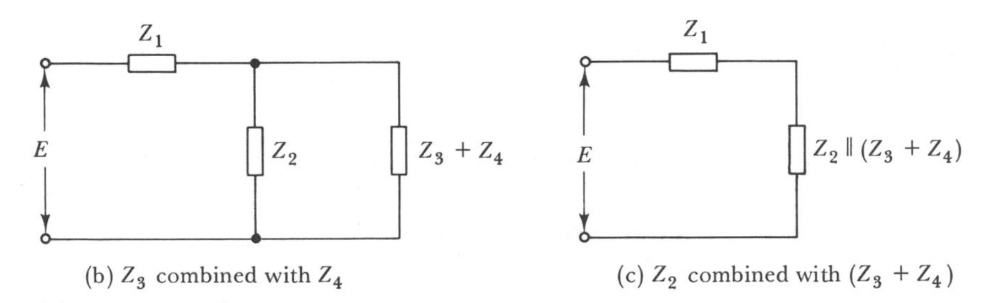

(b) Z_3 combined with Z_4

(c) Z_2 combined with $(Z_3 + Z_4)$

FIG. 21-6. Simplifying a series–parallel impedance circuit

equivalent impedance of Z_2 in parallel with (Z_3+Z_4) is found. At this stage the circuit has become two impedances in series, and these may be vectorially added to give the equivalent impedance for the whole circuit.

EXAMPLE 21-6 For the circuit shown in Figure 21-6(a): $Z_1=(560\ \Omega-j620\ \Omega)$, $Z_2=(330\ \Omega+j470\ \Omega)$, $Z_3=(390\ \Omega+j270\ \Omega)$, and $Z_4=(220\ \Omega-j220\ \Omega)$. If the supply voltage is 30 V, determine the current through Z_2.

SOLUTION

$$Z_3+Z_4=(390\ \Omega+j270\ \Omega)+(220\ \Omega-j220\ \Omega)$$
$$=610\ \Omega+j50\ \Omega$$
$$=\sqrt{610^2+50^2}\ \angle \arctan\frac{50\ \Omega}{610\ \Omega}$$
$$\mathbf{Z_3+Z_4=612\ \Omega\ \angle 4.7^\circ}$$

$$Y_{3,4}=\frac{1}{Z_3+Z_4}=\frac{1}{612\ \Omega\ \angle 4.7^\circ}$$
$$\mathbf{Y_{3,4}=1.63\ mS\ \angle -4.7^\circ}$$

$$Z_2=\sqrt{330^2+470^2}\ \angle \arctan\frac{470}{330}$$
$$Z_2=574\ \angle 54.9^\circ$$
$$Y_2=\frac{1}{Z_2}=\frac{1}{574\ \angle 54.9^\circ}$$
$$\mathbf{Y_2=1.74\ mS\ \angle -54.9^\circ}$$

$$Y_{3,4}=1.63\ \text{mS}[\cos(-4.7^\circ)+j\sin(-4.7^\circ)]$$
$$\mathbf{Y_{3,4}=1.62\ mS-j0.134\ mS}$$

$$Y_2=1.74\ \text{mS}[\cos(-54.9^\circ)+j\sin(-54.9^\circ)]$$
$$\mathbf{Y_2=1\ mS-j1.42\ mS}$$

$$Y_{3,4}+Y_2=(1.62\ \text{mS}-j0.134\ \text{mS})+(1\ \text{mS}-j1.42\ \text{mS})$$
$$=2.62\ \text{mS}-j1.55\ \text{mS}$$
$$=\sqrt{(2.62\ \text{mS})^2+(1.55\ \text{mS})^2}\ \angle \arctan\frac{-1.55\ \text{mS}}{2.62\ \text{mS}}$$
$$\mathbf{Y_{3,4}+Y_2=3.04\ mS\ \angle -30.6^\circ}$$

$$Z_2 \| (Z_3 + Z_4) = \frac{1}{Y_{3,4} + Y_2}$$

$$= \frac{1}{3.04\ \text{mS} \angle -30.6^\circ}$$

$$Z_2 \| (Z_3 + Z_4) = 329\ \Omega \angle 30.6^\circ$$

$$= 329\ \Omega [\cos(30.6^\circ) + j\sin(30.6^\circ)]$$

$$\mathbf{Z_2 \| (Z_3 + Z_4) = 283\ \Omega + j167\ \Omega}$$

$$Z_T = Z_1 + [Z_2 \| (Z_3 + Z_4)] = (560\ \Omega - j620\ \Omega) + (283\ \Omega + j167\ \Omega)$$

$$= 843\ \Omega - j453\ \Omega$$

$$Z_T = \sqrt{843^2 + 453^2} \angle \arctan \frac{-453\ \Omega}{843\ \Omega}$$

$$\mathbf{Z_T = 957\ \Omega \angle -28.3^\circ}$$

$$I = \frac{E}{Z_T}$$

$$= \frac{30\ \text{V}}{957\ \Omega \angle -28.3^\circ}$$

$$\mathbf{I = 31.3\ \text{mA} \angle -28.3^\circ}$$

The total current I flows through Z_1 and splits up into i_2 and i_3, as indicated in Figure 21-6(a).

Using the current divider rule:

$$i_2 = I \frac{Z_3 + Z_4}{Z_2 + (Z_3 + Z_4)}$$

$$Z_2 + (Z_3 + Z_4) = (330\ \Omega + j470\ \Omega) + (610\ \Omega + j50\ \Omega)$$

$$= 940 + j520$$

$$= \sqrt{940^2 + 520^2} \angle \arctan \frac{520\ \Omega}{940\ \Omega}$$

$$\mathbf{Z_2 + (Z_3 + Z_4) = 1074 \angle 29^\circ}$$

Therefore,

$$i_2 = 31.3\ \text{mA} \angle 28.3^\circ \left(\frac{612\ \Omega \angle 4.7^\circ}{1074\ \Omega \angle 29^\circ} \right)$$

$$\mathbf{i_2 = 17.8\ \text{mA} \angle 4^\circ}$$

21-6 SERIES AND PARALLEL EQUIVALENT CIRCUITS

Sometimes it is convenient to replace a series *RL* circuit with a parallel *RL* circuit which offers exactly the same impedance characteristics, and vice versa. In this case each circuit is the *equivalent circuit* of the other, and if each were contained in a sealed box with only their terminals showing, it may be difficult to distinguish between them.

Figure 21-7 shows a series *RL* circuit consisting of R_s and X_s, and its equivalent parallel circuit, R_p and X_p. From Figure 21-7(a),

$$Z_s = R_s + jX_s$$

and from Figure 21-7(b),

$$Y_p = \frac{1}{R_p} - j\frac{1}{X_p}$$

Also,

$$Z_s = Z_p$$

or

$$\frac{1}{Z_s} = Y_p$$

Therefore,

$$\frac{1}{R_s + jX_s} = \frac{1}{R_p} - j\frac{1}{X_p}$$

$$\frac{1}{R_s + jX_s} \times \frac{R_s - jX_s}{R_s - jX_s} = \frac{1}{R_p} - j\frac{1}{X_p}$$

$$\frac{R_s - jX_s}{R_s^2 + X_s^2} = \frac{1}{R_p} - j\frac{1}{X_p}$$

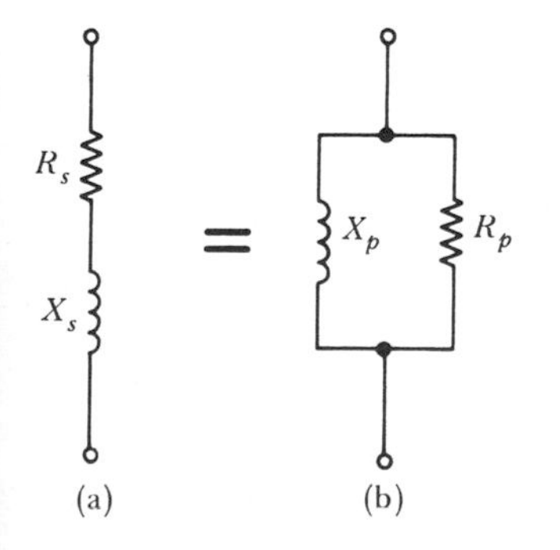

FIG. 21-7. Equivalent series and parallel *RL* circuits

Equating the real and imaginary terms,

$$\frac{R_s}{R_s^2+X_s^2}=\frac{1}{R_p}$$

or

$$\boxed{R_p=\frac{R_s^2+X_s^2}{R_s}} \tag{21-6}$$

and

$$\frac{X_s}{R_s^2+X_s^2}=\frac{1}{X_p}$$

or

$$\boxed{X_p=\frac{R_s^2+X_s^2}{X_s}} \tag{21-7}$$

Equations (21-6) and (21-7) also apply to the series *CR* circuit and its parallel equivalent shown in Figure 21-8. These equations enable the component values of the parallel *CR* or *LR* equivalent circuit to be calculated for any given series *CR* or *LR* circuit.

To derive equations for calculating the series *CR* or *LR* equivalent circuit for a given parallel circuit, Y_p is written in terms of a conductance and susceptance. Z_s is then equated to the reciprocal of Y_p.

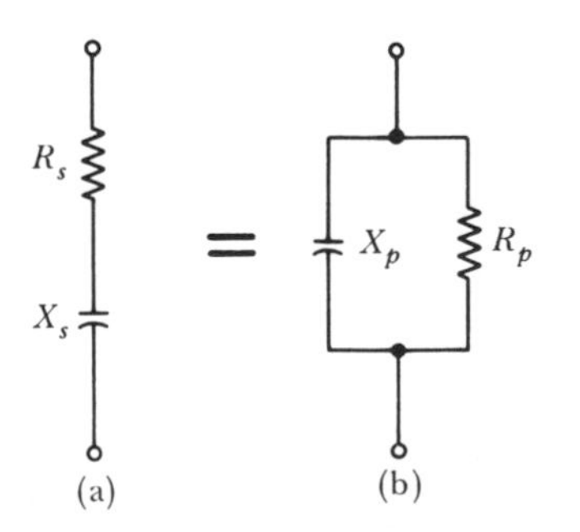

FIG. 21-8. Equivalent series and parallel *RC* circuits

For the circuits in Figure 21-8,

$$Z_s = R_s - jX_s$$

and

$$Y_p = G_p + jB_p$$

$$Z_s = \frac{1}{Y_p}$$

Therefore,

$$R_s - jX_s = \frac{1}{G_p + jB_p}$$

$$R_s - jX_s = \frac{1}{G_p + jB_p} \times \frac{G_p - jB_p}{G_p - jB_p}$$

$$R_s - jX_s = \frac{G_p - jB_p}{G_p^2 + B_p^2}$$

Equating the real and imaginary terms,

$$\boxed{R_s = \frac{G_p}{G_p^2 + B_p^2}} \tag{21-8}$$

and

$$\boxed{X_s = \frac{B_p}{G_p^2 + B_p^2}} \tag{21-9}$$

Since it may be more convenient to use R_p and X_p rather than G_p and B_p, the equations can be rewritten:
Eq. (21-8):

$$R_s = \frac{1/R_p}{1/R_p^2 + 1/X_p^2}$$

$$= \frac{1/R_p}{1/R_p^2 + 1/X_p^2} \times \frac{R_p^2 X_p^2}{R_p^2 X_p^2}$$

or

$$\boxed{R_s = \frac{R_p X_p^2}{X_p^2 + R_p^2}} \tag{21-10}$$

and from Eq. (21-9),

$$\boldsymbol{X_s = \frac{X_p R_p^2}{X_p^2 + R_p^2}} \qquad \textbf{(21-11)}$$

EXAMPLE 21-7

At a frequency of 1 kHz, an unknown impedance behaves as a 0.01-μF capacitor in series with a resistance of 10 kΩ. A dc measurement at the terminals of the impedance gives a resistance of 35.3 kΩ. Determine the actual components and how they are connected.

SOLUTION

$$X_s = \frac{1}{2\pi f C} = \frac{1}{2\pi \times 1\ \text{kHz} \times 0.01\ \mu\text{F}}$$

$$X_s = 15.9\ \text{k}\Omega$$

$$R_s = 10\ \text{k}\Omega$$

Eq. (21-6):

$$R_p = \frac{R_s^2 + X_s^2}{R_s} = \frac{(10\ \text{k}\Omega)^2 + (15.9\ \text{k}\Omega)^2}{10\ \text{k}\Omega}$$

$$\boldsymbol{R_p = 35.3\ \text{k}\Omega}$$

Eq. (21-7):

$$X_p = \frac{R_s^2 + X_s^2}{X_s} = \frac{(10\ \text{k}\Omega)^2 + (15.9\ \text{k}\Omega)^2}{15.9\ \text{k}\Omega}$$

$$X_p = 22.2\ \text{k}\Omega$$

$$C_p = \frac{1}{2\pi f X_p} = \frac{1}{2\pi \times 1\ \text{kHz} \times 22.2\ \text{k}\Omega}$$

$$\boldsymbol{C_p = 0.007\ \mu\text{F}}$$

If the actual impedance consisted of a series-connected resistor and capacitor, the dc resistance measurement would give a value greater than R_s; that is, it would be $R_s +$ (the capacitor dielectric resistance). Since the measured resistance corresponds with the value of R_p, the actual impedance consists of a 35.3-kΩ *resistor connected in parallel with a* 0.007-μF *capacitor.*

GLOSSARY OF FORMULAS

AC voltage divider:

$$V_3 = \frac{EZ_3}{Z_1 + Z_2 + Z_3}$$

Resolving polar form admittances into rectangular form:

$$Y = Y\cos\theta + jY\sin\theta$$

AC current divider:

$$i_1 = I\frac{Z_2}{Z_1 + Z_2}$$

$$i_2 = I\frac{Z_1}{Z_1 + Z_2}$$

Series–parallel equivalent circuits:

$$R_p = \frac{R_s^2 + X_s^2}{R_s}$$

$$X_p = \frac{R_s^2 + X_s^2}{X_s}$$

$$R_s = \frac{G_p}{G_p^2 + B_p^2} = \frac{R_p X_p^2}{X_p^2 + R_p^2}$$

$$X_s = \frac{B_p}{G_p^2 + B_p^2} = \frac{X_p R_p^2}{X_p^2 + R_p^2}$$

PROBLEMS

21-1 A coil having a resistance of 100 Ω and an inductance of 300 mH is connected in series with a 10-μF capacitor, another inductor of 150 mH, and a resistance of 180 Ω. The supply voltage is 115 V with a frequency of 60 Hz. Calculate the current taken from the supply.

21-2 For the circuit described in Problem 21-1, calculate the terminal voltages of the 10-μF capacitor and the 300-mH inductor. Also, draw a phasor diagram showing the supply voltage and current and the voltage across the capacitor and inductor.

21-3 An impedance of (3.3 kΩ + j3.9 kΩ) is connected in series with another impedance having a value of (6.8 kΩ − j5.6 kΩ). The supply is 55 V. Using the voltage divider rule, determine the voltage across each impedance.

21-4 The two impedances described in Problem 21-3 are connected in parallel across a supply. If the total current taken from the supply is 22 mA, use the current divider rule to calculate the current through each impedance.

21-5 Three impedances connected in parallel have the following values: $Z_1 = (5.6\ k\Omega - j3.3\ k\Omega)$, $Z_2 = (1.8\ k\Omega + j2.2\ k\Omega)$, $Z_3 = (8.2\ k\Omega - j6.2\ k\Omega)$. Determine the total equivalent impedance and the current taken from the supply. The supply voltage is 19 V.

21-6 Calculate the equivalent impedance of the circuit shown in Figure 21-9.

21-7 For the circuit shown in Figure 21-9, determine the voltage drops across L_2 and C_2.

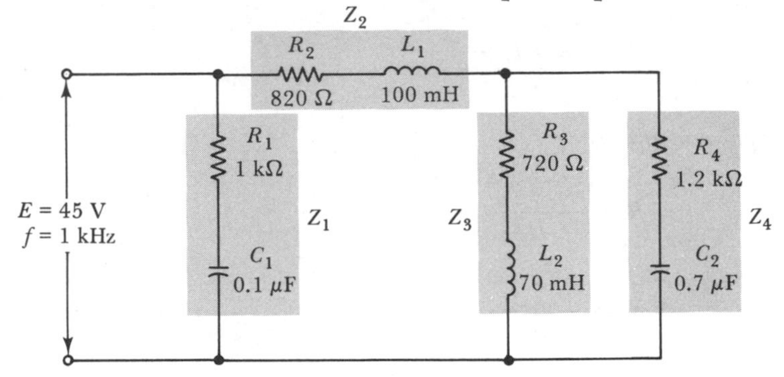

FIG. 21-9.

21-8 For the circuit shown in Figure 21-10, determine the total equivalent impedance.

21-9 Calculate the current that flows through R_2 and L_1 in the circuit of Figure 21-10.

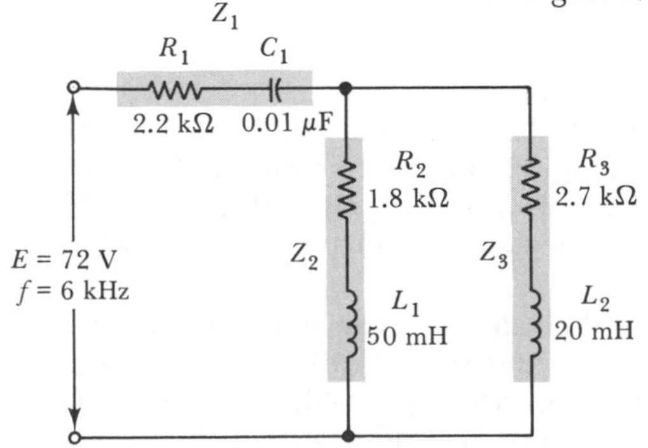

FIG. 21-10.

21-10 Derive the equations for converting a series-connected inductance and resistance into the equivalent parallel-connected inductance and resistance. Also, derive the equations for converting a parallel-connected capacitor and resistor into the equivalent series-connected capacitor and resistor.

21-11 An unknown impedance measured at a frequency of 5 kHz behaves as a 500-mH inductor in series with a 10-kΩ resistor. A dc measurement gives a terminal resistance of 10 kΩ. Determine the parallel equivalent circuit of the impedance and decide what the actual impedance consists of.

22

POWER IN AC CIRCUITS

INTRODUCTION

In an alternating-current circuit, power is dissipated in a resistor but not in an inductor or a capacitor. Because the current in an LR circuit lags the supply voltage by an angle ϕ, the amount of useful power supplied to the circuit is proportional to $\cos \phi$. Similarly, in a CR circuit, the useful power is proportional to the angle by which the current leads the supply voltage. The presence of the reactive components causes the supply of a *reactive power* to the circuit. This gives an *apparent power*, which is greater than the *true power* utilized in the circuit. The ratio of true power to apparent power is termed the *power factor* of the circuit.

22-1 POWER DISSIPATED IN A RESISTANCE

In Section 18-4 it was shown that when an alternating current flows in a resistance, the power dissipated can be calculated as follows:
Eq. (18-12):

$$P = \tfrac{1}{2} P_m \qquad \text{watts}$$

where

$$P_m = E_m I_m$$

Therefore,

$$P = \tfrac{1}{2} E_m I_m$$

$$= \frac{E_m}{\sqrt{2}} \times \frac{I_m}{\sqrt{2}}$$

or

$$P = EI \qquad \text{watts} \tag{22-1}$$

where E and I are rms values. This is exactly the same as for Eq. (3-5), which was derived for a dc circuit. As long as rms quantities are used for voltage and current, the power calculations for a resistor in an ac circuit are exactly the same as those for a dc circuit. Thus, from Eqs. (3-8) and (3-9),

$$P = \frac{E^2}{R} \qquad \text{watts} \tag{22-2}$$

and

$$P = I^2R \qquad \text{watts} \tag{22-3}$$

where E and I are again rms values.

Figure 18-8 shows that the instantaneous power dissipated in a resistance alternates between zero and a peak level as the current rises from zero to its positive and negative peak values. If the resistance were the tungsten filament of a lamp, and if the frequency of the alternating current were 1 Hz, the filament could clearly be seen to go bright and dim at a frequency of 2 Hz. The lamp would be bright when the current is $+I_p$, dim when the current is zero, and bright again when the current is $-I_p$. The waveform of power in Figure 18-8 shows that two positive peaks of power occur during each cycle of current. Since the normal domestic and industrial power frequency is 60 Hz (in North America), and since the tungsten filament does not lose its heat very rapidly, an electric filament lamp supplied with ac normally does not fluctuate in brightness.

EXAMPLE 22-1

A 100-W electric lamp is supplied from a 115-V 60-Hz source. Calculate:

a. The level of current that flows.

b. The resistance of the filament.

c. The peak power dissipated in the lamp filament.

SOLUTION

a. Eq. (22-1):

$$P = EI$$

Therefore,
$$I=\frac{P}{E}$$
$$\mathbf{I=\frac{100\ W}{115\ V}\simeq 870\ mA}$$

b. Eq. (22-2):
$$P=\frac{E^2}{R}$$

Therefore,
$$R=\frac{E^2}{P}$$
$$\mathbf{R=\frac{(115\ V)^2}{100\ W}=132\ \Omega}$$

c. Eq. (18-12):
$$P=\tfrac{1}{2}P_m$$

Therefore,
$$P_m=2\times P$$
$$\mathbf{P_m=2\times 100\ W=200\ W}$$

22-2 POWER IN AN INDUCTANCE

When an alternating voltage is applied to a pure inductance, the current lags the applied voltage by 90°. This was discussed in Section 20-1. The waveforms of current and voltage for an inductance are reproduced in Figure 22-1. Since the instantaneous power supplied to any component is calculated as

$$p=e\times i$$

the waveform of power can easily be derived from the current and voltage waves. At *time* t_1 on Figure 22-1,

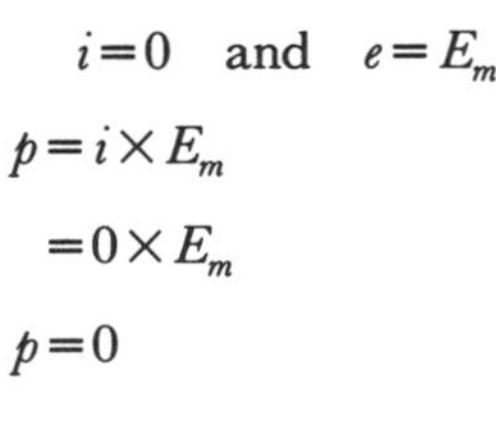

$$i=0 \quad \text{and} \quad e=E_m$$

Therefore,
$$p=i\times E_m$$
$$=0\times E_m$$
$$p=0$$

At t_2,
$$i=I_m\sin 45°$$
$$i=\frac{I_m}{\sqrt{2}}=I \quad (\textit{rms value})$$

and
$$e=E_m\sin 135°$$
$$e=\frac{E_m}{\sqrt{2}}=E \quad (\textit{rms value})$$

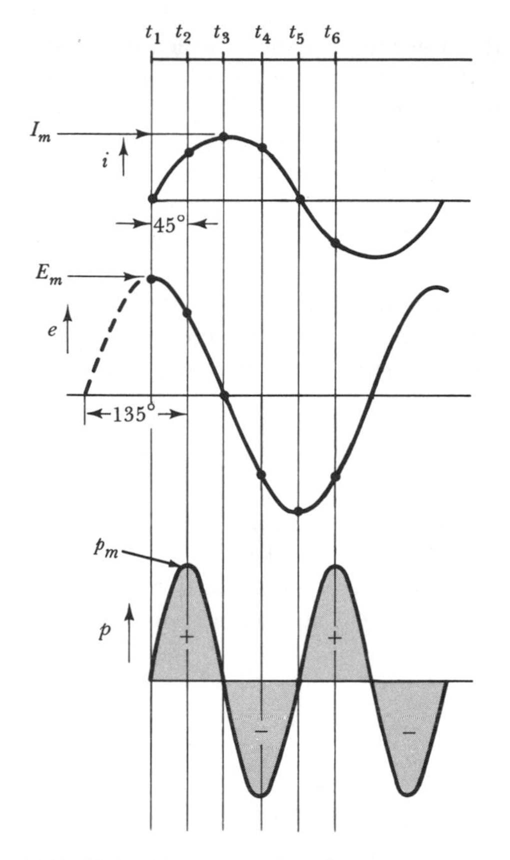

FIG. 22-1. Power in an inductance

Therefore, $p_m = IE$

At t_3,

$$i = I_m \quad \text{and} \quad e = 0$$
$$= I_m \times 0$$
$$p = 0$$

At t_4,

$$i = \frac{I_m}{\sqrt{2}} \quad \text{and} \quad e = \frac{-E_m}{\sqrt{2}}$$

$$p_m = \frac{I_m}{\sqrt{2}} \times \frac{-E_m}{\sqrt{2}}$$

giving $p_m = -IE$

At this point the power supplied is a negative quantity, which means that the inductance is not absorbing power but *supplying power*. Continuing the process of calculating and plotting the instantaneous power levels in Figure 22-1, it is seen that the waveform of power supplied is sinusoidal, with a frequency which is twice that of the voltage and current. Also, since the *negative* half-cycles of power supplied are equal to the positive half-cycles, the average power supplied to the inductance is zero.

In Chapter 15 it was shown that energy can be stored in an inductor. Therefore, referring to Figure 22-1 again, it can be said that the energy supplied to the inductor is stored during the time that energy input is a positive quantity, and that the stored energy is returned to the source of supply when the energy input is negative.

22-3 POWER IN A CAPACITANCE

In the case of a pure capacitance supplied with an alternating voltage, the current leads the voltage by 90° (see Section 20-3). Figure 22-2 shows the current and voltage waveforms for a capacitance, and the waveform of power supplied, as derived from the current and voltage waves.

At t_1,

$$p = i \times e$$
$$= 0 \times (-E_m)$$
$$= 0$$

At t_2,

$$i = I_m \sin 45°$$

$$i = \frac{I_m}{\sqrt{2}} = I \quad \text{(rms value)}$$

and

$$e = E_m \sin 315°$$

$$e = \frac{-E_m}{\sqrt{2}} = -\text{E} \quad \text{(rms value)}$$

Therefore,

$$p_m = -IE$$

Continuing the process, exactly as was done for the inductive circuit, it is seen that the waveform of power supplied to a capacitor is sinusoidal, and that its frequency is twice that of the voltage and current. It is also seen that the power supplied to the capacitor is alternatively positive and negative, meaning that the capacitor stores the power supplied to it, then returns the stored power to the source of supply. The result of this is that the average power supplied to the capacitance is zero.

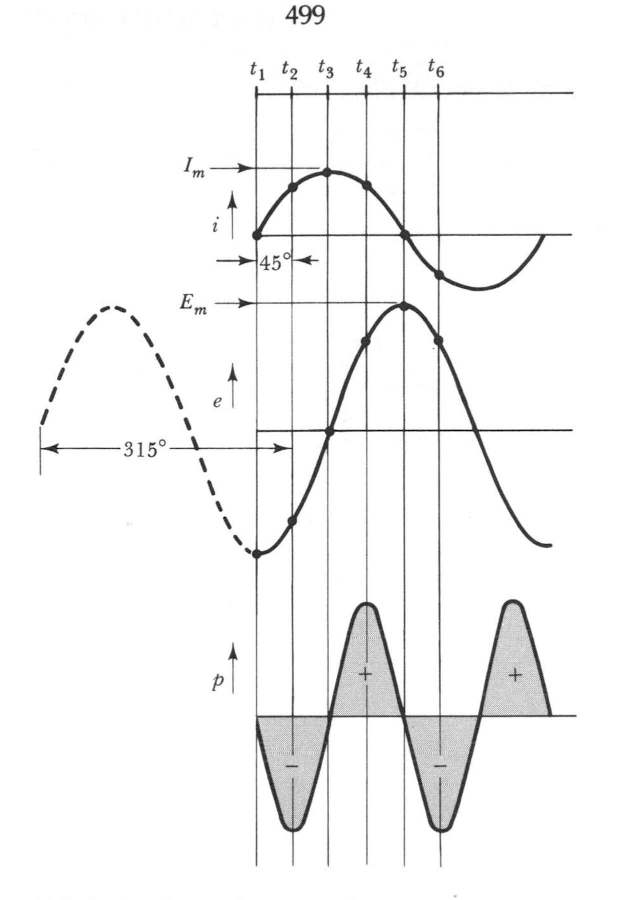

FIG. 22-2. Power in a capacitance

22-4 TRUE POWER AND REACTIVE POWER

The power supplied to a resistance is sometimes referred to as *resistive power*. Similarly, the power supplied to a reactance (inductive or capacitive) can be termed *reactive power*. Also, because the power supplied to a resistance is actually dissipated in the form of heat, while that supplied to a reactance averages out to zero, the *resistive power* is known as *true power*.

It is very important to distinguish between true power and reactive power. Consequently, true power is always measured in *watts*, while reactive power is given the name *volt-amps reactive*, abbreviated *var*. For a pure inductance or capacitance, the reactive power is calculated from

$$\boxed{Q = EI \qquad \text{vars}} \tag{22-4}$$

where Q = reactive power and E and I are rms voltage and current, respectively.

The reactive power can also be calculated as

$$\boxed{Q = I^2X \qquad \text{vars}} \tag{22-5}$$

or

$$\boxed{Q = \frac{E^2}{X} \qquad \text{vars}} \tag{22-6}$$

where X in Eqs. (22-5) *and* (22-6) *can be* X_C *or* X_L.

EXAMPLE 22-2 Calculate the power supplied when a 120-V 60-Hz source is connected to:

a. A 60-Ω resistor.

b. A 50-mH inductor.

c. A 33-μF capacitor.

SOLUTION

a.

$$P = \frac{E^2}{R}$$

$$= \frac{(120 \text{ V})^2}{60 \ \Omega}$$

$$\mathbf{P = 240 \text{ W}} \quad (\textit{true power})$$

b.

$$X_L = 2\pi fL$$

$$= 2 \times \pi \times 60 \text{ Hz} \times 50 \text{ mH}$$

$$X_L \cong 18.8 \ \Omega$$

From Eq. (22-6),

$$Q = \frac{E^2}{X_L}$$

$$= \frac{(120 \text{ V})^2}{18.8 \ \Omega}$$

$$\mathbf{Q = 766 \text{ vars}} \quad (\textit{reactive power})$$

c.

$$X_C = \frac{1}{2\pi fc}$$

$$= \frac{1}{2 \times \pi \times 60\ \text{Hz} \times 33\ \mu\text{F}}$$

$$X_C \cong 80.4\ \Omega$$

From Eq. (22-6),

$$Q = \frac{E^2}{X_C}$$

$$= \frac{(120\ \text{V})^2}{80.4\ \Omega}$$

$$\mathbf{Q = 179\ vars} \quad \textbf{(\textit{reactive power})}$$

22-5 POWER IN *LR* AND *CR* CIRCUITS

In a series circuit consisting of inductance and resistance, the current lags the voltage by an angle ϕ which is less than 90° (see Sections 20-5 and 20-6). Figure 22-3 shows typical voltage and current waveforms for an *RL* circuit, along with the waveform of power in the circuit. The power waveform is derived in the usual way by multiplying together instantaneous values of voltage and current. During the period from t_1 to t_2, both i and e are positive quantities; therefore, p remains positive throughout that time. From t_2 to t_3, i is positive while e is a negative quantity; consequently, the product of i and e is negative, and the power waveform is below the zero line. After t_3, the current and voltage are both negative until t_4, so from t_3 to t_4 the power is positive again.

The power waveform clearly shows that more positive power than negative power is supplied to the circuit. This is to be expected, of course, since power is dissipated in the resistance, while the average power supplied to the inductance remains at zero. The average power supplied to the circuit can now be represented by the dashed line shown on the power waveform.

Now consider the phasor diagram for a series *RL* circuit, as reproduced in Figure 22-4. The current I is shown lagging the applied voltage E by the angle ϕ. The voltage across the resistance E_R is in phase with I, so E_R also lags E by angle ϕ. The voltage across the inductance is E_L, and it leads the current by an angle of 90°. The phasor sum of E_R and E_L gives the supply voltage E, as illustrated.

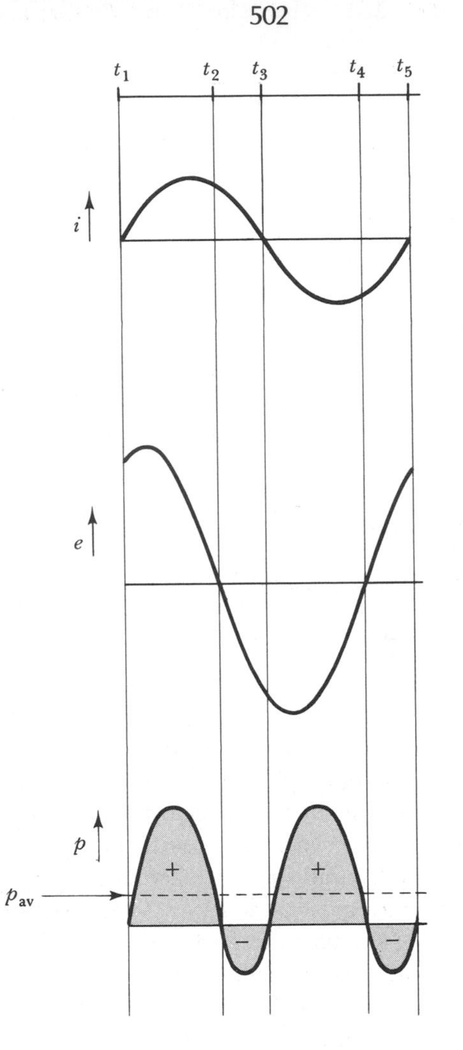

FIG. 22-3. Power in an *LR* circuit

The true power dissipated in the *RL* circuit is, of course, the power dissipated in the resistor. Therefore, true power,

$$P = E_R \times I$$

and from Figure 22-4, $E_R = E\cos\phi$

giving $P = (E\cos\phi) \times I$

or

$$\boxed{\mathbf{P = EI\cos\phi \quad watts}} \tag{22-7}$$

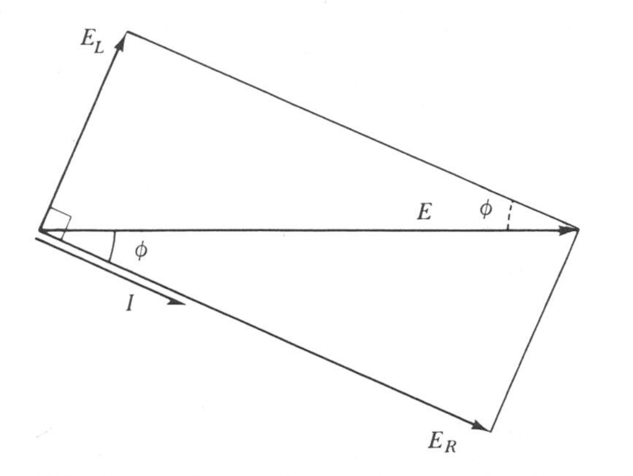

FIG. 22-4. Phasor diagram for a series LR circuit

The reactive power is the product of E_L and I. Also, from Figure 22-4,

$$E_L = E \sin\phi$$

Therefore,

$$Q = (E \sin\phi) \times I$$

or

$$\boxed{Q = EI \sin\phi \qquad \text{vars}} \tag{22-8}$$

If the supply voltage E and the measured current I in an RL circuit were multiplied together, the product would be neither the true power nor the reactive power. However, it would give a quantity that might appear to be the power supplied to the circuit. Therefore, the term *apparent power* is applied to this quantity, and the units of apparent power are *volt-amps*:

$$\boxed{S = EI \qquad \text{volt-amps}} \tag{22-9}$$

where S is the apparent power and E and I are rms values of supply voltage and current.

EXAMPLE 22-3 In a series RL circuit supplied with 50 V, the current is measured as 100 mA, with a phase angle of 25°. Calculate the apparent power, reactive power, and true power supplied to the circuit.

SOLUTION

Eq. (22-9):

$$S = EI$$
$$= 50\text{ V} \times 100\text{ mA}$$
$$\boldsymbol{S = 5}\textbf{ volt-amps}$$

Eq. (22-8):

$$Q = EI \sin\phi\text{ vars}$$
$$= 50\text{ V} \times 100\text{ mA} \times (\sin 25°)$$
$$\boldsymbol{Q \cong 2.1}\textbf{ vars}$$

Eq. (22-7):

$$P = EI \cos\phi\text{ watts}$$
$$= 50\text{ V} \times 100\text{ mA} \times (\cos 25°)$$
$$\boldsymbol{P \cong 4.5}\textbf{ W}$$

When the current and voltage waveforms in a series *CR* circuit are employed to derive the waveform of power supplied (Figure 22-5), it is seen that a similar result is obtained as in the case of the *LR* circuit. The average power supplied to the circuit is a positive quantity, and this represents the power dissipated in the resistance. The phasor diagram for the *CR* circuit, as shown in Figure 22-6, gives the same equations for true power, reactive power, and apparent power as those derived for the *LR* circuit.

The power supplied to *parallel CR* and *LR* circuits can also be resolved into true power, reactive power, and apparent power components. Using the supply voltage and current, and the phase angle between them, Eqs. (22-7), (22-8), and (22-9) are just as valid for parallel circuits as they are for series circuits. However, in the next section it is shown that capacitors connected in parallel with a load can be employed in correcting problems that arise because of the phase difference between supply current and voltage.

EXAMPLE 22-4

A series circuit consisting of $R = 1.2\text{ k}\Omega$ and $C = 0.1\ \mu\text{F}$ is supplied with 45 V at a frequency of 1 kHz. Determine the apparent power, true power and reactive power in the circuit.

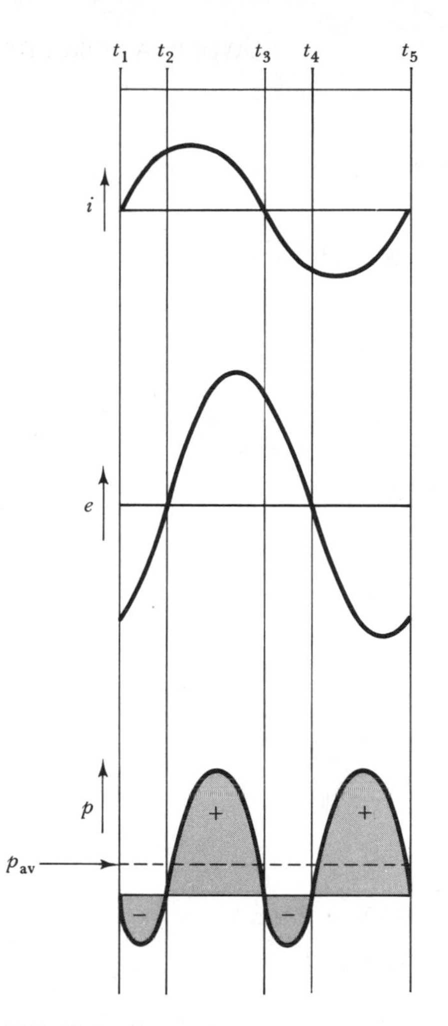

FIG. 22-5. Power in a *CR* circuit

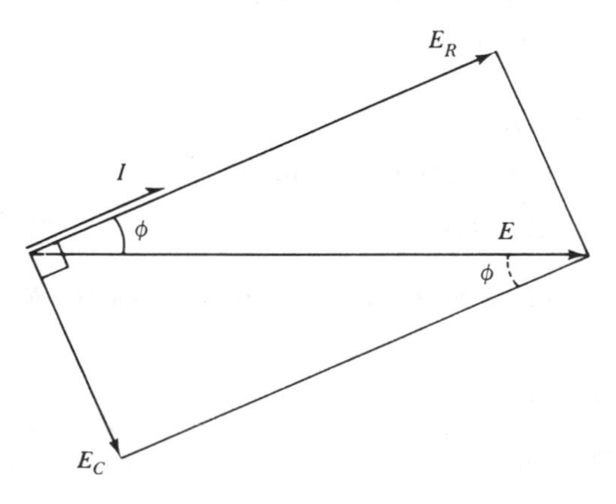

FIG. 22-6. Phasor diagram for a series *CR* circuit

SOLUTION

$$X_C = \frac{1}{2\pi fc}$$

$$= \frac{1}{2 \times \pi \times 1\ \text{kHz} \times 0.1\ \mu\text{F}}$$

$$\mathbf{X_C = 1.59\ k\Omega}$$

$$|Z| = \sqrt{R^2 + X_C^2}$$

$$= \sqrt{(1.2\ \text{k}\Omega)^2 + (1.59\ \text{k}\Omega)^2}$$

$$\mathbf{|Z| = 1.99\ k\Omega}$$

$$\phi = \arctan \frac{X_C}{R}$$

$$= \arctan\left(\frac{1.59\ \text{k}\Omega}{1.2\ \text{k}\Omega}\right)$$

$$\boldsymbol{\phi \cong 53^\circ}$$

$$|I| = \frac{E}{|Z|} = \frac{45\ \text{V}}{1.99\ \text{k}\Omega}$$

$$\mathbf{|I| = 22.6\ mA}$$

apparent power, $S = EI = 45\ \text{V} \times 22.6\ \text{mA}$

$$\mathbf{S \cong 1\ volt\text{-}amp}$$

true power, $P = EI \cos\phi$

$$= 45\ \text{V} \times 22.6\ \text{mA} \times (\cos 53^\circ)$$

$$\mathbf{P = 0.61\ W}$$

reactive power, $Q = EI \sin\phi$

$$= 45\ \text{V} \times 22.6\ \text{mA} \times (\sin 53^\circ)$$

$$\mathbf{Q = 0.81\ vars}$$

Note that since the true power is the power dissipated in the resistor, it can also be calculated as I^2R,

$$P = I^2R$$

$$= (22.6\ \text{mA})^2 \times 1.2\ \text{k}\Omega$$

$$\mathbf{P = 0.61\ W}$$

Also note that the reactive power can be calculated as I^2X_C,

$$Q = I^2X_C$$
$$= (22.6\text{ mA})^2 \times 1.59\text{ k}\Omega$$
$$\mathbf{Q = 0.81\text{ vars}}$$

22-6 POWER FACTOR

Ideally, all the supply volt-amps should be converted into true power in a load. When this is not the case, a certain kind of inefficiency occurs, and this will be discussed further. The ratio of *true power* to *apparent power* is termed the *power factor* of the load,

$$\text{power factor} = \frac{\text{true power}}{\text{apparent power}}$$

As explained in the previous section, the true power in a load is calculated from Eq. (22-7) as

$$P = EI\cos\phi$$

where E and I are the supply voltage and current and ϕ is the phase angle between them. Equation (22-7) can be rewritten to give

$$\cos\phi = \frac{P}{EI} = \frac{\text{true power}}{\text{apparent power}}$$

Therefore,

$$\boxed{\textbf{power factor} = \cos\phi = \frac{P}{EI}} \qquad \textbf{(22-10)}$$

If the power factor was 1, the phase angle ϕ would be zero and all the volt-amps from the supply would be dissipated as true power in the load. Where ϕ is greater than zero, $\cos\phi$ is less than 1 and only a portion of the supply volt-amps is converted into useful power. Thus, it is seen that the power factor always has a maximum value of 1, and is normally less than 1.

The power factor can be expressed as a ratio or as a percentage, and is also usually defined as *leading* or *lagging*. A 60% *lagging power factor* implies an inductive load in which the supply current lags the voltage by an angle with a cosine of 0.6 (i.e. by approximately 53°). A 90% *leading power factor* would indicate a capacitive load in which the current leads the voltage by approximately 26°.

Most industrial loads consist of electric motors and thus are inductive and have lagging power factors. Consider the situation illustrated in Figure 22-7(a), and assume that the supply voltage and current are measured as $E = 120$ V and $I = 100$ A, with the current lagging the voltage by an angle of 33.5°. Using the appropriate equations, the apparent power, true power, and reactive power are calculated as:

$$\textit{apparent power},\ S = EI$$
$$= 120\ \text{V} \times 100\ \text{A}$$
$$\mathbf{S = 12\ kVA}$$

$$\textit{true power},\ P = EI\cos\phi$$
$$= 12\ \text{kVA} \times (\cos 33.5°)$$
$$\mathbf{P = 10\ kW}$$

$$\textit{reactive power},\ Q = EI\sin\phi$$
$$= 12\ \text{kVA} \times (\sin 33.5°)$$
$$\mathbf{Q = 6.6\ kvars}$$

A *power triangle* can be drawn for the circuit, as illustrated in Figure 22-7(b). The true power P is represented by a horizontal vector, and the vector for apparent power S is drawn lagging the true power by the phase angle ϕ. The reactive power vector Q_L completes the triangle.

If the power factor were 1 (i.e., $\phi = 0$), then to supply a power of 10 kW to the load would require a current of

$$I = \frac{P}{E} = \frac{10\ \text{kW}}{120\ \text{V}}$$
$$I \cong 83\ \text{A}$$

This shows that if the power factor could be reduced to unity, the generators supplying the load would have to produce a current of only 83 A instead of 100 A, while still supplying the required amount of true power. Also, the conducting cables could be selected to carry 83 A instead of 100 A, and consequently they would be less expensive. It is always best to have a power factor as near unity as possible, and in fact the power

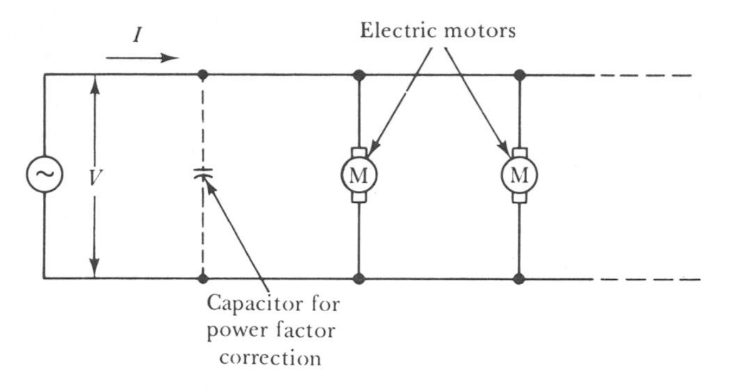

(a) Generator with an inductive load

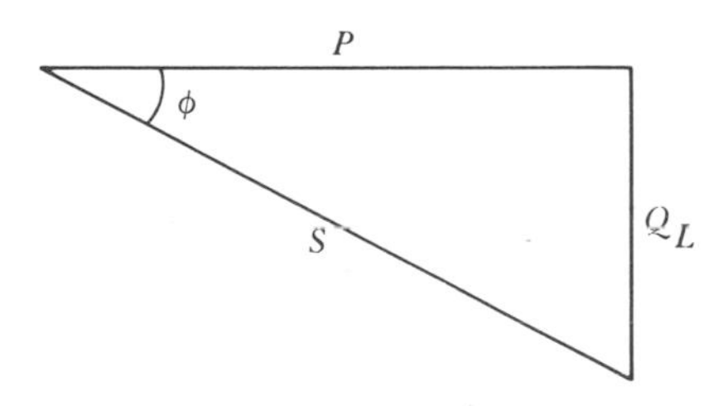

(b) Power triangle for an inductive load

FIG. 22-7. Inductive load and power triangle

factor of an inductive load can be adjusted toward unity. The process is known as *power factor correction.*

Power factor correction for an inductive load consists simply of connecting capacitance in parallel with the load. Suppose that the 100-A load discussed above had been capacitive instead of inductive. The current would then lead the supply voltage by the phase angle, and the power triangle would be drawn as illustrated in Figure 22-8(a). The *apparent power* vector S is shown leading the *true power* vector P by the angle ϕ, and the *reactive power* vector Q_C is drawn vertically *up* from the true power vector, as shown in the figure.

When capacitance is connected in parallel with an inductive load, the power triangle has a *capacitive reactive power* component as well as *inductive reactive power*. The diagram in Figure 22-8(b) illustrates the situation. The capacitive reactive power is represented by the vector Q_C, drawn vertically *up*, while Q_L drawn *down* represents the inductive reactive power. The net reactive power is the difference between Q_L and Q_C, and in the diagram it is so small that it gives a very small phase angle ϕ, which results in a near-unity power factor. Unity power factor would, of course, be achieved when Q_C and Q_L are equal.

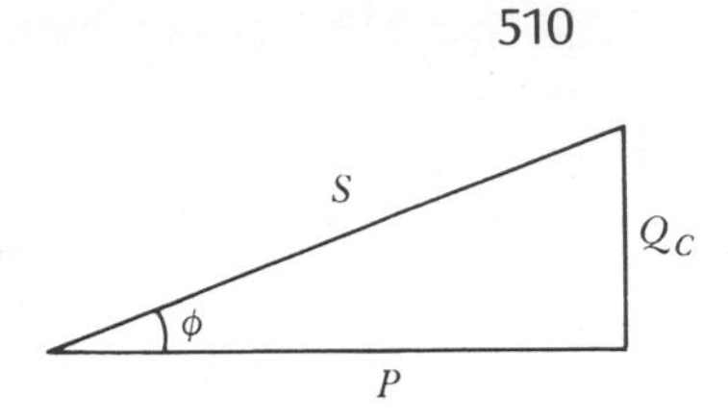

(a) Power triangle for a capacitive load

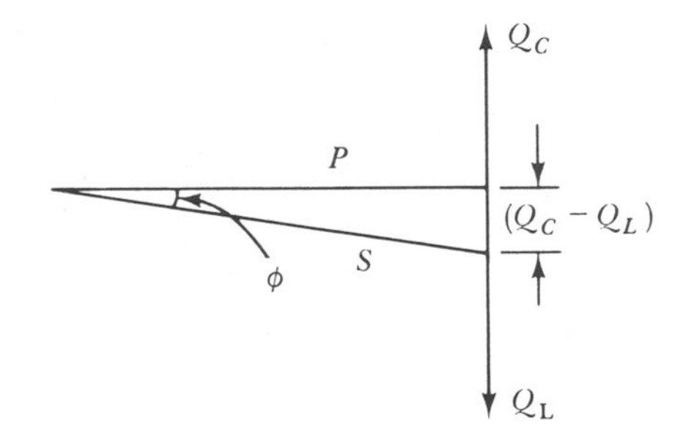

(b) Power triangle for a load with inductance and capacitance

FIG. 22-8. Power factor correction

EXAMPLE 22-5 The current taken from a 115-V 60-Hz supply is measured as 20 A with a lagging power factor of 75%. Calculate the apparent power, true power, and reactive power. Also determine the amount of capacitance that must be connected in parallel with the load to correct the power factor to 95% lagging.

SOLUTION

$$\textit{power factor} = 75\%$$

or

$$\cos\phi = 0.75$$

$$\phi = \arccos 0.75$$

$$\boldsymbol{\phi = 41.4°}$$

$$\textit{apparent power}, S = EI$$

$$= 115\text{ V} \times 20\text{ A}$$

$$\mathbf{S = 2.3\ kVA}$$

$$\textit{true power}, P = EI\cos\phi$$
$$= 2300 \text{ VA} \times 0.75$$
$$\mathbf{P = 1.725\ kW}$$

$$\textit{reactive power}, Q = EI\sin\phi$$
$$= 2300 \text{ VA} \times (\sin 41.4°)$$
$$\mathbf{Q = 1.52\ kvars}$$

For a 95% power factor,

$$\cos\phi = 0.95$$

and

$$\phi = \arccos 0.95$$
$$\phi = 18.2°$$

true power remains,

$$P = 1.725 \text{ W}$$

and

$$P = EI\cos\phi$$

new value of

$$S = EI = \frac{P}{\cos\phi}$$
$$EI = \frac{1.725 \text{ kW}}{0.95}$$
$$\mathbf{EI = 1.82\ kVA}$$

and new value of

$$Q = EI\sin\phi$$
$$= 1.82 \text{ kVA} \times (\sin 18.2°)$$
$$\mathbf{Q = 568\ vars}$$

and

$$Q = Q_L - Q_C$$

giving

$$Q_C = Q_L - Q$$
$$= 1.52 \text{ kvars} - 568 \text{ vars}$$
$$\mathbf{Q_C = 952\ vars}$$

Eq. (22-6):

$$Q_C = \frac{E^2}{X_C}$$

$$X_C = \frac{E^2}{Q_C}$$

$$= \frac{(115\ \text{V})^2}{952\ \text{vars}}$$

$$X_C = 13.9\ \Omega$$

$$X_C = \frac{1}{2\pi f C}$$

$$C = \frac{1}{2\pi f X_C}$$

$$= \frac{1}{2 \times \pi \times 60\ \text{Hz} \times 13.9\ \Omega}$$

$$\mathbf{C = 191\ \mu F}$$

GLOSSARY OF FORMULAS

True power:

$$P = EI \qquad \text{watts}$$

$$P = \frac{E^2}{R} \qquad \text{watts}$$

$$P = I^2 R \qquad \text{watts}$$

Reactive power:

$$Q = EI \qquad \text{vars}$$

$$Q = I^2 X \qquad \text{vars}$$

$$Q = \frac{E^2}{X} \qquad \text{vars}$$

True power:

$$P = EI \cos\phi$$

Reactive power:

$$Q = EI \sin\phi$$

Apparent power:

$$S = EI \text{ volt-amps}$$

REVIEW QUESTIONS

22-1 Sketch typical waveforms of current and voltage for a pure resistance connected in an alternating current circuit. From the voltage and current waveforms, derive the waveform of instantaneous power dissipation. Explain briefly.

22-2 Repeat Review Question 22-1 for a pure inductance connected in an ac circuit.

22-3 Repeat Review Question 22-1 for a pure capacitance connected in an ac circuit.

22-4 Repeat Review Question 22-1 for a series *LR* circuit.

22-5 Repeat Review Question 22-1 for a series *CR* circuit.

22-6 Define: reactive power, apparent power, and true power. Write equations for each.

22-7 Define power factor, write an equation for it, and discuss its importance.

22-8 Sketch typical power triangles for a series *LR* circuit and for a series *CR* circuit. Show the effect that a parallel capacitor can have on the power triangle for a series *LR* circuit.

PROBLEMS

22-1 A 2-kW heating element is supplied from a 220-V ac source. Determine:

a. The resistance of the element.

b. The current level.

c. The peak power dissipated in the element.

22-2 Calculate the power supplied by a 240-V 100-Hz supply when it is connected to:

a. A 200-mH inductance.

b. A 100-Ω resistor.

c. A 100-μF capacitor.

22-3 Calculate the apparent power, reactive power, and true power in a series *LR* circuit, when the supply voltage is 75 V, the current is 1 A, and the phase angle is 30° lagging. Also, draw a power triangle for the circuit.

22-4 A 200-mH choke is connected in series with a 600-Ω resistor. The ac supply voltage is 15 V with a frequency of 1 kHz. Determine the apparent power, true power, and reactive power in the circuit, and draw the power triangle.

22-5 The current taken by a certain load connected to a 220-V 100-Hz supply is measured as 4 A with a power factor of 69% lagging. Determine the apparent power, true power, and reactive power. Calculate the parallel capacitance required to correct the power factor to 97% lagging. Draw the power triangle for the circuit before and after correction.

22-6 An inductive circuit with a 250-μF parallel capacitor has a power factor of 93% lagging. The supply voltage is 120 V with a frequency of 60 Hz, and the supply current is 15 A. Determine the apparent power, true power, and reactive power when the capacitor is disconnected.

22-7 A 24-V 400-Hz supply is connected to a 4-kW load with a 65% lagging power factor. Calculate the current that must be carried by the conductors. Also calculate the new level of conductor current when the power factor is corrected to 85% lagging, and determine the capacitor value for power factor correction.

22-8 A 120-mH choke with a series resistance of 70 Ω is connected to a 50-V 400-Hz supply. Determine the power supplied to the circuit. Also calculate the load current when a 1-μF capacitor is connected in parallel with the load.

23

AC NETWORK ANALYSIS

INTRODUCTION

The analysis techniques and theorems employed to solve dc network problems in Chapters 10 and 11 can also be applied to ac networks. However, all impedances and alternating voltages and currents must be treated as phasor quantities. Consequently, there are a lot more calculations involved in analyzing an ac impedance network compared to a similar dc resistance circuit. As in the case of purely resistive circuits, the calculations can be greatly simplified by application of the various network theorems.

23-1 AC SOURCES AND KIRCHHOFF'S LAWS FOR AC CIRCUITS

In analyzing an ac network, it is necessary to know the phase angle of each source as well as the amplitude of the voltage or current generated at each source. Also, although alternating voltages are continuously reversing their polarity, + and − terminals must be identified at each generator. This is necessary because if a generator is connected in reverse, the phase angle of the generator output is altered by 180°.

Figure 23-1(a) shows a circuit that has two voltage sources: $E_1 = 6\ \mathrm{V}\angle 0^\circ$ and $E_2 = 12\ \mathrm{V}\angle -20^\circ$. The waveforms in the illustration show the phase relationship between these two sources. Because the phase angle of E_2 is -20° and that of E_1 is 0°, the zero level of E_2 occurs 20° after the zero level of E_1, as illustrated. Now assume that the generator which produces E_2 is reconnected in reverse, as shown in Figure 23-1(b). Then

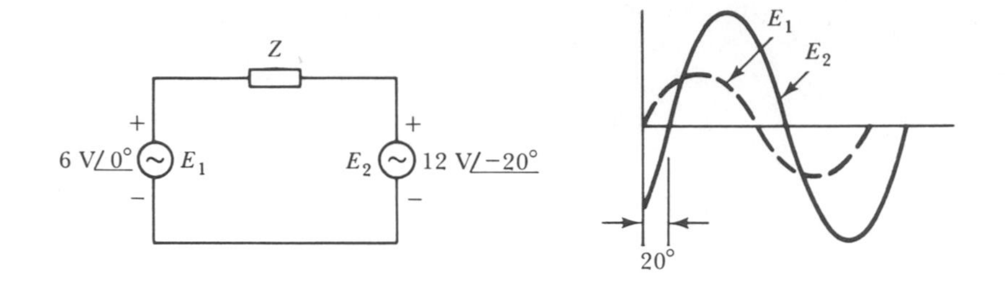

(a) Two signal sources 20° out of phase

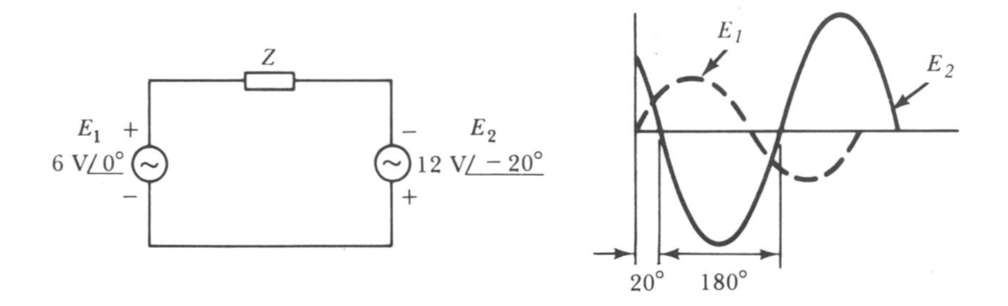

(b) E_2 connected in reverse adds a further 180° phase shift

FIG. 23-1. ac sources should have a terminal polarity identification

E_2 still has its zero level 20° after E_1, but instead of growing positively, its output grows *negatively*, as illustrated. Thus, the reversal of the output terminals has added a further 180° to the phase of E_2 with respect to E_1. It is for this reason that the terminals of all ac sources in a circuit diagram must have a polarity identification.

Kirchhoff's Voltage Law can be stated for ac circuits as follows:

Kirchhoff's Voltage Law (ac circuits)

In any closed electric circuit, the phasor sum of the voltage drops must equal the phasor sum of the applied voltages.

The statement differs from that made for dc networks (see Section 7-2) only in that it refers to the *phasor sum* of the voltages instead of the *algebraic sum*. The same remarks apply to the currents referred to in Kirchhoff's Current Law (Section 8-1 for dc), which for ac circuits is stated thus:

Kirchhoff's Current Law (ac circuits)

The phasor sum of the currents entering a point in an electric circuit must equal the phasor sum of the currents leaving that point.

It is obvious that in solving ac networks, the phase angles of all impedances, and of all voltages and currents, must be carefully considered. Where two quantities are to be multiplied or divided, they should be stated in polar form. Where they are to be added or subtracted, they must be converted into rectangular form.

23-2 AC CIRCUIT LOOP EQUATIONS

The procedure for analysis of ac networks by loop currents is exactly the same as that for dc networks, with the exception that impedances are being dealt with instead of resistances. Loop currents are drawn first, usually in a clockwise direction, as for dc circuits. When the analysis is complete, those branch currents that come out as positive quantities are (instantaneously) in the same direction as that selected for loop currents. The branch currents that have negative signs have an additional 180° phase shift in relation to the loop currents. The procedure for ac network analysis by loop equations is as follows:

1. Draw all loop currents in clockwise direction and identify them by number.
2. Identify all impedance voltage drops as + to − in the direction of the loop current.
3. Identify all voltage sources with their correct polarity.
4. Write the equations for the voltage drops around each loop.
5. Solve the equations to find the required branch current.

Because of the necessity of converting from polar to rectangular form, and vice versa, during calculations, there is much more work involved in analyzing an ac network than for a similar dc network. Consequently, the possibility of calculation errors is increased many times. To minimize the calculations, the equations for the unknown quantities should be reduced to their very simplest state. As noted in Section 10-3, the use of *determinants* can simplify the derivation of equations for the unknown quantities. The following example illustrates the approach that should be taken.

EXAMPLE 23-1

Using loop equations, analyze the impedance network shown in Figure 23-2 to determine the current through Z_3.

SOLUTION

Loop currents I_1 and I_2 are drawn clockwise, as shown on the circuit diagram. For loop 1:

$$E_1 = I_1(Z_1 + Z_3) - I_2 Z_3 \qquad (1)$$

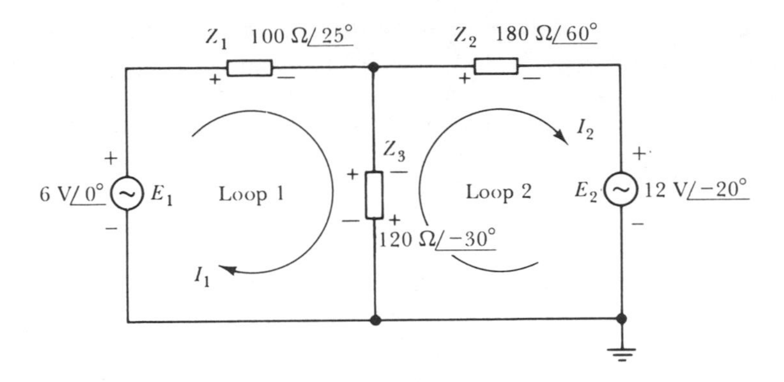

FIG. 23-2. ac circuit analysis by loop equations

For loop 2:

$$0\ \text{V} = I_2(Z_2 + Z_3) - I_1 Z_3 + E_2$$

Therefore,

$$-E_2 = I_2(Z_2 + Z_3) - I_1 Z_3 \tag{2}$$

Eq. $(1) \times \dfrac{(Z_2 + Z_3)}{Z_3}$:

$$\frac{E_1(Z_2 + Z_3)}{Z_3} = \frac{I_1(Z_1 + Z_3)(Z_2 + Z_3)}{Z_3} - I_2(Z_2 + Z_3) \tag{3}$$

Adding Eq. (2) *to* Eq. (3) *gives*

$$\frac{E_1(Z_2 + Z_3)}{Z_3} - E_2 = I_1\left[\frac{(Z_1 + Z_3)(Z_2 + Z_3)}{Z_3} - Z_3\right] \tag{4}$$

Multiplying through by Z_3:

$$E_1(Z_2 + Z_3) - E_2 Z_3 = I_1[(Z_1 + Z_3)(Z_2 + Z_3) - Z_3^2]$$

and

$$I_1 = \frac{E_1(Z_2 + Z_3) - E_2 Z_3}{(Z_1 + Z_3)(Z_2 + Z_3) - Z_3^2}$$

$$= \frac{E_1 Z_2 + E_1 Z_3 - E_2 Z_3}{Z_1 Z_2 + Z_1 Z_3 + Z_2 Z_3 + Z_3^2 - Z_3^2}$$

$$I_1 = \frac{E_1 Z_2 + E_1 Z_3 - E_2 Z_3}{Z_1 Z_2 + Z_1 Z_3 + Z_2 Z_3} \tag{5}$$

Eq. $(2) \times \dfrac{Z_1 + Z_3}{Z_3}$:

$$-\frac{E_2(Z_1 + Z_3)}{Z_3} = \frac{I_2(Z_2 + Z_3)(Z_1 + Z_3)}{Z_3} - I_1(Z_1 + Z_3) \tag{6}$$

Adding Eq. (1) *to* Eq. (6) *gives*

$$E_1 - \frac{E_2(Z_1+Z_3)}{Z_3} = I_2\left[\frac{(Z_2+Z_3)(Z_1+Z_3)}{Z_3} - Z_3\right]$$

Multiplying through by Z_3:

$$E_1Z_3 - E_2(Z_1+Z_3) = I_2[(Z_2+Z_3)(Z_1+Z_3) - Z_3^2]$$

and

$$I_2 = \frac{E_1Z_3 - E_2(Z_1+Z_3)}{(Z_2+Z_3)(Z_1+Z_3) - Z_3^2}$$

$$= \frac{E_1Z_3 - E_2Z_1 - E_2Z_3}{Z_1Z_2 + Z_2Z_3 + Z_1Z_3 + Z_3^2 - Z_3^2}$$

$$I_2 = \frac{E_1Z_3 - E_2Z_1 - E_2Z_3}{Z_1Z_2 + Z_1Z_3 + Z_2Z_3} \qquad (7)$$

$$I_3 = I_1 - I_2$$

$$= \frac{E_1Z_2 + E_1Z_3 - E_2Z_3 - E_1Z_3 + E_2Z_1 + E_2Z_3}{Z_1Z_2 + Z_1Z_3 + Z_2Z_3}$$

$$I_3 = \frac{E_1Z_2 + E_2Z_1}{Z_1Z_2 + Z_1Z_3 + Z_2Z_3} \qquad (8)$$

The equation above for the current through Z_3 *is stated in its simplest possible form. The component values for the equation can now be calculated separately, and then used for determination of* I_3.

$$Z_1 = 100\ \Omega\ \underline{/\ 25^\circ} = 90.6\ \Omega + j42.3\ \Omega$$
$$Z_2 = 180\ \Omega\ \underline{/\ 60^\circ} = 90\ \Omega + j156\ \Omega$$
$$Z_3 = 120\ \Omega\ \underline{/\ -30^\circ} = 104\ \Omega - j60\ \Omega$$
$$E_1 = 6\ \text{V}\ \underline{/\ 0^\circ} = 6\ \text{V} + j0\ \text{V}$$
$$E_2 = 12\ \text{V}\ \underline{/\ -20^\circ} = 11.3\ \text{V} - j4.1\ \text{V}$$
$$Z_1Z_2 = 18\ \text{k}\Omega\ \underline{/\ 85^\circ} = 1.57\ \text{k}\Omega + j17.9\ \text{k}\Omega$$
$$Z_1Z_3 = 12\ \text{k}\Omega\ \underline{/\ -5^\circ} = 11.95\ \text{k}\Omega - j1.05\ \text{k}\Omega$$
$$Z_2Z_3 = 21.6\ \text{k}\Omega\ \underline{/\ 30^\circ} = 18.7\ \text{k}\Omega + j10.8\ \text{k}\Omega$$
$$Z_1Z_2 + Z_1Z_3 + Z_2Z_3 = 32.22\ \text{k}\Omega + j27.65\ \text{k}\Omega$$
$$= 42.46\ \text{k}\Omega\ \underline{/\ 40.6^\circ}$$

$$E_1Z_2 = 1080\ \underline{/\ 60^\circ} = 540 + j935$$
$$E_2Z_1 = 1200\ \underline{/\ 5^\circ} = 1195 + j105$$
$$E_1Z_2 + E_2Z_1 = 1735 + j1040$$
$$= 2023\ \underline{/\ 30.9^\circ}$$

Substituting into Eq. (8):

$$I_3 = \frac{2023 \angle 30.9^\circ}{42.46 \times 10^3 \angle 40.6^\circ}$$

$$\mathbf{I_3 = 47.6\ mA \angle -9.7^\circ}$$

Although the circuit in Figure 23-2 is not an extremely complex network, the analysis of it (in Example 23-1) is quite lengthy. In following sections it is shown that the analysis can be simplified by application of one of the circuit theorems. In general, the solution of more complex networks requires use of the circuit theorems.

23-3 SUPERPOSITION THEOREM APPLIED TO AC NETWORKS

The superposition theorem allows a complex network involving several voltage and/or current sources to be reduced to several simpler networks each of which has only one voltage or current source. For application to ac networks, the *superposition theorem* is stated:

Superposition Theorem (ac circuits)

In a network containing more than one source of voltage or current, the current through any branch is the phasor sum of the currents produced by each source acting independently.

The procedure for determining an ac network branch current by the superposition theorem is as follows:

1. Select one source, and replace all other sources with their internal impedances.
2. Determine the amplitude and phase angle of the current that flows through the desired branch, as a result of the single source acting alone.
3. Repeat steps 1 and 2 using each source in turn until the branch current components have been calculated for all sources.
4. Determine the phasor sum of the current components to obtain the actual branch current.

EXAMPLE 23-2

Using the superposition theorem, analyze the impedance network shown in Figure 23-2 to determine the current through Z_3.

SOLUTION

Selecting source E_1, and replacing source E_2 with its internal impedance, gives the circuit shown in Figure 23-3.

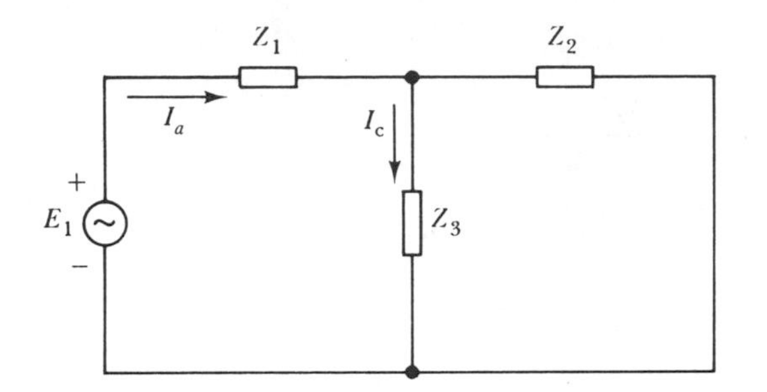

FIG. 23-3. Example 23-2

From Figure 23-3:

$$I_a = \frac{E_1}{Z_1 + Z_2 \| Z_3}$$

$$= \frac{E_1}{Z_1 + \dfrac{Z_2 Z_3}{Z_2 + Z_3}}$$

By the current divider rule:

$$I_c = I_a \times \frac{Z_2}{Z_2 + Z_3}$$

Therefore,

$$I_c = \frac{E_1}{Z_1 + \dfrac{Z_2 Z_3}{Z_2 + Z_3}} \times \frac{Z_2}{Z_2 + Z_3}$$

$$= \frac{E_1 Z_2}{Z_1 Z_2 + Z_1 Z_3 + Z_2 Z_3} \qquad (1)$$

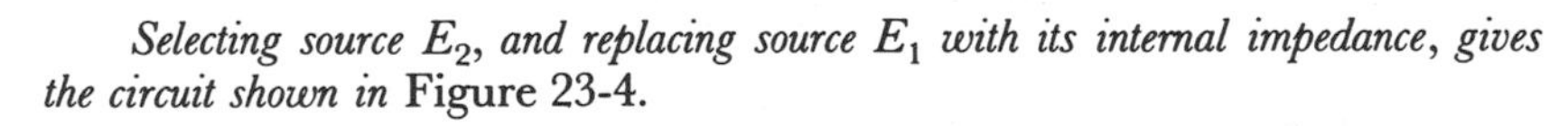

Selecting source E_2*, and replacing source* E_1 *with its internal impedance, gives the circuit shown in* Figure 23-4.

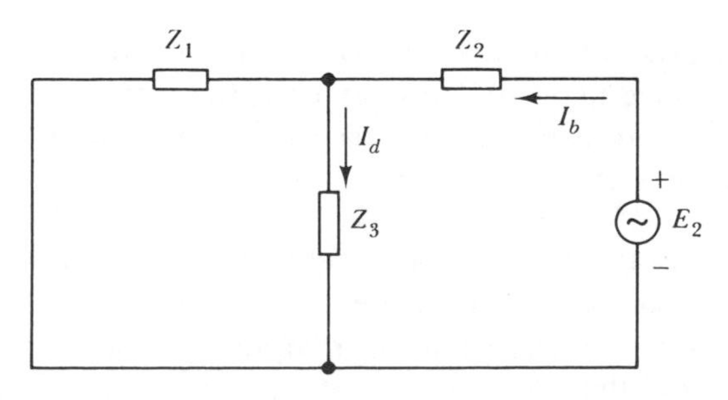

FIG. 23-4. Example 23-2

From Figure 23-4:

$$I_b = \frac{E_2}{Z_2 + \dfrac{Z_1 Z_3}{Z_1 + Z_3}}$$

and

$$I_d = I_b \times \frac{Z_1}{Z_1 + Z_3}$$

Therefore,

$$I_d = \frac{E_2}{Z_2 + \dfrac{Z_1 Z_3}{Z_1 + Z_3}} \times \frac{Z_1}{Z_1 + Z_3}$$

$$I_d = \frac{E_2 Z_1}{Z_2 Z_1 + Z_2 Z_3 + Z_1 Z_3} \qquad (2)$$

$$I_3 = I_c + I_d$$

$$I_3 = \frac{E_1 Z_2 + E_2 Z_1}{Z_1 Z_2 + Z_1 Z_3 + Z_2 Z_3} \qquad (3)$$

Equation (3) *is exactly the same equation obtained for* I_3 *in* Example 23-1. *Therefore, substituting the values of impedances and voltages into the equation, exactly as in* Example 23-1, *gives*

$$\mathbf{I_3 = 47.6\ mA \angle -9.7°}$$

When Examples 23-1 and 23-2 are compared, it is seen that the equation for the unknown branch current was arrived at much more quickly by the use of the superposition theorem than when the loop currents method was used. However, in the case of a more complex circuit, the superposition theorem may not simplify the derivation of the equations.

23-4 NODAL ANALYSIS FOR AC CIRCUITS

The procedure for nodal analysis of an impedance network is exactly the same as that for a dc resistance network, with the exception that once again the phase angle of all quantities must be taken into consideration.

Procedure for nodal anaysis of ac networks:

1. Convert all voltage sources to current sources and redraw the circuit.
2. Identify all nodes and choose a reference node.
3. Write the equations for the currents flowing into and out of each node, with the exception of the reference node.
4. Solve the equations to determine the node voltage and the required branch currents.

EXAMPLE 23-3

Using nodal analysis on the impedance network shown in Figure 23-2, determine the current through Z_3.

SOLUTION

Converting the voltage sources to current sources gives the circuit shown in Figure 23-5(a) *and* (b).

The current sources are:

$$I_a = \frac{E_1}{Z_1} = \frac{6\text{ V} \angle 0^\circ}{100\ \Omega \angle 25^\circ}$$

$$I_a = 60\text{ mA} \angle -25^\circ = 54.4\text{ mA} - j25.4\text{ mA}$$

$$I_b = \frac{E_2}{Z_2} = \frac{12\text{ V} \angle -20^\circ}{180\ \Omega \angle 60^\circ}$$

$$I_b = 66.7\text{ mA} \angle -80^\circ = 11.58\text{ mA} - j65.69\text{ mA}$$

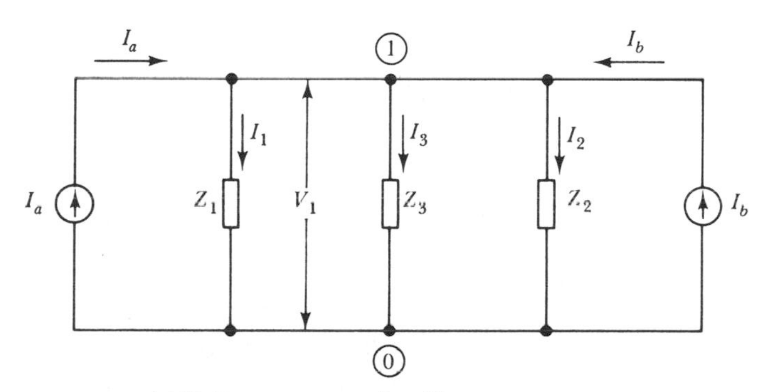

(a) Voltage sources replaced by current sources

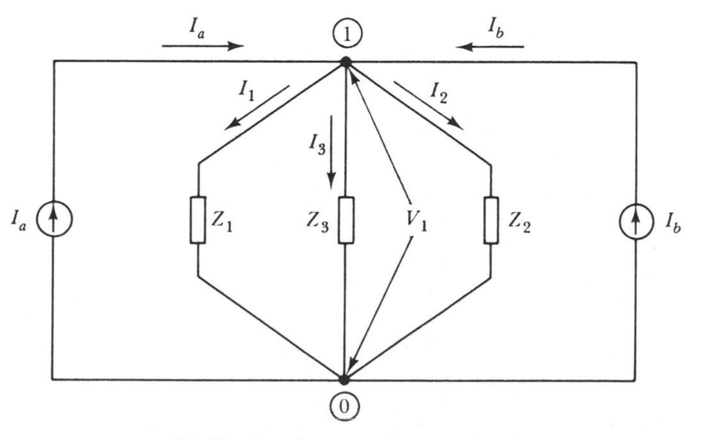

(b) Circuit redrawn to show currents entering and leaving node 1

FIG. 23-5. Example 23-3

The current equation at node 1 *in* Figure 23-5(b) *is*:

$$I_a + I_b = I_1 + I_2 + I_3 \tag{1}$$

$$= \frac{V_1}{Z_1} + \frac{V_1}{Z_2} + \frac{V_1}{Z_3}$$

$$I_a + I_b = V_1\left(\frac{1}{Z_1} + \frac{1}{Z_2} + \frac{1}{Z_3}\right)$$

Therefore,

$$V_1 = \frac{I_a + I_b}{\dfrac{1}{Z_1} + \dfrac{1}{Z_2} + \dfrac{1}{Z_3}} \tag{2}$$

and

$$I_3 = \frac{V_1}{Z_3}$$

$$I_3 = \frac{I_a + I_b}{Z_3\left(\dfrac{1}{Z_1} + \dfrac{1}{Z_2} + \dfrac{1}{Z_3}\right)} \tag{3}$$

$$Z_1 = 100\ \Omega\ \underline{/\ 25^\circ}$$

$$\frac{1}{Z_1} = 0.01\ \text{S}\ \underline{/\ -25^\circ} = (0.009 - j0.0042)\ \text{S}$$

$$Z_2 = 180\ \Omega\ \underline{/\ 60^\circ}$$

$$\frac{1}{Z_2} = 0.0056\ \text{S}\ \underline{/\ -60^\circ} = (0.0028 - j0.0048)\ \text{S}$$

$$Z_3 = 120\ \Omega\ \underline{/\ -30^\circ}$$

$$\frac{1}{Z_3} = 0.0083\ \text{S}\ \underline{/\ 30^\circ} = (0.0072 + j0.0042)\ \text{S}$$

$$\frac{1}{Z_1} + \frac{1}{Z_2} + \frac{1}{Z_3} = (0.019 - j0.0048)\ \text{S}$$

$$= 0.0196\ \text{S}\ \underline{/\ -14.2^\circ}$$

and

$$I_a + I_b = 65.98\ \text{mA} - j91.09\ \text{mA}$$

$$= 112.5\ \text{mA}\ \underline{/\ -54.1^\circ}$$

Substituting into Eq. (3):

$$I_3 = \frac{112.5\ \text{mA}\ \underline{/-54.1^\circ}}{120\ \Omega\ \underline{/\ -30^\circ} \times 0.0196\ \text{S}\ \underline{/\ -14.2^\circ}}$$

$$\mathbf{I_3 = 47.8\ mA}\ \underline{/\ \mathbf{-9.9^\circ}}$$

Note that because of approximations in the impedance calculations, I_3 is slightly different from that determined in Example 23-1.

The nodal analysis method of finding the current through the unknown impedance in Figure 23-2 is seen to be simpler than the use of either loop currents or the superposition theorem. Again, this method may not be the simplest for more complex circuits.

23-5 THÉVENIN'S THEOREM APPLIED TO AC CIRCUITS

Thévenin's theorem allows any single impedance in a network to be isolated. The rest of the network is replaced by a single impedance and a single voltage source. In this way the entire network is reduced to one voltage source in series with two impedances. Calculation of the current level through the desired impedance is then quite simple.

For application to ac networks, Thévenin's theorem is stated:

Thévenin's Theory (ac circuits)

Any two-terminal network containing impedances and voltage and/or current sources may be replaced by a single voltage source in series with a single impedance. The emf of the voltage source is the open-circuit emf at the network terminals, and the series impedance is the impedance between the network terminals when all sources are replaced by their internal impedances.

The procedure for *thevenizing* an ac network is as follows:

1. Calculate the open-circuit terminal voltage of the network.
2. Redraw the network with each voltage source replaced by a short circuit in series with its internal impedance, and each current source replaced by an open circuit in parallel with its internal impedance.
3. Calculate the impedance of the redrawn network as *seen* from the output terminals.

To apply Thévenin's theorem to the problem of finding the current through impedance Z_3 in the circuit of Figure 23-2, Z_3 is first removed as shown in Figure 23-6(a). The open-circuit terminal voltage (V) is then calculated. Next, the circuit is redrawn with the voltage sources replaced by their internal impedances [Figure 23-6(b)], and the impedance *looking into* the output terminals is calculated. Finally, as illustrated in Figure 23-6(c), the open-circuit output voltage is shown in series with the internal impedance and Z_3, and the output current is calculated.

EXAMPLE 23-4 Use Thévenin's theorem to determine the current through Z_3 in the circuit shown in Figure 23-2.

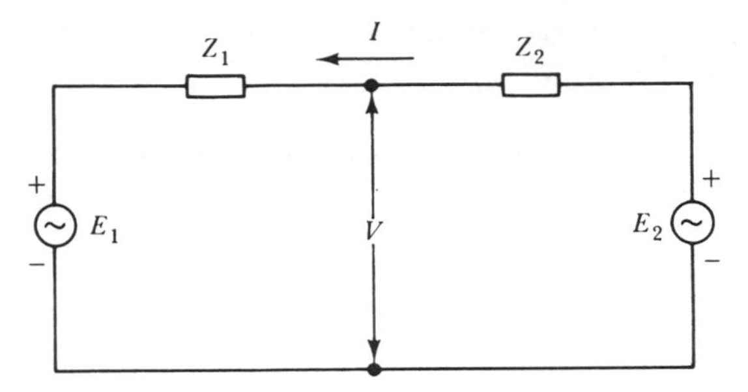

(a) Open-circuit terminal voltage

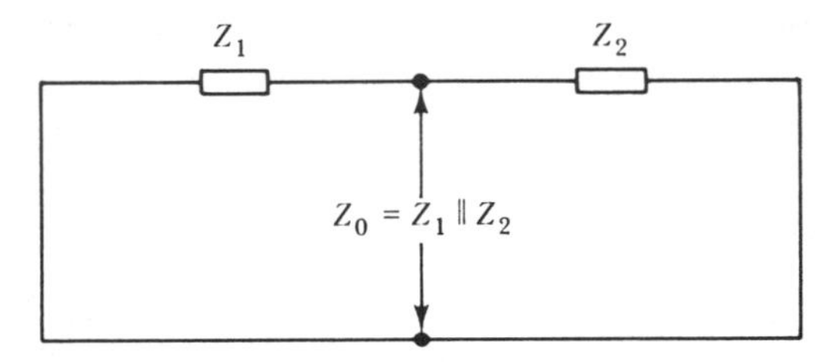

(b) Internal impedance

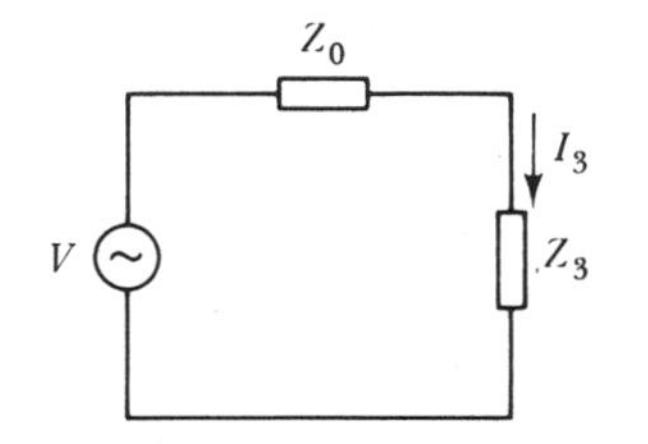

(c) Thévenin equivalent circuit

FIG. 23-6. Thevenizing an ac network

SOLUTION

Referring to Figure 23-6(a):

$$I = \frac{E_2 - E_1}{Z_1 + Z_2}$$

and

$$V = E_1 + IZ_1$$

$$= E_1 + \frac{Z_1 (E_2 - E_1)}{Z_1 + Z_2}$$

$$= \frac{E_1 (Z_1 + Z_2) + Z_1 (E_2 - E_1)}{Z_1 + Z_2}$$

$$V = \frac{E_1 Z_2 + E_2 Z_1}{Z_1 + Z_2} \tag{1}$$

From Figure 23-6(b):

$$Z_0 = Z_1 \| Z_2$$

$$Z_0 = \frac{Z_1 Z_2}{Z_1 + Z_2} \tag{2}$$

and from Figure 23-6(c):

$$I_3 = \frac{V}{Z_3 + Z_0}$$

$$= \frac{V}{Z_3 + \dfrac{Z_1 Z_2}{Z_1 + Z_2}}$$

$$= \frac{V(Z_1 + Z_2)}{Z_3(Z_1 + Z_2) + Z_1 Z_2}$$

$$I_3 = \frac{V(Z_1 + Z_2)}{Z_1 Z_2 + Z_1 Z_3 + Z_2 Z_3} \tag{3}$$

Substituting Eq. (1) *into* Eq. (3):

$$I_3 = \frac{Z_1 + Z_2}{Z_1 Z_2 + Z_1 Z_3 + Z_2 Z_3} \times \frac{E_1 Z_2 + E_2 Z_1}{Z_1 + Z_2}$$

$$I_3 = \frac{E_1 Z_2 + E_2 Z_1}{Z_1 Z_2 + Z_1 Z_3 + Z_2 Z_3} \tag{4}$$

Once again, this is the same equation for the current through Z_3 *as derived in* Examples 23-1 *and* 23-2. *Therefore,*

$$\mathbf{I_3 = 47.6\ mA\ \underline{/-9.7^\circ}}$$

Comparing Example 23-4 to Example 23-2, it is seen that the solution of this particular problem by Thévenin's theorem is approximately as simple (or as complex) as the use of the superposition theorem.

23-6 NORTON'S THEOREM APPLIED TO AC CIRCUITS

By application of Norton's theorem, any impedance in a network can be isolated, and the rest of the network replaced by a single current source in parallel with a single impedance.

For ac networks applications, Norton's theorem is stated:

Norton's Theorem (ac circuits)

Any two-terminal network containing impedances and voltage sources, and/or current sources, may be replaced by a single current source in parallel with a single impedance. The output from the current source is the short-circuit current at the network terminals, and the parallel impedance is the impedance between the network terminals when all sources are replaced by their internal impedances.

The procedure for *nortonizing* an ac network is as follows:

1. Calculate the short-circuit current at the network terminals.
2. Redraw the network with each voltage source replaced by a short circuit in series with its internal impedance, and each current source replaced by an open circuit in parallel with its internal impedance.
3. Calculate the impedance of the redrawn networks as *seen* from its output terminals.

EXAMPLE 23-5

Using Norton's theorem, determine the current through Z_3 in the circuit shown in Figure 23-2.

SOLUTION

From Figure 23-7(a):

$$I_1 = \frac{E_1}{Z_1}$$

and

$$I_2 = \frac{E_2}{Z_2}$$

$$I_{sc} = I_1 + I_2$$

$$= \frac{E_1}{Z_1} + \frac{E_2}{Z_2} \qquad (1)$$

From Figure 23-7(b):

$$Z_0 = Z_1 \| Z_2$$

$$Z_0 = \frac{Z_1 Z_2}{Z_1 + Z_2} \qquad (2)$$

In Figure 23-7(c):

$$I_N = I_{sc} = \frac{E_1}{Z_1} + \frac{E_2}{Z_2}$$

and

$$Z_N = Z_0 = Z_1 \| Z_2$$

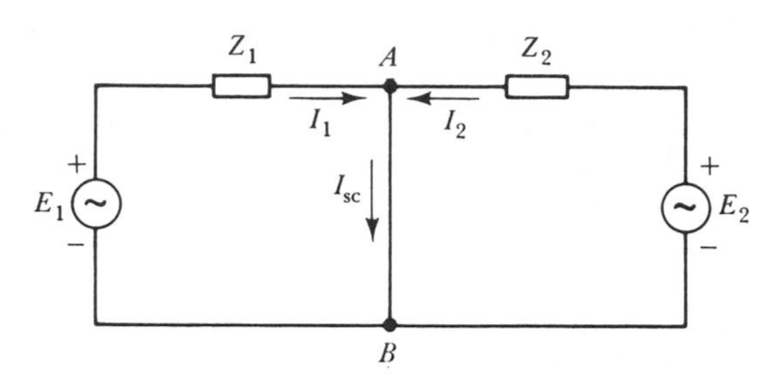

(a) Short-circuit output current

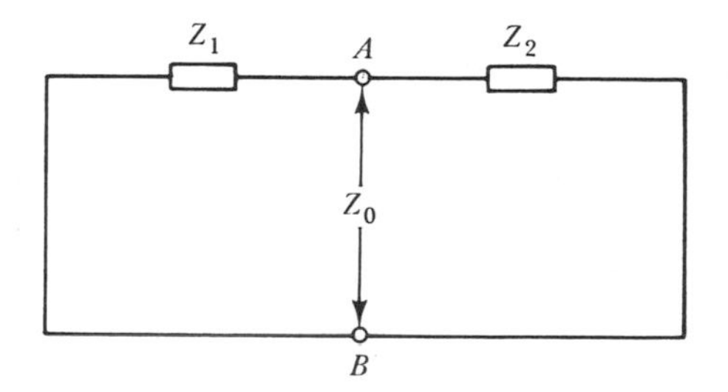

(b) Internal impedance

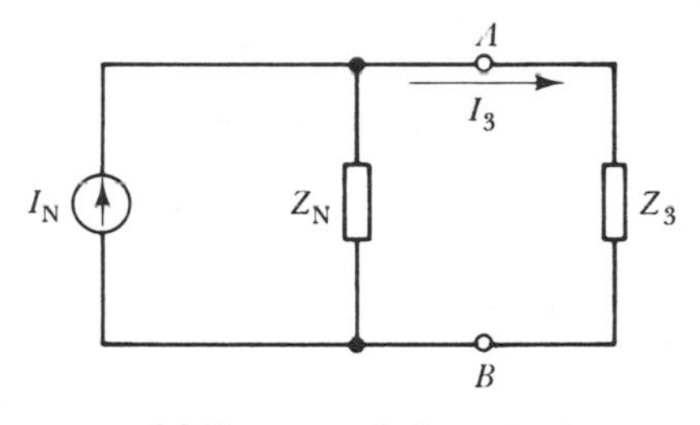

(c) Norton equivalent circuit

FIG. 23-7. Nortonizing an ac network

Using the current divider rule:

$$I_3 = \frac{I_N \times Z_N}{Z_3 + Z_N}$$

Substituting for Z_N:

$$I_3 = \frac{I_N \times \dfrac{Z_1 Z_2}{Z_1 + Z_2}}{Z_3 + \dfrac{Z_1 Z_2}{Z_1 + Z_2}}$$

$$I_3 = \frac{I_N Z_1 Z_2}{Z_1 Z_3 + Z_2 Z_3 + Z_1 Z_2} \qquad (3)$$

Substituting for I_N:

$$I_3 = \frac{Z_1Z_2\left(\frac{E_1}{Z_1}+\frac{E_2}{Z_2}\right)}{Z_1Z_3+Z_2Z_3+Z_1Z_2}$$

$$I_3 = \frac{E_1Z_2+E_2Z_1}{Z_1Z_3+Z_2Z_3+Z_1Z_2} \qquad (4)$$

Once again this is the same equation for I_3 as derived in other examples, and it gives

$$\mathbf{I_3 = 47.6\ mA\ \angle -9.7^\circ}$$

The use of Norton's theorem to determine the current through impedance Z_3 in Figure 23-2 is seen to simplify the solution of the problem to the same degree as Thévenin's theorem.

23-7 MAXIMUM POWER TRANSFER THEOREM APPLIED TO AC CIRCUITS

When considering purely resistive circuits, it was found that maximum power is transferred from a voltage source (or current source) when the load resistance equals the source resistance (see Section 11-5). To understand how the maximum power transfer theorem applies to impedance networks, consider the circuit shown in Figure 23-8. The source impedance is

$$Z_s = R_1 + jX_1$$

and the load impedance is

$$Z_L = R_2 - jX_2$$

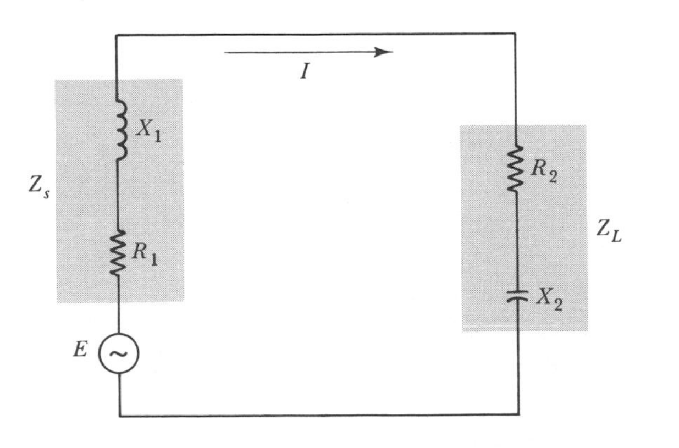

FIG. 23-8. For maximum power transfer $X_1 = -X_2$ and $R_1 = R_2$

The output current is

$$I=\frac{E}{Z_s+Z_L}$$

$$=\frac{E}{(R_1+jX_1)+(R_2-jX_2)}$$

or

$$|I|=\frac{E}{\sqrt{(R_1+R_2)^2+(X_1-X_2)^2}}$$

The power dissipation in the load impedance occurs in the resistive portion of the load. Thus,

$$P_o=I^2R_2$$

or

$$\boxed{P_o=\frac{E^2R_2}{(R_1+R_2)^2+(X_1-X_2)^2}} \tag{23-1}$$

Maximum power output occurs when the denominator in Eq. (23-1) has its minimum value (i.e., when $X_1=X_2$) which gives

$$\boxed{P_{o(\max)}=\frac{E^2R_2}{(R_1+R_2)^2}} \tag{23-2}$$

When this condition is obtained, the circuit behaves as if there were no reactive components, and therefore requires $R_1=R_2$ for maximum power output.

It is seen that maximum power output is derived from an ac source when the resistive components of the source impedance and load impedance are equal, and when the reactive components of the source impedance and load impedance are equal in magnitude but opposite in sign. Thus, for maximum power transfer, an *LR* source must have a *CR* load in which the capacitive reactance has the same magnitude as the inductive reactance. Similarly, a *CR* source must have an *LR* load, with $|X_L|=|X_C|$. Another way of stating this requirement is that the load impedance must be the *conjugate* of the source impedance.

For application to ac impedance networks and ac sources, the maximum power transfer theorem is stated as follows:

Maximum Power Transfer Theorem (ac circuits)

Maximum output power is obtained from a network or source when the load impedance is the conjugate of the output impedance of the network or source, as seen from the terminals of the load.

EXAMPLE 23-6

A voltage source has an equivalent circuit consisting of $R_s = 100\ \Omega$ in series with $L = 20\ \mu\text{H}$. Calculate the optimum load for maximum output power at a frequency of 500 kHz.

SOLUTION

Reactive component of source impedance:

$$X_1 = 2\pi fL$$

$$= 2\pi \times 500\ \text{kHz} \times 20\ \mu\text{H}$$

$$X_1 = 62.83\ \Omega$$

Reactive component of load impedance:

$$X_2 = -X_1$$

$$= -62.83\ \Omega \quad (\textit{capacitive})$$

Therefore,

$$\frac{1}{2\pi fC} = 2\pi fL$$

and

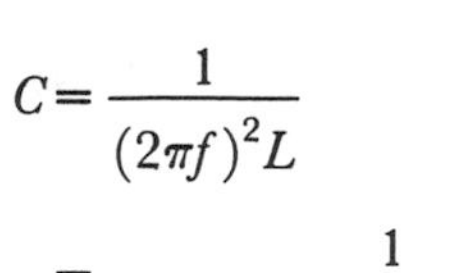

$$C = \frac{1}{(2\pi f)^2 L}$$

$$= \frac{1}{(2\pi \times 500\ \text{kHz})^2 \times 20\ \mu\text{H}}$$

$$\mathbf{C = 0.005\ \mu F}$$

Resistive component of load impedance:

$$\mathbf{R_L = R_s = 100\ \Omega}$$

EXAMPLE 23-7

For the voltage source described in Example 23-6, a parallel CR circuit is to be used as a load. Determine the required component values.

SOLUTION

From Example 23-6, *a parallel CR equivalent circuit is required for the series connected* $R = 100\ \Omega$ *and* $C = 0.005\ \mu\text{F}$.

Series circuit is

$$R - jX = 100\ \Omega - j62.83\ \Omega$$

from Eq. (21-6),

$$R_p = \frac{R_s^2 + X_s^2}{R_s}$$

$$= \frac{100^2 + 62.83^2}{100}$$

$$\mathbf{R_p = 139.5\ \Omega}$$

from Eq. (21-7),

$$X_p = \frac{R_s^2 + X_s^2}{X_s}$$

$$= \frac{100^2 + 62.83^2}{62.83}$$

$$X_p = 222\ \Omega$$

$$X_p = \frac{1}{2\pi f C_p}$$

Therefore,

$$C_p = \frac{1}{2\pi f X_p}$$

$$= \frac{1}{2\pi \times 500\ \text{kHz} \times 222\ \Omega}$$

$$\mathbf{C_p = 0.0014\ \mu F}$$

23-8 DELTA–WYE TRANSFORMATIONS FOR AC NETWORKS

Conversion between *delta* (Δ) and *wye* (Y) impedance networks can be performed using the formulas derived in Section 10-5 for delta–wye conversion of resistance networks. Figure 23-9 shows Δ and Y impedance networks, and the conversion formulas for the networks as illustrated are

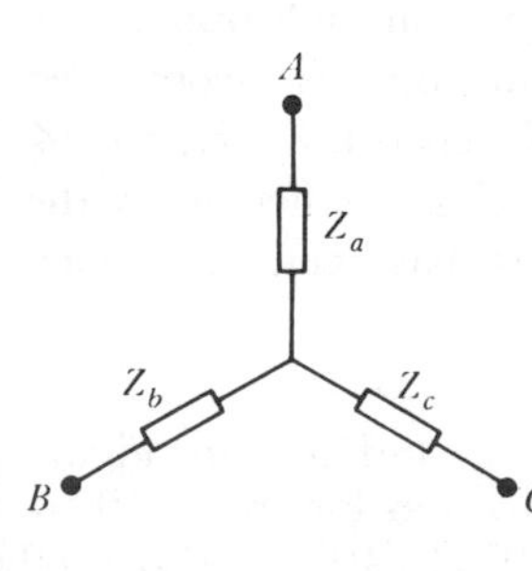

(a) Y impedance network

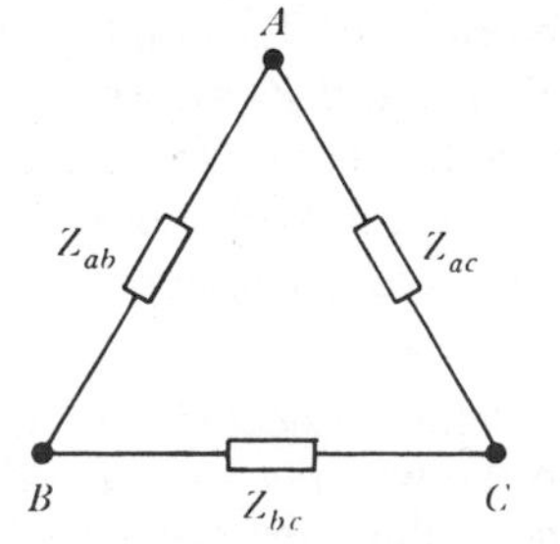

(b) Δ impedance network

FIG. 23-9. Y and Δ impedance networks

listed below. As always, the impedances must be treated as phasor quantities.

Converting from Δ to Y:

$$Z_a = \frac{Z_{ab}Z_{ac}}{Z_{ab}+Z_{ac}+Z_{bc}} \tag{23-3}$$

$$Z_b = \frac{Z_{ab}Z_{bc}}{Z_{ab}+Z_{ac}+Z_{bc}} \tag{23-4}$$

$$Z_c = \frac{Z_{ac}Z_{bc}}{Z_{ab}+Z_{ac}+Z_{bc}} \tag{23-5}$$

Converting from Y to Δ:

$$Z_{ab} = \frac{Z_aZ_b+Z_aZ_c+Z_bZ_c}{Z_c} \tag{23-6}$$

$$Z_{ac} = \frac{Z_aZ_b+Z_aZ_c+Z_bZ_c}{Z_b} \tag{23-7}$$

$$Z_{bc} = \frac{Z_aZ_b+Z_aZ_c+Z_bZ_c}{Z_a} \tag{23-8}$$

A typical application of Δ–Y conversion is illustrated in Figure 23-10. In the circuit shown in Figure 23-10(a), the current through impedance Z_3 is to be determined. To simplify the calculations, the delta-connected impedances Z_4, Z_5, and Z_6 are converted to the Y network Z_a, Z_b, and Z_c shown in Figure 23-10(b). Note that the value of Z_3 is unaffected by the conversion. Figure 23-10(c) shows that the network has been very much simplified by the Δ to Y conversion.

EXAMPLE 23-8

Convert the delta-connected impedances, Z_4, Z_5, and Z_6, in Figure 23-10(a), into the wye network, Z_a, Z_b, and Z_c, shown in Figure 23-10(b). The impedance values are $Z_4 = 100\ \Omega\ \angle 30°$, $Z_5 = 95\ \Omega\ \angle 40°$, and $Z_6 = 60\ \Omega\ \angle 20°$.

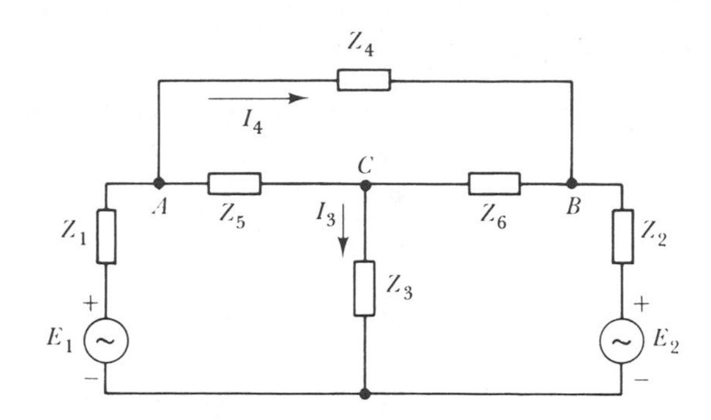

(a) Impedance network with Δ connected impedances Z_4, Z_5 and Z_6

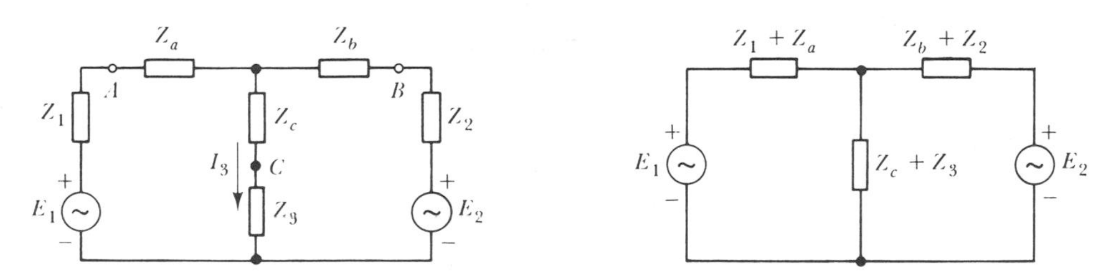

(b) Δ connected impedances replaced by equivalent Y network

(c) Simplified impedance network

FIG. 23-10. Use of Δ–Y transformation to simplify a complex impedance network

SOLUTION

$$Z_{ab} = Z_4 = 100\ \Omega\ \angle 30^\circ = 86.6\ \Omega + j50\ \Omega$$

$$Z_{ac} = Z_5 = 95\ \Omega\ \angle 40^\circ = 72.8\ \Omega + j61.1\ \Omega$$

$$Z_{bc} = Z_6 = 60\ \Omega\ \angle 20^\circ = 56.4\ \Omega + j20.5\ \Omega$$

$$Z_{ab} + Z_{ac} + Z_{bc} = 215.8\ \Omega + j131.6\ \Omega$$

$$= 253\ \Omega\ \angle 31.4^\circ$$

From Eq. (23-3):

$$Z_a = \frac{100\ \Omega\ \angle 30^\circ \times 95\ \Omega\ \angle 40^\circ}{253\ \Omega\ \angle 31.4^\circ}$$

$$\mathbf{Z_a = 37.5\ \Omega\ \angle 38.6^\circ}$$

From Eq. (23-4):

$$Z_b = \frac{100\ \Omega\ \angle 30^\circ \times 60\ \Omega\ \angle 20^\circ}{253\ \Omega\ \angle 31.4^\circ}$$

$$\mathbf{Z_b = 23.7\ \Omega\ \angle 18.6^\circ}$$

From Eq. (23-5):

$$Z_c = \frac{95\ \Omega\ \angle 40^\circ \times 60\ \Omega\ \angle 20^\circ}{253\ \Omega\ \angle 31.4^\circ}$$

$$\mathbf{Z_c = 22.5\ \Omega\ \angle 28.6^\circ}$$

GLOSSARY OF FORMULAS

Power output from an ac source to an impedance:

$$P_o = \frac{E^2 R_2}{(R_1 + R_2)^2 + (X_1 - X_2)^2}$$

$$P_{o(\max)} = \frac{E^2 R_2}{R_1 + R_2}$$

Δ to Y network conversion:

$$Z_a = \frac{Z_{ab} Z_{ac}}{Z_{ab} + Z_{ac} + Z_{bc}}$$

$$Z_b = \frac{Z_{ab} Z_{bc}}{Z_{ab} + Z_{ac} + Z_{bc}}$$

$$Z_c = \frac{Z_{ac} Z_{bc}}{Z_{ab} + Z_{ac} + Z_{bc}}$$

Y to Δ network conversion:

$$Z_{ab} = \frac{Z_a Z_b + Z_a Z_c + Z_b Z_c}{Z_c}$$

$$Z_{ac} = \frac{Z_a Z_b + Z_a Z_c + Z_b Z_c}{Z_b}$$

$$Z_{bc} = \frac{Z_a Z_b + Z_a Z_c + Z_b Z_c}{Z_a}$$

REVIEW QUESTIONS

23-1 State Kirchhoff's voltage law and current law as applied to ac impedance networks.

23-2 Explain why ac voltage and current sources must have their terminals identified as + and −.

23-3 State the superposition theorem as applied to ac impedance networks.

23-4 State Thévenin's theorem as applied to ac impedance networks.

23-5 State Norton's theorem as applied to ac impedance networks.

23-6 State the maximum power transfer theorem as applied to ac sources. Sketch the circuit of an ac source and load, and derive the appropriate equation to show the truth of the theorem.

PROBLEMS

23-1 Using loop equations, determine the current through Z_4 in the circuit shown in Figure 23-11.

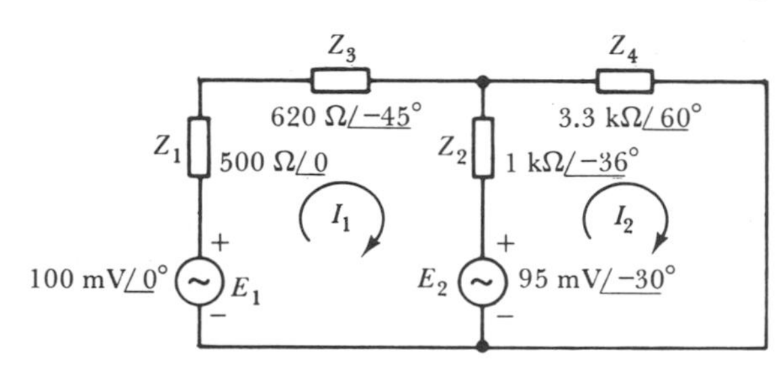

FIG. 23-11.

23-2 For the circuit shown in Figure 23-12, use loop equations to calculate the voltage across R_2.

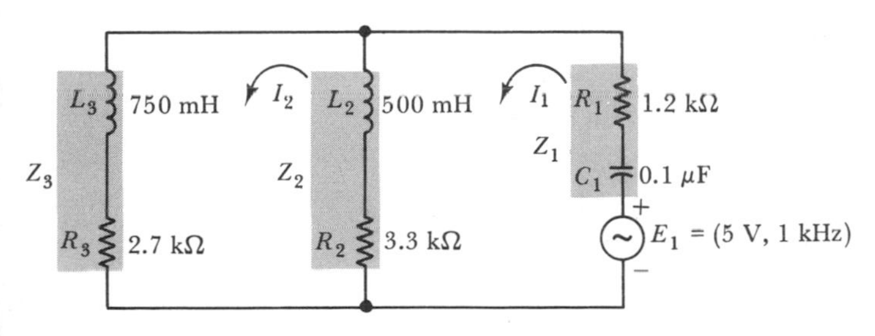

FIG. 23-12.

23-3 Determine the current taken from voltage source E_1 in Figure 23-13. Use the loop equation method.

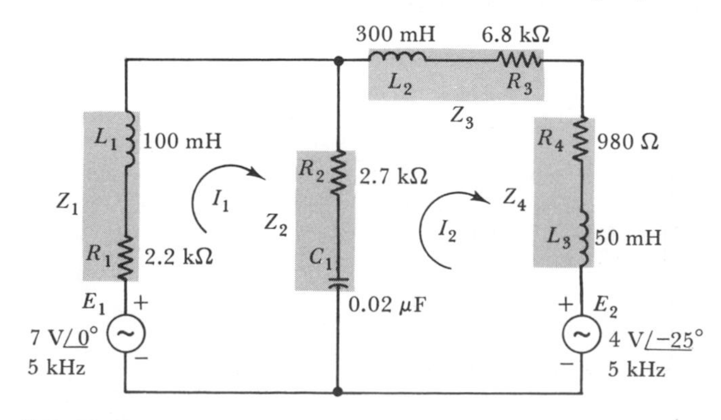

FIG. 23-13.

23-4 Using the superposition theorem, repeat Problem 23-1.

23-5 Apply the superposition theorem to solve Problem 23-3.

23-6 By applying the superposition theorem, determine the current through Z_4 in the circuit shown in Figure 23-14.

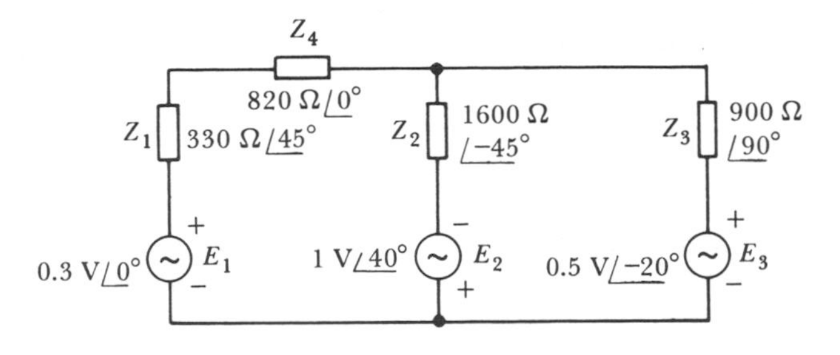

FIG. 23-14.

23-7 Use the superposition theorem to calculate the voltage drop across Z_4 in Figure 23-15.

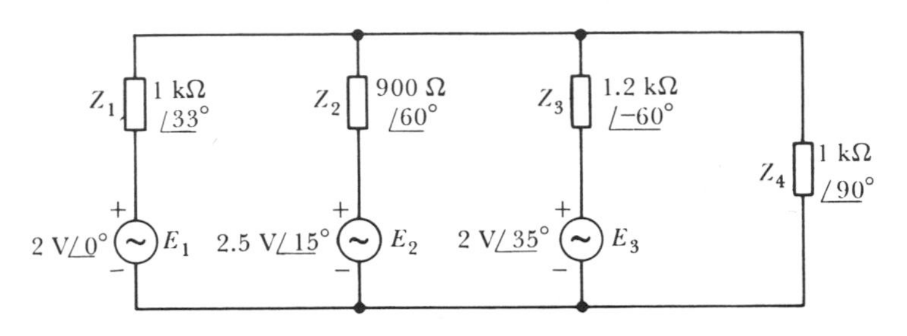

FIG. 23-15.

23-8 Apply the nodal analysis method to solve Problem 23-2.

23-9 Using nodal analysis, solve Problem 23-3.

23-10 Solve Problem 23-7 by nodal analysis.

23-11 Use Thévenin's theorem to determine the current through Z_5 in Figure 23-16.

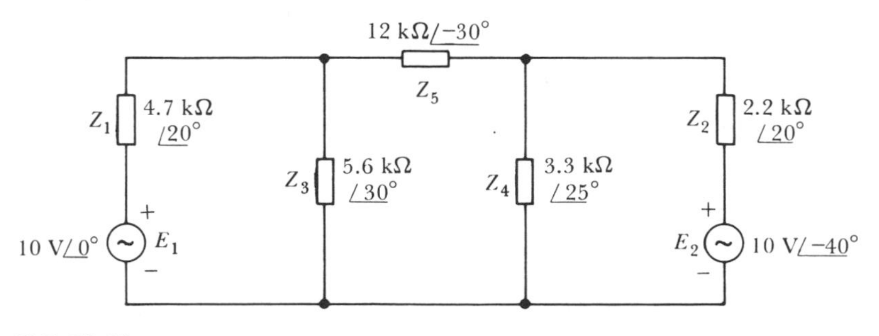

FIG. 23-16.

23-12 Apply delta–wye transformation and nodal analysis to determine the current through Z_4 in the circuit of Figure 23-17.

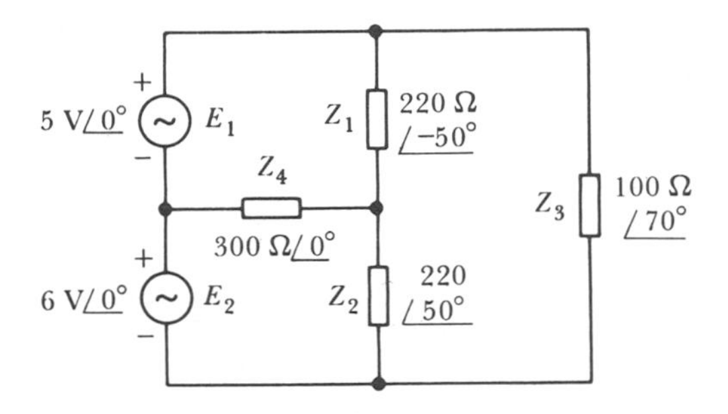

FIG. 23-17.

23-13 Apply Norton's theorem to solving Problem 23-7.

23-14 Calculate the current that flows through resistor R_3, in the circuit shown in Figure 23-18. Do the calculation:

a. By the use of Norton's theorem.

b. By applying Thévenin's theorem.

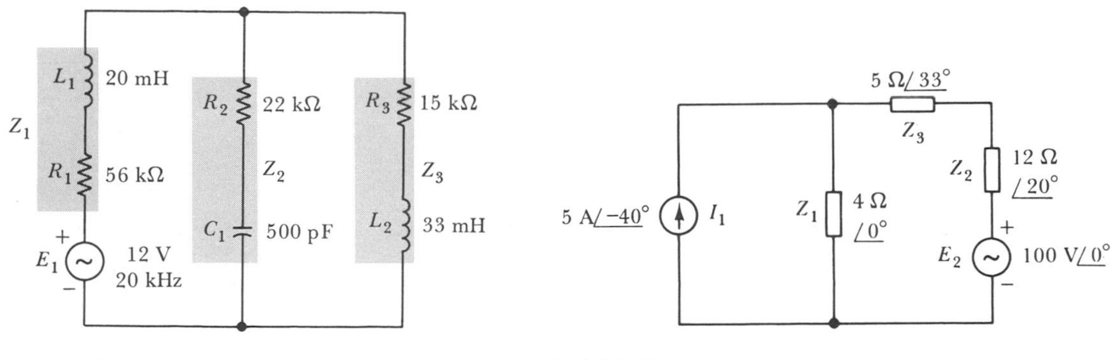

FIG. 23-18.

FIG. 23-19.

23-15 For the circuit shown in Figure 23-19, calculate the current through Z_3:

a. By nodal analysis.

b. By Norton's theorem.

23-16 Apply the delta–wye transformation method to determine the current taken from the voltage source in the circuit of Figure 23-20.

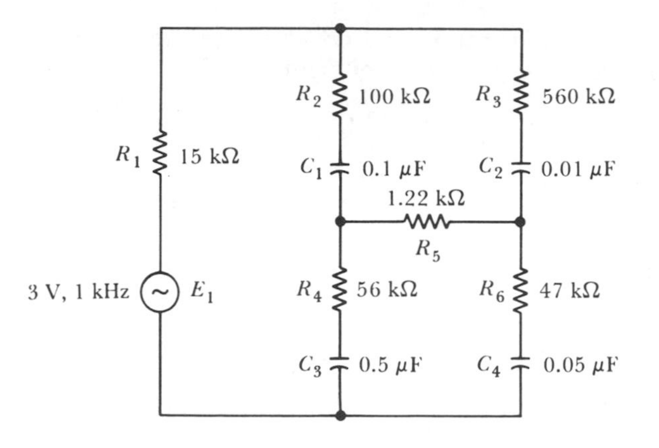

FIG. 23-20.

23-17 A voltage source has an equivalent circuit consisting of $R=500\ \Omega$ in series with $C=0.01\ \mu F$. Calculate the required component values of a *parallel LR* circuit that will draw maximum power from the source when the signal frequency is 1 MHz.

24

RESONANCE

INTRODUCTION

When a series connected *LCR* circuit has the frequency of its alternating supply voltage varied, it is found that $X_L = X_C$ at a particular frequency. The two reactances cancel, and the result is that the series impedance of the circuit has a minimum value of $Z = R$. The circuit is said to be in a state of *resonance*, and the frequency at which this occurs is termed the *resonance frequency*. Since the impedance is a minimum, the series current is a maximum, and there is a rise in the voltage developed across *L* and *C*.

A parallel *LC* circuit can also be in a state of resonance, and again this occurs when $X_L = X_C$. In this case, the circuit impedance has a maximum value, and there is a rise in current through each component. A circuit can be tuned to resonate over a range of frequencies by making *L* or *C* adjustable.

24-1 SERIES RESONANCE

FREQUENCY EFFECT ON CIRCUIT IMPEDANCE. The series *LCR* circuit shown in Figure 24-1(a) has an impedance of

$$\boxed{Z = R + j(X_L - X_C)} \qquad (24\text{-}1)$$

Because $X_L = 2\pi fL$ and $X_C = 1/2\pi fC$, the actual value of the impedance obviously depends upon the frequency of the alternating supply. Figure

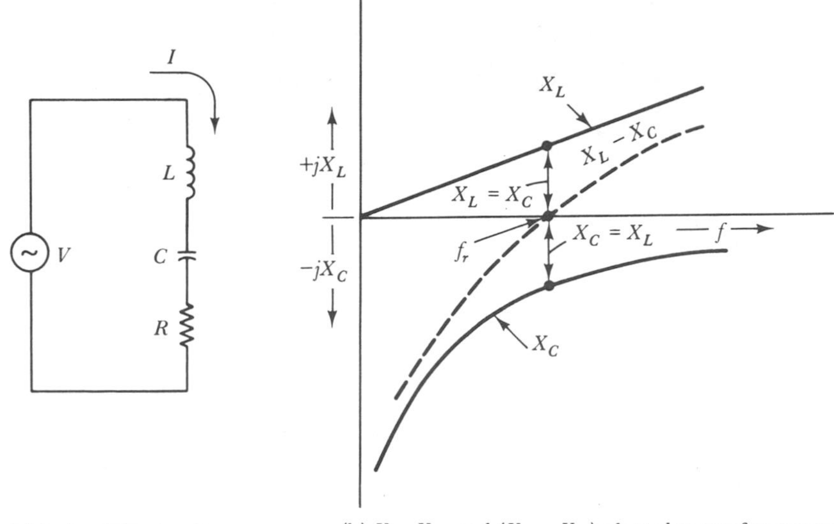

(a) Series *LCR* circuit (b) X_L, X_C, and $(X_L - X_C)$ plotted versus frequency

FIG. 24-1. Series *LCR* circuit and reactances plotted versus frequency

24-1(b) shows the values of X_L and X_C plotted versus supply frequency. It is seen that the inductive reactance increases linearly from zero with increase in frequency, because X_L is directly proportional to f. However, the capacitive reactance is inversely proportional to the frequency, and at low frequency X_C is infinitely large. As f increases, X_C rapidly decreases, resulting in a curved graph, as illustrated.

The total reactance is $(X_L - X_C)$, and the graph of this quantity is shown by the dashed line in Figure 24-1(b). At low frequencies, it is clear that X_L is much smaller than X_C, and thus the total reactance is largely capacitive. At the higher frequencies, where X_L is much larger than X_C, the total reactance becomes largely inductive. At one particular frequency, identified as f_r on the graph, the values of X_L and X_C are equal. Consequently, the impedance of the series *LCR* circuit becomes

$$Z = R + j(0)$$

or

$$Z = R$$

The frequency at which this phenomenon occurs is known as the *resonance frequency* (f_r), and at this frequency the circuit is said to be in a state of *electrical resonance*.

The resonance frequency for a series LCR circuit is defined as the frequency at which $X_L = X_C$.

Referring again to Figure 24-1(b), it is seen that at frequencies above and below resonance, the reactance has a large value. Therefore, when the series LCR circuit impedance is plotted versus frequency [Figure 24-2(a)], it is found that the impedance has a high value above and below the resonance frequency, and that at resonance, it dips to its minimum value of $Z=R$. Thus, *a series LCR circuit has a minimum impedance at the resonance frequency.*

CURRENT AT RESONANCE. The current in the series LCR circuit is determined from

$$\boxed{I=\frac{V}{R+j(X_L-X_C)}} \tag{24-2}$$

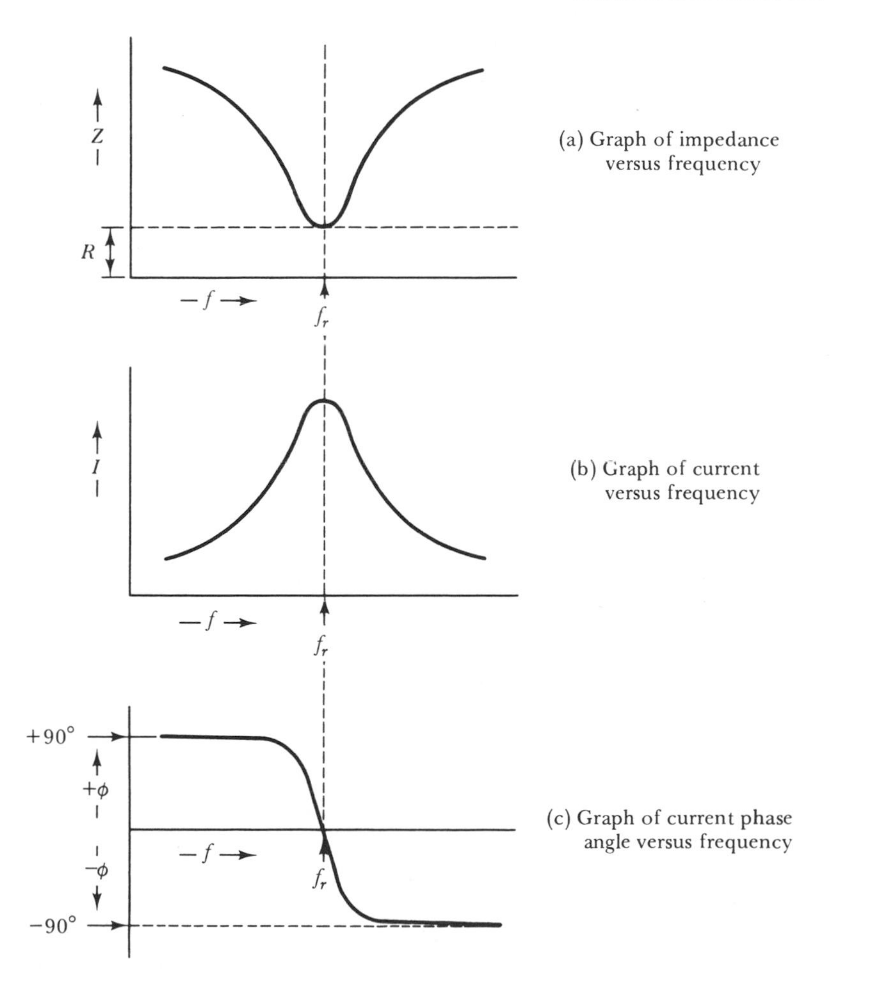

FIG. 24-2. Effect of frequency variation on impedance, current, and phase angle for an *LCR* circuit

or
$$|I| = \frac{V}{\sqrt{R^2 + (X_L - X_C)^2}}$$

where $|I|$ is the numerical value or amplitude of the current, without reference to its phase angle.

Thus, at resonance, where $X_L = X_C$, the current equation becomes

$$I = \frac{V}{R}$$

A typical graph of current versus frequency for the series LCR circuit is shown in Figure 24-2(b). The current is a minimum at frequencies above and below resonance, but peaks sharply at resonance as the impedance falls to a minimum. This I versus f graph is sometimes referred to as the *frequency response curve* of the circuit.

As already noted from Figure 24-1(b), the impedance of the series LCR circuit is largely capacitive at frequencies well below resonance. This means that the circuit current *leads* the applied voltage by a phase angle of approximately 90°. Conversely, because the impedance is largely inductive at frequencies much greater than the resonance frequency, the phase angle of the current above resonance is approximately −90°. The graph of phase angle versus frequency for the series LCR circuit [Figure 24-2(c)] shows the 90° leading phase angle at low frequencies, changing to 0° at the resonance frequency, and moving to a 90° lagging phase angle above f_r.

RESONANCE FREQUENCY. The frequency at which resonance occurs is easily calculated by equating X_L and X_C:

$$X_L = 2\pi f_r L$$

$$X_C = \frac{1}{2\pi f_r C}$$

Therefore,
$$2\pi f_r L = \frac{1}{2\pi f_r C}$$

$$f_r^2 = \frac{1}{(2\pi)^2 LC}$$

and

$$\boxed{f_r = \frac{1}{2\pi\sqrt{LC}}} \quad \textbf{(24-3)}$$

When L and C are in henrys and farads, respectively, Eq. (24-3) *gives f_r in hertz.*

EXAMPLE 24-1

If the series LCR circuit shown in Figure 24-1(a) has: $L=85\ \mu H$, $C=298$ pF, $R=100\ \Omega$, and $V_s=10$ V, determine the resonance frequency. Also, calculate the circuit currents at: $0.25f_r$, $0.5f_r$, $0.8f_r$, f_r, $1.25f_r$, $2f_r$, and $4f_r$. Plot a graph of I versus f to a logarithmic base.

SOLUTION

Eq. (24-3):

$$f_r=\frac{1}{2\pi\sqrt{LC}}$$

$$=\frac{1}{2\pi\sqrt{85\ \mu H\times 298\ pF}}$$

$$\mathbf{f_r=1\ MHz}$$

Eq. (24-2):

$$I=\frac{V}{R+j(X_L-X_C)}$$

or

$$|I|=\frac{V}{\sqrt{R^2+(X_L-X_C)^2}}$$

The values of the various quantities are tabulated below for each frequency, and the graph of current versus frequency is plotted in Figure 24-3.

f	X_L (Ω)	X_C (Ω)	(X_L-X_C) (Ω)	$\|z\|$ (Ω)	I (mA)
250 kHz	134	2136	$-j2$ kΩ	≅2 kΩ	5
500 kHz	267	1068	$-j801$	≅807	12.4
800 kHz	427	667.5	$-j241$	≅261	38.3
1 MHz	534	534	0	100	100
1.25 MHz	667.5	427	$+j241$	≅261	38.3
2 MHz	1068	267	$+j801$	≅807	12.4
4 MHz	2136	134	$+j2$ kΩ	≅2 kΩ	5

The graph of current versus frequency plotted in Figure 24-3 clearly shows that the current reaches a maximum value at the resonance frequency, and that it falls off sharply at frequencies above and below resonance. Note from the table that when the frequency is doubled from 250 kHz to 500 kHz, the current goes from 5 mA to 12.4 mA. Also note that above resonance the current goes from 12.4 mA to 5 mA when the frequency is doubled from 2 MHz to 4 MHz. The same sort of corresponding current ratios are evident for other frequency changes above and below the resonance frequency. It is seen that the ratio of I to f is *not*

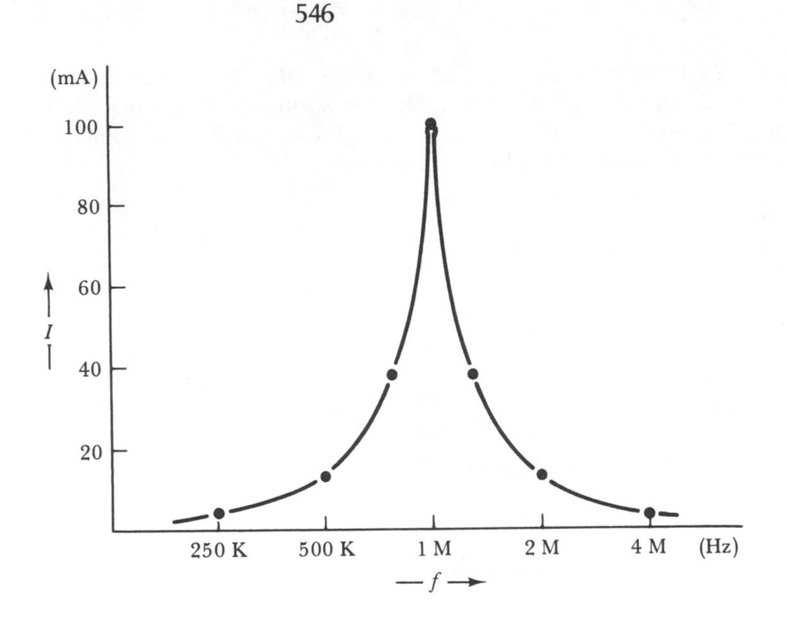

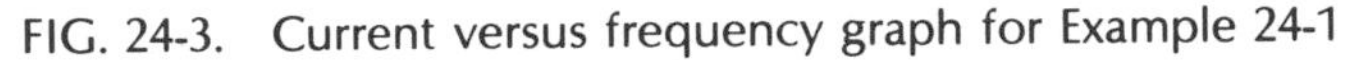
FIG. 24-3. Current versus frequency graph for Example 24-1

linear but logarithmic, and this is why a logarithmic frequency scale is employed on the graph.

RESONANCE RISE IN VOLTAGE. The voltage across the resistance of the series *LCR* circuit is $V_R = I \times R$, and at resonance this is $V_R =$ (the supply voltage), V in Figure 24-1(a).

The voltage across the capacitor and the inductor is

$$\boxed{V_C = IX_C} \qquad (24\text{-}4)$$

and

$$\boxed{V_L = IX_L} \qquad (24\text{-}5)$$

When the capacitor voltage and inductor voltage values are calculated for various frequencies, and plotted versus a logarithmic frequency scale, it is found that their *frequency response curves* are similar in shape to that of the I versus f graph. It is also found that V_C and V_L at resonance can be *many times greater than the supply voltage*. This effect is termed the *resonant rise in voltage*.

EXAMPLE 24-2 For the series *LCR* circuit referred to in Example 24-1, determine the values of capacitor voltage and inductor voltage at each frequency. Also plot V_R, V_C, and V_L versus frequency.

SOLUTION

$$V_C = IX_C \qquad V_L = IX_L \qquad V_R = IR$$

The values of the various quantities are tabulated below for each frequency, and the graphs are plotted in Figure 24-4.

f	IR (V)	IX_L (V)	IX_C (V)
250 kHz	0.5	0.67	10.7
500 kHz	1.24	3.3	13.2
800 kHz	3.83	16.4	25.5
1 MHz	10	53.4	53.4
1.25 MHz	3.83	25.5	16.4
2 MHz	1.24	13.2	3.3
4 MHz	0.5	10.7	0.67

Example 24-2 demonstrates that at resonance the voltages across the capacitor and inductor become much greater than the supply voltage. The voltage across the resistance (V_R) is, of course, in phase with the circuit current, whereas V_L leads the current by 90°, and V_C lags the current phasor by 90°. This is illustrated in the phasor diagram in Figure 24-5(a). From the phasor diagram it is clear that V_L and V_C cancel, and the supply voltage is equal to the voltage across the resistance.

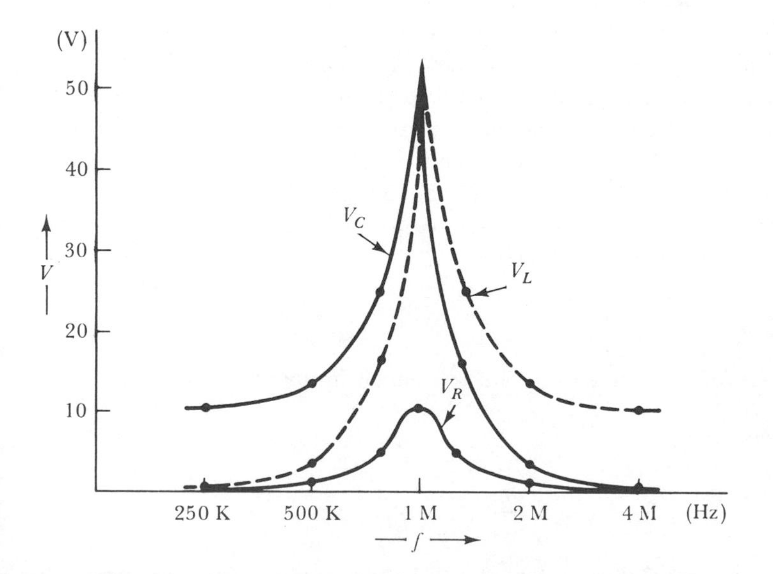

FIG. 24-4. V_C, V_L, and V_R plotted versus frequency for Example 24-2

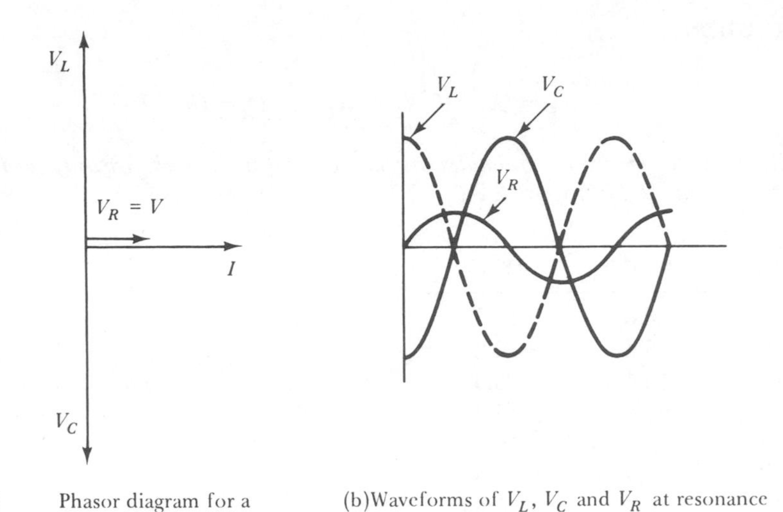

(a) Phasor diagram for a series LCR circuit at resonance

(b) Waveforms of V_L, V_C and V_R at resonance

FIG. 24-5. Phasor diagram and voltage waveforms for a series LCR circuit at resonance

ENERGY TRANSFER BETWEEN L AND C. When the waveforms of V_L, V_R, and V_C are plotted for the resonance frequency [see Figure 24-5(b)], it is evident that as V_L grows positively, V_C is becoming more negative, and vice versa. The positive peak of V_L occurs at the same instant as the negative peak of V_C, and the positive peak of V_C corresponds in time with the negative peak of V_L. Note also that the positive and negative peaks of voltage across the resistance coincide with the instant when V_L and V_C are zero. The information to be derived from these waveforms is that there is energy stored in the resonant circuit. The energy is continuously being transferred from the inductor to the capacitor and back again at the resonance frequency. The resonance frequency, therefore, is the only frequency at which this transfer of energy can take place for the values of capacitor and inductor involved. Note, however, that *there can be no energy stored and no transfer of energy between the reactive components unless there is an energy input at the resonance frequency.*

24-2 TUNING FOR RESONANCE

The equation for calculating the resonance frequency has already been derived as
Eq. (24-3):

$$f_r = \frac{1}{2\pi\sqrt{LC}}$$

When L and C have fixed values, the frequency of the supply must be made equal to f_r to achieve resonance. Suppose, instead, that f had a fixed value; then either L or C (or perhaps both) could be adjusted until the circuit resonates at the supply frequency. In fact, this is exactly what is done in a radio receiver. Figure 24-6 shows a circuit in which V_s represents a small voltage induced from the atmosphere into the inductance L. Resistance R represents the resistive component of L, and the capacitor C is variable. The electromagnetic signals sent into the atmosphere by different radio stations have different frequencies. Thus, by adjusting capacitor C the circuit may be made to resonate at any one of a wide range of radio frequencies. Adjusting the value of C (or L) is termed *tuning* the circuit, and C is referred to as a *tuning capacitor*. The resonance frequency causes a *resonance rise in voltage* across the tuning capacitor. The signal developed across the capacitor can then be passed on to the other circuits of the radio receiver.

EXAMPLE 24-3

The circuit shown in Figure 24-6 has $L = 100\ \mu\text{H}$ and is to be tuned to resonate at frequencies ranging from 500 kHz to 1 MHz. Calculate the range of adjustment of the tuning capacitor.

SOLUTION

From Eq. (24-3),

$$C = \frac{1}{4\pi^2 f_r^2 L}$$

At $f_r = 500$ kHz:

$$C = \frac{1}{4\pi^2 \times (500\ \text{kHz})^2 \times 100\ \mu\text{H}}$$

$$\cong 1000\ \text{pF}$$

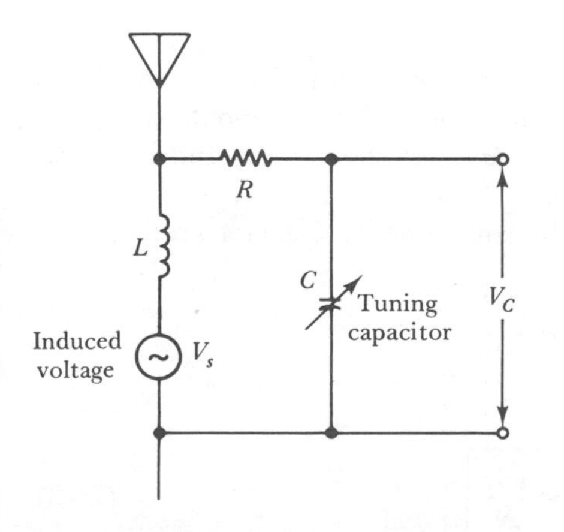

FIG. 24-6. Resonance circuit with tuning capacitor

At $f_r = 1$ MHz:

$$C = \frac{1}{4\pi^2 \times (1\ \text{MHz})^2 \times 100\ \mu\text{H}}$$

$$\cong 253\ \text{pF}$$

Capacitor range is from **253 pF** *to* **1000 pF.**

24-3 Q FACTOR OF A SERIES RESONANT CIRCUIT

Reconsider the equations for I, V_L, and V_C at resonance:

$$V_L = IX_L$$

and

$$I = \frac{V}{R}$$

Therefore,

$$V_L = \frac{V}{R} X_L$$

or

$$\frac{\text{inductor voltage at resonance}}{\text{supply voltage}} = \frac{V_L}{V}$$

and

$$\boxed{\frac{V_L}{V} = \frac{X_L}{R}} \tag{24-6}$$

Similarly,

$$\boxed{\frac{V_C}{V} = \frac{X_C}{R}} \tag{24-7}$$

A measure of the *quality* of a resonance circuit is the ratio of the voltage developed across the capacitor (or inductor) at resonance, to the supply voltage. This is termed the *Q factor* of the circuit, and it is also known as the *voltage magnification factor.*

From Eqs. (24-6) and (24-7), the equations for Q factor are:

$$Q = \frac{X_L}{R}$$

or

$$\boxed{Q = \frac{\omega L}{R}} \tag{24-8}$$

and

$$Q = \frac{X_C}{R}$$

giving

$$\boxed{Q = \frac{1}{\omega CR}} \tag{24-9}$$

Since the coil resistance is often the only resistance in a series resonance circuit, the Q is sometimes referred to as the *Q factor of the coil.* Rewriting Eq. (24-8),

$$Q = \frac{2\pi f_r L}{R}$$

and substituting for f_r from Eq. (24-3),

$$Q = \frac{2\pi L\left(\dfrac{1}{2\pi\sqrt{LC}}\right)}{R}$$

which reduces to

$$\boxed{Q = \frac{1}{R}\sqrt{\frac{L}{C}}} \tag{24-10}$$

It is seen that the Q factor of a series resonance circuit may be increased either by reducing R or by increasing the L/C ratio.

The Q factor can also be defined in terms of the ratio of the reactive power to the power dissipated in the circuit resistance. Using this definition, the equations for Q come out exactly as derived above.

EXAMPLE 24-4

A series LCR circuit which resonates at $f_r \cong 500$ kHz has $L = 100\ \mu$H, $R = 25\ \Omega$, and $C = 1000$ pF. Determine the Q factor of the circuit. Also determine the new value of C required for resonance at 500 kHz when the value of L is doubled, and calculate the new Q factor.

SOLUTION

Eq. (24-10):

$$Q = \frac{1}{R}\sqrt{\frac{L}{C}}$$

$$= \frac{1}{25\ \Omega}\sqrt{\frac{100\ \mu\text{H}}{1000\ \text{pF}}}$$

$$Q_1 \cong 12.6$$

When L is doubled:
From Eq. (24-3):

$$C = \frac{1}{4\pi^2 f_r^2 L}$$

$$= \frac{1}{4\pi^2 \times (500\ \text{kHz})^2 \times 200\ \mu\text{H}}$$

$$\mathbf{C \cong 500\ pF}$$

$$Q_2 = \frac{1}{25\ \Omega}\sqrt{\frac{200\ \mu\text{H}}{500\ \text{pF}}}$$

$$\mathbf{Q_2 \cong 25}$$

24-4 BANDWIDTH OF A SERIES RESONANT CIRCUIT

Consider the current/frequency response graph once again, as reproduced in Figure 24-7. It is clear from the graph that the current reaches a maximum level (I_m) at resonance. It is also clear that at frequencies close to resonance, the current level is only a little below its maximum value. Thus, the resonant circuit is said to select a *band* of frequencies rather than just one frequency. The lowest and highest frequencies of the band are identified as f_1 and f_2, respectively, in Figure 24-7, and the *bandwidth* is

$$\boxed{\Delta f = f_2 - f_1} \qquad \textbf{(24-11)}$$

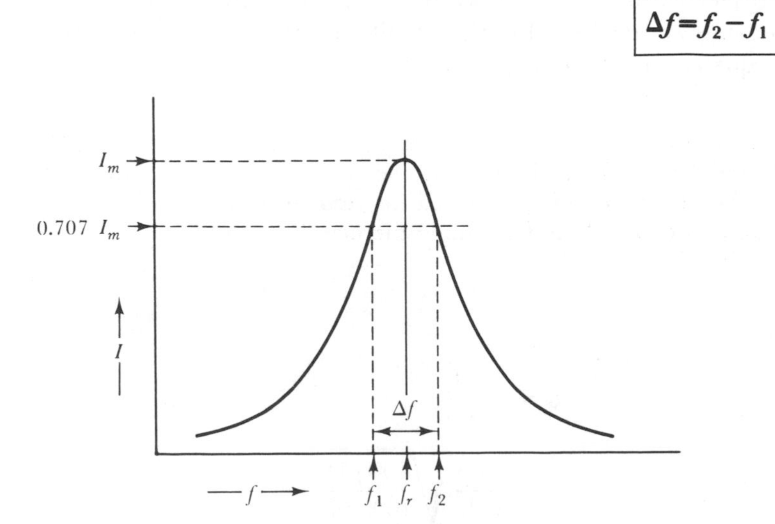

FIG. 24-7. Bandwidth of a series resonant circuit

The frequencies f_1 and f_2 are arbitrarily defined as those frequencies at which the power delivered to the circuit is half the power delivered at resonance. For this reason, f_1 and f_2 are termed the *half-power points*. The power delivered to the circuit at resonance is

$$P = I_m^2 R$$

Thus, for half-power,

$$\tfrac{1}{2}P = \frac{I_m^2 R}{2}$$

which gives

$$\boxed{\tfrac{1}{2}P = \left(\frac{I_m}{\sqrt{2}}\right)^2 R} \qquad (24\text{-}12)$$

Equation (24-12) shows that at the half-power points, the current in the series *LCR* circuit is

$$I = \frac{I_m}{\sqrt{2}} = 0.707 I_m$$

From Eq. (24-2), the current in the circuit is

$$|I| = \frac{V}{\sqrt{R^2 + (X_L - X_C)^2}}$$

and this can be rewritten as

$$|I| = \frac{V}{R\sqrt{1 + \left(\frac{X_L - X_C}{R}\right)^2}}$$

or

$$|I| = \frac{I_m}{\sqrt{1 + \left(\frac{X_L - X_C}{R}\right)^2}}$$

and from Eq. (24-12) at the half-power points,

$$I = \frac{I_m}{\sqrt{2}}$$

Therefore, at the half-power points,

$$\frac{X_L - X_C}{R} = 1$$

or

$$\boxed{X_L - X_C = R} \tag{24-13}$$

But at f_r, $X_L - X_C = 0$. Therefore, when the frequency increases from f_r to f_2, X_L must increase by $R/2$ and X_C must decrease by $R/2$, to fulfill Eq. (24-13). Therefore,

$$2\pi f_2 L - 2\pi f_r L = \tfrac{1}{2} R$$

which gives

$$\boxed{f_2 - f_r = \frac{R}{4\pi L}} \tag{24-14}$$

Similarly, when the frequency decreases from f_r to f_1, X_C must increase by $R/2$ and X_L must decrease by $R/2$. Consequently,

$$2\pi f_r L - 2\pi f_1 L = \tfrac{1}{2} R$$

and

$$\boxed{f_r - f_1 = \frac{R}{4\pi L}} \tag{24-15}$$

Adding Eq. (24-14) to Eq. (24-15):

$$f_2 - f_1 = \frac{R}{2\pi L}$$

or the circuit bandwidth is

$$\boxed{\Delta f = \frac{R}{2\pi L}} \tag{24-16}$$

Multiplying the right-hand side of Eq. (24-16) by f_r/f_r gives

$$\Delta f = \frac{R f_r}{2\pi f_r L}$$

or

$$\boxed{\Delta f = \frac{f_r}{Q}} \qquad (24\text{-}17)$$

The ability of a resonant circuit to select one particular frequency and to discriminate against other frequencies is termed the *selectivity* of the circuit. The circuits with the narrowest bandwidths obviously have the greatest selectivity. Equation (24-17) clearly shows that the greater the Q of a circuit, the better its selectivity.

EXAMPLE 24-5

For the circuit described in Example 24-4, calculate the half-power frequencies and the bandwidths:

a. When $L = 100\ \mu\text{H}$.

b. When the value of L is doubled.

SOLUTION

a. for $L = 100\ \mu\text{H}$:
From Eq. (24-14):

$$f_2 = \frac{R}{4\pi L} + f_r$$

$$= \frac{25\ \Omega}{4\pi \times 100\ \mu\text{H}} + 500\ \text{kHz}$$

$$\mathbf{f_2 \cong 520\ kHz}$$

From Eq. (24-15):

$$f_1 = f_r - \frac{R}{4\pi L}$$

$$= 500\ \text{kHz} - \frac{25\ \Omega}{4\pi \times 100\ \mu\text{H}}$$

$$\mathbf{f_1 \cong 480\ kHz}$$

$$\Delta f = f_2 - f_1$$

$$\mathbf{\Delta f = 40\ kHz}$$

Alternatively, *from* Eq. (24-17):

$$\Delta f = \frac{f_r}{Q}$$

From Example 24-4, $Q = 12.6$ *when* $L = 100\ \mu H$. *Therefore,*

$$\Delta f = \frac{500\ \text{kHz}}{12.6}$$

$$\mathbf{\Delta f \simeq 40\ kHz}$$

b. *for* $L = 200\ \mu H$,

$$f_2 = \frac{R}{4\pi L} + f_r$$

$$= \frac{25\ \Omega}{4\pi \times 200\ \mu H} + 500\ \text{kHz}$$

$$\mathbf{f_2 \simeq 510\ kHz}$$

$$f_1 = f_r - \frac{R}{4\pi L}$$

$$\mathbf{f_1 \simeq 490\ kHz}$$

$$\mathbf{\Delta f = 20\ kHz}$$

Alternatively, from Example 24-4, $Q = 25$ *when* $L = 200\ \mu H$:

$$\Delta f = \frac{f_r}{Q}$$

$$= \frac{500\ \text{kHz}}{25}$$

$$\mathbf{\Delta f = 20\ kHz}$$

24-5 PARALLEL RESONANCE

IDEAL PARALLEL RESONANT CIRCUIT. Consider the parallel *LCR* circuit shown in Figure 24-8(a). The admittance of the circuit is

$$Y = \frac{1}{R_P} - j\frac{1}{X_L} + j\frac{1}{X_C}$$

If the supply frequency is adjusted until X_L and X_C are equal, the admittance becomes

$$Y = \frac{1}{R_P}$$

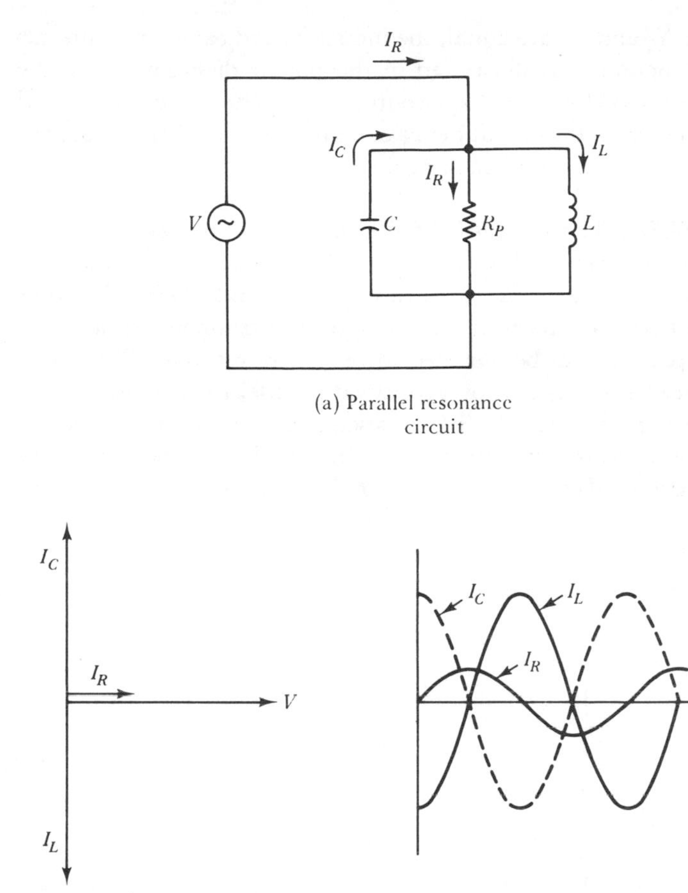

(a) Parallel resonance circuit

(b) Phasor diagram for parallel resonance circuit

(c) Waveforms of I_C, I_L, and I_R at resonance

FIG. 24-8. Parallel resonance circuit, phasor diagram, and waveforms

and the circuit impedance is

$$Z = R_P$$

Consequently, the current taken from the supply source is

$$I = \frac{V}{R_P}$$

The current through R_P is in phase with the supply voltage, while the current through L lags the supply voltage by 90°, and the capacitor current leads the supply voltage by 90°. This is illustrated by the phasor diagram in Figure 24-8(b), and by the current waveforms in Figure

24-8(c). When X_L and X_C are equal, the inductive and capacitive currents are equal and opposite, as illustrated in the phasor diagram. Thus, the total current supplied by the voltage source is I_R. I_C and I_L are the result of the energy stored in the circuit being continuously transferred from the inductor to the capacitor, and back again.

PRACTICAL PARALLEL RESONANT CIRCUIT. Since a practical inductor is not purely reactive but has a winding resistance, it must be represented as a resistance R_S in series with an inductance L. Some capacitors must also be shown as having a resistive component; however, in general, capacitors can be assumed to be purely reactive. The circuit shown in Figure 24-9(a) is that of a practical parallel LC circuit, and the phasor diagram in Figure 24-9(b) represents the circuit conditions at resonance. It is seen that the presence of the resistive component in the inductive branch makes the I_L phase angle ϕ slightly less than 90°.

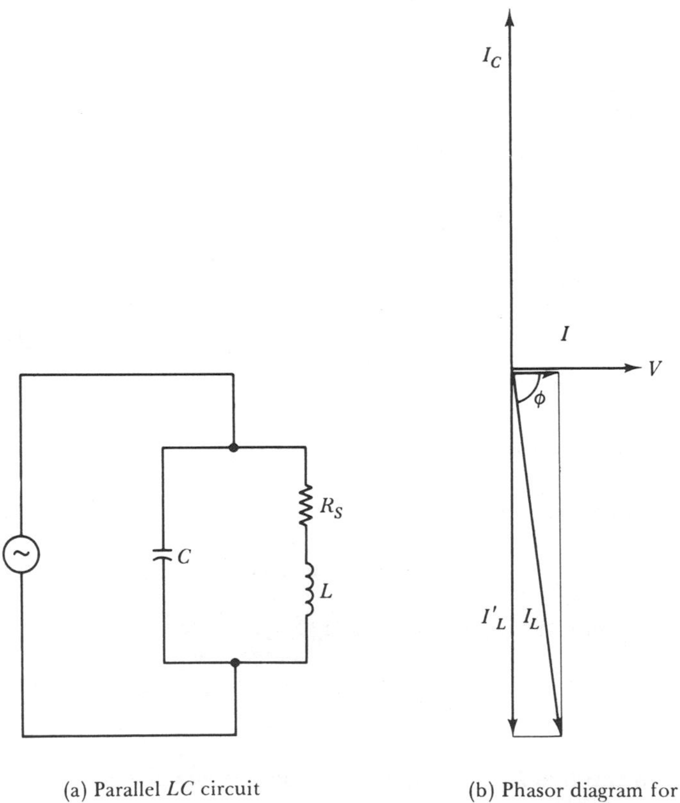

(a) Parallel LC circuit

(b) Phasor diagram for parallel LC circuit

FIG. 24-9. Practical parallel resonance circuit and phasor diagram

Resonance is achieved when the vertical component of I_L [I'_L in Figure 24-9(b)] is equal to the capacitor current I_C.

The admittance of the parallel circuit in Figure 24-9(a) is

$$Y=\frac{1}{R_S+jX_L}+j\frac{1}{X_C}$$

Multiplying

$$\frac{1}{R_S+jX_L} \text{ by } \frac{R_S-jX_L}{R_S-jX_L}$$

gives

$$\boxed{Y=\frac{R_S}{R_S^2+X_L^2}-j\frac{X_L}{R_S^2+X_L^2}+j\frac{1}{X_C}} \qquad (24\text{-}18)$$

At resonance:

$$\frac{1}{X_C}=\frac{X_L}{R_S^2+X_L^2}$$

or

$$\boxed{X_C=\frac{R_S^2+X_L^2}{X_L}} \qquad (24\text{-}19)$$

which when substituted into Eq. (24-18) gives the admittance at resonance as

$$Y=\frac{R_S}{R_S^2+X_L^2}$$

Thus,

$$Z=\frac{R_S^2+X_L^2}{R_S}$$

From Eq. (24-19):

$$X_LX_C=R_S^2+X_L^2$$

Therefore,

$$Z=\frac{X_LX_C}{R_S}$$

$$=\frac{2\pi f_rL}{2\pi f_rCR_S}$$

or at resonance

$$\boxed{Z = \frac{L}{CR_S}} \qquad (24\text{-}20)$$

If the impedance of the parallel LC circuit [as represented by the inverse of Eq. (24-18)] is plotted to a logarithmic frequency base, it is found that the impedance peaks at the resonance frequency (see Figure 24-10). *Thus, a parallel LC circuit has a maximum impedance at the resonance frequency*, while, as already discussed, a series resonant circuit has a minimum impedance at the resonance frequency. As a consequence of the peak in the impedance value of a parallel resonant circuit, there is a dip in the current drawn from the supply at the resonance frequency, as illustrated in Figure 24-10. Once again, this is the opposite of the case with series resonance.

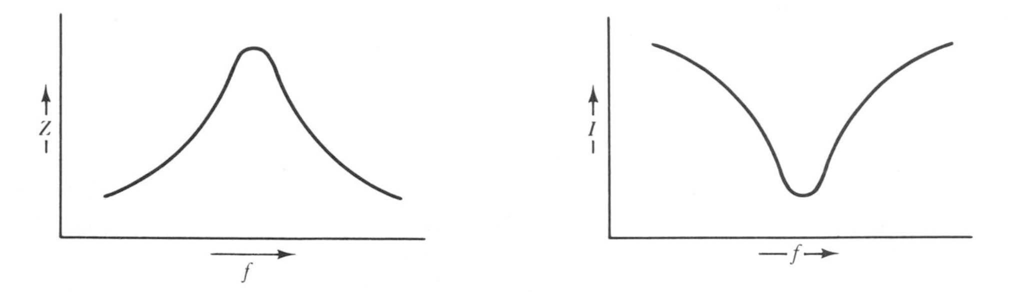

FIG. 24-10. Impedance and current graphs plotted versus frequency for a parallel resonance circuit

24-6 Q FACTOR FOR PARALLEL LC CIRCUIT

Referring again to Figure 24-9(b) it is seen that the total current drawn from the supply is I, which is in phase with V. Also, I is very much less than I_C and I_L. Thus, in the parallel resonant LC circuit, there is a *current magnification*, which is analogous to the voltage magnification that occurs in a series resonant circuit. The Q factor, *or current magnification factor* of the parallel resonant circuit, can be determined as the ratio of I_L (or I_C) to the supply current I.

From Figure 24-9(b), the supply current is

$$I = I_L \cos\phi = I_L \frac{R_S}{X_L}$$

and

$$Q = \frac{I_L}{I} = \frac{X_L}{R_S}$$

or

$$Q = \frac{\omega_L}{R_S} \qquad \text{[Eq. (24-8)]}$$

This equation is exactly the same as the Q factor equation for a series resonant circuit; that is, the Q is once again the Q factor of the inductance.

24-7 RESONANCE FREQUENCY FOR PARALLEL CIRCUIT

Returning to Eq. (24-19) which gives the value of X_C at resonance:
Eq. (24-19):

$$X_C = \frac{R_S^2 + X_L^2}{X_L}$$

When $Q > 10$,

$$X_L^2 \gg R_S^2$$

and therefore,

$$X_C \cong X_L$$

This gives the resonance frequency as

$$f_r = \frac{1}{2\pi\sqrt{LC}} \qquad \text{[Eq. (24-3)]}$$

The equation above is also the expression for the resonance frequency of a series LCR circuit. This equation does not apply in situations where Q is less than 10. Where $Q < 10$, the value of f_r can be shown to be

$$f_r = \frac{1}{2\pi\sqrt{LC}}\sqrt{1 - \frac{CR_S^2}{L}} \qquad (24\text{-}21)$$

The bandwidth of a parallel resonant circuit is determined in exactly the same way as that for a series resonant circuit:

$$\Delta f = \frac{f_r}{Q} \qquad \text{[Eq. (24-17)]}$$

EXAMPLE 24-6

A parallel LC circuit has an inductance $L=100\ \mu H$, which has a coil resistance of 12 Ω. The capacitor C is adjustable over the range 200 pF to 300 pF. Determine the maximum and minimum resonance frequency for the circuit. Also, calculate the Q and bandwidth of the circuit at the two resonance frequency extremes.

SOLUTION

Eq. (24-3):

$$f_r = \frac{1}{2\pi\sqrt{LC}}$$

When $c=200$ pF,

$$f_r = \frac{1}{2\pi\sqrt{100\ \mu H \times 200\ pF}}$$

$$\mathbf{f_{r(max)} = 1.13\ MHz}$$

When $c=300$ pF,

$$f_r = \frac{1}{2\pi\sqrt{100\ \mu H \times 300\ pF}}$$

$$\mathbf{f_{r(min)} = 919\ kHz}$$

Eq. (24-8):

$$Q = \frac{\omega_L}{R_S}$$

At $f=1.13$ MHz,

$$Q = \frac{2\pi \times 1.13\ MHz \times 100\ \mu H}{12\ \Omega}$$

$$\mathbf{Q = 59.2}$$

At $f=919$ kHz,

$$Q = \frac{2\pi \times 919\ kHz \times 100\ \mu H}{12\ \Omega}$$

$$\mathbf{Q = 48}$$

Eq. (24-16):

$$\Delta f = \frac{R_S}{2\pi L} = \frac{12\ \Omega}{2\pi \times 100\ \mu H}$$

$$\mathbf{\Delta f = 19.1\ kHz}$$

24-8 RESISTANCE DAMPING OF PARALLEL *LC* CIRCUIT

Figure 24-11 shows a parallel LC circuit with a resistor R_P connected in parallel with the capacitor and inductance. R_S is the resistive component of the inductance. Since the impedance of the parallel LC circuit is a maximum at resonance, the presence of R_P reduces the total impedance of the circuit. As will be shown, it also reduces the Q factor of the circuit.

When R_P is not present, the supply current is determined as:

$$I = I_L \times \frac{R_S}{X_L} \qquad (\textit{see}\ \text{Section 24-6})$$

$$= \frac{V}{X_L} \times \frac{R_S}{X_L}$$

$$I = \frac{VR_S}{X_L^2}$$

The current through R_P is in phase with I and is

$$I_R = \frac{V}{R_P}$$

Therefore, the total supply current is

$$I_T = I + I_R$$

$$I_T = \frac{VR_S}{X_L^2} + \frac{V}{R_P}$$

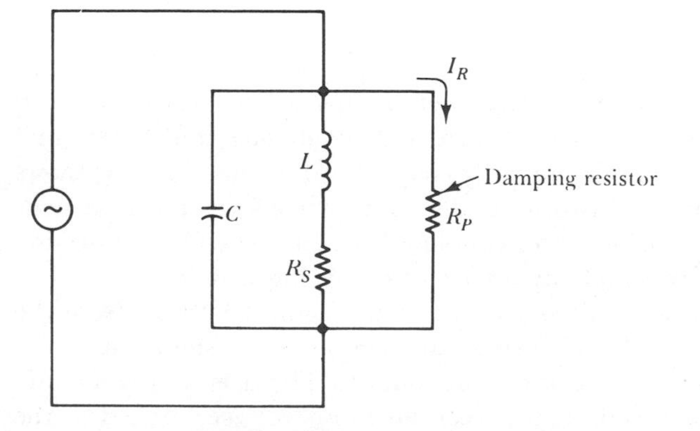

FIG. 24-11. Resistance damping of a parallel LC circuit

The Q factor for the parallel resonance LCR circuit is determined as

$$Q = \frac{I_L}{I_T}$$

$$= \frac{V/X_L}{\dfrac{VR_S}{X_L^2} + \dfrac{V}{R_P}}$$

giving

$$Q = \frac{1}{\dfrac{R_S}{\omega L} + \dfrac{\omega L}{R_P}}$$

When

$$\frac{R_S}{\omega L} \ll \frac{\omega L}{R_P}$$

$$\boxed{Q = \frac{R_P}{\omega L}} \qquad (24\text{-}22)$$

Note that this equation for the Q of a parallel LCR circuit is inverted by comparison with the equation for the Q of the inductance ($Q = \omega L / R_S$). It is seen that when R_P is made smaller, the circuit Q is reduced. The effect is known as *damping*, and R_P is referred to as a *damping resistor*. Damping may be employed where resonance of LC circuits is undesirable, for example, to avoid picking up radio frequencies in audio equipment. The Q factor for the inductance is given as $\omega L / R_S$. Thus, to obtain Eq. (24-22), $\omega L / R_S \gg R_P / \omega L$; that is, the inductance Q factor must be at least 10 times the Q factor of the LCR circuit.

24-9 TUNED COUPLED COILS

In Chapter 15 it was shown that mutual inductance exists between adjacent coils, and that when a current flows in one coil, a terminal voltage may be induced in a mutually coupled coil. Figure 24-12(a) shows a typical arrangement of two mutually coupled coils wound on a non-magnetic core. The coefficient of coupling between the coils (see Section 15-4) can be adjusted simply by altering the spacing between them. The circuit diagram for coupled coils is shown in Figure 24-12(b). Note the dots at one end of each coil. These are employed to show the phase relationship between input and output voltages. The dots simply identify in-phase ends of each coil. Thus, when an input voltage applied to the left-hand coil has the polarity shown (positive at the dotted end of the

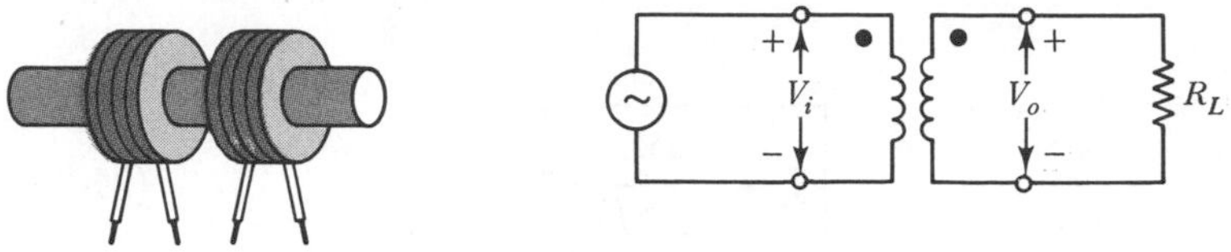

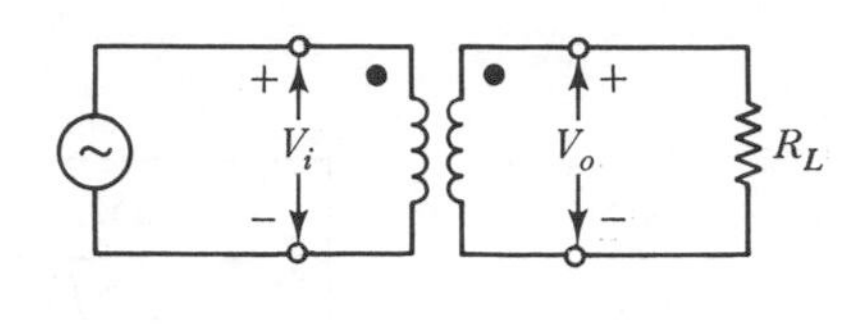

(a) Coupled coils (b) Circuit diagram for two coupled coils

FIG. 24-12. Two coupled coils and their circuit diagram

coil), the output voltage from the right-hand coil is also positive at the dotted end of the coil.

The equivalent circuit for the coupled coils is shown in Figure 24-13. R_1 represents the resistive component of the *primary* winding (the left-hand winding in Figure 24-13), and L_1 is the inductive component. In the *secondary* winding R_2 and L_2 are the total resistive and inductive components, respectively (R_2 including the load resistance R_L). When the current I_1 flows in the primary winding, a voltage ωMI_1 is induced in the secondary winding [see Eq. (20-6)], causing current I_2 to flow. The flow of I_2 also induces a voltage ωMI_2 in the primary winding. This induced voltage must be shown as having a polarity that opposes the primary current. Two voltage generators are included in the equivalent circuit to represent the voltages induced in each winding.

The equation for the voltage drops around the primary circuit may be written

$$\boxed{V = I_1R_1 + jI_1X_1 + \omega MI_2} \tag{24-23}$$

and the equation for the voltage drops around the secondary circuit is

$$\omega MI_1 = I_2R_2 + jI_2X_2$$

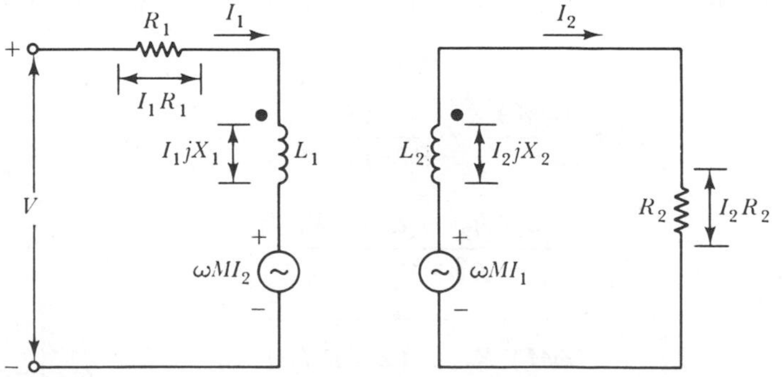

FIG. 24-13. Equivalent circuit of two coupled coils

giving

$$I_2 = \frac{\omega M I_1}{R_2 + jX_2} \tag{24-24}$$

Substituting for I_2 in Eq. (24-23),

$$V = I_1 R_1 + jI_1 X_1 + \frac{(\omega M)^2 I_1}{R_2 + jX_2}$$

from which the primary impedance is

$$Z_1 = \frac{V}{I_1} = R_1 + jX_1 + \frac{(\omega M)^2}{R_2 + jX_2} \tag{24-25}$$

The primary circuit impedance is seen to consist of a *self-impedance* Z and a *coupled impedance* Z_2', where

$$\text{self-impedance } Z = R_1 + jX_1$$

and

$$\textbf{coupled impedance } Z_2' = \frac{(\omega M)^2}{R_2 + jX_2} \tag{24-26}$$

The coupled impedance can be resolved into resistive and reactive components by multiplying by the conjugate of the denominator (see Section 19-4):

$$Z_2' = \frac{(\omega M)^2}{R_2 + jX_2} \times \frac{R_2 - jX_2}{R_2 - jX_2}$$

$$= \frac{(\omega M)^2 R_2 - j(\omega M)^2 X_2}{R_2^2 + X_2^2}$$

$$Z_2' = \frac{(\omega M)^2 R_2}{R_2^2 + X_2^2} - j\frac{(\omega M)^2 X_2}{R_2^2 + X_2^2} \tag{24-27}$$

It is seen that the coupled impedance is made up of a coupled resistance R_2' and a coupled reactance X_2':

$$\boxed{\textit{coupled resistance, } R_2' = \frac{(\omega M)^2 R_2}{R_2^2 + X_2^2}} \qquad \text{(24-28)}$$

and

$$\boxed{\textit{coupled reactance, } X_2' = -j\frac{(\omega M)^2 X_2}{R_2^2 + X_2^2}} \qquad \text{(24-29)}$$

Now consider the coupled circuits shown in Figure 24-14(a), in which the primary and secondary circuits include capacitors and are tuned to resonate at the same frequency. At the resonance frequency, the series reactance in each circuit is zero. Thus, when X_2 is zero, only the resistive component of the impedance is coupled from the secondary into the primary. Also, because of the presence of X_2 in the denominator of the expression for the coupled resistance [Eq. (24-28)] the coupled resistance is a maximum when X_2 becomes zero (i.e., at the resonance frequency). This (maximum) coupled resistance adds to the coil resistance of the primary, and thus has the effect of reducing the primary current at resonance.

Figure 24-14(b) shows the equivalent circuit of the coupled circuits at resonance. R_1 and R_2 are the primary and secondary coil resistances, respectively, and R_2' is the secondary resistance coupled into the primary, where at resonance

$$\boxed{R_2' = \frac{(\omega M)^2}{R_2}} \qquad \text{(24-30)}$$

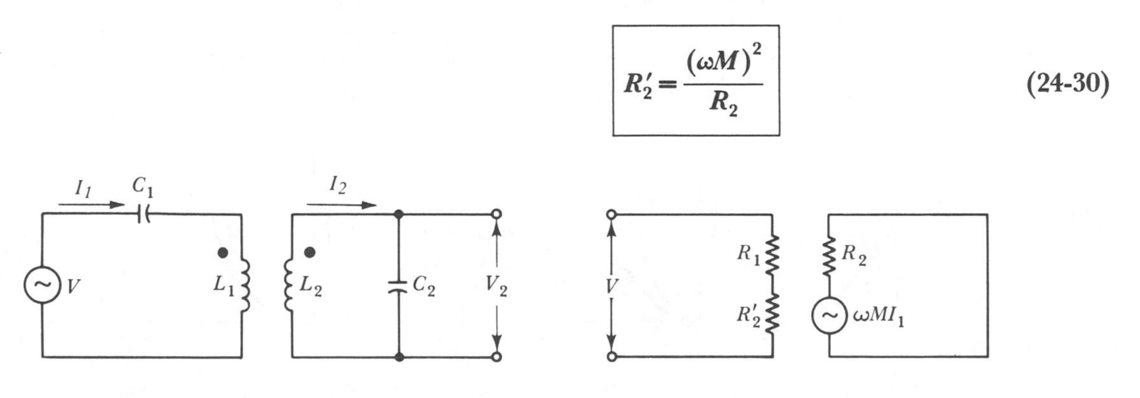

(a) Coupled coils tuned to resonate

(b) Equivalent circuit at resonance

FIG. 24-14. Tuned coupled coils

Viewed from the input terminals, the primary circuit (operating alone) is a series resonant circuit, with a maximum impedance at resonance:

$$Z = R_1$$

The secondary circuit behaves as a series resonant circuit with a supply voltage of $\omega M I_1$. The coupled resistance R_2' causes the maximum (resonance) impedance of the primary circuit to become

$$\boxed{Z_1 = R_1 + R_2'} \qquad (24\text{-}31)$$

It is clear that the effect of R_2' is to increase Z, and consequently reduce the current flowing through the primary winding. The degree to which the primary current is affected by the secondary depends upon how tightly the coils are coupled. That is, upon the coefficient of coupling k.

The frequency response curves of I_1 and I_2 for *loosely coupled coils* are shown in Figure 24-15(a). The shape of the I_1 response curve is largely unaffected by the coupled impedance, and the I_2 curve is seen to be very much smaller than I_1 but with the same shape as the I_1 curve.

With *tight coupling*, Figure 24-15(b) the coupled resistance has a considerable effect at the resonance frequency. The total primary impedance is increased, and the primary current is significantly reduced. Also, the primary response curve is distorted by the occurrence of two *humps*, one above and one below the resonance frequency. These humps result from the fact that the secondary impedance is capacitive below the resonance frequency and inductive above f_r. When coupled into the primary, the coupled impedances are inductive below f_r and capacitive above f_r. These result in minimum primary impedances and therefore maximum currents, at two frequencies above and below resonance. Because of the tight coupling, the response curve of the secondary current is simply a slightly amplified version of the primary.

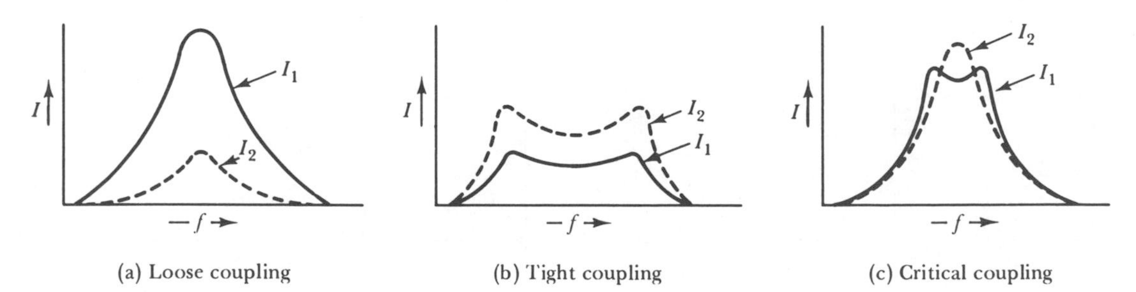

FIG. 24-15. Frequency response of coupled coils for various degrees of coupling

With *critical coupling* [Figure 24-15(c)], maximum power transfer occurs from the primary to the secondary, and thus the secondary current is greater than in either of the other two cases. The condition for maximum power transfer is that the source impedance and load impedance be equal (see Section 23-7). The two humps still occur in the primary response curve but are absent in the secondary. For maximum power transfer and critical coupling:

$$R_1 = R_2'$$

$$R_1 = \frac{(\omega M)^2}{R_2}$$

or

$$\boxed{M^2 = \frac{R_1 R_2}{\omega^2}} \qquad (24\text{-}32)$$

and from Eq. (15-13),

$$M^2 = k^2 L_1 L_2$$

Therefore,

$$k^2 L_1 L_2 = \frac{R_1 R_2}{\omega^2}$$

$$k^2 = \frac{R_1}{\omega L_1} \times \frac{R_2}{\omega L_2}$$

or

$$\boxed{\textit{critical coupling coefficient } k = \frac{1}{\sqrt{Q_1 Q_2}}} \qquad (24\text{-}33)$$

EXAMPLE 24-7 A 100-μH coil with a resistance of $R_1 = 20\ \Omega$ is connected in series with a capacitor and a supply voltage of 2.5 V rms. The supply frequency is 1 MHz and the capacitor is tuned to resonate at this frequency. A second coil with $L_2 = 50\ \mu$H and $R_2 = 8\ \Omega$ is coupled to the first coil and is also tuned to resonate at 1 MHz. Calculate the critical value of the coefficient of coupling, and determine the levels of the currents in each coil when critically coupled.

SOLUTION

$$Q_1 = \frac{\omega L_1}{R_1}$$

$$= \frac{2\pi \times 1\ \text{MHz} \times 100\ \mu\text{H}}{20\ \Omega}$$

$$\mathbf{Q_1 = 31.4}$$

$$Q_2 = \frac{\omega L_2}{R_2}$$

$$= \frac{2\pi \times 1\ \text{MHz} \times 50\ \mu\text{H}}{8\ \Omega}$$

$$\mathbf{Q_2 = 39.3}$$

Eq. (24-33):

$$\textit{critical } k = \frac{1}{\sqrt{Q_1 Q_2}}$$

$$= \frac{1}{\sqrt{31.4 \times 39.3}}$$

$$\textbf{\textit{critical} } \mathbf{k = 0.028}$$

$$I_1 = \frac{V}{R_1 + R_2'}$$

When critically coupled:

$$R_1 = R_2'$$

$$I_1 = \frac{V}{2R_1}$$

$$= \frac{2.5\ \text{V}}{2 \times 20\ \Omega}$$

$$\mathbf{I_1 = 62.5\ mA}$$

From Eq. (24-32):

$$(\omega M)^2 = R_1 R_2$$

$$\omega M = \sqrt{R_1 R_2}$$

$$= \sqrt{20\ \Omega \times 8\ \Omega}$$

$$\omega M = 12.65$$

$$\textit{secondary voltage, } e_2 = \omega M I_1$$

$$\textit{and secondary current, } I_2 = \frac{e_2}{R_2}$$

$$= \frac{\omega M I_1}{R_2}$$

$$= \frac{12.65 \times 62.5\ \text{mA}}{8\ \Omega}$$

$$\mathbf{I_2 = 98.8\ mA}$$

24-10 RESONANCE FILTERS

In Sections 20-9 and 20-10 it was shown that *CR* and *LC* circuits can be employed to *filter out* unwanted ac frequencies. These circuits were designated *low-pass filters* and *high-pass filters*. Resonance circuits can also be employed very effectively as filters. Because resonance occurs over a narrow band of frequencies, resonance filters are normally either *band-pass filters* or *band-stop filters*.

The circuit of a resonance band-pass filter is shown in Figure 24-16(a), and its equivalent circuit is drawn in Figure 24-16(b). The impedance of the series *LC* circuit is

$$Z = R_s + j(\omega L - 1/\omega C)$$

where R_s is the resistance of the inductor.

As shown in Figure 24-16(b), the input voltage V_i is potentially divided across Z and R_1 to give an output voltage of

$$V_o = \frac{V_i R_1}{R_1 + Z}$$

At frequencies far above and below resonance, Z is much greater than R_1, and consequently the output voltage V_o is very much smaller

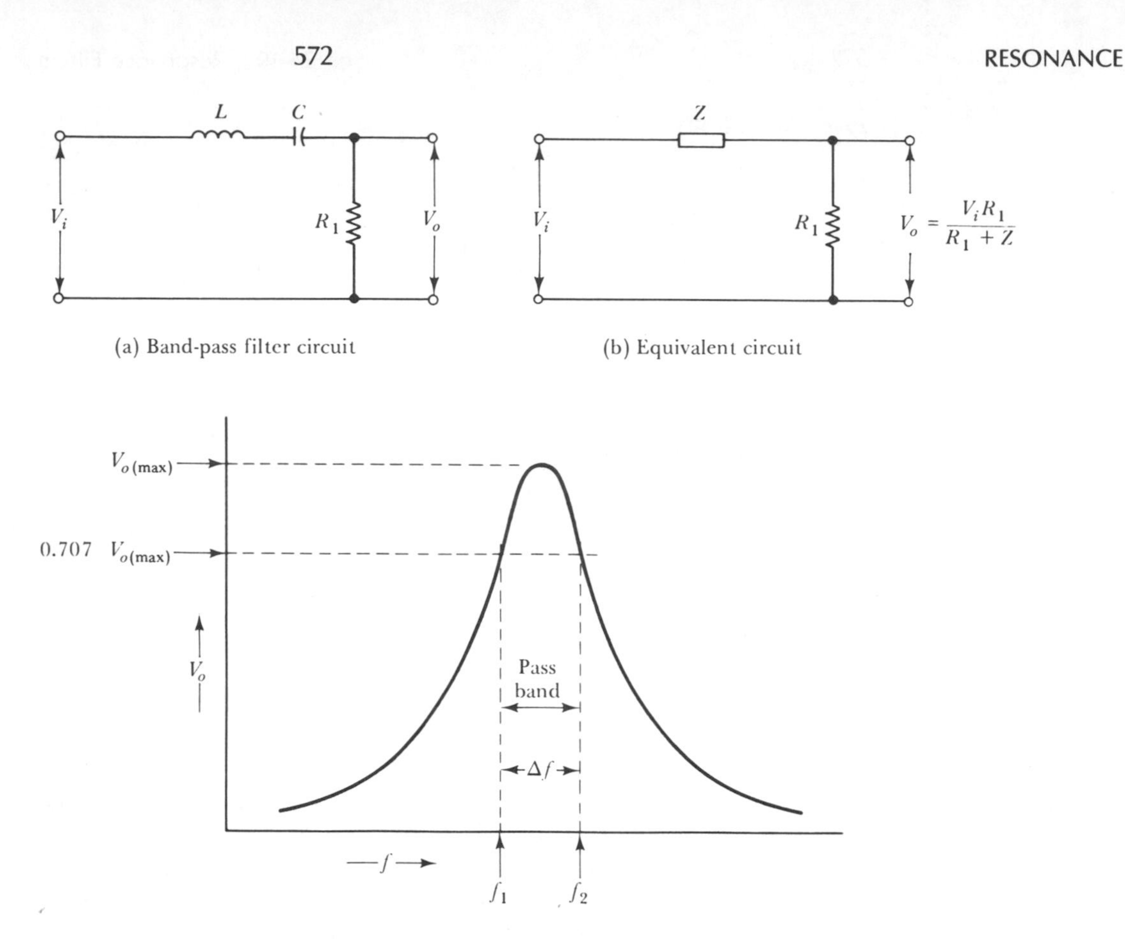

FIG. 24-16. Band-pass filter using a series *LC* circuit

than V_i. However, at resonance $\omega L = 1/\omega C$ and the impedance of the series *LC* circuit becomes

$$Z = R_s$$

and the output voltage is

$$V_o = \frac{V_i R_1}{R_1 + R_s}$$

If R_s is very much smaller than R_1, very little of the input voltage is lost across R_s and the output voltage at resonance is almost equal to the input voltage.

The graph of V_o plotted versus frequency is shown in Figure 24-16(c). As discussed in Section 24-4, there is a bandwidth of frequencies over which the circuit is close to resonance. Thus, for the band-pass filter circuit of Figure 24-16(a), the output voltage remains high for the band of frequencies between f_1 and f_2 on Figure 24-16(c). This is said to be the *pass band* of the filter circuit. Recall that f_1 and f_2 are those frequencies at which V_o is 0.707 $V_{o(max)}$. The width of the pass band is inversely proportional to the Q factor of the circuit (see Section 24-4). To calculate the circuit Q factor, the total resistance of the circuit must be taken as $R_1 + R_s$.

EXAMPLE 24-8

The band-pass filter circuit shown in Figure 24-16(a) has $L=1$ mH, $C=100$ pF, $R_1=150\ \Omega$, and $R_s=15\ \Omega$. Determine the resonance frequency, Q factor, and pass band of the circuit.

SOLUTION

From Eq. (24-3):

$$f_r = \frac{1}{2\pi\sqrt{LC}} = \frac{1}{2\pi\sqrt{1\text{ mH}\times 100\text{ pF}}}$$

$$\mathbf{f_r \cong 500\ kHz}$$

From Eq. (24-10):

$$Q = \frac{1}{R_s + R_1}\sqrt{\frac{L}{C}}$$

$$= \frac{1}{15\ \Omega + 150\ \Omega}\sqrt{\frac{1\text{ mH}}{100\text{ pF}}}$$

$$\mathbf{Q = 19.2}$$

From Eq. (24-17):

$$\Delta f = \frac{f_r}{Q} = \frac{500\text{ kHz}}{19.2}$$

$$\mathbf{\Delta f = 26\ kHz}$$

$$f_1 = f_r - \frac{\Delta f}{2} = 500\text{ kHz} - \frac{26\text{ kHz}}{2}$$

$$\mathbf{f_1 = 487\ kHz}$$

and

$$f_2 = f_r + \frac{\Delta f}{2}$$

$$\mathbf{f_2 = 513\ kHz}$$

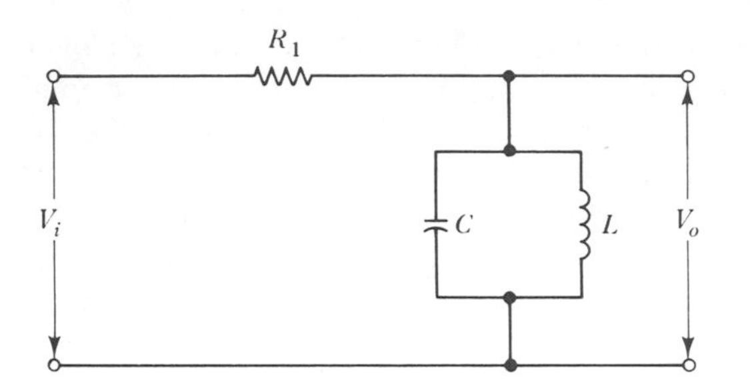

(a) Band-pass filter circuit

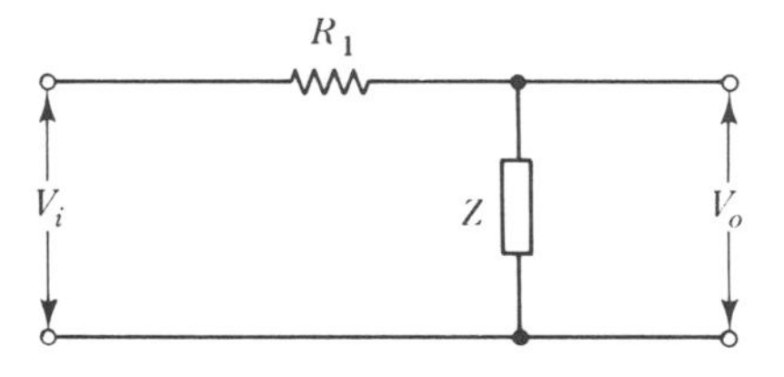

(b) Equivalent circuit

FIG. 24-17. Band-pass filter using a parallel *LC* circuit

Another band-pass filter circuit, this time using a parallel resonance circuit, is shown in Figure 24-17(a). In this case the output voltage is developed across the resonance circuit. As illustrated in Figure 24-17(b), the input voltage is potentially divided across R_1 and Z, so that

$$V_o = \frac{V_i \times Z}{R_1 + Z}$$

At frequencies above and below resonance, Z is much smaller than R_1. At resonance Z is a maximum for a parallel LC circuit. If Z becomes much larger than R_1 at resonance, then V_o is just a little smaller than V_i. The frequency response for this filter circuit is exactly as shown for the previous circuit in Figure 24-16(c).

Two *band-stop filter* circuits and their frequency response graph are shown in Figure 24-18. Consider the circuit in Figure 24-18(a). The output voltage V_o is developed across the series LC circuit. At frequencies away from resonance, the impedance of this series LC circuit will be much larger than R_1, while at resonance the impedance will be much smaller than R_1. Thus, at resonance, V_o becomes much smaller than V_i, and off resonance, V_o is approximately equal to V_i.

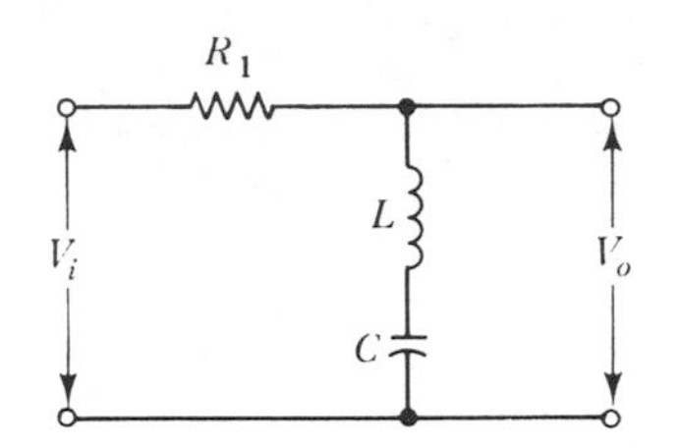

(a) Band-stop filter using series LC

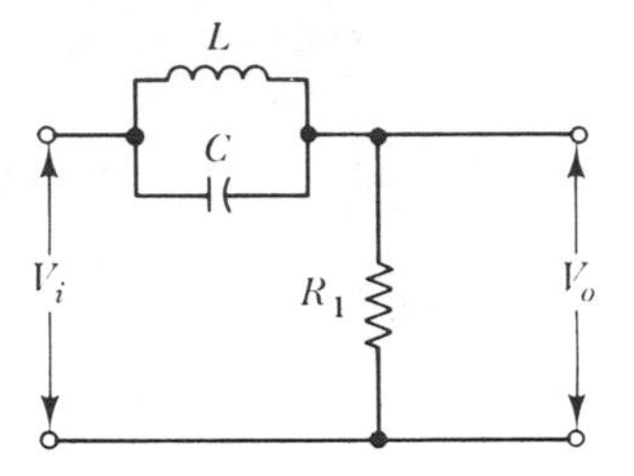

(b) Band-stop filter using parallel LC

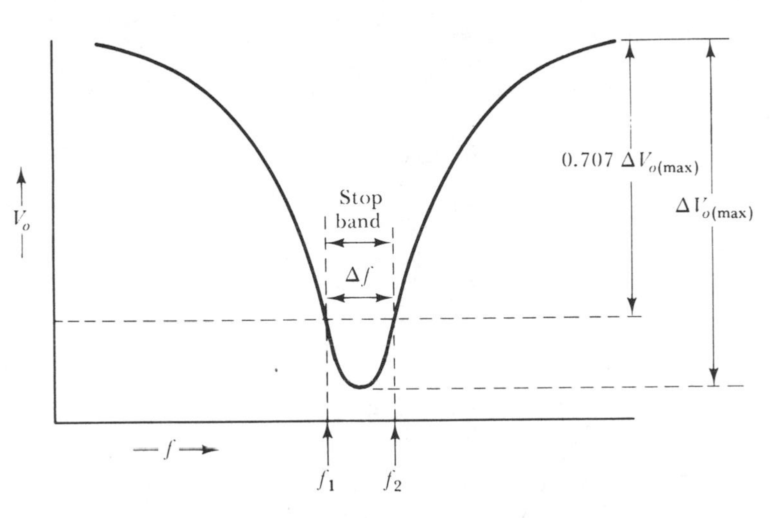

(c) Frequency response of band-stop filter

FIG. 24-18. Band-stop filter circuits and frequency response graph

For the circuit of Figure 24-18(b), the parallel LC circuit has an impedance that is smaller than R_1 at frequencies off resonance, and much larger than R_1 at the resonance frequency. Again, V_o is much smaller than V_i over a band of frequencies close to resonance. As shown in Figure 24-18(c), all frequencies except those in the *stop band* are passed by a band-stop filter. The frequency response graph shows $\Delta V_{o(\max)}$ as the voltage reduction of the output at resonance. Frequencies f_1 and f_2 are those frequencies at which the output reduction is 0.707 of $\Delta V_{o(\max)}$.

GLOSSARY OF FORMULAS

Impedance of series LCR circuit:

$$Z = R + j(X_L - X_C)$$

At resonance:

$$Z = R$$

Resonance frequency for series resonance:

$$f_r = \frac{1}{2\pi\sqrt{LC}}$$

Q factor for series resonance:

$$Q = \frac{\omega L}{R}$$

$$Q = \frac{1}{\omega CR}$$

$$Q = \frac{1}{R}\sqrt{\frac{L}{C}}$$

Bandwidth:

$$\Delta f = f_2 - f_1$$

$$\Delta f = \frac{R}{2\pi L}$$

$$\Delta f = \frac{f_r}{Q}$$

For parallel LC circuit:

$$Y = \frac{1}{R_S + jX_L} + j\frac{1}{X_C}$$

At resonance:

$$Z = \frac{L}{CR_S}$$

Q factor for parallel resonance:

$$Q = \frac{\omega L}{R_S}$$

$$Q = \frac{1}{\omega CR_S}$$

Resonance frequency for parallel resonance:

When $Q > 10$:

$$f_r = \frac{1}{2\pi\sqrt{LC}}$$

When $Q < 10$:

$$f_r = \frac{1}{2\pi\sqrt{LC}}\sqrt{1 - \frac{CR_S^2}{L}}$$

For parallel LCR circuit:

$$Q = \frac{R_P}{\omega L}$$

For coupled coils:

Secondary current:

$$I_2 = \frac{\omega M I_1}{R_2 + jX_2}$$

Coupled impedance:

$$Z_2' = \frac{(\omega M)^2}{R_2 + jX_2}$$

Coupled resistance:

$$R_2' = \frac{(\omega M)^2 R_2}{R_2^2 + X_2^2}$$

At resonance:

$$R_2' = \frac{(\omega M)^2}{R_2}$$

Coupled reactance:

$$X_2' = -j\frac{(\omega M)^2 X_2}{R_2^2 + X_2^2}$$

At resonance:

$$Z_1 = R_1 + R_2'$$

For critical coupling:

$$R_1 = R_2'$$

$$M^2 = \frac{R_1 R_2}{\omega^2}$$

$$k = \frac{1}{\sqrt{Q_1 Q_2}}$$

REVIEW QUESTIONS

24-1 Sketch a series *LCR* circuit with an ac supply. Write the equation for the impedance of the circuit and explain what occurs at the resonance frequency.

24-2 Draw a sketch to show the graphs of inductive reactance and capacitive reactance plotted versus frequency. Also show the graph of total impedance for a series *LCR* circuit. Briefly explain.

24-3 Sketch a phasor diagram for a series LCR circuit at resonance. Also, draw typical graphs showing the variation of impedance, current, and the current phase angle above and below resonance.

24-4 Derive an equation for the resonance frequency of a series LCR circuit.

24-5 Define Q factor with respect to a series resonance circuit. Derive an expression for the Q factor of the inductance coil in a series resonance circuit. Also, derive an equation for the circuit Q factor in terms of R, L, and C.

24-6 Using illustrations, define half-power points, bandwidth, and selectivity as applied to a resonance circuit. Derive an equation for the bandwidth of a resonance circuit in terms of the resonance frequency and the circuit Q factor.

24-7 Sketch a parallel LC circuit with some coil resistance and an ac supply. Write the equation for the circuit admittance and explain what occurs at resonance. Derive an expression for the impedance of the circuit at resonance.

24-8 Sketch the phasor diagram for the component currents in a parallel LC circuit at resonance. Briefly explain.

24-9 For a parallel LC circuit, sketch the typical graphs of circuit impedance, inductor current, and supply current plotted versus frequency. Explain the shape of the graphs.

24-10 Derive equations for the Q factor and resonance frequency of a parallel LC circuit. Discuss any approximations made.

24-11 A coil supplied with an alternating voltage V is inductively coupled to another coil with a load resistance R_L. Sketch the circuit diagram for the coils, and sketch the equivalent circuit showing the voltage drops around the primary and secondary circuits. Derive equations for the resistance and reactance coupled from the secondary into the primary.

24-12 Two coupled coils each have a capacitor connected in series with them and are tuned to resonate at the same frequency. Explain the effect of resonance on the coupled reactance and resistance, and write the equation for the total primary circuit resistance at resonance.

24-13 Define the term critical coupling as applied to two coupled coils that are tuned to resonate at the same frequency. Derive an equation for the critical coupling factor, and sketch the graphs of primary current and secondary current versus frequency for various degrees of coupling between the coils. Explain the shapes of the graphs.

24-14 Sketch the circuits of a band-pass filter and a band-stop filter using series resonance. Sketch the frequency response graphs and explain briefly the operation of each circuit.

24-15 Sketch the circuits of a band-pass filter and a band-stop filter using parallel resonance. Sketch the frequency response graphs and explain briefly the operation of each circuit.

PROBLEMS

24-1 A series LCR circuit has $L=506$ μH, $C=200$ pF, $R=32$ Ω, and supply voltage $V_S=5$ V. Calculate the resonance frequency and determine the value of the supply current at resonance.

24-2 For the circuit described in Problem 24-1, calculate the values of the supply current at several frequencies above and below resonance. Plot a graph of current to a logarithmic frequency base.

24-3 For the circuit described in Problem 24-1, plot the graphs of V_C, V_L, and V_R to a logarithmic frequency base. Discuss the shape of the graphs and explain resonance rise in voltage.

24-4 A 750-μH inductance is to be connected in series with a variable capacitor and tuned to resonate at frequencies ranging from 100 kHz to 900 kHz. Determine the required range of adjustment of the capacitor.

24-5 A series LCR circuit supplied with 7 V rms has $L=600$ μH, $C=150$ pF, and $R=20$ Ω. Plot the graphs of I, V_L, V_C, and V_R for the circuit versus a logarithmic frequency base.

24-6 The voltage across the capacitor in Problem 24-5 is monitored on an oscilloscope. If the oscilloscope has an input capacitance of 40 pF, determine its effect on the resonance frequency of the circuit.

24-7 Calculate the Q factor for the circuits described in Problems 24-1 and 24-5. Also, determine the new Q factor for the circuit described in Problem 24-5 when an oscilloscope is connected to the circuit as explained in Problem 24-6.

24-8 Determine the half-power frequencies and the bandwidth for the circuit described in Problem 24-1.

24-9 For the circuit in Problem 24-5, calculate the half-power frequencies and the bandwidth. Also, determine the new bandwidth value when an oscilloscope is connected to the circuit as described in Problem 24-6.

24-10 An inductance with $L=300$ μH and $R=5$ Ω is connected in parallel with a capacitor having a value of $C=300$ pF. Determine the resonance frequency of the circuit and calculate the circuit Q factor and bandwidth.

24-11 The circuit described in Problem 24-10 is connected to an oscilloscope with an input capacitance of 30 pF. Determine the new values of resonance frequency, Q, and half-power frequencies.

24-12 Explain what is meant by resistance damping of a parallel resonance circuit. Derive an expression for the Q factor of a parallel resonance circuit and calculate the value of damping resistor required to give a Q of 15 to the circuit described in Problem 24-10.

24-13 A 300-μH coil with a resistance of $R_1 = 15\ \Omega$ is connected in series with a supply of 3.3 V rms and a 300-pF capacitor. A second coil with $L = 100\ \mu$H and $R_2 = 10\ \Omega$ is tuned to resonate at the same frequency as the first coil. Determine the critical value of the coefficient of coupling for the coils and calculate the current in each coil when critically coupled.

24-14 A band-stop filter using a parallel resonance circuit has $L = 5$ mH, $C = 250$ pF, $R_s = 12\ \Omega$, and $R_1 = 200\ \Omega$. Calculate the stop band of the circuit.

25

TRANSFORMERS

INTRODUCTION

A transformer basically consists of two coils wound on a single iron core. When an alternating voltage is applied to one of the coils, the mutual inductance causes an alternating voltage to be generated in the other coil. The ratio of the voltage amplitudes depends upon the number of turns on each coil. A transformer may be used either to increase or decrease an applied voltage, or to increase or decrease a current.

The emf induced in each winding of a transformer can be calculated from the *transformer emf equation*. The behavior of the transformer under *no-load* and *full-load* conditions is best understood by drawing the appropriate phasor diagrams and by studying the transformer equivalent circuit. Simplification of the equivalent circuit is possible by the process of *referring* the secondary resistance and reactance to the primary winding.

The performance of a transformer is described in terms of its *voltage regulation* and its efficiency. The performance can be predicted from the results of two tests known as the *open-circuit test* and the *short-circuit test*.

25-1 PRINCIPLE OF TRANSFORMER OPERATION

The transformer is an application of mutual inductance. An alternating voltage applied to a *primary* winding generates an alternating magnetic flux which links with a *secondary* winding and induces an alternating emf in the secondary.

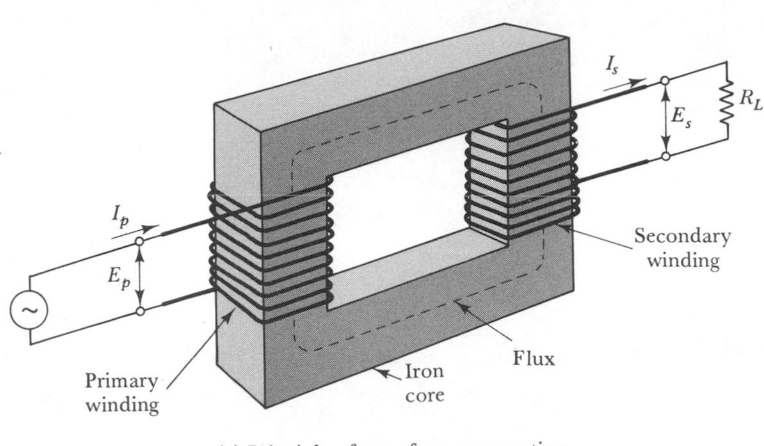

(a) Principle of transformer operation

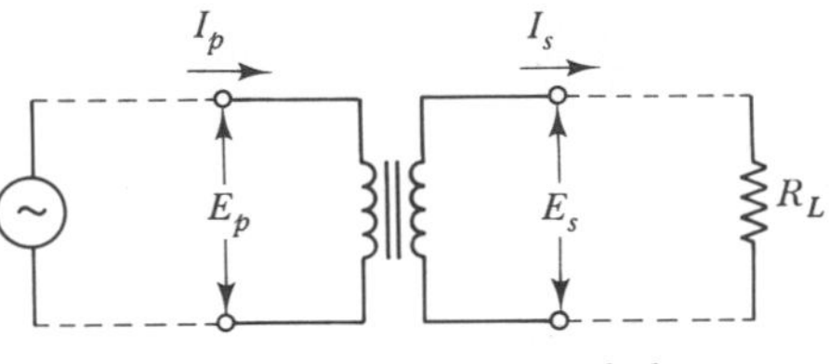

(b) Transformer circuit symbol

FIG. 25-1. Principle of operation of a transformer and transformer circuit symbol

Figure 25-1(a) illustrates the basic construction of the transformer. The iron core forms a closed magnetic circuit, and the windings are simply coils of insulated wire wrapped around the core. The presence of the iron core causes virtually 100% of the magnetic flux generated by the primary to be linked with the secondary. The transformer is therefore similar to two *coupled coils* (see Section 15-4) in which the coefficient of coupling is 1. In fact, coupled coils are sometimes referred to as an *air-cored transformer*. The circuit symbol for the transformer is shown in Figure 25-1(b). The two lines between the coils indicate the presence of an iron core.

The alternating magnetic flux in the iron core of the transformer links with both the primary winding and the secondary winding. The flux linking with the primary winding generates a counter-emf in the primary input voltage. From Eq. (15-2),

$$E_p = \frac{\Delta\Phi}{\Delta t} N_p$$

where E_p is the primary induced voltage and N_p is the number of turns on the primary winding. Also,

$$E_s = \frac{\Delta\Phi}{\Delta t} N_s$$

where E_s and N_s are the secondary induced voltage and the secondary turns, respectively. Combining the two equations,

$$\frac{E_p}{E_s} = \frac{\Delta\Phi/\Delta t}{\Delta\Phi/\Delta t} \times \frac{N_p}{N_s}$$

or

$$\frac{E_p}{E_s} = \frac{N_p}{N_s}$$

It is seen that the ratio of primary voltage to secondary voltage is the same as the ratio of primary coil turns to secondary turns. Where the primary (input) voltage and *turns ratio* are known, the secondary (output) voltage can be calculated from

$$\boxed{E_s = E_p \times \frac{N_s}{N_p}} \qquad \textbf{(25-1)}$$

When the primary and secondary windings have an equal number of turns, the output voltage is equal to the input voltage, and the transformer is referred to as a *a 1:1 transformer*. If the secondary has more turns than the primary, E_s is greater than E_p, and the device is termed a *step-up* transformer. This means that the transformer *steps up* the input voltage to produce a higher output level. It is also possible to have a secondary winding with fewer turns than the primary. In this case the output voltage is less than the input, and the device is a *step-down* transformer.

Figure 25-2 shows the circuit diagram of a transformer with three separate secondary windings. The output voltage from each winding depends upon the number of turns on that winding as well as upon the primary voltage and number of primary turns.

EXAMPLE 25-1

The transformer in Figure 25-2 has an alternating input of 100 V. Determine the output voltage from each secondary winding. Also determine the total output voltage if all three secondary windings were connected in series. The numbers of turns on each winding are $N_p = 375$, $N_{s1} = 750$, $N_{s2} = 500$, and $N_{s3} = 75$.

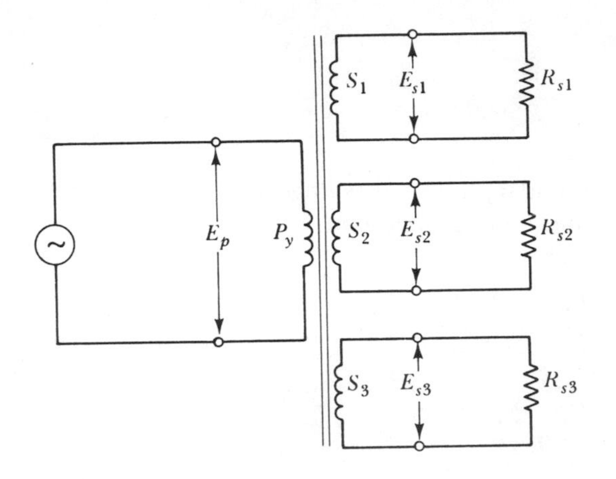

FIG. 25-2. Transformer with three secondary windings

SOLUTION

Eq. (25-1):

$$E_s = E_p \times \frac{N_s}{N_p}$$

$$E_{s1} = E_p \times \frac{N_{s1}}{N_p}$$

$$= 100\text{ V} \times \frac{750}{375}$$

$$\mathbf{E_{s1} = 200\ V}$$

$$E_{s2} = E_p \times \frac{N_{s2}}{N_p}$$

$$= 100\text{ V} \times \frac{500}{375}$$

$$\mathbf{E_{s2} = 133\ V}$$

$$E_{s3} = E_p \times \frac{N_{s3}}{N_p}$$

$$= 100\text{ V} \times \frac{75}{375}$$

$$\mathbf{E_{s3} = 20\ V}$$

$$\textit{total secondary turns} = N_{s1} + N_{s2} + N_{s3}$$

$$N_s = 750 + 500 + 75$$

$$= 1325$$

$$\textit{total secondary voltage, } E_s = E_p \times \frac{N_s}{N_p}$$

$$= 100 \text{ V} \times \frac{1325}{375}$$

$$\mathbf{E_s = 353 \text{ V}}$$

or

$$E_s = E_{s1} + E_{s2} + E_{s3}$$

$$= 200 \text{ V} + 133 \text{ V} + 20 \text{ V}$$

$$\mathbf{E_s = 353 \text{ V}}$$

It is important to note that the total output voltage calculation in Example 25-1 assumes that the secondary windings are connected *series-aiding*, that is, such that all the winding voltages are in phase. If one of the windings is connected so that its voltage is in antiphase to the others, this voltage will subtract from the others, and the connection is termed *series-opposing*. Figure 25-3(a) and (b) shows a transformer that has two outputs. One of these gives a waveform with a peak level of 20 V, and the other has a peak level of 15 V. When the two are connected series-aiding, as illustrated in Figure 25-3(a), the resultant waveform has a peak output of (20 V + 15 V) (i.e., 35 V). When connected series-opposing, as shown in Figure 25-3(b), the peak level of the bottom winding is −15 V when the top winding has a peak value of +20 V. Thus, the resultant output peak level is (20 V − 15 V) = 5 V.

The transformer shown diagramatically in Figure 25-3(c) has a secondary that is said to be *center-tapped*. This means that the secondary winding consists of two equal turn windings connected series-aiding. If the center tap is grounded, as shown dashed, the output from one winding is positive-going with respect to ground, while the output from the other is negative-going, and vice versa. The total output voltage from the two windings is the sum of the outputs from each.

From Eq. (22-7), the input power to a transformer is:

$$P_p = E_p I_p \cos\phi_p \quad \text{watts}$$

and the output power is

$$P_s = E_s I_s \cos\phi_s \quad \text{watts}$$

The primary and secondary circuit phase angles are closely equal, and as an approximation the transformer efficiency can be taken as 100%. Therefore,

$$P_p \cong P_s$$

or

$$E_p I_p = E_s I_s$$

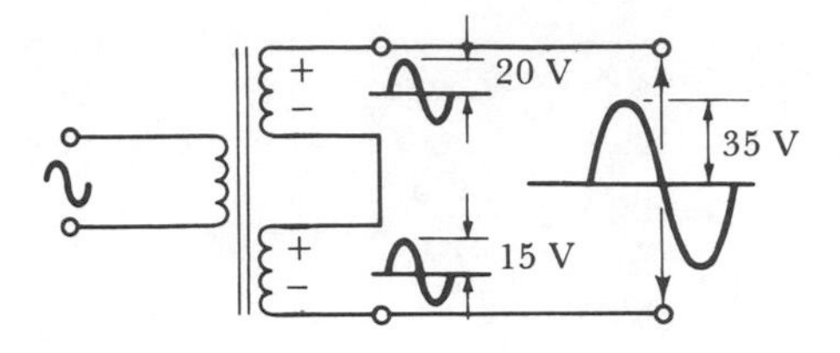

(a) Transformer with two secondary windings connected series-aiding

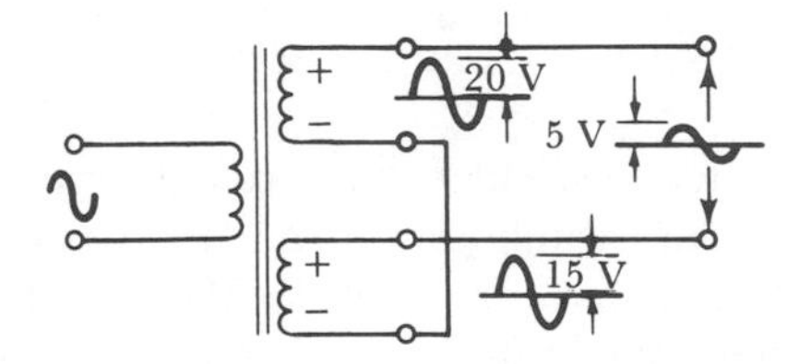

(b) Secondary windings connected series-opposing

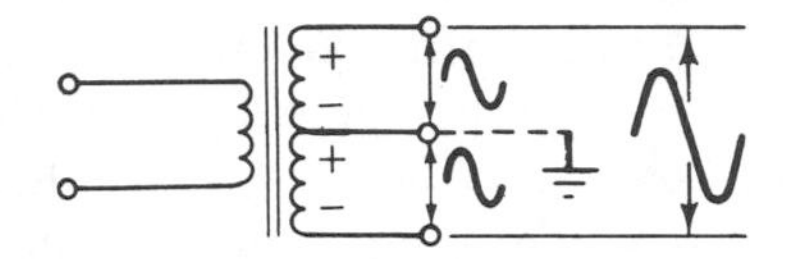

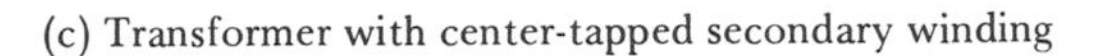

(c) Transformer with center-tapped secondary winding

FIG. 25-3. Transformers with various secondary winding arrangements

which gives

$$\frac{I_p}{I_s} = \frac{E_s}{E_p} = \frac{N_s}{N_p}$$

or

$$\boxed{I_s = I_p \times \frac{N_p}{N_s}} \qquad (25\text{-}2)$$

Comparing Eqs. (25-2) and (25-1), it is seen that the ratio of primary and secondary currents in terms of the turns ratio is the inverse of the ratio of primary and secondary voltages. This shows that if a transformer *steps up* the voltage from the primary, it may be said to *step down* the current, and vice versa.

EXAMPLE 25-2

The three secondary windings on the transformer described in Example 25-1 have the following loads: $R_{s1}=1\ \text{k}\Omega$, $R_{s2}=500\ \Omega$, and $R_{s3}=100\ \Omega$. Assuming that the transformer is 100% efficient, determine the total primary current.

SOLUTION

$$I_{s1}=\frac{E_{s1}}{R_{s1}}=\frac{200\ \text{V}}{1\ \text{k}\Omega}$$

$$\mathbf{I_{s1}=200\ mA}$$

From Eq. (25-2),

$$I_p=I_s\times\frac{N_s}{N_p}$$

Therefore, the primary current required to supply I_{s1} *is*:

$$I_{p1}=I_{s1}\times\frac{N_{s1}}{N_p}$$

$$=200\ \text{mA}\times\frac{750}{375}$$

$$\mathbf{I_{p1}=400\ mA}$$

$$I_{s2}=\frac{E_{s2}}{R_{s2}}=\frac{133\ \text{V}}{500\ \Omega}$$

$$\mathbf{I_{s2}=266\ mA}$$

To supply I_{s2}:

$$I_{p2}=I_{s2}\times\frac{N_{s2}}{N_p}$$

$$=266\ \text{mA}\times\frac{500}{375}$$

$$\mathbf{I_{p2}=355\ mA}$$

$$I_{s3}=\frac{E_{s3}}{R_{s3}}=\frac{20\ \text{V}}{100\ \Omega}$$

$$\mathbf{I_{s3}=200\ mA}$$

To supply I_{s3}:

$$I_{p3} = I_{s3} \times \frac{N_{s3}}{N_p}$$

$$= 200 \text{ mA} \times \frac{75}{375}$$

$$\boldsymbol{I_{p3} = 40 \text{ mA}}$$

Total primary current:

$$I_p = I_{p1} + I_{p2} + I_{p3}$$

$$= 400 \text{ mA} + 355 \text{ mA} + 40 \text{ mA}$$

$$\boldsymbol{I_p = 795 \text{ mA}}$$

25-2 EMF EQUATION

As discussed in the preceding section, the emfs induced in the primary and secondary windings are

$$E_p = \frac{\Delta\Phi}{\Delta t} N_p \quad \text{and} \quad E_s = \frac{\Delta\Phi}{\Delta t} N_s$$

The flux in the transformer core has a sinusoidal waveform, because the (primary) current producing it is sinusoidal. Therefore, as illustrated in Figure 25-4, the flux increases from zero to its maximum value Φ_m in a

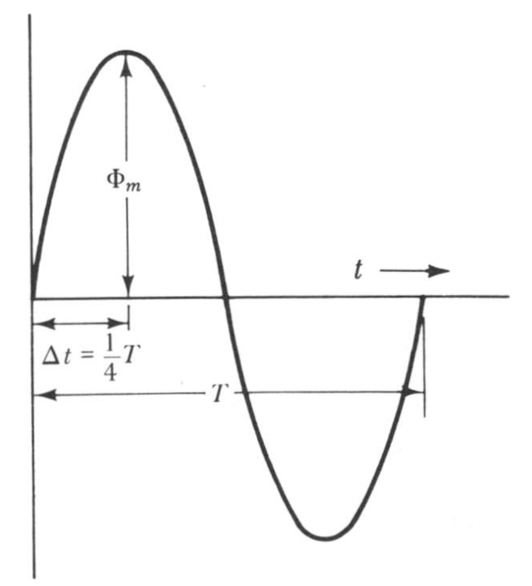

FIG. 25-4. The flux in a transformer core has a sinusoidal waveform

time period of

$$\Delta t = \tfrac{1}{4} T$$

or

$$\Delta t = \frac{1}{4f}$$

where f is the frequency of the flux waveform (i.e., the frequency of the primary voltage and current). Consequently, the average induced voltage is

$$E_{\text{av}} = \frac{\Phi_m}{1/(4f)} N$$

or

$$E_{\text{av}} = 4\Phi_m f N$$

For a sine wave:

$$E_{\text{rms}} = \frac{0.707}{0.637} E_{\text{av}} \qquad \text{(see Section 18-5)}$$

$$E_{\text{rms}} = 1.11 E_{\text{av}}$$

and

$$E_{\text{rms}} = 1.11 \times 4\Phi_m f N$$

$$E_{\text{rms}} = 4.44\Phi_m f N$$

This gives the primary rms voltage as:

$$\boxed{E_p = 4.44\ \Phi_m f N_p} \qquad \textbf{(25-3)}$$

and the secondary as:

$$\boxed{E_s = 4.44\ \Phi_m f N_s} \qquad \textbf{(25-4)}$$

where Φ_m is in webers, f is in hertz, and E_s and E_p are in volts.

Equations (25-3) and (25-4) enable the maximum value of flux in a transformer core to be calculated from a knowledge of the coil turns, coil voltage, and supply frequency. Once the maximum flux is known, the flux density in the core can be determined using the core dimensions.

EXAMPLE 25-3 The input voltage to the transformer described in Example 25-1 has a frequency of 400 Hz. Determine the peak value of the flux.

SOLUTION

From Eq. (25-3):

$$\Phi_m = \frac{E_p}{4.44 f N_p}$$

$$= \frac{100\ \text{V}}{4.44 \times 400\ \text{Hz} \times 375}$$

$$\mathbf{\Phi_m = 0.15 \times 10^{-3}\ Wb}$$

25-3 TRANSFORMER ON NO-LOAD

A transformer is said to be on *no-load* when the output (secondary) terminals are open-circuited. In this condition there is no current flowing in the secondary windings. However, with an alternating voltage applied to the input (primary) terminals, a small primary current flows in order to create the magnetic flux in the core. This current is termed the *magnetizing current*.

In Section 13-8 it was explained that when the magnetic flux in a core is continuously increasing to a peak level in one direction and then reversing to a peak in the opposite direction, energy is absorbed by the core, owing to *hysteresis* loss. The alternating magnetic flux also induces *eddy currents* in the transformer core, as discussed in Section 13-9. The eddy currents cause additional energy to be dissipated in the core. Hysteresis loss is kept to a minimum by the use of a magnetic material with a narrow hysteresis loop (see Section 13-8), and eddy current loss is minimized by constructing the core of *laminations* (see Section 13-9). Because of the laminated construction, there are air gaps in the core, as illustrated in Figure 25-5. The magnetizing current must also create a

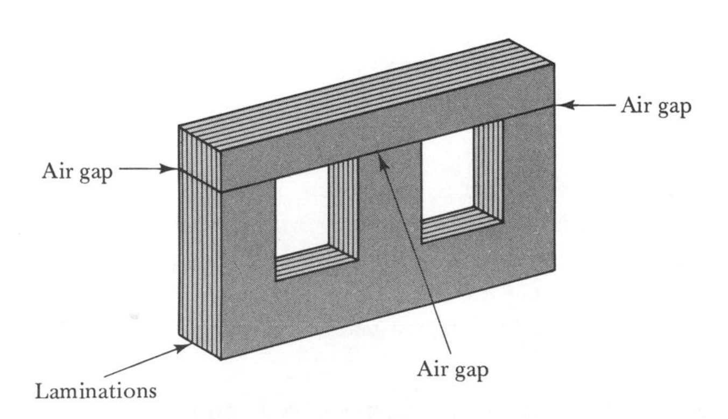

FIG. 25-5. Laminated transformer core

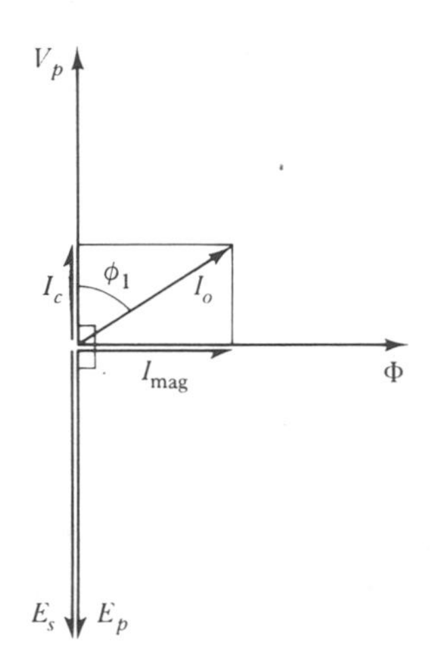

FIG. 25-6. Approximate phasor diagram for a transformer on no-load

flux in the air gaps. A further energy loss that must be supplied is the power dissipated in the transformer primary windings (the I^2R loss). The total primary *no-load current* is composed of the magnetizing current and the current required to supply the core losses.

An approximate phasor diagram for a transformer under no-load conditions is shown in Figure 25-6. Since the transformer windings are inductive, the input voltage V_p leads the magnetizing current I_{mag} by 90°. The flux Φ increases and decreases as the magnetizing current rises and falls; consequently, the flux phasor is shown in phase with I_{mag}. The current I_c is the component of the primary current which supplies the core losses and the small power loss in the primary winding. The total power losses are equal to $(I_c \times V_p)$, and so I_c is in phase with V_p. The no-load primary current I_o is the phasor sum of I_c and I_{mag}. Since I_c is normally much smaller than I_{mag}, the no-load power factor ($\cos\phi_1$) is very small. The voltages induced in the secondary and primary windings, E_s and E_p, respectively, lag the flux by 90°, and thus the E_s and E_p phasors are drawn opposite to the V_p phasor. In Figure 25-6, E_s and E_p are shown as equal voltages (i.e., assuming a 1 : 1 transformer). An approximation in the phasor diagram of Figure 25-6 occurs because E_p has been taken as exactly equal and opposite to V_p. In fact, E_p is equal and opposite to the phasor sum of V_p and the winding voltage drops due to I_o. This becomes more evident when a transformer *on-load* is considered.

The no-load equivalent circuit for the transformer is shown in Figure 25-7. The transformer is replaced by an ideal (no-loss) transformer, with a resistance R_o and an inductive reactance X_o in parallel with its primary. R_o represents the core losses, and so the current I_c which supplies the core losses is shown passing through R_o. The inductive reactance X_o represents

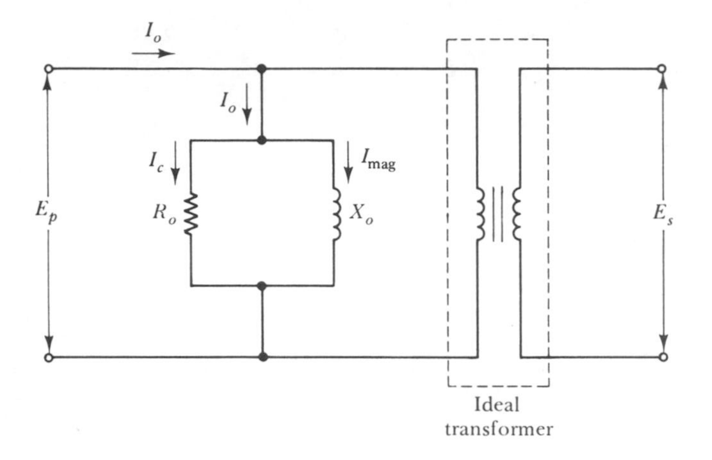

FIG. 25-7. No-load equivalent circuit for a transformer

a loss-free coil which passes the magnetizing current I_{mag}. Thus, the combination of R_o, X_o, and the ideal transformer simulates the actual transformer under no-load conditions.

25-4 TRANSFORMER ON LOAD

When a load is connected to a transformer secondary terminals, a secondary winding (load) current flows. As illustrated in Figure 25-8, the secondary current tends to generate a flux Φ_2 in the transformer core. To supply the secondary current, current must flow in the primary winding. The primary current generates a flux Φ_1 which is exactly equal and opposite to Φ_2. Thus, Φ_1 and Φ_2 cancel each other out, and the core flux remains at the level set up by the magnetizing current.

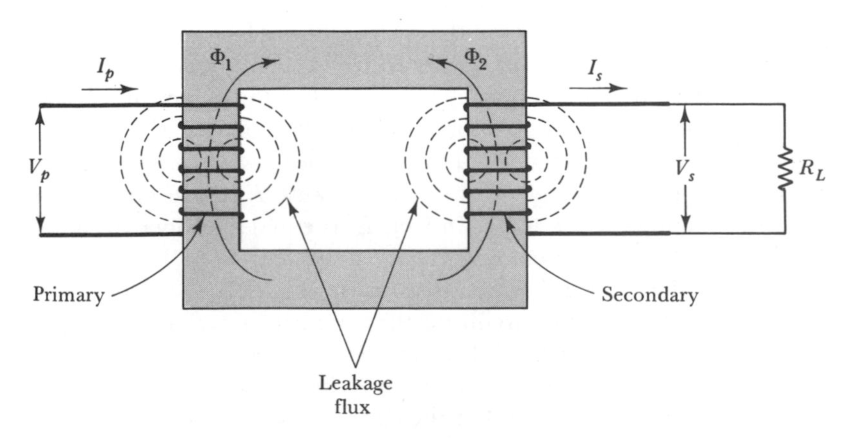

FIG. 25-8. Transformer core flux and leakage flux

Figure 25-8 also shows that when primary and secondary currents flow, not all the flux set up by the currents pass through the iron core. Instead, there is some *leakage flux* passing through the air surrounding each coil. Since the magnetic path through the iron core has a very much smaller reluctance than the air path around each coil, the leakage flux is normally quite small. However, the leakage flux links with the winding turns in each coil, and sets up emfs that oppose the flow of current through each coil. Thus, the leakage flux produces the same effect as an unwanted inductance connected in series with each winding. The effect is termed the *leakage inductance*.

The complete equivalent circuit for a transformer is shown in Figure 25-9. Inductive reactances X_p and X_s represent the leakage inductance of the primary and secondary windings, respectively, and R_p and R_s represent the winding resistances. An ideal (no-loss) transformer is shown with R_s and X_s connected in series with the secondary, so an output (load) current causes a voltage drop across R_s and X_s. Similarly, R_p and X_p are connected in series with the primary windings, and voltage drops are produced across them when a primary current flows. As before, R_o and X_o are shown in parallel with the primary, to simulate the no-load losses and the magnetizing current.

The phasor diagrams for the secondary circuit of a transformer under load is shown in Figure 25-10. Referring to Figure 25-9 and Figure 25-10(a), V_o is the voltage at the transformer output terminals, and I_s is the secondary (load) current. For a load with a lagging phase angle ϕ_o, the I_s phasor is shown lagging the V_o phasor by angle ϕ_o. The secondary current flows through R_s and X_s and produces voltage drops across them.

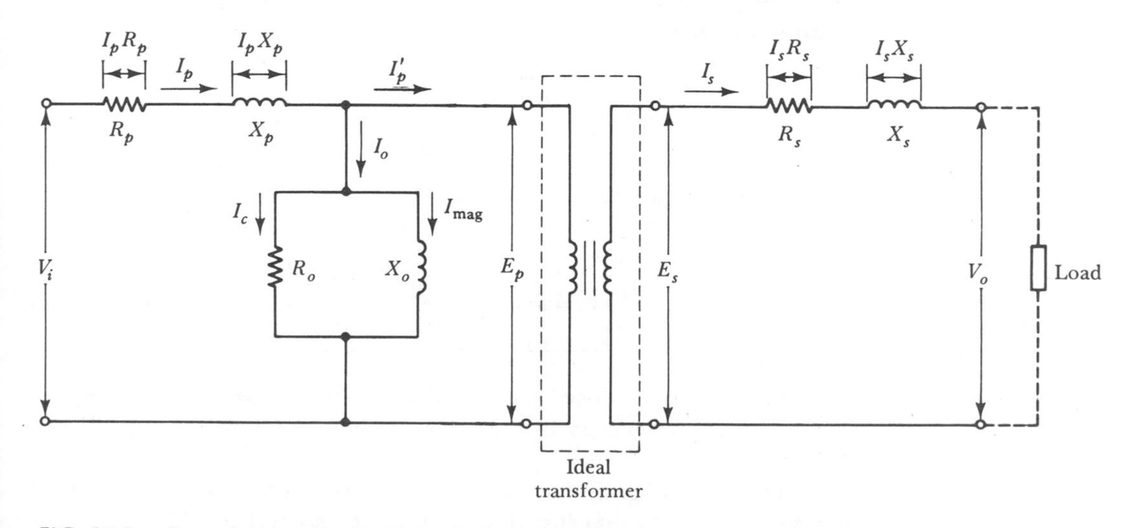

FIG. 25-9. Complete equivalent circuit for a transformer

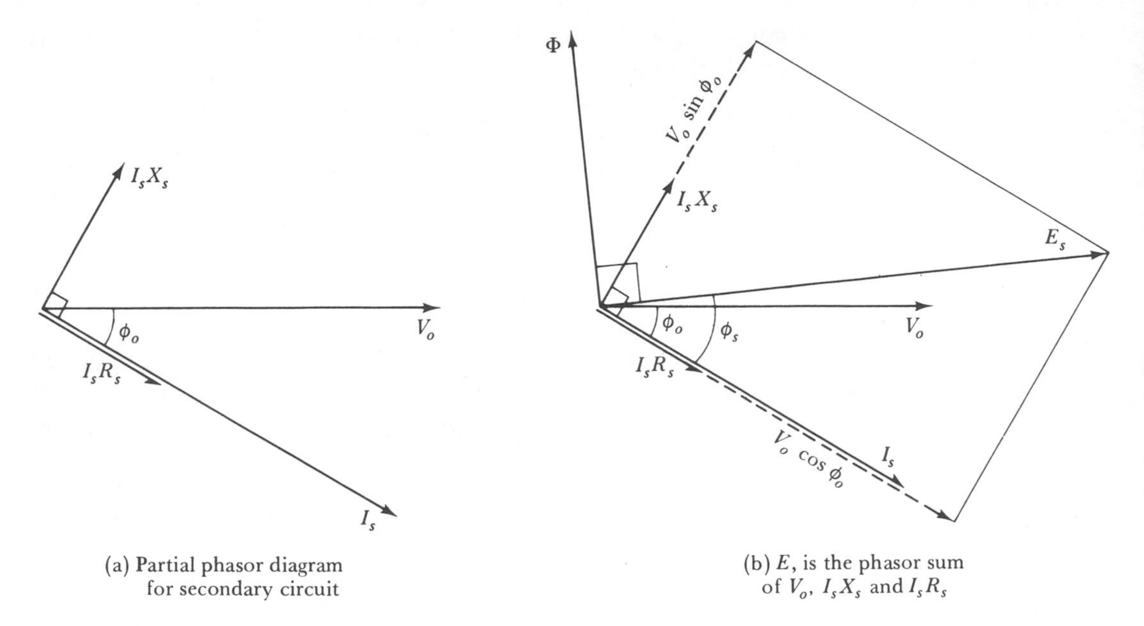

(a) Partial phasor diagram for secondary circuit

(b) E_s is the phasor sum of V_o, I_sX_s and I_sR_s

FIG. 25-10. Phasor diagram for the secondary circuit of a transformer under load

I_sR_s is the resistive voltage drop; consequently, its phasor is in phase with I_s. I_sX_s is an inductive voltage drop, so it leads the current by 90°, as shown in Figure 25-10(a).

Turning to Figure 25-10(b), the phasors of V_o, I_s, I_sX_s, and I_sR_s are reproduced. The secondary induced voltage E_s (also see Figure 25-9) is the phasor sum of V_o, I_sR_s, and I_sX_s. From the discussion of phasor addition in Chapter 19:

$$E_s = \sqrt{(V_o \cos\phi_o + I_sR_s)^2 + (V_o \sin\phi_o + I_sX_s)^2} \quad (25\text{-}5)$$

and the phase angle between E_s and I_s is

$$\phi_s = \arctan\left(\frac{V_o \sin\phi_o + I_sX_s}{V_o \cos\phi_o + I_sR_s}\right) \quad (25\text{-}6)$$

The secondary induced voltage always lags 90° behind the core flux Φ, so the flux phasor may be drawn 90° ahead of E_s, as shown in Figure 25-10(b).

The phasor diagram for the transformer primary may be constructed in a similar way to that just discussed for the secondary. This time it is

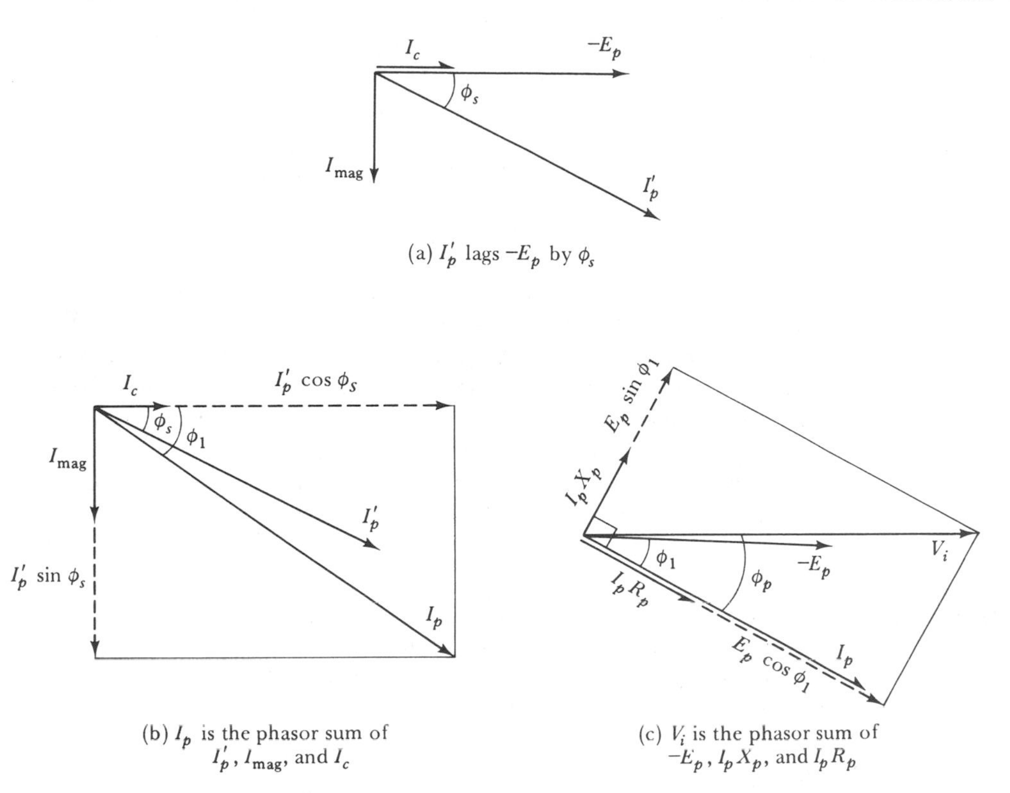

(a) I_p' lags $-E_p$ by ϕ_s

(b) I_p is the phasor sum of I_p', I_{mag}, and I_c

(c) V_i is the phasor sum of $-E_p$, I_pX_p, and I_pR_p

FIG. 25-11. Phasor diagram for the primary circuit of a transformer under load

necessary to commence with the voltage E_p and current I_p' right at the primary winding (see Figure 25-9). E_p and I_p' may be calculated from a knowledge of the turns ratio and the secondary current and voltage. The phase angle between them is ϕ_s, the phase angle of the secondary circuit.

Referring to Fig. 25-11(a), the $-E_p$ phasor is first drawn horizontally, and the I_p' current phasor is drawn at an angle of ϕ_s lagging $-E_p$. Note that because E_p is the voltage induced in the primary winding by the changing core flux, it is equal and opposite to the component of the applied voltage at the (ideal) transformer primary winding. Consequently, the applied voltage component is $-E_p$.

As well as I_p', the no-load current I_o (composed of I_c and I_{mag}) must be supplied. The actual current drawn from the supply is I_p, which is the phasor sum of I_p', I_c, and I_{mag} [see Figures 25-9 and 25-11(b)].

$$I_p = \sqrt{(I_p' \cos\phi_s + I_c)^2 + (I_p' \sin\phi_s + I_{mag})^2} \qquad (25\text{-}7)$$

and

$$\phi_1 = \arctan\left(\frac{I_p' \sin\phi_s + I_{mag}}{I_p' \cos\phi_s + I_c}\right) \tag{25-8}$$

I_p causes a voltage drop along R_p and X_p (Figure 25-9). The I_pR_p phasor is in phase with I_p, and the I_pX_p phasor leads I_p by 90°, as shown in Figure 25-11(c). The phasor sum of I_pR_p, I_pX_p, and $-E_p$ gives the supply voltage V_i. The primary input phase angle is then ϕ_p, which is the angle between V_i and I_p, as illustrated.

Given the secondary load and the parameters of the transformer equivalent circuit, the primary input voltage and current can be calculated by means of the phasor diagram.

EXAMPLE 25-4

A transformer has the following parameters: $R_p = 2.5\ \Omega$, $X_p = 6\ \Omega$, $R_o = 5$ kΩ, $X_o = 2$ kΩ, $R_s = 0.25\ \Omega$, $X_s = 1\ \Omega$, and $N_p/N_s = 2$. If the output voltage is 50 V and the load at the output terminals is $Z_L = 25\ \Omega\ \angle 30°$, determine the supply voltage and current.

SOLUTION

Secondary circuit calculations (*refer to* Figure 25-10):
load current,

$$\begin{aligned} I_s &= V_o/Z_L = 50\ \text{V}/(25\ \Omega\ \angle 30°) \\ &= 2\ \text{A}\ \angle -30° \end{aligned}$$

R_s *volts drop*,

$$\begin{aligned} I_sR_s &= 2\text{A} \times 0.25\ \Omega \\ &= 0.5\ \text{V} \end{aligned}$$

X_s *volts drop*,

$$\begin{aligned} I_sX_s &= 2\ \text{A} \times 1\ \Omega \\ &= 2\ \text{V} \end{aligned}$$

Phasor sum of V_o, I_sR_s, and I_sX_s:
Eq. (25-5)

$$E_s = \sqrt{(V_o \cos\phi_o + I_sR_s)^2 + (V_o \sin\phi_o + I_sX_s)^2}$$

$$= \sqrt{(50 \text{ V} \cos 30° + 0.5 \text{ V})^2 + (50 \text{ V} \sin 30° + 2 \text{ V})^2}$$

$$\mathbf{E_s = 51.5 \text{ V}}$$

Eq. (25-6):

$$\phi_s = \arctan\left(\frac{V_o \sin\phi_o + I_sX_s}{V_o \cos\phi_o + I_sR_s}\right)$$

$$= \arctan\left(\frac{50 \text{ V} \sin 30° + 2 \text{ V}}{50 \text{ V} \cos 30° + 0.5 \text{ V}}\right)$$

$$\mathbf{\phi_s = 31.7°}$$

Primary circuit calculations (refer to Figure 25-11):

$$I_p' = \frac{N_s}{N_p} \times I_s = \tfrac{1}{2} \times 2 \text{ A}$$

$$= 1 \text{ A}$$

$$E_p = \frac{N_p}{N_s} \times E_s = \frac{2}{1} \times 51.5 \text{ V} = 103 \text{ V}$$

$$I_c = \frac{E_p}{R_o} = \frac{103 \text{ V}}{5 \text{ k}\Omega} = 20.6 \text{ mA} \qquad (\textit{see} \text{ Figure 25-9})$$

$$I_{\text{mag}} = \frac{E_p}{X_o} = \frac{103 \text{ V}}{2 \text{ k}\Omega} = 51.5 \text{ mA}$$

Eq. (25-7):

$$I_p = \sqrt{(I_p' \cos\phi_s + I_c)^2 + (I_p' \sin\phi_s + I_{\text{mag}})^2}$$

$$= \sqrt{(1 \text{ A} \cos 31.7° + 20.6 \text{ mA})^2 + (1 \text{ A} \sin 31.7° + 51.5 \text{ mA})^2}$$

$$\mathbf{I_p = 1.05 \text{ A}}$$

Eq. (25-8):

$$\phi_1 = \arctan\left(\frac{1\ \text{A} \sin 31.7° + 51.5\ \text{mA}}{1\ \text{A} \cos 31.7° + 20.6\ \text{mA}}\right)$$

$$\boldsymbol{\phi_1 = 33.5°}$$

$$I_pR_p = 1.05\ \text{A} \times 2.5\ \Omega = 2.63\ \text{V} \qquad (\textit{see}\ \text{Figure 25-9})$$

$$I_pX_p = 1.05\ \text{A} \times 6\ \Omega = 6.3\ \text{V}$$

$$V_i = \sqrt{(E_p \cos\phi_1 + I_pR_p)^2 + (E_p \sin\phi_1 + I_pX_p)^2}$$

$$= \sqrt{(103\ \text{V} \cos 33.5° + 2.63\ \text{V})^2 + (103\ \text{V} \sin 33.5° + 6.3\ \text{V})^2}$$

$$\boldsymbol{V_i = 108.7\ \text{V}}$$

$$\phi_p = \arctan\left(\frac{103\ \text{V} \sin 33.5° + 6.3\ \text{V}}{103\ \text{V} \cos 33.5° + 2.63\ \text{V}}\right)$$

$$\boldsymbol{\phi_p = 35.5°}$$

25-5 REFERRED RESISTANCE AND REACTANCE

The equivalent circuit of a transformer can be considerable simplifed by replacing the secondary circuit resistive and reactive components by primary circuit components that have the same effect. Consider Figure 25-12(a). The secondary load resistance R_L can obviously be written as

$$R_L = \frac{E_s}{I_s}$$

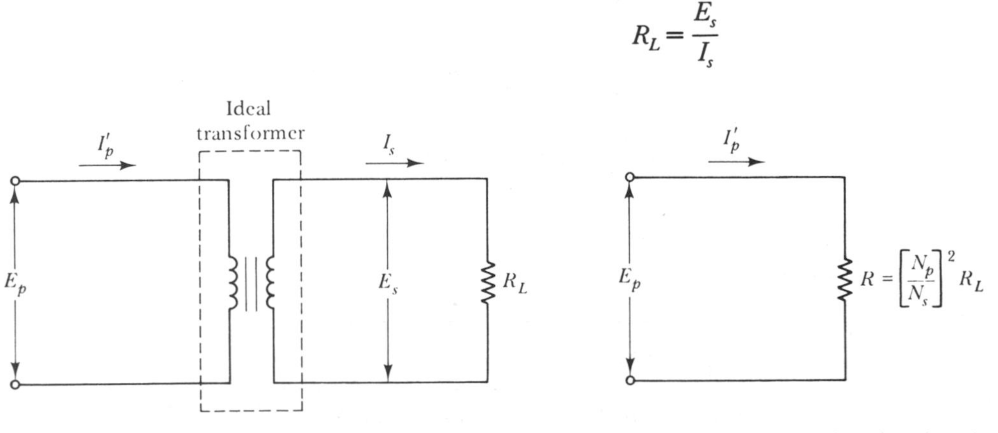

(a) Ideal transformer with load resistance R_L

(b) Load resistance R_L referred to the primary

FIG. 25-12. A load connected to the secondary of a transformer may be referred to the primary

Also, the primary circuit *sees* a resistance of

$$R = \frac{E_p}{I'_p}$$

but

$$E_p = E_s\left(\frac{N_p}{N_s}\right) \quad \text{and} \quad I'_p = I_s\left(\frac{N_s}{N_p}\right)$$

Therefore,

$$R = \frac{E_s\left(\frac{N_p}{N_s}\right)}{I_s\left(\frac{N_s}{N_p}\right)}$$

$$R = R_L\left(\frac{N_p}{N_s}\right)^2$$

Writing,

$$\frac{N_p}{N_s} = a$$

$$\boxed{R = a^2 R_L} \qquad (25\text{-}9)$$

Thus, a primary resistance of a^2R_L would have the same effect on the transformer primary circuit as a secondary resistance of R_L has on the transformer primary circuit [Figure 25-12(b)]. The equivalent primary resistance of the secondary calculated in this way is termed *the referred resistance* (i.e., the resistance of the secondary is said to be *referred* to the primary).

The secondary resistance, reactance, and load impedance may all be referred to the primary:

$$\boxed{\textit{referred reactance} = a^2 X_s} \qquad (25\text{-}10)$$

$$\boxed{\textit{referred load} = a^2 Z_L} \qquad (25\text{-}11)$$

Figure 25-13 illustrates the simplification of the transformer equivalent circuit by the technique of referring everything to the primary. R_s, X_s, and Z_L, become a^2R_s, a^2X_s, and a^2Z_L, respectively, when referred to the primary circuit [see Figure 25-13(a) and (b)]. Note that these referred

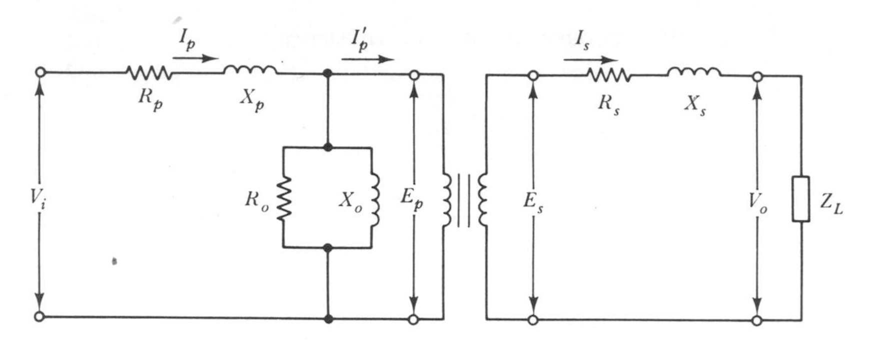

(a) Complete equivalent circuit

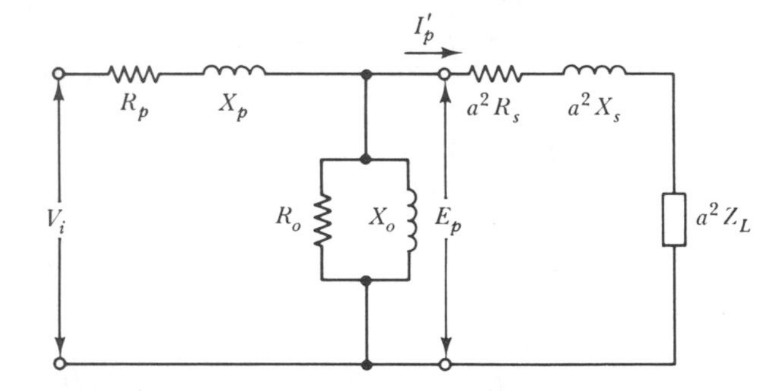

(b) Secondary circuit components referred to the primary

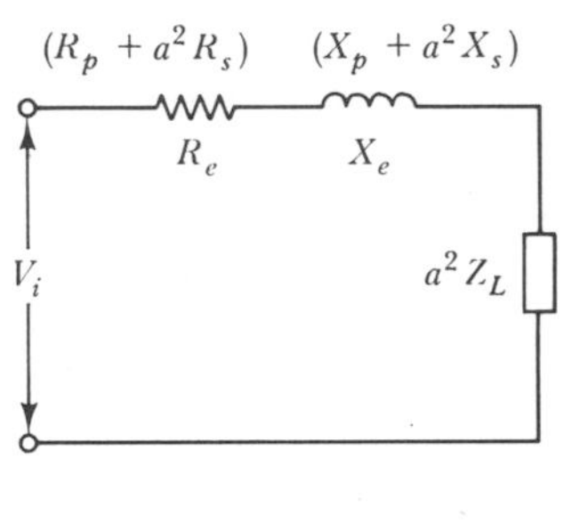

(c) Equivalent circuit simplified by neglecting R_o and X_o, and by combining R_p with $a^2 R_s$ and X_p with $a^2 X_s$

FIG. 25-13. Simplification of transformer equivalent circuit by referring secondary circuit components to the primary

components are *seen looking into* the primary winding of the ideal transformer. Consequently, the three of them (in series) must be shown in parallel with R_o and X_o in the primary equivalent circuit.

As an approximation to further simplify the circuit, R_o and X_o may be omitted, since for a transformer on full load, I_c and I_{mag} are a very small percentage of the total load current I_p. The total primary resistive and reactive components may now be added together [Figure 25-13(c)] to give a primary equivalent resistance and equivalent reactance:

$$\boxed{R_e = R_p + a^2 R_s} \tag{25-12}$$

and

$$\boxed{X_e = X_p + a^2 X_s} \tag{25-13}$$

The referred load impedance remains a^2Z_L; however, it may also be taken care of by adding its resistive and reactive components to R_e and X_e.

EXAMPLE 25-5

For the transformer circuit described in Example 25-4, refer all secondary components to the primary. Neglecting R_o and X_o, calculate I_p and ϕ_p when V_i is 108.7 V.

SOLUTION

$$
\begin{aligned}
Z_L &= 25\ \Omega\angle 30^\circ \\
&= R_L + jX_L \\
&= 25\cos 30^\circ + j25\sin 30^\circ \\
Z_L &= 21.65\ \Omega + j12.5\ \Omega
\end{aligned}
$$

total secondary resistance,

$$
\begin{aligned}
R_s' &= R_s + R_L \\
&= 0.5\ \Omega + 21.65\ \Omega \\
R_s' &= 22.15\ \Omega
\end{aligned}
$$

total secondary reactance,

$$
\begin{aligned}
X_s' &= X_s + X_L \\
&= 1\ \Omega + 12.5\ \Omega \\
X_s' &= 13.5\ \Omega
\end{aligned}
$$

$$
\begin{aligned}
R_s' \textit{ referred to primary} &= a^2R_s' \\
&= 2^2 \times 22.15\ \Omega \\
&= 88.6\ \Omega
\end{aligned}
$$

$$
\begin{aligned}
X_s' \textit{ referred to primary} &= a^2X_s' = 2^2 \times 13.5\ \Omega \\
&= 54\ \Omega
\end{aligned}
$$

total equivalent primary resistance,

$$
\begin{aligned}
R_e &= R_p + a^2R_s' = 2.5\ \Omega + 88.6\ \Omega \\
&= 91.1\ \Omega
\end{aligned}
$$

total equivalent primary reactance,

$$X_e = X_p + a^2 X_s' = 6\ \Omega + 54\ \Omega$$

$$= 60\ \Omega$$

$$R_e + jX_e = 91.1\ \Omega + j60\ \Omega$$

$$|I_p| = \frac{V_i}{\sqrt{R_e^2 + X_e^2}}$$

$$= \frac{108.7\ \text{V}}{\sqrt{(91.1\ \Omega)^2 + (60\ \Omega)^2}}$$

$$\mathbf{|I_p| = 0.996\ A}$$

$$\phi_p = \arctan\left(\frac{X_e}{R_e}\right) = \arctan\left(\frac{60\ \Omega}{91.2\ \Omega}\right)$$

$$\boldsymbol{\phi_p = 33.37°}$$

Note that I_p and Φ_p are slightly different from the values calculated in Example 25-4. *This is because R_o and X_o have been neglected.*

25-6 VOLTAGE REGULATION

It is clear from the transformer equivalent circuit in Figure 25-13(a) that the secondary current I_s produces voltage drops I_sR_s and I_sX_s across the resistive and reactive components. Also, the primary current I_p causes primary circuit voltage drops I_pR_p and I_pX_p. Consequently, the induced primary voltage E_p is less than the input V_i, and the output voltage V_o is less than E_s.

When there is no load connected to the output terminals of the transformer, no secondary current flows, and therefore, no voltage drops occur across R_s and X_s. With zero secondary current, the primary current drops to the no-load current I_o, and the voltage drops across R_p and X_p become very small. Thus, in the no-load situation, E_p is almost equal to V_i, and V_o equals E_s.

It appears that the transformer output voltage is greatest on no-load, and that under loaded conditions the voltage drops across the resistive and reactive components of the equivalent circuit cause V_o to drop below its no-load level. (Note that, depending upon the power factor of the load, the output full-load voltage may actually be larger than the no-load voltage.) The percentage change in output voltage from no-load to full load is termed the *voltage regulation* of the transformer. Ideally, there should be no change in V_o from no-load to full-load, (i.e., regulation = 0%). For best possible performance, the transformer should have the lowest possible regulation.

$$\textbf{\textit{voltage regulation}} = \frac{V_{o(NL)} - V_{o(FL)}}{V_{o(FL)}} \times 100\% \qquad (25\text{-}14)$$

where $V_{o(NL)}$ *is the transformer no-load output voltage, and* $V_{o(FL)}$ *is the full-load output voltage.*

EXAMPLE 25-6

Neglecting the no-load current, calculate the voltage regulation for the transformer described in Example 25-4.

SOLUTION

$$\textit{full-load output voltage, } V_{o(FL)} = 50 \text{ V}$$

$$\textit{input voltage, } V_i = 108.7 \text{ V}$$

At no-load:

$$E_p \cong V_i$$

and

$$E_s = \frac{N_s}{N_p} \times E_p$$

$$= \tfrac{1}{2} \times 108.7 \text{ V}$$

$$E_s = 54.35 \text{ V}$$

no-load output voltage, $V_{o(NL)} = E_s = 54.35$ V

From Eq. (25-14):

$$\textit{voltage regulation} = \frac{54.35 \text{ V} - 50 \text{ V}}{50 \text{ V}} \times 100\%$$

voltage regulation = 8.7%

25-7 TRANSFORMER EFFICIENCY

The efficiency of a transformer, like that of any other piece of equipment, is the output power expressed as a percentage of the input power.

$$\textbf{\textit{efficiency}}, \eta = \frac{P_o}{P_i} \times 100\% \qquad (25\text{-}15)$$

or $$\eta = \frac{V_o I_s \cos\phi_s}{V_i I_p \cos\phi_p} \times 100\%$$

Since $$P_i = P_o + \text{losses}$$

$$\boxed{\eta = \frac{P_o}{P_o + \text{losses}} \times 100\%} \tag{25-16}$$

The power losses in a transformer consist of core losses due to hysteresis and eddy currents and copper losses due to the currents flowing in the primary and secondary windings (see Section 25-3). As long as the supply frequency remains constant, the core losses tend to be a constant quantity. The copper losses have two components:

$$\text{primary winding copper loss} = I_p^2 R_p$$

$$\text{secondary winding copper loss} = I_s^2 R_s$$

The two copper losses can be lumped into one power loss proportional to the load current, if R_p is referred to the secondary winding, and added to R_s. Thus, where R_{es} is the equivalent of (R_p referred $+R_s$):

$$\text{total copper losses in both windings} = I_s^2 R_{es}$$

Now, rewriting Eq. (25-16):

$$\eta = \frac{V_o I_s \cos\phi_s}{V_o I_s \cos\phi_s + I_s^2 R_{es} + P_c} \times 100\%$$

where $V_o I_s \cos\phi_s$ *is the output power,* $I_s^2 R_{es}$ *is the total copper losses, and* P_c *is the core loss*. Dividing the numerator and denominator by I_s,

$$\eta = \frac{V_o \cos\phi_s}{V_o \cos\phi_s + I_s R_{es} + (P_c/I_s)} \times 100\%$$

Since the output voltage V_o remains substantially constant (within the limits of the regulation), the maximum value of η is obtained from the equation above when the denominator has its minimum value. If I_s were zero, $I_s R_{es}$ becomes zero, but P_c/I_s becomes infinity. Similarly, if I_s were made very large, P_c/I_s might be made very small, but $I_s R_{es}$ would become very large. So, neither $I_s = 0$, nor $I_s =$ (a very large current) gives maximum efficiency. It can be shown by differential calculus, or by substituting practical values into $I_s R_{es}$ and P_c/I_s, that maximum transformer efficiency is obtained when

$$I_s R_{es} = \frac{P_c}{I_s}$$

or

$$\boxed{I_s^2 R_{es} = P_c} \qquad \textbf{(25-17)}$$

That is, *for maximum efficiency,*

$$(\textit{winding copper losses}) = (\textit{core loss})$$

EXAMPLE 25-7

Determine the efficiency of the transformer described in Example 25-4. Also calculate the level of output current at which maximum efficiency occurs.

SOLUTION

$$\begin{aligned} P_o &= V_s I_s \cos\phi_s \\ &= 50\ \text{V} \times 2\ \text{A} \times \cos 31.7° \\ &= 85.3\ \text{W} \end{aligned}$$

$$\begin{aligned} P_i &= V_i I_p \cos\phi_p \\ &= 108.7\ \text{V} \times 1.05\ \text{A} \times \cos 35.5° \\ &= 92.9\ \text{W} \end{aligned}$$

Eq. (25-15):

$$\begin{aligned} \eta &= \frac{P_o}{P_i} \times 100\% \\ &= \frac{85.3\ \text{W}}{92.9\ \text{W}} \times 100\% \\ &\cong 92\% \end{aligned}$$

$$\begin{aligned} \textit{Core losses},\ P_c &= \frac{V_i^2}{R_o} \\ &= \frac{108.7^2}{5\ \text{k}\Omega} \\ P_c &= 2.36\ \text{W} \end{aligned}$$

Referring R_p to the secondary, the equivalent secondary resistance is

$$R_{es} = R_s + R_p \left(\frac{N_s}{N_p} \right)^2$$

$$= 0.25\ \Omega + 2.5\ \Omega (\tfrac{1}{2})^2$$

$$R_{es} = 0.875\ \Omega$$

For maximum efficiency:
Eq. (25-17):

$$I_s^2 R_{es} = P_c$$

Therefore,

$$I_s = \sqrt{\frac{P_c}{R_{es}}}$$

$$= \sqrt{\frac{2.36\ \text{W}}{0.875\ \Omega}}$$

$$\boldsymbol{I_s \cong 1.64\ \text{A}}$$

25-8 OPEN-CIRCUIT AND SHORT-CIRCUIT TESTS

A transformer could be tested under no-load and full-load conditions, to determine its turns ratio, regulation, and efficiency. However, without fully loading the transformer it is possible to perform two tests from which all important data can be derived.

Figure 25-14 shows the circuit for the transformer *open-circuit test*. The alternating input voltage is set to the normal primary level for the

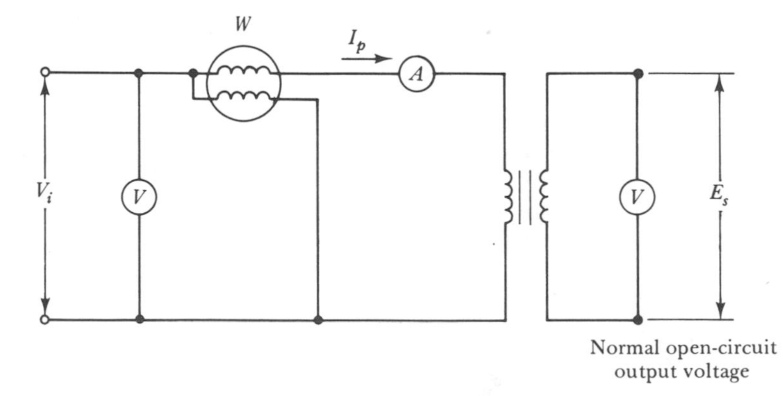

FIG. 25-14. Open-circuit test on a transformer

transformer, and the voltage at the open-circuited output terminals is monitored on a voltmeter, as shown. The input power is measured by the wattmeter (see Section 26-5), and the ammeter measures the primary current.

Since the secondary is effectively open-circuited, the primary current is very small, and the voltage drops across the ammeter and wattmeter can be assumed to be negligible. In this case the input voltage can be taken as the transformer primary voltage, and thus the ratio of the two voltmeter readings gives the turns ratio:

$$\frac{E_p}{E_s} = \frac{N_p}{N_s}$$

With a very small primary current, and near-zero secondary current (i.e., the voltmeter current), the copper loss in the windings can be taken as negligible. The input power measured on the wattmeter is then the total transformer core losses, and the ammeter indicates the no-load primary current I_o.

From the measured values of input voltage, current, and power, the components of the no-load equivalent circuit can be determined:

$$\textit{true power}, P = \frac{E_p^2}{R_o}$$

$$\boxed{R_o = \frac{E_p^2}{P}} \tag{25-18}$$

$$\textit{apparent power}, S = E_p I_o$$

and $S = \sqrt{(\text{true power})^2 + (\text{reactive power})^2}$ [see Figure 22-7(b)]

$S = \sqrt{P^2 + Q^2}$

or $Q = \sqrt{S^2 - P^2}$

or

$$\boxed{\textit{reactive power}, Q = \sqrt{(E_p I_o)^2 - P^2}} \tag{25-19}$$

and from Eq. (22-6),

$$Q = \frac{E_p^2}{X_o}$$

Therefore,

$$\boxed{X_o = \frac{E_p^2}{Q}} \tag{25-20}$$

EXAMPLE 25-8

An open-circuit test on a certain transformer produced the following measurements: $E_p = 115$ V, $E_s = 57.5$ V, $P = 9.5$ W, and $I_o = 180$ mA. Determine the transformer turns ratio and the values of R_o and X_o.

SOLUTION

$$\frac{N_s}{N_p} = \frac{E_s}{E_p} = \frac{57.5\text{ V}}{115\text{ V}}$$

$$\frac{N_s}{N_p} = \frac{1}{2}$$

Eq. (25-18):

$$R_o = \frac{E_p^2}{P} = \frac{(115\text{ V})^2}{9.5\text{ W}}$$

$$\mathbf{R_o = 1.39\ k\Omega}$$

Eq. (25-19):

$$Q = \sqrt{(E_p I_o)^2 - (P)^2}$$

$$= \sqrt{(115\text{ V} \times 180\text{ mA})^2 - (9.5\text{ W})^2}$$

$$\mathbf{Q = 18.39\ vars}$$

Eq. (25-20):

$$X_o = \frac{E_p^2}{Q} = \frac{(115\text{ V})^2}{18.39\text{ vars}}$$

$$\mathbf{X_o = 719\ \Omega}$$

The transformer *short-circuit test* is performed with the secondary terminals short-circuited, as illustrated in Figure 25-15. Note that the primary voltage E_p is measured right at the transformer primary terminals, to avoid the error due to the voltage drops across the ammeter and wattmeter. The input voltage is increased from zero until the ammeter in

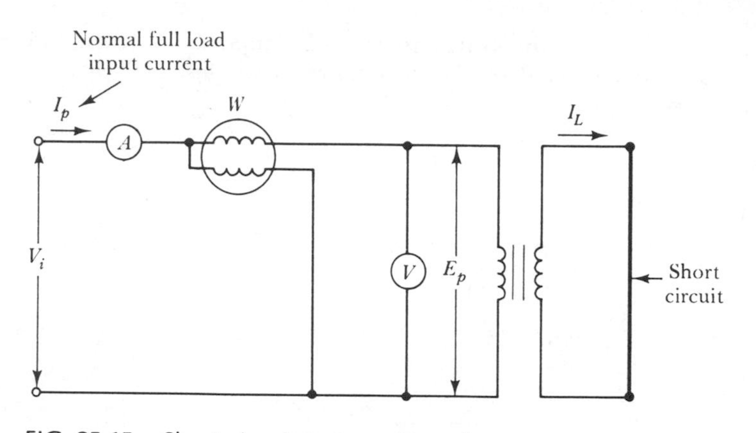

FIG. 25-15. Short-circuit test on a transformer

the primary circuit indicates normal full-load primary current. When this occurs, the normal full-load secondary current is circulating in the secondary winding. Because the secondary terminals are short-circuited, the input voltage required to produce full-load primary and secondary currents is around 3% of the normal input voltage level. With such a low level of input voltage, the core flux is a minimum, and consequently the core losses are so small that they can be neglected. However, the windings are carrying normal full-load current, and thus the input is supplying the normal full-load copper losses.

The wattmeter measuring true input power indicates the full-load copper losses. The product of the ammeter and voltmeter readings gives the apparent input power. From these quantities, calculation may be made of the resistive and reactive components of the full-load equivalent circuit referred to the primary [see Figure 25-13(c)]:

$$\textit{true power}, \ P = I_p^2 R_e$$

$$\boxed{R_e = \frac{P}{I_p^2}} \tag{25-21}$$

$$\textit{apparent power}, \ S = E_p I_p$$

$$\textit{reactive power}, \ \boxed{Q = \sqrt{(E_p I_p)^2 - P^2}} \tag{25-22}$$

and

$$Q = I_p^2 X_e$$

Therefore,

$$\boxed{X_e = \frac{Q}{I_p^2}} \tag{25-23}$$

EXAMPLE 25-9 Determine R_e and X_e for the transformer in Example 25-8 when the following measurements were made on a short-circuit test: $E_p = 5.5$ V, $I_p = 1$ A, and $P = 5.25$ W.

SOLUTION

Eq. (25-21):

$$R_e = \frac{P}{I_p^2} = \frac{5.25 \text{ W}}{(1 \text{ A})^2}$$

$$\mathbf{R_e = 5.25\ \Omega}$$

Eq. (25-22):

$$Q = \sqrt{(E_p I_p)^2 - P^2}$$

$$= \sqrt{(5.5 \text{ V} \times 1 \text{ A})^2 - (5.25)^2}$$

$$\mathbf{Q = 1.64\ vars}$$

Eq. (25-23):

$$X_e = \frac{Q}{I_p^2} = \frac{1.64 \text{ vars}}{(1 \text{ A})^2}$$

$$\mathbf{X_e = 1.64\ \Omega}$$

EXAMPLE 25-10 From the open-circuit and short-circuit test results given in Examples 25-8 and 25-9, determine the transformer regulation for a full load with a lagging phase angle of $\phi_s = 10°$.

SOLUTION

Eq. (25-14):

$$\text{voltage regulation} = \frac{V_{o(\text{NL})} - V_{o(\text{FL})}}{V_{o(\text{FL})}} \times 100\%$$

$$\mathbf{V_{o(NL)} = 57.5\ V} \qquad (\textit{from O.C. test})$$

From S.C. test:

$$V_{i(\text{sc})} = 5.5 \text{ V} \qquad \textbf{when } \mathbf{V_o = 0}$$

Therefore, the S.C. V_i is the full-load volt drop in the transformer equivalent circuit when the load has a zero phase angle.

Referring to Figure 25-11, $\phi_1 = \phi_s$ *when core losses are neglected* and

$$\phi_p = \text{primary phase angle on S.C.}$$

$$= \arctan\left(\frac{Q}{P}\right)$$

$$= \arctan\left(\frac{1.64 \text{ vars}}{5.25 \text{ W}}\right)$$

$$\boldsymbol{\phi_p \cong 17.3°}$$

Again referring to Figure 25-11(c), *the phase angle between* V_i *and* $-E_p$ *is*

$$\phi = (\phi_p - \phi_1)$$

$$\cong (\phi_p - \phi_s)$$

The component of full-load voltage drop which is in phase with $-E_p$ *is*

$$E = V_{i(\text{sc})} \cos(\phi_p - \phi_s)$$

Therefore, full-load primary voltage

$$E_{p(\text{FL})} = E_{p(\text{NL})} - [V_{i(\text{sc})} \cos(\phi_p - \phi_s)]$$

$$= 115 \text{ V} - [5.5 \text{ V} \cos(17.3° - 10°)]$$

$$\boldsymbol{E_{p(\text{FL})} = 109.5 \text{ V}}$$

and full-load output voltage

$$V_{o(\text{FL})} = E_{p(\text{FL})} \times \frac{N_s}{N_p}$$

$$= 109.5 \text{ V} \times \frac{1}{2}$$

$$\boldsymbol{V_{o(\text{FL})} = 54.75 \text{ V}}$$

$$\text{voltage regulation} = \frac{57.5 \text{ V} - 54.75 \text{ V}}{54.75 \text{ V}} \times 100\%$$

$$\textbf{voltage regulation} \cong \mathbf{5\%}$$

EXAMPLE 25-11 From the open-circuit and short-circuit test results given in Examples 25-8 and 25-9, calculate the transformer efficiency on full-load with a load phase angle of $\phi_s = 10°$ lagging.

SOLUTION

$$\textit{full-load output current, } I_{s(\mathrm{FL})} = I_{p(\mathrm{FL})} \times \frac{N_p}{N_s}$$

$$= 1\ \mathrm{A} \times \frac{2}{1}$$

$$\mathbf{I_{s(FL)} = 2\ A}$$

$$P_o = V_{o(\mathrm{FL})} \times I_{s(\mathrm{FL})} \times \cos\phi_s$$

$$= 54.75\ \mathrm{V} \times 2\ \mathrm{A} \times \cos 10°$$

$$\mathbf{P_o = 107.8\ W}$$

$$\textit{core losses} = 9.5\ \mathrm{W} \qquad (\textit{from O.C. test})$$

$$\textit{full-load copper losses} = 5.25\ \mathrm{W} \qquad (\textit{from S.C. test})$$

Eq. (25-15):

$$\textit{efficiency, } \eta = \frac{P_o}{P_i} \times 100\%$$

$$= \frac{P_o}{P_o + \text{losses}} \times 100\%$$

$$= \frac{107.8\ \mathrm{W}}{107.8\ \mathrm{W} + 9.5\ \mathrm{W} + 5.25\ \mathrm{W}} \times 100\%$$

$$\boldsymbol{\eta = 88\%}$$

25-9 AUTOTRANSFORMER

The *autotransformer* has a single winding on an iron core. One of the coil terminals is common to both input and output, and the other output terminal is movable so that it can make contact with any turn on the winding. The principle is illustrated in Figure 25-16(a). Several fixed output terminals (or *taps*) are sometimes used instead of a continuously variable output.

The induced primary voltage, as in the case of a transformer with two windings, is approximately

$$E_p = \frac{V_p}{N_p} \qquad \text{volts/turn} \qquad [\text{see Figure 25-16(a)}]$$

and the total induced secondary voltage is

$$E_s = N_s\left(\frac{E_p}{N_p}\right) \qquad \text{volts}$$

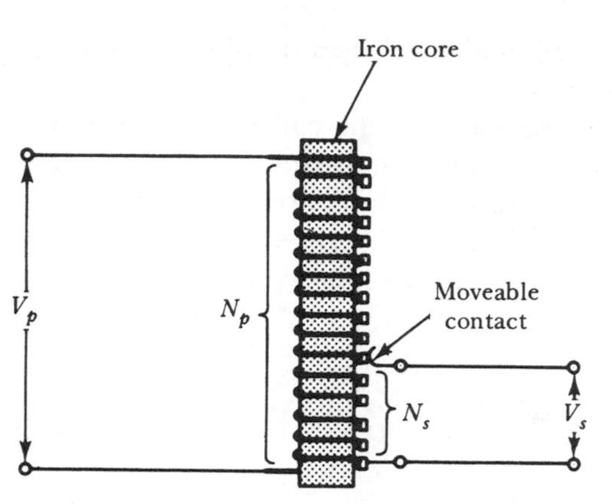

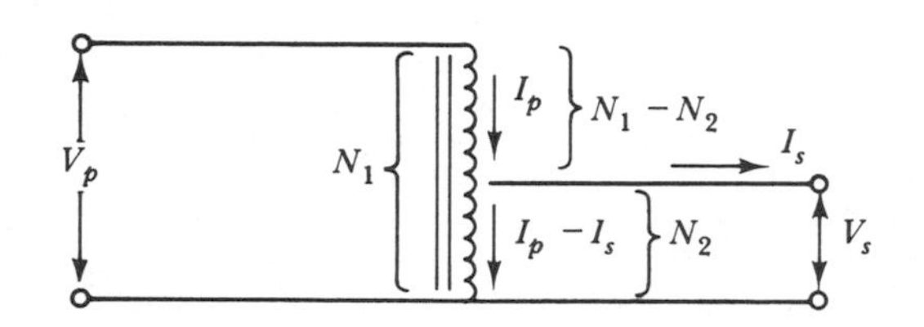

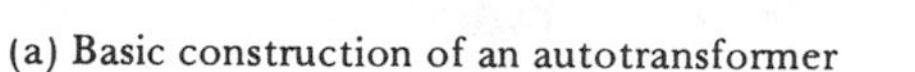
(a) Basic construction of an autotransformer

(b) Currents in an autotransformer

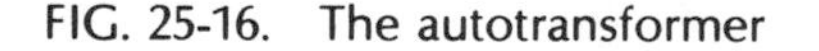
FIG. 25-16. The autotransformer

Therefore,
$$\frac{E_s}{E_p} = \frac{N_s}{N_p}$$

which is the same as in the case of a double-wound transformer.

The obvious advantage of the autotransformer is the facility for adjusting the output voltage to any desired level. The obvious disadvantage is that the output is no longer dc isolated from the input.

Another advantage of the autotransformer is illustrated in Figure 25-16(b). The input current I_p is seen to flow through turns $(N_1 - N_2)$. Because the secondary current flows out at the output terminal, the current flowing through N_2 turns is $(I_p - I_s)$. When I_p and I_s are almost equal, $(I_p - I_s)$ can be a very small quantity, and the N_2 portion of the winding can be constructed of relatively thin copper wire. This cannot apply in the case of a continuously variable output transformer designed to supply a wide range of loads. However, in the case of a fixed output autotransformer, the reduced thickness of the copper windings can result in a significant saving of copper compared to a double-wound transformer designed to do the same job.

25-10 CURRENT TRANSFORMER

A high-level alternating current can be most easily measured by accurately transforming the current to a much lower level. Also, because conductors carrying large currents are frequently at high voltage levels, a measuring instrument directly connected to the conductor would have to be very well insulated. Another problem that arises when rectifier ammeters are to be used (see Section 26-4) is the relatively large voltage drop across each rectifier. The dc isolation afforded by a transformer solves the insulation and rectifier problems.

Figure 25-17 illustrates the principle of the current transformer. A conductor carrying a large current passes through a circular laminated iron core. The conductor constitutes a one-turn primary winding. The secondary winding consists of a number of turns of much finer wire wrapped around the core as shown. The secondary current is given by Eq. (25-2):

$$I_s = I_p \times \frac{N_p}{N_s}$$

For example, suppose that $I_p = 100$ A in Figure 25-17, and let the ammeter be capable of measuring a maximum of 1 A. Then

$$N_s = N_p \times \frac{I_p}{I_s}$$

$$= 1 \times \frac{100 \text{ A}}{1 \text{ A}}$$

$$N_s = 100$$

It is very important to note that a current transformer must never be operated with its secondary winding open-circuited. This is because, when there is no secondary current to oppose the core flux generated by the primary, serious overheating of the core can occur. It is also possible that the secondary open-circuit voltage could reach a dangerously high level.

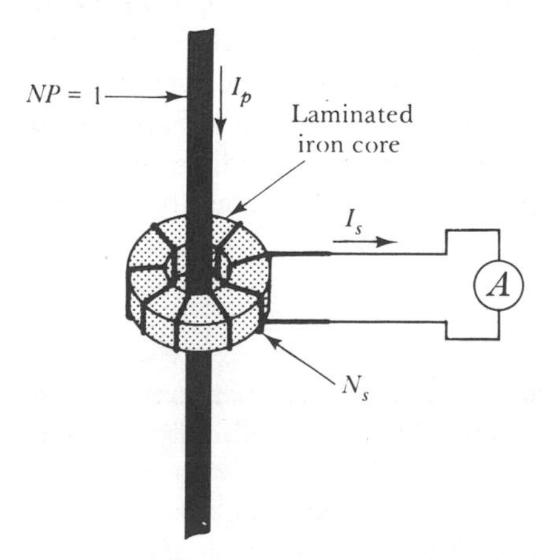

FIG. 25-17. Current transformer

25-11 AUDIO TRANSFORMER

An *audio transformer* is designed to operate over the range of audible frequencies. These devices are usually employed in coupling an *audio amplifier* to the *speaker* of a sound system. The audio range is approximately from 50 Hz to 12 kHz; however, because other factors reduce the frequency range of the system, most audio amplifiers are designed to amplify frequencies ranging from about 20 Hz to 20 kHz, or higher.

As discussed in Section 25-5, a load Z_L connected to the secondary appears as an impedance of (a^2Z_L) in series with the primary input terminals. Most speakers used in sound-reproduction systems have impedance values on the order of 8 Ω to 16 Ω. By careful selection of the transformer turns ratio, the low-value speaker impedance can be made to appear in the transformer primary as a more convenient larger-value impedance. The process is referred to as *impedance matching*.

Over most of the audio range of frequencies, the load impedance referred to the primary dominates the transformer equivalent circuit. Thus the input impedance remains approximately a^2Z_L, and a constant level of ac input voltage to the transformer primary produces a constant level of output voltage. Over this range, the transformer is said to have a *flat frequency response*. A typical transformer frequency response is shown in Figure 25-18, and it is seen that the output voltage level is constant from about 100 Hz to 10 kHz.

Remembering that the transformer is an ac device which does not give an output change for direct input voltages, it is obvious that neither will it respond to very low input frequencies. Thus, as the frequency of the input voltage is decreased below a lower limit, the transformer output voltage falls off, as illustrated in Figure 25-18. Similarly, when the input frequency becomes high enough, energy losses in the iron core increase to the point at which the output voltage falls off as the input frequency increases. Resonance of the leakage reactance and the capacitance between the turns of the winding causes the frequency response to rise initially at the high frequency end before falling off.

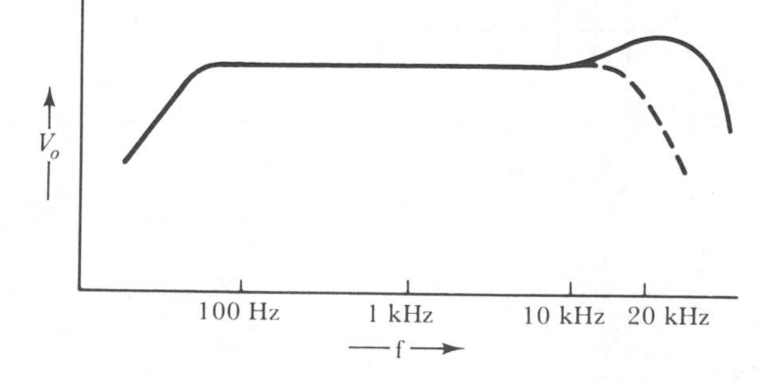

FIG. 25-18. Frequency response of an audio transformer

GLOSSARY OF FORMULAS

Primary / secondary voltage ratio:

$$\frac{E_s}{E_p} = \frac{N_s}{N_p}$$

Primary / secondary current ratio:

$$\frac{I_s}{I_p} = \frac{N_p}{N_s}$$

emf equations:

$$E_p = 4.44\Phi_m f N_p$$
$$E_s = 4.44\Phi_m f N_s$$

Secondary induced voltage:

$$E_s = \sqrt{(V_o \cos\phi_o + I_s R_s)^2 + (V_o \sin\phi_o + I_s X_s)^2}$$

Phase angle between E_s and I_s:

$$\phi_s = \arctan\left(\frac{V_o \sin\phi_o + I_s X_s}{V_o \cos\phi_o + I_s R_s}\right)$$

Primary current:

$$I_p = \sqrt{(I_p' \cos\phi_s + I_c)^2 + (I_p' \sin\phi_s + I_{\text{mag}})^2}$$

Phase angle between I_p and E_p:

$$\phi_1 = \arctan\left(\frac{I_p' \sin\phi_s + I_{\text{mag}}}{I_p' \cos\phi_s + I_c}\right)$$

Referred resistance:

$$R = \left(\frac{N_p}{N_s}\right)^2 R_L = a^2 R_L$$

Referred reactance:

$$X = a^2 X_s$$

Referred impedance:

$$Z = a^2 Z_L$$

Primary equivalent resistance:

$$R_e = R_p + a^2 R_s$$

Primary equivalent reactance:

$$X_e = X_p + a^2 X_s$$

Voltage regulation:

$$\frac{V_{o(\mathrm{NL})} - V_{o(\mathrm{FL})}}{V_{o(\mathrm{FL})}} \times 100\%$$

Efficiency:

$$\eta = \frac{P_o}{P_i} \times 100\%$$

For maximum efficiency:

$$I_s^2 R_{es} = P_c$$

From open-circuit test:

$$R_o = \frac{E_p^2}{P}$$

$$Q = \sqrt{(E_p I_o)^2 - P^2}$$

$$X_o = \frac{E_p^2}{Q}$$

From short-circuit test:

$$R_e = \frac{P}{I_p^2}$$

$$Q = \sqrt{(E_p I_p)^2 - P^2}$$

$$X_e = \frac{Q}{I_p^2}$$

REVIEW QUESTIONS

25-1 Draw a sketch to show the basic construction of a transformer and explain how it functions. Sketch the circuit diagram for a simple two-coil transformer and derive the approximate relationship between the numbers of coil turns and the primary and secondary voltages.

25-2 Define primary winding, secondary winding, turns ratio, 1:1 transformer, step-up transformer, and step-down transformer.

25-3 Derive an expression relating the number of primary and secondary turns on a transformer to the primary and secondary currents.

25-4 Derive the transformer emf equation relating primary or secondary voltage to the core flux, the supply frequency, and the number of turns on the winding.

25-5 Discuss the power losses that occur in a transformer, and explain any measures that may be taken to minimize the losses.

25-6 Sketch the approximate phasor diagram of transformer primary voltage and currents under no-load conditions. Also sketch the no-load equivalent circuit for the transformer. Explain briefly the phasor diagram and equivalent circuit.

25-7 Explain the terms *leakage flux* and *leakage inductance*. Sketch the complete equivalent circuit for a transformer under load. Identify all components of the circuit and all voltages and currents. Explain the origin of each quantity.

25-8 Sketch a complete phasor diagram for the secondary circuit of a transformer under load. Briefly explain the diagram.

25-9 Sketch a complete phasor diagram for the primary circuit of a transformer under load. Briefly explain the diagram.

25-10 Explain the principle of *referred resistance* and *referred reactances*. Show how the transformer equivalent circuit may be simplified by referring the secondary circuit components to the primary.

25-11 Explain what is meant by the *voltage regulation* of a transformer, and write the equation for calculating the voltage regulation.

25-12 Discuss the efficiency of transformers, and derive an equation for transformer efficiency in terms of the output voltage and current and core losses.

25-13 Sketch the circuit diagrams for performing open- and short-circuit tests on a transformer. Explain the testing procedure and develop the necessary equation from which R_o, X_o, R_e, and X_e may be calculated from the test results.

25-14 Sketch the circuit diagram for an *autotransformer* and explain its operation. Also, discuss the advantages and disadvantages of an autotransformer.

25-15 Describe a *current transformer*. Explain its application and discuss any precautions necessary when using a current transformer.

25-16 Explain the purpose of an audio transformer. Sketch the typical frequency response graph of an audio transformer, and explain its shape.

PROBLEMS

25-1 A transformer with a primary winding of 250 turns has an input of 115 V rms. There are two secondary windings; one with 65 turns and the other with 80 turns. Determine the output voltage from each winding. Also, calculate the combined output voltage:

a. When the secondary windings are connected series-aiding.

b. When the secondary windings are connected series-opposing.

25-2 A transformer with a primary winding consisting of 250 turns has four secondary windings with the following numbers of turns: $N_{s1}=13$, $N_{s2}=52$, $N_{s3}=125$, and $N_{s4}=1042$. The secondary windings have the following loads connected to them: $R_{s1}=31.5\ \Omega$, $R_{s2}=50\ \Omega$, $R_{s3}=1.2\ \text{k}\Omega$, and $R_{s4}=500\ \text{k}\Omega$. If the primary voltage is 120 V, determine the output voltage from each secondary winding. Also, calculate the total primary current. Assume that the transformer is 100% efficient.

25-3 If the transformer described in Problem 25-2 has a supply frequency of 60 Hz, calculate the peak value of the core flux.

25-4 A transformer with an output voltage of 75 V supplies a load consisting of $R_L=33.5\ \Omega$ and $X_L=22\ \Omega$. Determine the supply voltage and current if the transformer has the following parameters: $R_P=2\ \Omega$, $X_P=5\ \Omega$, $R_o=7.5\ \text{k}\Omega$, $X_o=3\ \text{k}\Omega$, $R_S=0.25\ \Omega$, $X_S=1.2\ \Omega$, and $N_p/N_s=3/2$.

25-5 Refer all the secondary components of the transformer described in Problem 25-4 to the primary winding. Neglecting the components representing the core loss, calculate the primary current and its phase angle with respect to the input voltage when the input voltage is 120 V.

25-6 Neglecting the transformer losses, determine the voltage regulation of the transformer specified in Problem 25-4.

25-7 Calculate the efficiency of the transformer specified in Problem 25-4, and calculate the output current for maximum efficiency.

25-8 An open-circuit test on a transformer gave the following results: $E_p=120$ V, $E_s=35$ V, $P=5$ W, and $I_o=125$ mA. A short-circuit test on the same transformer produced: $E_p=4$ V, $I_p=0.8$ A, and $P=3$ W. Calculate R_o, X_o, N_p/N_s, R_e, and X_e.

25-9 From the open-circuit and short-circuit test results given in Problem 25-8, calculate the transformer full-load regulation when the load has a phase angle of 15° lagging.

25-10 Using the open-circuit and short-circuit test results given in Problem 25-8, calculate the efficiency of the transformer under full load with a load phase angle of 15° lagging.

26

AC MEASURING INSTRUMENTS

INTRODUCTION

The permanent magnet moving coil instrument operates only on dc. For ac measurements, the alternating quantity must be converted to dc by rectification, before being applied to a PMMC instrument. The dynamometer instrument can be used directly to measure *dc* quantities, or *rms ac* quantities. Its most important application is as a wattmeter, and on *ac* the dynamometer wattmeter measures *true power*. Where very high input impedances or measurement of very low level quantities are required, an electronic instrument must be used. This may give the measurement either in digital form (a *digital instrument*) or by means of a pointer moving over a scale (an *analog instrument*). Measurements of inductance and capacitance are usually made by the use of an *ac bridge*.

26-1 PMMC INSTRUMENT ON AC

In Chapter 14 it is pointed out that the *permanent magnet moving coil* (PMMC) instrument is *polarized*. This means that the terminals of the instrument are identified as + and −, and that for correct deflection the meter must be connected with that polarity. When incorrectly connected, the pointer attempts to deflect to the left of zero (i.e., *off-scale*).

Now consider what occurs when a PMMC instrument is connected directly to an alternating current source, as shown in Figure 26-1. For the illustration showing instantaneous pointer position [Figure 26-1(b)] it is assumed that the frequency of the alternating current is very low, on the

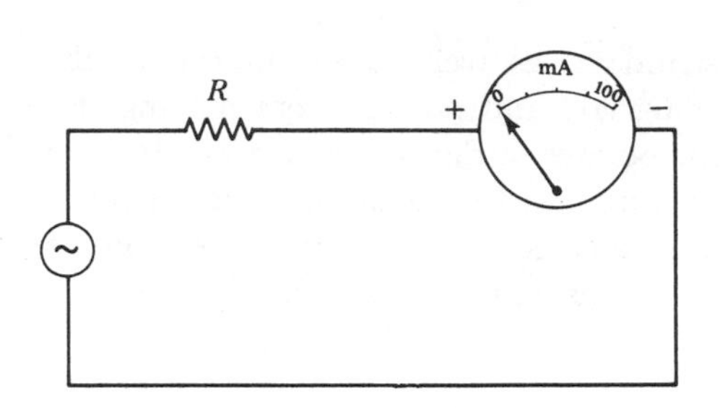

(a) PMMC ammeter connected to alternating current source

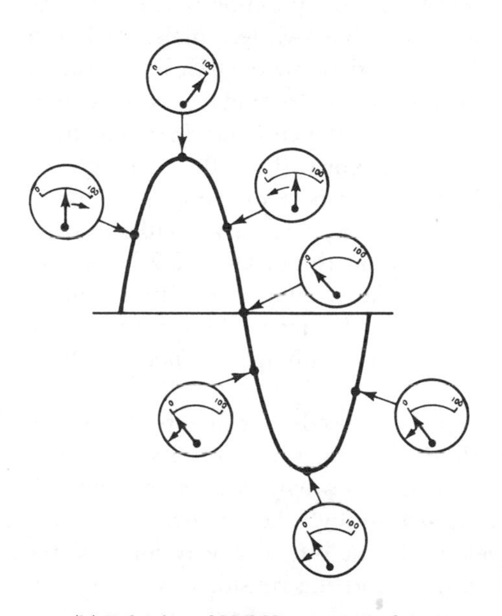

(b) Behavior of PMMC ammeter when measuring a very low frequency alternating current

FIG. 26-1. PMMC instrument on ac

order of 0.1 Hz. As the current level increases positively from zero, the meter pointer moves to the right until it indicates peak current. Then the pointer falls back toward zero again as the current level drops. When the current direction reverses during the second half-cycle of the waveform, the meter attempts to deflect to the left, which of course, is not possible. Consequently, the pointer remains at (or just below) the zero mark, during the negative half-cycle of the current.

Most ac frequencies are much greater than 0.1 Hz, 60 Hz being the normal ac supply frequency in North America. At this frequency, the meter pointer would have to rise and fall 60 times in every second. The damping mechanism of the instrument, as well as the inertia of the moving system, prevents the moving coil and pointer from moving this fast. Consequently, the instrument pointer settles at the average level of the alternating current passing through the windings. For normal sinusoidal alternating current, the average level is zero. Therefore, a PMMC instrument used directly to measure 60 Hz ac indicates zero.

26-2 RECTIFICATION

A rectifier is (usually) a *semiconductor diode*, although it may also be a vacuum tube. Without going into the theory of operation of the semiconductor diode, it is sufficient to note that it has two terminals, *anode* and *cathode*, as illustrated in Figure 26-2(a), and that current can flow through the device in only one direction. When the anode terminal is positive with respect to the cathode, current flows (in the conventional direction) from anode to cathode [see Figure 26-2(b)]. Current will not flow through the rectifier when the cathode is positive with respect to the anode [Figure 26-2(c)]. When *forward-biased* (i.e., conducting), there is a small volt drop from anode to cathode. This is normally on the order of 0.7 V or less, depending upon the semiconductor material employed in manufacturing the rectifier. The conventional direction of current flow through the device is indicated by the arrowhead in the circuit symbol, pointing from anode to cathode.

Figure 26-3 shows how a rectifier is used to convert alternating current into a series of pulses which all have the same polarity. In Figure 26-3(a) a single rectifier is connected in series with a resistor. Only the positive half-cycles of current are passed through the rectifier, so that the negative portion of the ac waveform is cut off. The waveform of the current that passes through the rectifier and the resistor is a continuous series of positive pulses with intervening spaces (see the illustration). This waveform is termed *half-wave-rectified*.

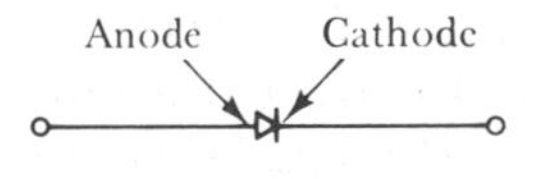

(a) Circuit symbol for a rectifier

Conventional current direction

I_F

+ −

Arrowhead indicates conventional current direction

(b) Current flows when anode is + and cathode is −

(c) Effectively zero current flows when the cathode is + and the anode is −

FIG. 26-2. The semiconductor rectifier

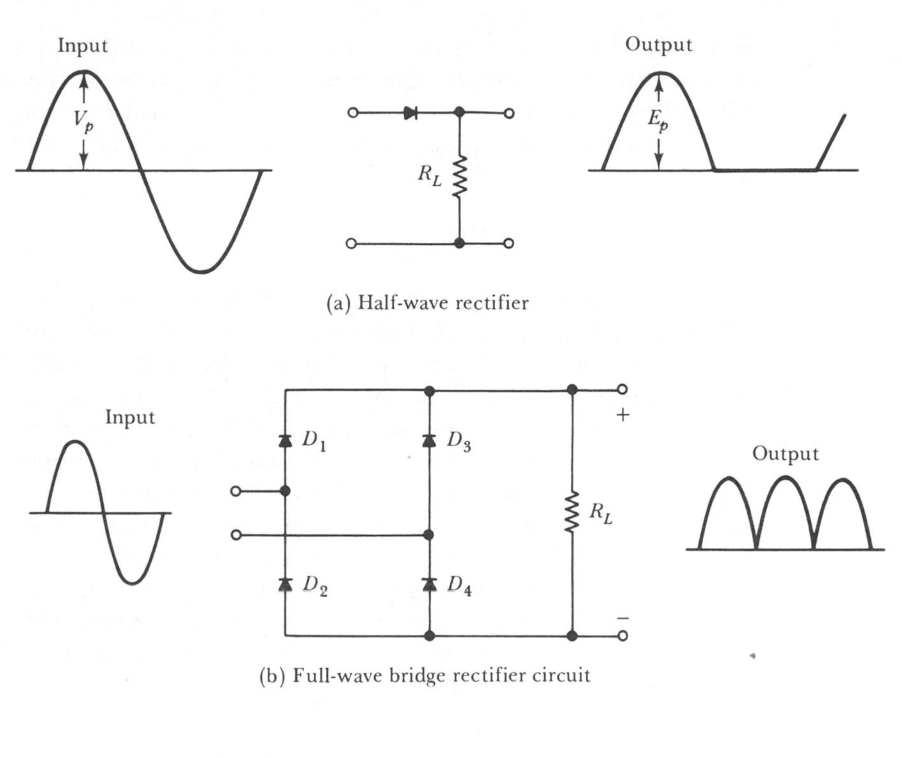

(a) Half-wave rectifier

(b) Full-wave bridge rectifier circuit

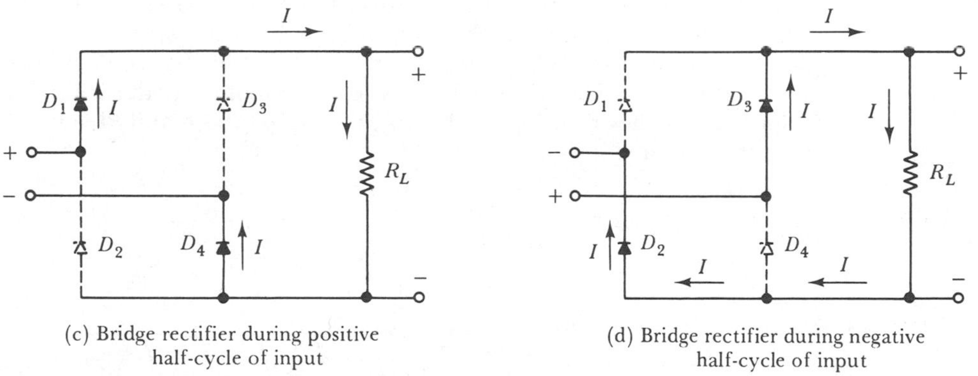

(c) Bridge rectifier during positive half-cycle of input

(d) Bridge rectifier during negative half-cycle of input

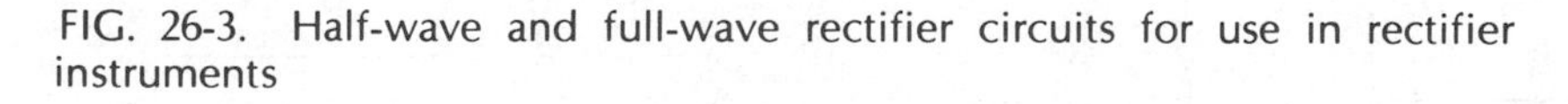

FIG. 26-3. Half-wave and full-wave rectifier circuits for use in rectifier instruments

The circuit shown in Figure 26-3(b) employs four rectifiers and is termed a *bridge rectifier circuit*. In this case, when the input waveform is positive [see Figure 26-3(c)], rectifiers D_1 and D_4 are forward-biased, while D_2 and D_3 are reverse-biased. Thus, current flows through D_1, resistor R_L, and D_4, as illustrated. Conversely, when the input waveform is negative [Figure 26-3(d)], D_2 and D_3 are forward-biased and D_1 and D_4

are reverse-biased. Current now flows through D_2, resistor R_L, and D_3, and it is seen that the current direction through the resistor is the same as before. The result of this is that the output waveform is a continuous series of unidirectional pulses, and the process is termed *full-wave rectification.*

26-3 RECTIFIER VOLTMETER

The circuit shown in Figure 26-4(a) is that of an ac voltmeter using a PMMC instrument and a full-wave rectifier circuit. As in the case of a dc voltmeter, a *multiplier* resistance must be included in the circuit to limit the current through the instrument. The actual current that flows through the deflection instrument has the full-wave rectified waveform shown in Figure 26-4(b). Since the deflection of a PMMC instrument is proportional to the average level of the current through its coil, the pointer tends to indicate 0.637 of V_p. However, the effective value, or rms value of the alternating waveform, is the quantity that is normally required in any measurement. Because there is a direct relationship between rms and average values, the instrument can be designed to have its scale marked to indicate rms volts. The following design example shows how this is accomplished.

EXAMPLE 26-1

An ac voltmeter to indicate 100 V rms at full scale is to be constructed using a deflection instrument which has full-scale deflection (FSD) at 500 μA. The coil resistance of the instrument is 1 kΩ, and the rectifiers used have a forward voltage drop of 0.7 V each. Employing the full-wave rectifier circuit shown in Figure 26-4(a), determine the required value of multiplier resistance.

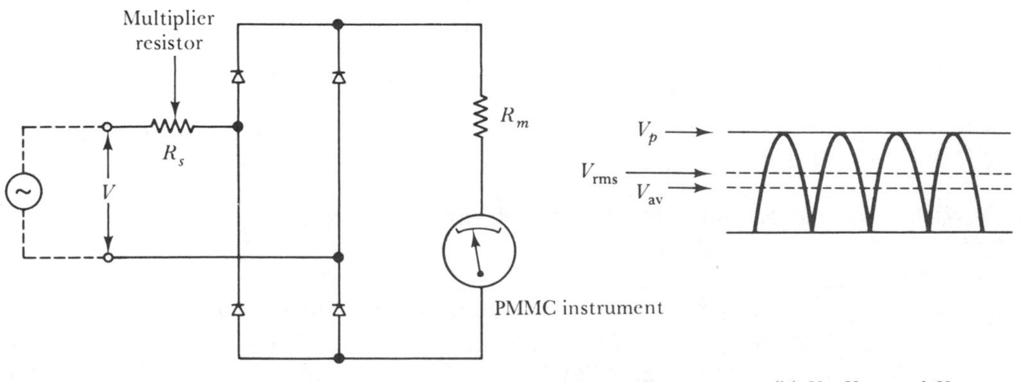

(a) Circuit of ac voltmeter

(b) V_p, V_{rms}, and V_{av}

FIG. 26-4. ac voltmeter employing PMMC instrument and full-wave bridge rectifier circuit

SOLUTION

At FSD, the average current through the deflection instrument is

$$I_{av} = 500\ \mu A$$

For a sinusoidal waveform:

$$I_{rms} = 1.11\ I_{av}$$

$$= 1.11 \times 500\ \mu A$$

$$I_{rms} = 555\ \mu A$$

and

$$I_{rms} = \frac{(\text{applied rms voltage}) - (\text{rectifier voltage drops})}{\text{total circuit resistance}}$$

$$I_{rms} = \frac{V - 2V_r}{R_s + R_m} \qquad (2V_r \textit{ for two rectifiers in series})$$

Therefore,

$$R_s = \frac{V - 2V_r}{I_{rms}} - R_m$$

$$= \frac{100\ V - (2 \times 0.7\ V)}{555\ \mu A} - 1\ k\Omega$$

$$\mathbf{R_s \cong 177\ k\Omega}$$

A half-wave rectifier circuit can also be employed with a deflection instrument, to produce an ac voltmeter (see the circuit in Figure 26-5). The procedure for calculating the value of multiplier resistor for the half-wave instrument is not exactly the same as in the full-wave case. This is because, as shown in the illustration, the average current through the

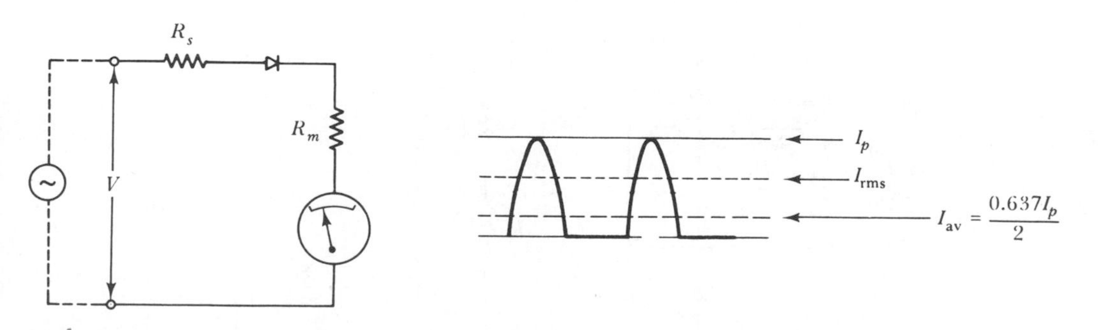

FIG. 26-5. Half-wave rectifier circuit employed with a deflection instrument to produce an ac voltmeter

meter is now

$$I_{av} = \frac{1}{2} \times 0.637 \times I_p$$

It is important to note that rectifier voltmeters designed for sine-wave operation *can be used only where pure sine waves are involved.* Where the voltage to be measured has any other waveform, the voltmeter will *not* correctly indicate rms voltage. This is because the 1.11 relationship between rms and average current levels applies only to pure sine waves.

For multirange ac voltmeters, a rotary switch is used to select one of several values of multiplier resistor, exactly as in the case of dc voltmeters.

26-4 RECTIFIER AMMETER

Recall from Chapter 14 that a dc ammeter must have a very low resistance, because it is always connected in series with the circuit in which current is to be measured. For the same reasons, an ac ammeter must also have a very low resistance. The low resistance requirement implies that when measuring current, the voltage drop across the ammeter must be very small. In most circumstances this voltage drop should be not greater than about 100 mV. But the voltage drop across one rectifier is typically 0.7 V, and for a bridge rectifier circuit in which there are two rectifiers in series, the total rectifier voltage drop becomes typically 1.4 V. Clearly, the ordinary rectifier instrument cannot be directly applied as an ac ammeter.

The use of a *current transformer* (see Chapter 25) gives the ammeter a low terminal resistance and low voltage drop while providing sufficient voltage to operate the rectifier instrument. Figure 26-6 shows a full-wave rectifier ammeter using a current transformer. The transformer currents might typically be 0.1 A in the primary and 500 μA in the secondary, as illustrated. In this case, from Eq. (25-2), the ratio of secondary to primary

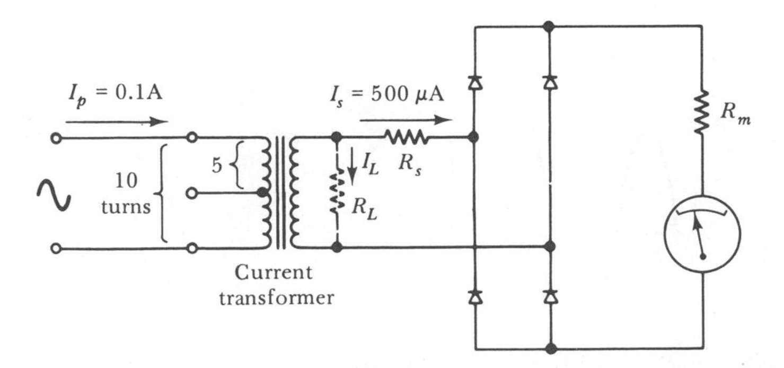

FIG. 26-6. Rectifier ammeter circuit

turns is

$$\frac{N_s}{N_p} = \frac{I_p}{I_s}$$

$$= \frac{0.1 \text{ A}}{500 \ \mu\text{A}} = 200$$

This would normally mean a one-turn primary, with 200 turns on the secondary. The meter scale would, of course, be marked as 0.1 A for full-scale deflection.

When a load resistance R_L (shown dashed in Figure 26-6) is connected across the secondary winding, a larger secondary current is drawn from the transformer. This allows the use of a larger number of primary turns, thus giving a more accurate current transformation. For example, if the inclusion of R_L increased the total secondary current to 5 mA, then with 200 turns on the secondary the number of primary turns would be

$$N_p = \frac{I_s}{I_p} \times N_s$$

$$= \frac{5 \text{ mA}}{0.1 \text{ A}} \times 200$$

$$N_p = 10 \text{ turns}$$

If *taps* are provided on the primary winding so that its number of turns can be altered, as illustrated in Figure 26-6, the range of the ammeter can be changed. When N_p is 5 turns and I_s is to be 5 mA to give full-scale deflection on the instrument, then

$$I_p = \frac{N_s}{N_p} \times I_s$$

$$= \frac{200}{5} \times 5 \text{ mA}$$

$$I_p = 0.2 \text{ A}$$

It is seen that halving the number of primary turns doubled the range of current that the ammeter can measure. Thus, the provision of several taps on the primary winding allows the selection of several ac current ranges. Note that the range of the instrument may also be changed by switching to different values of secondary load resistance (R_L). For range switching, make-before-break rotary switches should be employed, as already discussed for dc ammeters.

Like the rectifier voltmeter, the rectifier ammeter is suitable only for use with pure sinusoidal waveforms.

EXAMPLE 26-2

A rectifier ammeter has the circuit shown in Figure 26-6. The deflection instrument has full-scale deflection for a coil current of 1 mA and has a coil resistance of 100 Ω. The series resistance R_s is 10 kΩ and R_L is 1.42 kΩ. If the current transformer has $N_s=250$ turns and $N_p=5$ turns, determine the primary current required to give full-scale deflection on the instrument.

SOLUTION

For FSD:

$$I_{av}=1 \text{ mA}$$

and

$$I_{rms}=1.11\times I_{av}$$
$$=1.11 \text{ mA}$$

Transformer secondary voltage:

$$V_s=I(R_s+R_m)+2V_r$$
$$=1.11 \text{ mA }(10 \text{ k}\Omega+100\ \Omega)+(2\times 0.7 \text{ V})$$
$$=12.6 \text{ V}$$

Current through R_L:

$$I_L=\frac{V_s}{R_L}=\frac{12.6 \text{ V}}{1.42 \text{ k}\Omega}$$
$$=8.87 \text{ mA}$$

Total secondary current:

$$I_s=I+I_L$$
$$=1.11 \text{ mA}+8.87 \text{ mA}$$
$$\cong 10 \text{ mA}$$
$$I_p=\frac{N_s}{N_p}\times I_s$$
$$=\frac{250}{5}\times 10 \text{ mA}$$
$$\mathbf{I_p=500 \text{ mA}}$$

26-5 DYNAMOMETER INSTRUMENT ON AC

As mentioned in Section 14-7, the dynamometer instrument can be employed as a voltmeter or ammeter or in its major application as a wattmeter. Consider the illustrations in Figure 26-7, which shows the field coils and moving coil of a dynamometer instrument connected in series. With current flowing in the direction shown in Figure 26-7(a), the field coils set up fluxes which have their N poles on the right-hand side and their S poles on the left-hand side. Also, the moving coil flux has the polarity illustrated, S at the bottom and N at the top. The N pole of the moving coil flux is repelled from the adjacent N pole of the field coil flux, and the two adjacent S poles also repel each other. The result of this is

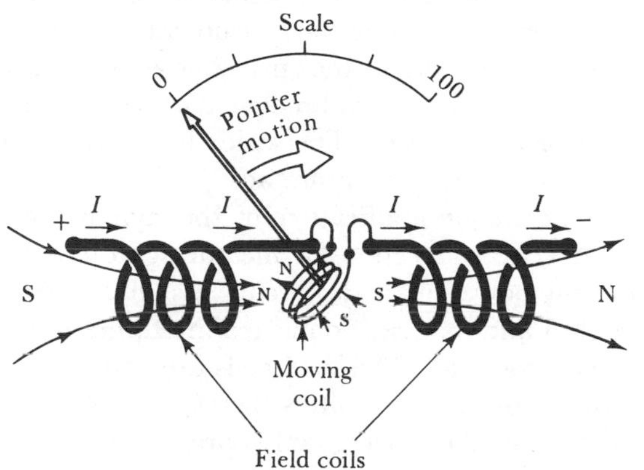

(a) Current flowing from left to right produces positive deflection

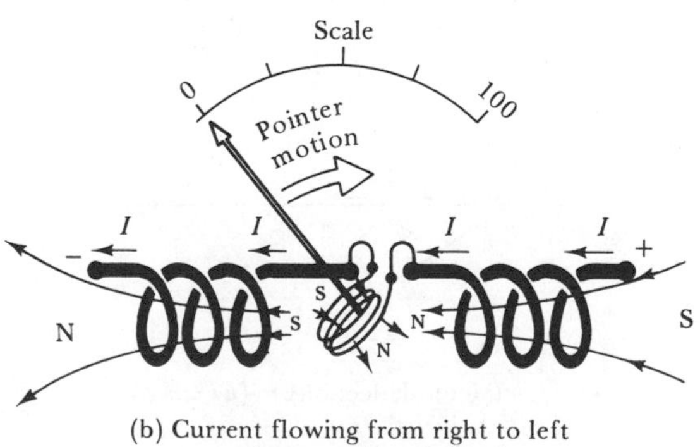

(b) Current flowing from right to left produces positive deflection

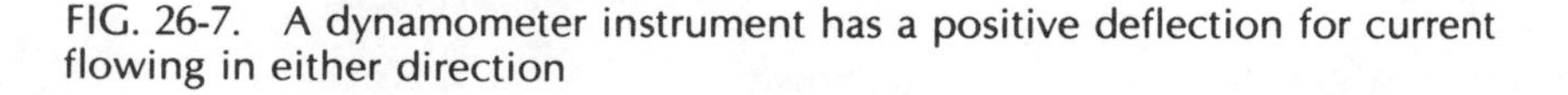

FIG. 26-7. A dynamometer instrument has a positive deflection for current flowing in either direction

that the moving coil is deflected clockwise, causing the pointer to move over the scale from left to right.

Now consider the effect of reversing the direction of the current through the coils. As illustrated in Figure 26-7(b), the field coil fluxes are now S at the right-hand side, and N at the left-hand side. The moving coil flux is also reversed, having S at the top and N at the bottom. Once again there are like poles adjacent to each other, and the pointer moves from left to right over the scale.

It is seen that dynamometer instrument has a positive deflection, no matter what the direction of the current through the meter. This means that the meter terminals are not marked + and −; that is, the dynamometer instrument is *unpolarized*. It also means that the instrument gives a positive deflection when either direct or alternating current flows through the coils. As an ammeter, the scale of the instrument can be read as direct current when measuring dc, and as the rms value of alternating current when measuring ac. Similarly, as a voltmeter, the dynamometer instrument indicates dc volts or rms ac volts. The scale of the instrument can be calibrated on dc and then used to measure ac.

For ac as well as dc, the major application of the dynamometer instrument is as a wattmeter. The connection for measuring ac power is similar to that for measuring power in a dc circuit. As illustrated in Figure 26-8(a), the moving coil and its series-connected multiplier resistor are connected in parallel with the load. The field coils are connected in series with the load. The flux from the field coils is therefore proportional to the load current, and that from the moving coil is proportional to the load voltage. As the supply voltage reverses, the load current and voltage both reverse; consequently, both the fluxes are reversed and the instrument deflection remains positive.

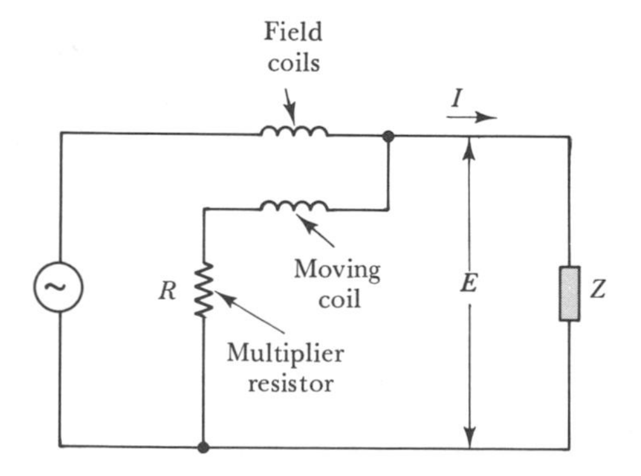

(a) Dynamometer instrument as an ac wattmeter

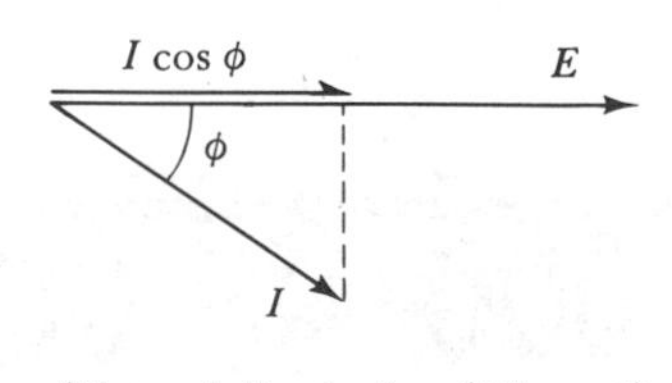

(b) Phasor diagram of load current and voltage

FIG. 26-8. A dynamometer wattmeter measures true ac power

In alternating-current applications, the load current could lead or lag the load voltage by a phase angle ϕ. The deflection of the instrument is proportional to the in-phase components of current and voltage. Thus, as shown in Figure 26-8(b), the instrument indication is proportional to $EI\cos\phi$.

As explained in Chapter 22, the *true power* dissipated in a load with an ac supply is

$$P = EI\cos\phi$$

Therefore, when used as an ac wattmeter, the dynamometer instrument measures the true power supplied to the load.

EXAMPLE 26-3

A wattmeter measures the ac power delivered to a certain load as 100 W, and an ammeter and voltmeter monitor the load current as 1.5 A and 100 V, respectively. Calculate the phase angle between the load current and voltage.

SOLUTION

$$\text{true power} = P = EI\cos\phi$$

Therefore,

$$\cos\phi = \frac{P}{EI}$$

$$P = 100\text{ W}$$

$$EI = 100\text{ V} \times 1.5\text{ A}$$

Therefore,

$$\cos\phi = \frac{100\text{ W}}{100\text{ V} \times 1.5\text{ A}}$$

$$= 0.667$$

and

$$\phi = 48.2^\circ$$

26-6 MULTIMETERS

The *multimeter* is an instrument that can function as a voltmeter, ammeter, or ohmmeter; thus it is sometimes referred to as a *VOM* (volt-ohmmeter). Basically the multimeter contains a sensitive PMMC deflection instrument and the necessary ammeter shunts and voltmeter multipliers for it to function as either a multirange voltmeter or a multirange ammeter. Rectifiers and a current transformer are also usually included so that it may be employed to measure ac quantities as well as dc. For the ohmmeter function, a battery and several standard resistors are provided.

One of the disadvantages of the usual type of multimeter is that in functioning as a voltmeter, its resistance sometimes affects the voltage being measured (see Section 14-4). Also, when functioning as either a voltmeter or an ammeter, the power to operate the instruments has to come from the source being measured. A further disadvantage is the inability to measure very low levels of voltage or current. These problems are overcome by the use of *electronic multimeters*.

The two types of electronic multimeters are *analog instruments*, in which the quantity being measured is indicated by a pointer moving over a scale, and *digital instruments*, which display the measurement in digital form. Digital instruments are discussed in Section 26-7. Analog electronic instruments could be subdivided into *vacuum-tube voltmeters* (VTVM) and *transistor voltmeters* (TVM).

Both transistor and vacuum-tube multimeters basically employ an *amplifier*, as shown in Figure 26-9. This presents a high input resistance (or impedance) to a voltage source being measured. The input impedance might typically be as high as 10 MΩ.

As well as having a high input impedance, the amplifier is easily able to provide enough power to drive a PMMC deflection instrument. The power to operate the amplifier and the deflection instrument is derived from the *dc power supply* contained within the multimeter (see Figure 26-9). This power supply converts a 115-V ac supply to perhaps ±30-V dc supply. Thus, there is no need to draw power from the source being measured.

The amplifier also has an accurate controllable gain. A 100-mV input can easily be amplified to a level that produces full-scale deflection on the PMMC instrument. Alternatively, instead of amplifying the input, the amplifier can be made to function as an *attenuator*. For example, a 1000-V input might be divided down so that full scale-deflection of the meter represents 1000 V.

For functioning as an ammeter, the electronic multimeter contains precise low-value resistors. The current to be measured is passed through one of these resistors, and the voltage across it is measured by the electronic voltmeter circuitry. The pointer indication is then read from a

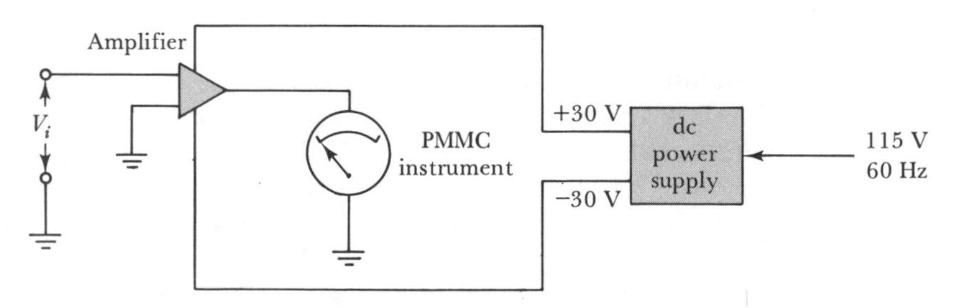

FIG. 26-9. Analog electronic voltmeter

scale calibrated to indicate current level. Operation as an ohmmeter is also easily achieved, and a battery is not required, since the electronic multimeter contains its own power supply.

Measurement of ac quantities require rectification. In the case of the electronic instrument, low-level ac voltages can be amplified before rectification, and then measured. Because of the voltage drop across the rectifiers, ordinary (nonelectronic) instruments cannot measure very low level voltages and currents. So this is another advantage that electronic instruments have over nonelectronic instruments.

The accuracy of measurements made with electronic analog instruments obviously depends upon the accuracy of the PMMC instrument used, and upon the accuracy of the amplifier gain (or attenuation). Electronic multimeter accuracies of $\pm 1\%$ to $\pm 3\%$ are typical.

26-7 DIGITAL INSTRUMENTS

A thorough understanding of digital instruments can be achieved only when the operation of semiconductor devices and the theory of pulse circuits have been studied. However, a basic knowledge of the operation of digital instruments is useful for their intelligent application.

DISPLAYS. The devices that constitute the basic component of many digital displays are the *light-emitting diode* (LED) and the *digital indicator tube*. The light-emitting diode can be thought of as a semiconductor rectifier, which, in addition to passing current in only one direction, glows brightly (usually red) when current is flowing through it. The circuit symbol for the LED is the same as that for an ordinary diode, with two arrows included to represent light emission [Figure 26-10(a)].

For numerical display purposes, LEDs are arranged in the *seven-segment* format shown in Figure 26-10(b). When segments *b* and *c* are energized, the display represents the numeral *1*. To display numeral *2*, segment *a, b, g, e,* and *d* are energized, while numeral *3* requires *a, b, g, c,* and *d* to be glowing. Thus, by selection of the appropriate segments, any numeral from *0* to *9* can be displayed.

The digital indicator tube is a much older device than the LED; however, it is still widely used in digital instruments. As illustrated in Figure 26-10(c), the device has one flat metal electrode termed the *anode*, and 10 separate wire *cathodes*, each in the shape of a numeral from 0 to 9. The electrodes are enclosed in a gas-filled glass tube with connecting terminals at the bottom. *Neon* gas is usually employed. When a voltage is applied, positive to the anode and negative to one of the cathodes, the gas breaks down and conducts current. When conduction occurs, ionization of the gas atoms takes place at the cathode and results in a visible glow around the cathode. Since the cathodes are in the shape of numerals, a

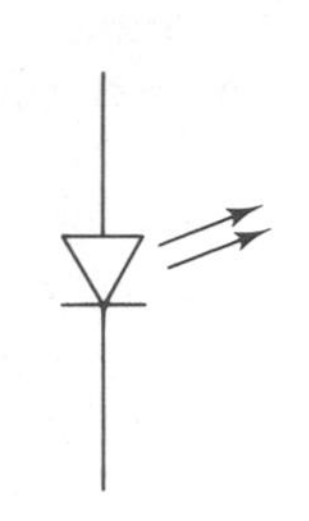

(a) Light-emitting diode

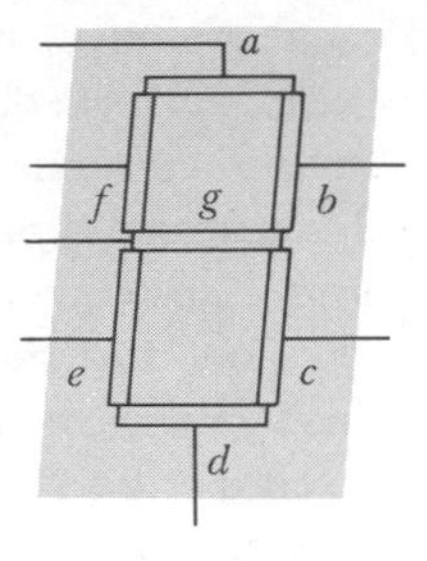

(b) 7-segment LED display

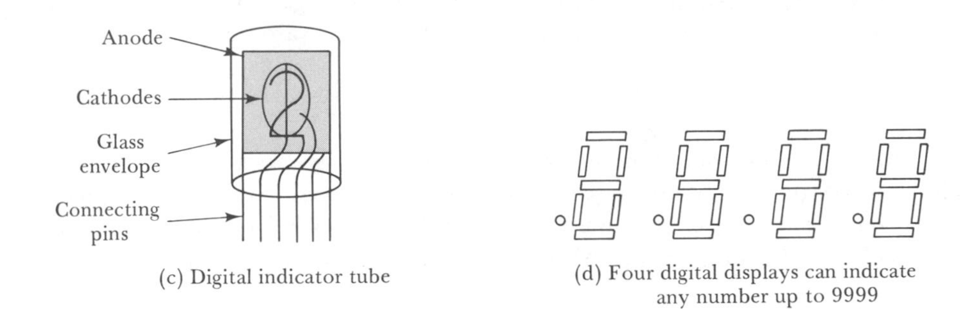

(c) Digital indicator tube

(d) Four digital displays can indicate any number up to 9999

FIG. 26-10. Digital display methods

glowing numeral is displayed, depending upon which cathode is selected.

Four digital display devices, placed side by side as illustrated in Figure 26-10(d), can be used to display any number from zero to 9999. When decimal points are included (individual LEDs or small gas tubes) the display can be used to read out a number to one, two, three, or four decimal places. Frequently, a small sign is illuminated alongside the display, so that the number is identified in units such as Hz, kHz, V, mV, and so on.

DIGITAL FREQUENCY METER. The block diagram in Figure 26-11 gives a basic idea of how a digital frequency meter operates. The signal that is to have its frequency measured is fed into an amplifier which offers a high input impedance. The input signal (that can have virtually any waveform) is amplified and fed to a *wave-shaping circuit*, in which it is converted to a train of pulses, as illustrated. The pulses have the same frequency as the original signal and have the necessary amplitude and shape for triggering the counting circuits. A *timer* permits counting to take place over an accurately measured time period. If the time period is 1 s, and if 1000 pulses are counted during that time, the display registers 1000 and is read as 1000 Hz or 1.000 kHz. Similarly, if 1293 pulses are counted

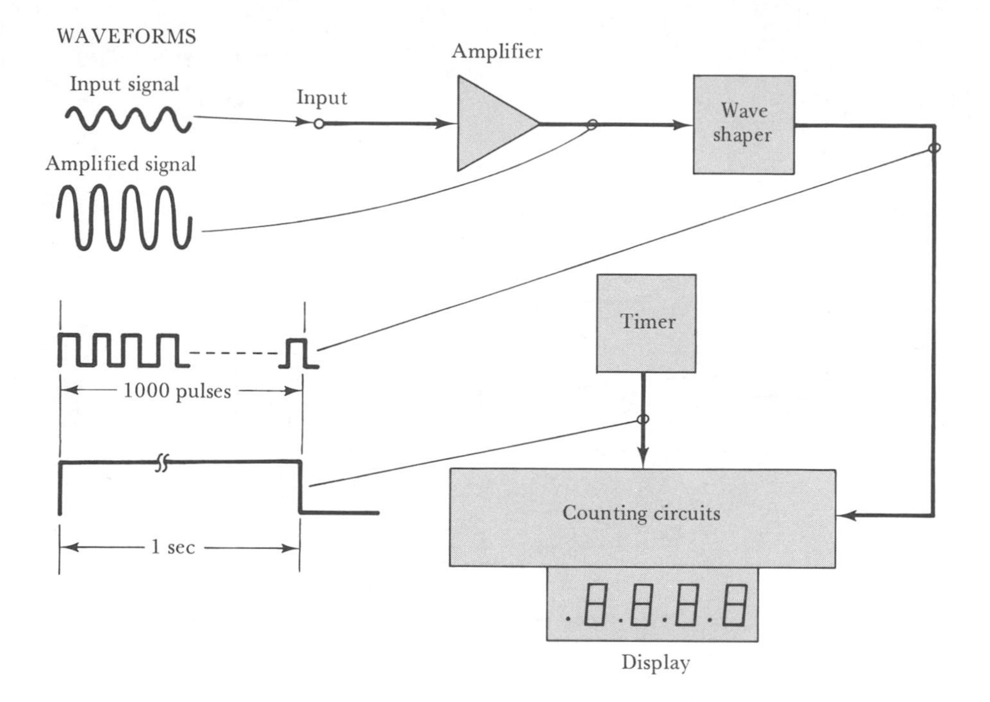

FIG. 26-11. Block diagram of a digital frequency meter

during the 1-s period, the display registers 1.293 kHz. To change the range of the instrument, the timer can be switched to one of several time bases. For example, if the time period is selected as 0.1 s, a count of 1950 pulses during this time would be the result of 19 500 pulses during 1 s. Hence, the display registers a frequency of 19.50 kHz.

DIGITAL VOLTMETER (DVM). The basic block diagram of a digital voltmeter, shown in Figure 26-12, is not very different from that of the digital frequency meter already discussed. Instead of the counting circuits being triggered by the pulse train from a wave shaper, a very stable *oscillator* is provided to generate the necessary triggering signals. The timer now generates an output time period which is proportional to the voltage to be measured. Suppose that the oscillator frequency is 1 kHz, and suppose that an input of exactly 1 V generates a time period of exactly 1 s. The timer switches the counting circuits on for 1 s, and during that time they receive 1000 pulses from the 1-kHz oscillator. The 1000 pulses are then registered on the digital displays as 1.000 V.

If the input voltage is 1.375 V, a time period of 1.375 s is generated. During that time the 1-kHz oscillator feeds 1375 pulses into the counting circuits, and the displays register 1.375 V. The timer usually has a range

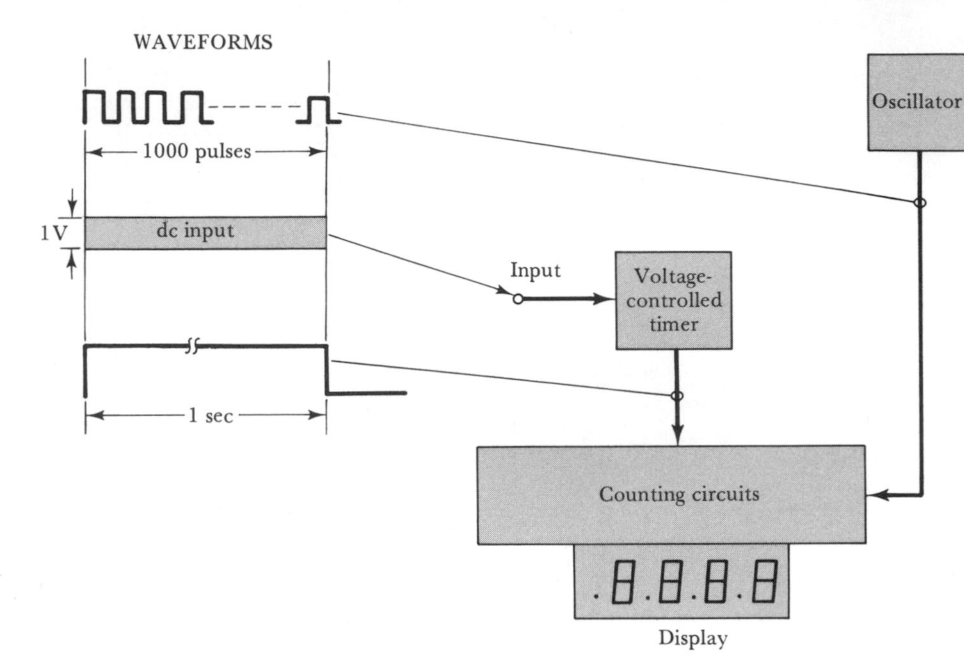

FIG. 26-12. Block diagram of one type of digital voltmeter

switch, so that a 100-V input might also produce a time period of only 1 s. In this case, however, the decimal point position would be switched when the timer range is changed, and the 1000 pulses counted would be displayed as 100.0 V.

For ac voltage measurements, the input might be first rectified and a time period generated proportional to its rms value. Then, the number of pulses counted are read directly from the display as *rms volts*.

The input impedance of a digital voltmeter depends upon the electronic circuitry employed. Usually, the input impedance is on the order of 1 MΩ resistive, and may easily be as high as 10 MΩ.

It is important to note that the basic block diagram shown in Figure 26-12 is only one possible digital voltmeter system. There are many DVM systems in general use.

DIGITAL AMMETER. As in the case of the analog electronic instruments, a digital ammeter is simply a digital voltmeter which includes several precise values of low resistance. The current to be measured is passed through one of these resistors, and the voltage drop across it is measured by the DVM. For a 1-Ω resistor with a 1-V drop across it, the display would be registered as 1.000 A. Similarly, for a 100-mV drop across the 1-Ω resistor, 100.0 mA would be displayed.

26-8 AC BRIDGES

AC bridges are used for measurement of inductance and capacitances, and all ac bridge circuits are based on the Wheatstone bridge (see Section 14-8).

Figure 26-13(a) shows the circuit of a *simple capacitance bridge*. C_s is a precise standard capacitor, while Q and P are standard resistors, one or both of which is adjustable. An ac supply is used, and the *null detector* (D) must be an ac instrument. A rectifier ammeter is frequently employed as null detector. Q is adjusted until the null detector indicates zero, and when this is obtained, the bridge is said to be *balanced*.

When the detector indicates null, the voltage drop across C_s must equal that across C_x, and similarly the voltage across Q must be equal to the voltage across P. Therefore,

$$V_{cs} = V_{cx}$$

or

$$\boxed{i_1 X_{cs} = i_2 X_{cx}} \qquad (26\text{-}1)$$

and

$$V_Q = V_p$$

or

$$\boxed{i_1 Q = i_2 P} \qquad (26\text{-}2)$$

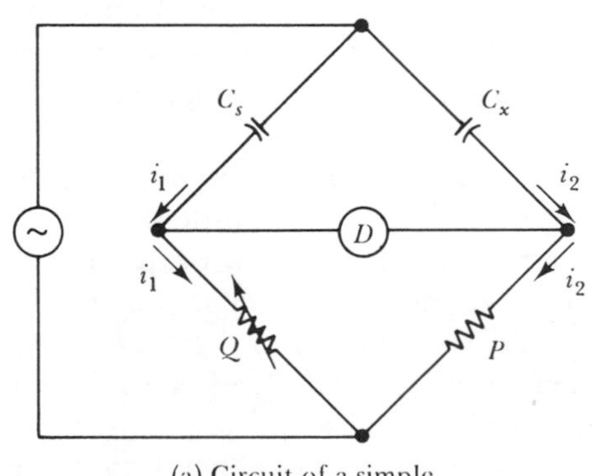

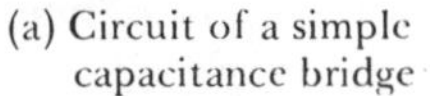
(a) Circuit of a simple capacitance bridge

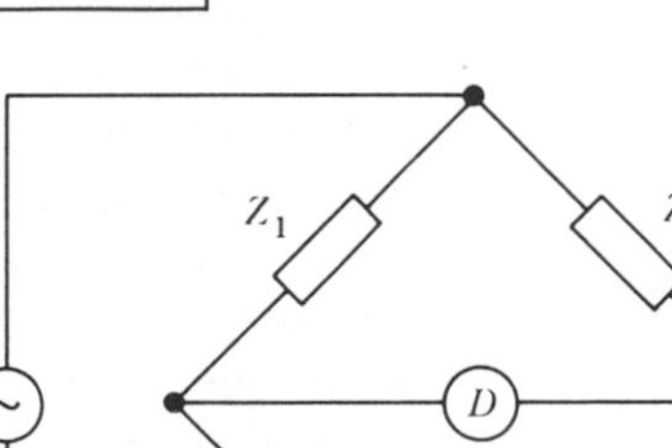

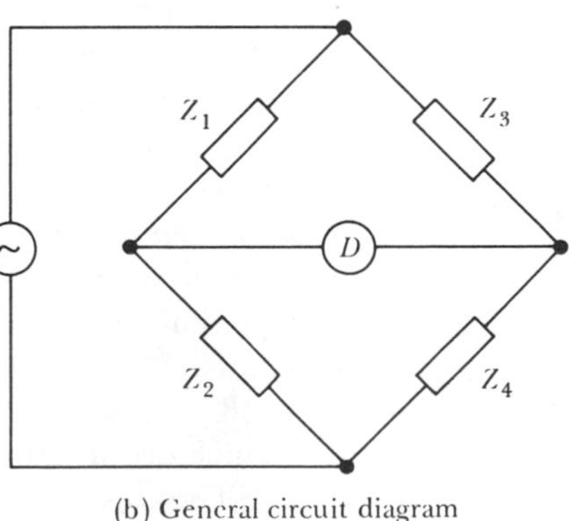

(b) General circuit diagram for an ac bridge

FIG. 26-13. Simple capacitance bridge and general ac bridge diagram

Dividing Eq. (26-1) by Eq. (26-2):

$$\frac{X_{cs}}{Q} = \frac{X_{cx}}{P} \tag{26-3}$$

Referring to Figure 26-13(b), the general *balance equation* for all ac bridges can be written

$$\frac{Z_1}{Z_2} = \frac{Z_3}{Z_4} \tag{26-4}$$

Substituting $1/\omega C_s$ for X_{cs}, and $1/\omega C_x$ for X_{cx} in Eq. (26-3),

$$\frac{1}{\omega C_s Q} = \frac{1}{\omega C_x P}$$

or

$$C_x = \frac{Q\omega C_s}{P\omega}$$

$$C_x = \frac{QC_s}{P} \tag{26-5}$$

It is seen that the unknown capacitance C_x can now be calculated from the known values of Q, C_s, and P.

One disadvantage of the simple capacitance bridge is that perfect balance of the bridge is obtained only when C_s and C_x are both pure capacitances (i.e., there is no resistance component). In general, this occurs only with capacitors that have air or mica dielectrics. Capacitors with other types of dielectric have a leakage current, and consequently the equivalent circuits for the capacitors have resistive components that must be included in the bridge circuit.

The circuit of the *series resistance capacitance bridge* shown in Figure 26-14 eliminates the balance problems that can occur with the simple capacitance bridge. Resistance r_x in series with the unknown capacitance, represents the resistive component of the capacitor equivalent circuit. The standard capacitor C_s normally has mica dielectric, and thus has a very small resistive component. Consequently, the adjustable resistance S must be included in the circuit to balance the effect of r_x.

The *balance equations* for the series resistance capacitance bridge are derived as follows:

Eq. (26-4):

$$\frac{Z_1}{Z_2} = \frac{Z_3}{Z_4}$$

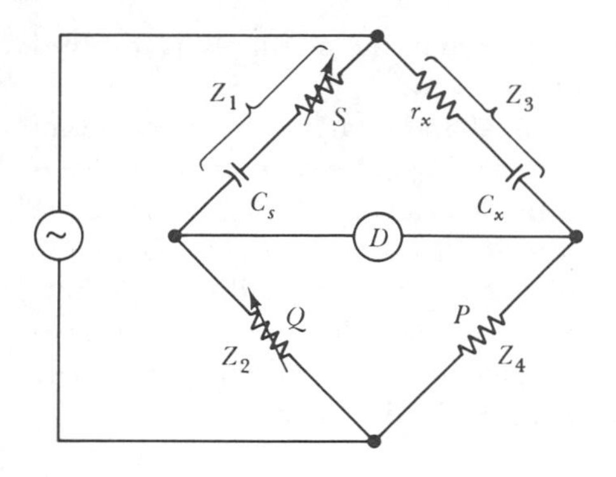

FIG. 26-14. Series resistance capacitance bridge

Therefore,

$$\frac{S - j\dfrac{1}{\omega C_s}}{Q} = \frac{r_x - j\dfrac{1}{\omega C_x}}{P}$$

$$\boxed{\frac{S}{Q} - j\frac{1}{\omega C_s Q} = \frac{r_x}{P} - j\frac{1}{\omega C_x P}} \qquad (26\text{-}6)$$

For the equation above to be correct, the real parts on each side must be equal, *and* the imaginary parts on each side must be equal.

Equating the real parts,

$$\frac{S}{Q} = \frac{r_x}{P}$$

and

$$\boxed{r_x = \frac{PS}{Q}} \qquad (26\text{-}7)$$

Equating the imaginary parts,

$$\frac{1}{\omega C_s Q} = \frac{1}{\omega C_x P}$$

and

$$\boxed{C_x = \frac{QC_s}{P}} \qquad (26\text{-}8)$$

The resistive and capacitive components of the unknown capacitor can now be calculated by means of Eqs. (26-7) and (26-8). Note that neither the supply voltage nor the frequency of the ac supply is involved in the balance equations for the bridge.

Because of the need to balance the real and imaginary components of the bridge impedances, the process of obtaining balance in an ac bridge is a little more complicated than with the Wheatstone bridge. One of the adjustable components Q or S in Figure 26-14 is first altered to obtain the lowest possible indication on the null meter. Then the other adjustable component is varied to obtain a lower reading. The process is repeated until further adjustment of either component cannot produce any lower reading on the null meter. At this point the bridge is balanced.

EXAMPLE 26-4

The capacitance bridge shown in Figure 26-14 has a 0.1-μF standard capacitor (C_s) and a standard resistor of $Q = 10.25$ kΩ. Zero deflection is obtained on the null detector when $P = 1$ kΩ and $S = 2.25$ kΩ. Calculate the value of the unknown capacitance and its resistive component.

SOLUTION

Eq. (26-8):

$$C_x = \frac{QC_s}{P}$$

$$= \frac{10.25\ \text{k}\Omega \times 0.1\ \mu\text{F}}{1\ \text{k}\Omega}$$

$$\mathbf{C_x = 1.025\ \mu F}$$

Eq. (26-7):

$$r_x = \frac{PS}{Q}$$

$$= \frac{1\ \text{k}\Omega \times 2.25\ \text{k}\Omega}{10.25\ \text{k}\Omega}$$

$$\mathbf{r_x = 219.5\ \Omega}$$

For measurement of inductance, the *Maxwell bridge* shown in Figure 26-15(a) can be employed. It is seen that the circuit of the Maxwell bridge is simply a repeat of the series resistance capacitance bridge, with the capacitors replaced by inductors. A disadvantage of this bridge is that standard inductors are larger and more difficult to manufacture than standard capacitors. Consequently, a variation of this circuit, known as the *Maxwell–Wein bridge*, is most often employed for inductance measurement.

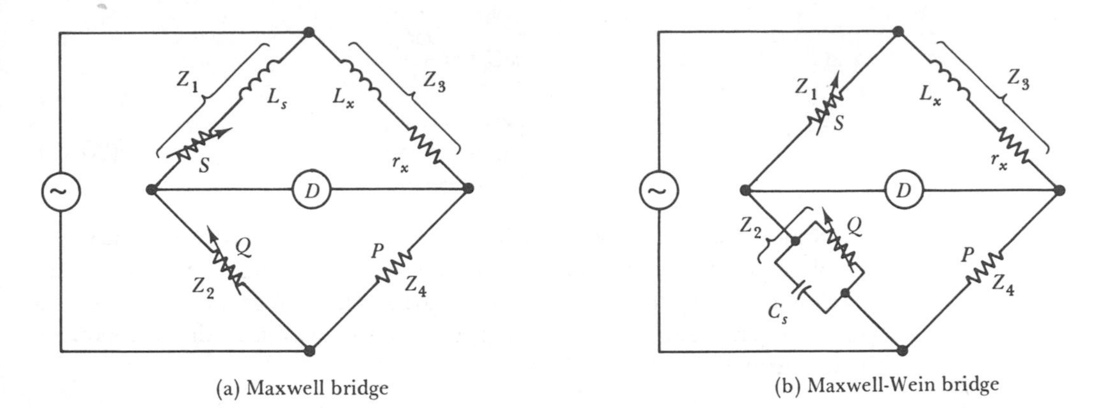

(a) Maxwell bridge (b) Maxwell-Wein bridge

FIG. 26-15. Inductance bridges

The circuit of the Maxwell–Wein bridge is shown in Figure 26-15(b). L_X is the unknown inductance to be measured and r_x is the resistance of its windings. C_s is again a precise standard capacitor, and P is a standard resistor. Q and S are accurate adjustable resistors. At balance:

Eq. (26-4):

$$\frac{Z_1}{Z_2} = \frac{Z_3}{Z_4}$$

Therefore

$$\boxed{\frac{S}{1/\left(\dfrac{1}{Q} + j\omega C_s\right)} = \frac{r_x + j\omega L_x}{P}} \tag{26-9}$$

or

$$S\left(\frac{1}{Q} + j\omega C_s\right) = \frac{r_x + j\omega L_x}{P}$$

$$\boxed{\frac{S}{Q} + j\omega C_s S = \frac{r_x}{P} + j\frac{\omega L_x}{P}} \tag{26-10}$$

Equating the real and imaginary terms:

$$\frac{S}{Q} = \frac{r_x}{P}$$

$$\boxed{r_x = \frac{SP}{Q}} \tag{26-11}$$

and $$\omega C_s S = \frac{\omega L_x}{P}$$

$$\boxed{L_x = PC_sS} \qquad (26\text{-}12)$$

Once again it is seen that the supply voltage and frequency are not involved in the balance equations for the bridge. This is *not* always the case with ac bridges; indeed, one particular bridge can be used to measure the frequency of the supply in terms of the bridge component values at balance.

EXAMPLE 26-5

The Maxwell–Wein bridge shown in Figure 26-15(b) uses a 0.1-μF standard capacitor (C_s) and a standard resistor of $Q = 1$ kΩ. Zero deflection of the null detector is obtained when $P = 1.33$ kΩ and $S = 870$ Ω. Calculate the inductance and resistance of impedance Z_3.

SOLUTION

Eq. (26-11):

$$r_x = \frac{SP}{Q}$$

$$= \frac{870\ \Omega \times 1.33\ \text{k}\Omega}{1\ \text{k}\Omega}$$

$$\boldsymbol{r_x = 1.16\ \text{k}\Omega}$$

Eq. (26-12):

$$L_x = PC_sS$$

$$= 1.33\ \text{k}\Omega \times 0.1\ \mu\text{F} \times 870\ \Omega$$

$$\boldsymbol{L_x = 115.7\ \text{mH}}$$

Comparing the circuits shown in Figures 26-14 and 26-15(b), it is seen that a series capacitance bridge and a Maxwell–Wein bridge could each be constructed from the same set of components. In fact, many commercial ac bridges use one set of components which are connected in the form of a series resistance capacitance bridge for capacitance measurement and are switched into the Maxwell–Wein bridge configuration for inductance measurement.

EXAMPLE 26-6

An ac bridge has the following components: $C_s = 0.2$ μF, $S = (500$ Ω to 1.6 kΩ), $Q = 500$ Ω to 15 kΩ, and $P = 100$ Ω to 1 MΩ. If the bridge can be switched between a series resistance capacitance configuration and a Maxwell–Wein circuit, determine the range of unknown C and L that can be measured.

SOLUTION

For capacitance measurements:
Eq. (26-8):

$$C_x = \frac{QC_s}{P}$$

$$\text{maximum } C_x = \frac{Q_{\max} \times C_s}{P_{\min}}$$

$$= \frac{15\ \text{k}\Omega \times 0.2\ \mu\text{F}}{100\ \Omega}$$

$$\textbf{maximum } \boldsymbol{C_x} \mathbf{= 30\ \mu F}$$

$$\text{minimum } C_x = \frac{Q_{\min} \times C_s}{P_{\max}}$$

$$= \frac{500\ \Omega \times 0.2\ \mu\text{F}}{1\ \text{M}\Omega}$$

$$\textbf{minimum } \boldsymbol{C_x} \mathbf{= 100\ pF}$$

For inductance measurements:
Eq. (26-12):

$$L_x = PC_sS$$

$$\text{maximum } L_x = P_{\max} \times C_s \times S_{\max}$$

$$= 1\ \text{M}\Omega \times 0.2\ \mu\text{F} \times 1.6\ \text{k}\Omega$$

$$\textbf{maximum } \boldsymbol{L_x} \mathbf{= 320\ H}$$

$$\text{minimum } L_x = P_{\min} \times C_s \times S_{\min}$$

$$= 100\ \Omega \times 0.2\ \mu\text{F} \times 500\ \Omega$$

$$\textbf{minimum } \boldsymbol{L_x} \mathbf{= 10\ mH}$$

REVIEW QUESTIONS

26-1 Explain what a rectifier is used for. Sketch the circuit symbol for a rectifier and show the current direction and voltage polarity when the rectifier is conducting.

26-2 Sketch half-wave and bridge rectifier circuits. Show the input and output waveform and explain briefly the operation of each circuit.

26-3 Sketch the circuit of a rectifier voltmeter that uses a bridge rectifier circuit. Explain how the instrument operates and discuss its limitations.

26-4 Sketch the circuit for an *ac* voltmeter using a PMMC instrument and half-wave rectification.

26-5 Sketch the circuit of rectifier ammeter. Explain how the instrument operates and discuss its limitations.

26-6 Using illustrations, explain why a dynamometer instrument gives a positive deflection for current flow through the meter in either direction. What changes are necessary to produce a negative deflection on a dynamometer instrument?

26-7 Show how a dynamometer wattmeter is connected to measure the power delivered to a load and explain why the instrument measures the true power in an ac circuit.

26-8 Explain what a VOM is. Describe its functions and list its disadvantages.

26-9 State the advantages of electronic multimeters over ordinary deflection instruments and define the major types of electronic multimeters. Also, discuss the measurement accuracy obtainable with electronic instruments.

26-10 Sketch a basic block diagram of an analog electronic voltmeter and explain how it operates. Also, discuss how an electronic instrument may be used for measuring current and resistance.

26-11 Using illustrations, describe two types of digital display devices used in digital electronic instruments.

26-12 Sketch a basic block diagram of a digital frequency meter. Show the waveforms at various points in the system and explain how the instrument operates.

26-13 Sketch a basic block diagram of a digital voltmeter. Show the waveforms at various points in the system and explain how the instrument operates. Also, discuss how the instrument may be used for measuring current.

26-14 Sketch the circuit of a simple capacitance bridge. Explain how it operates and derive an expression for the unknown capacitance.

26-15 Sketch the circuit of a Maxwell bridge. Explain the steps involved in balancing the bridge and derive the balance equations.

26-16 Sketch the circuits of two ac bridges that may be constructed from a standard capacitor and three variable standard resistors. Identify each bridge by name and derive the balance equations for the bridge.

PROBLEMS

26-1 A PMMC voltmeter having a full-scale deflection of 100 V is used to measure an alternating voltage with an rms value of 70.7 V. What will the meter reading be:

a. When the ac frequency is 60 Hz?

b. When the frequency is 0.05 Hz? Explain.

26-2 A PMMC instrument that gives full-scale deflection for 50 μA and has a coil resistance of 850 Ω is to be used as an ac voltmeter. A bridge rectifier circuit is to be used, with the rectifiers having a forward voltage drop of 0.7 V each. If the voltmeter is to have full-scale deflection at 150 V, calculate the value of the required multiplier resistance.

26-3 Calculate the value of multiplier resistance for a half-wave rectifier PMMC voltmeter to meet the requirements listed in Problem 26-2.

26-4 A 200-μA PMMC instrument with a coil resistance of 900 Ω is used in a rectifier ammeter which is to indicate 1-A rms at full scale. The ammeter circuit has a bridge rectifier and a current transformer with 20 primary turns and 180 secondary turns. If a 120-kΩ resistance is connected in series with the PMMC instrument, calculate the required value of the resistance that must be connected in parallel with the secondary winding.

26-5 The ammeter described in Problem 26-4 is to have its range changed to 100 mA.

a. Determine the additional number of transformer primary turns required.

b. If the transformer is not altered, determine the new value of secondary parallel load resistance.

26-6 A wattmeter measures the ac power delivered to a load as 250 W. The supply voltage is 115 V and the load is known to have a phase angle of $\phi = 33°$. Calculate the load current.

26-7 A series-resistance capacitance bridge (Figure 26-14) uses a 0.15-μF standard capacitor, with a resistance $S = 1.8$ kΩ connected in series with it. The standard resistor Q has a value of 12 kΩ. If the unknown capacitance is known to be between 2 μF and 2.5 μF, calculate the required range of adjustment of resistor P.

26-8 A 1-μF capacitor and three decade resistors which may be adjusted from 100 Ω to 5 kΩ are available for construction of the two bridges referred to in Review Question 26-16. Calculate the range of unknown capacitance and inductance that can be measured.

Appendix 1
CIRCUIT SYMBOLS

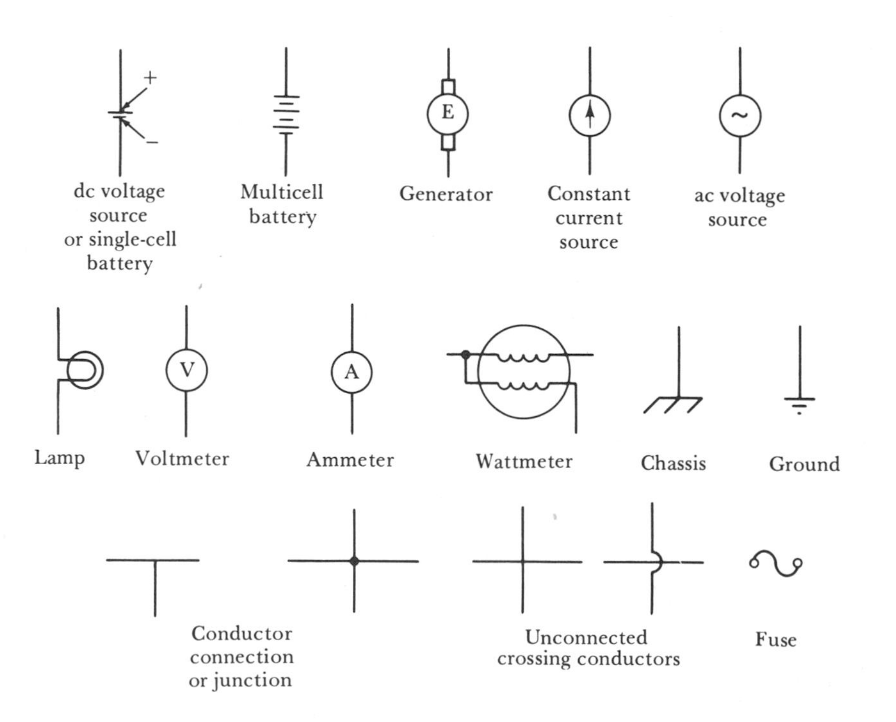

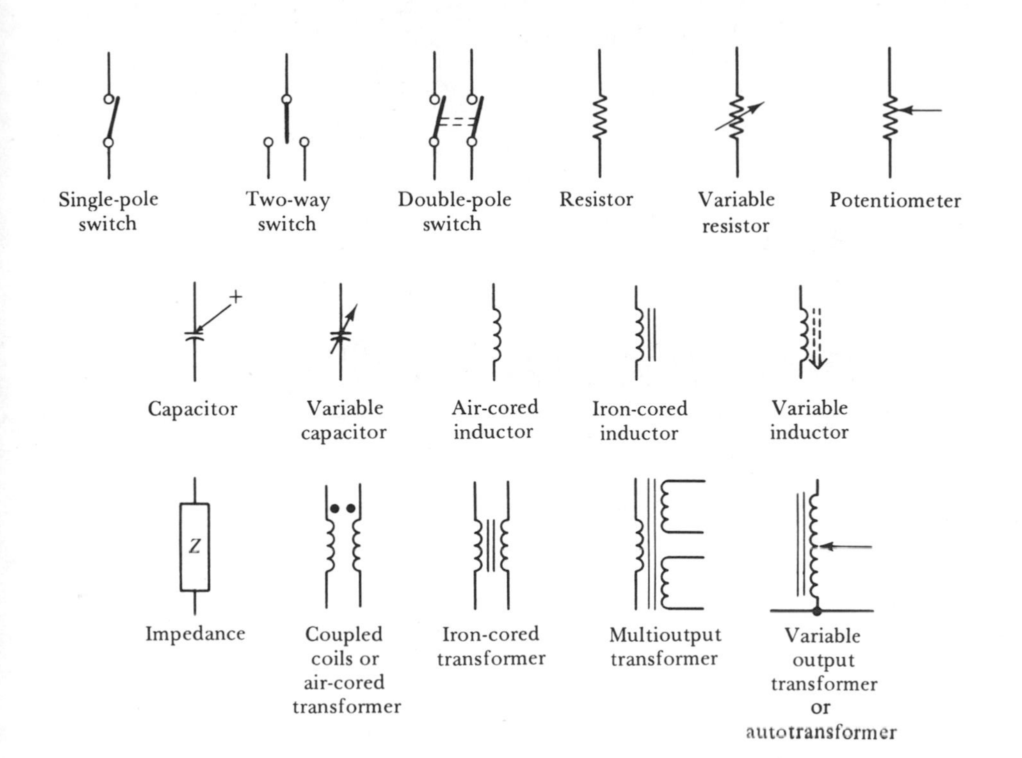
Single-pole switch
Two-way switch
Double-pole switch
Resistor
Variable resistor
Potentiometer
+
Capacitor
Variable capacitor
Air-cored inductor
Iron-cored inductor
Variable inductor
Z
Impedance
Coupled coils or air-cored transformer
Iron-cored transformer
Multioutput transformer
Variable output transformer or autotransformer

Appendix 2

UNIT CONVERSION FACTORS

The following factors may be used to convert non-SI units to SI units.

TO CONVERT	TO	MULTIPLY BY
acres	square meters	4047
acres	hectares	0.4047
amperes/inch	amperes/meter	39.37
angstroms	meters	1×10^{-10}
atmospheres	kg/m^2	10 332
bars	kg/m^2	1.020×10^4
Btu	joules	1054.8
Btu	kWh	2.928×10^{-4}
bushels	m^3	0.035 24
circular mils	m^2	5.067×10^{-2}
cubic feet	m^3	0.028 32
cubic inches	liters	0.016 39
dynes	grams	1.020×10^{-3}
dynes	newtons	10^{-5}
ergs	joules	10^{-7}
ergs	kWh	0.2778×10^{-13}

TO CONVERT	TO	MULTIPLY BY
Fahrenheit (°F)	°C	$(°F+32)\times\frac{5}{9}$
feet	meters	0.3048
foot-pounds	joules	1.356
foot-pounds	kgm	0.1383
gallons	m^3	3.785×10^{-3}
gallons	liters	3.785
gausses	Teslas	10^{-4}
gilberts	ampere (turns)	0.7958
gills	liters	0.1183
grams	newtons	9.807×10^{-3}
horsepower	watts	745.7
inches	cm	2.54
knots	km/h	1.853
lines/sq. in.	Teslas	1.55×10^{-5}
Maxwells	webers	10^{-8}
mhos	Siemens	1
microns	meters	10^{-6}
miles (nautical)	km	1.853
miles (statute)	km	1.609
miles/hour	km/h	1.609
mils	cm	2.54×10^{-3}
ounces	grams	28.35
pints	liters	0.4732
poundals	newtons	0.1383
pounds	grams	453.59
pounds-force	newtons	4.448
pounds/sq. ft.	kg/m^2	4.882
pounds/sq. in.	kg/m^2	703
quarts	liters	0.9463
Slug	kg	14.59
sq. ft.	m^2	0.0929
sq. in.	cm^2	6.452
sq. miles	km^2	2.59
tons (long)	kg	1016
tons (short)	kg	907.18

Appendix 3

METRIC EQUIVALENTS OF AMERICAN WIRE GAUGE SIZES

GAUGE	DIA. (mm)	RESISTANCE (Ω/km)	DIA. (mil)	RESISTANCE (Ω/1000 ft)
0000	11.68	0.160	460	0.049
000	10.40	0.203	409.6	0.062
00	9.266	0.255	364.8	0.078
0	8.252	0.316	324.9	0.098
1	7.348	0.406	289.3	0.124
2	6.543	0.511	257.6	0.156
3	5.827	0.645	229.4	0.197
4	5.189	0.813	204.3	0.248
5	4.620	1.026	181.9	0.313
6	4.115	1.29	162	0.395
7	3.665	1.63	144.3	0.498
8	3.264	2.06	128.5	0.628
9	2.906	2.59	114.4	0.792
10	2.588	3.27	101.9	0.999
11	2.30	4.10	90.7	1.26
12	2.05	5.20	80.8	1.59
13	1.83	6.55	72	2
14	1.63	8.26	64.1	2.52
15	1.45	10.4	57.1	3.18
16	1.29	13.1	50.8	4.02

GAUGE	DIA. (mm)	RESISTANCE (Ω/km)	DIA. (mil)	RESISTANCE (Ω/1000 ft)
17	1.15	16.6	45.3	5.06
18	1.02	21.0	40.3	6.39
19	0.912	26.3	35.9.	8.05
20	0.813	33.2	32	10.1
21	0.723	41.9	28.5	12.8
22	0.644	52.8	25.3	16.1
23	0.573	66.7	22.6	20.3
24	0.511	83.9	20.1	25.7
25	0.455	106	17.9	32.4
26	0.405	134	15.9	41
27	0.361	168	14.2	51.4
28	0.321	213	12.6	64.9
29	0.286	267	11.3	81.4
30	0.255	337	10	103
31	0.227	425	8.9	130
32	0.202	537	8	164
33	0.180	676	7.1	206
34	0.160	855	6.3	261
35	0.143	1071	5.6	329
36	0.127	1360	5	415
37	0.113	1715	4.5	523
38	0.101	2147	4	655
39	0.090	2704	3.5	832
40	0.080	3422	3.1	1044

ANSWERS TO PROBLEMS

CHAPTER 2	2-1	196×10^3 N
	2-2	117×10^3 N
	2-3	1.47×10^6 J
	2-4	24.5 kW
	2-5	196 J, 109 W
	2-6	2.84×10^4 MW
	2-7	2.16 kW, 3.89×10^6 J
CHAPTER 3	3-1	1.68×10^{22} 0.75 Ah
	3-2	(a) 20 000 V/m; (b) 2 000 eV; (c) 6×10^3 J
	3-3	(a) 20 mA; (b) 5.5 kΩ; (c) 33 V
	3-4	(a) 300 V; (b) 21 Ah
	3-5	5.45 mA, 6.55 C
	3-6	129.6×10^3 J
	3-7	10 V, 1.602×10^{-16} N
	3-8	1.8×10^{-3} S, 90 mA
	3-9	10.9 A, 6.25 kWh
	3-10	1.265 kW, 84%
	3-11	5.06 kWh

CHAPTER 4

4-1 320 mA, 570 mA, 820 mA
4-2 80 mA, 142 mA, 205 mA
4-3 (a) 20 V, 59 V, 82.5 V; (b) 64 mV, 185 mV, 260 mV
4-4 (a) $R \times 10$; (b) $R \times 1$ k; (c) $R \times 10$ k; (d) $R \times 1$
4-5 (a) 680 Ω, 250 Ω, 130 Ω; (b) 680 kΩ, 250 kΩ, 130 kΩ
4-6 0.43 mA, 1.67 mA, 2.38 mA
4-7 0.38 mA to 0.48 mA; 1.62 mA to 1.72 mA; 2.33 mA to 2.43 mA
4-8 (a) 220 Ω to 280 Ω; (b) 220 kΩ to 280 kΩ
4-9 (a) 9.398 V to 9.438 V; (b) 1.286 V to 1.294 V; (c) 3.06 V to 3.074 V
4-10 (a) 4.634 kΩ to 4.692 kΩ; (b) 556.6 Ω to 563.6 Ω; (c) 27.59 Ω to 27.95 Ω

CHAPTER 5

5-1 115 V, 2.5 W
5-2 2.19×10^{-2} Ω, 5.47×10^{-3} Ω, 2.43×10^{-3} Ω, 1.37×10^{-3} Ω, 8.76×10^{-4} Ω
5-3 0.06 Ω, 0.1 Ω
5-4 0.47 mm
5-5 0.516 mV
5-6 13.12 Ω
5-7 0.607 mV
5-8 440 Ω, 470 Ω, 530 Ω, 560 Ω, 590 Ω, 620 Ω, 650 Ω, 680 Ω, 710 Ω, 740 Ω
5-9 0.0038 silver, 0.0034 gold
5-10 5600 Ω ± 20%, 820 Ω ± 10%, 390 kΩ ± 5%
5-11 (a) 17.46 mA, 14.3 V; (b) 0.707 mA, 707 V
5-12 2.64 Ω
5-13 5646.2 Ω, 826.8 Ω, 393.2 kΩ

CHAPTER 6

6-1 7.5 Ah, 30 h
6-2 1.4 V, 33 A
6-3 8 V, 7.5 V
6-4 8 cells, 0.4 Ω, 11.54 V
6-5 10 cells
6-6 1.7 V, 6 A, 72 Ah, 8.75×10^{-3} Ω
6-7 16 cells
6-8 (a) 33.3 h; (b) 37.5 minutes; (c) 22.5 minutes
6-9 (a) 50 A; (b) 11.75 V
6-10 352 kΩ

CHAPTER 7

7-1 20 mA
7-2 3 V, 5 V, 2.5 V, 1.5 V
7-3 (a) 300 mA, 3.6 V, 6.6 V, 4.8 V; (b) 60 mA, 0.72 V, 1.32 V, 0.96 V
7-4 22.2 V, 52.8 V
7-5 (a) 1.08 W, 1.98 W, 1.44 W, 4.5 W; (b) 43.2 mW, 79.2 mW, 57.6 mW, 180 mW
7-6 28.8 Ω, 125 W
7-7 200 Ω, 45 mW
7-8 2.31 A, 30.9 W

CHAPTER 8

8-1 115 mA, 250 mA, 365 mA
8-2 555 mA, 221 mA, 455 mA, 1.23 A
8-3 3.9 kΩ
8-4 123.2 Ω, 365 mA
8-5 12.19 Ω, 1.23 A
8-6 115 mA, 250 mA, 365 mA
8-7 555 mA, 221 mA, 455 mA, 1.23 A
8-8 115 mA, 250 mA
8-9 5.19 mW, 11.25 W, 16.43 W
8-10 8.3 W, 3.3 W, 6.8 W, 18.4 W
8-11 870 mA, 348 mA, 522 mA, 217 mA, 225 W

CHAPTER 9

9-1 83.5 mA, 52.5 mA, 31 mA, 62.6 V, 17.4 V, 17.4 V
9-2 18.5 mA, 2.155 mA, 20.66 mA, 17.7 V, 32.3 V
9-3 4.4 V, 4.6 V, 0.38 mA, 0.68 mA, 0.26 mA
9-4 0.726 mA, 0.297 mA, 0.429 mA, 0.543 mA, 11.6 V, 3.4 V, 3 V, 12 V
9-5 0.427 mA, 0.384 mA, 3.8 V, 1.7 V, 26.1 V, 3.4 V
9-6 0.427 mA, 0.403 mA, 4 V, 0 V, 27.4 V, 3.6 V
9-7 0.427 mA, 0.36 mA, 3.6 V, 1.6 V, 24.5 V, 5.4 V
9-8 3.73 mA
9-9 0.89 mA

CHAPTER 10

10-1 75 V, 5 Ω, 62.5 V, 2.5 A
10-2 0.98 V
10-3 3.05 mA
10-4 0.528 V
10-5 0.536 mA
10-6 0.98 V
10-7 3.05 mA
10-8 0.528 V

10-9 0.536 mA
10-10 0.98 V
10-11 0.536 mA
10-12 39 V
10-13 2.38 mA
10-14 20.9 mA

CHAPTER 11

11-1 1.207 mA
11-2 0.383 mA
11-3 0.53 mA
11-4 1.207 mA
11-5 0.84 V
11-6 3.64 V
11-7 4.1 mA
11-8 0.26 mA
11-9 0.544 mA
11-10 0.65 mA
11-11 5.7 V
11-12 0.939 mA, 0.664 mA

CHAPTER 12

12-1 3.3×10^{-3} T, 2×10^{-4} T
12-2 6.16×10^{-3} Wb
12-3 4.5×10^{-5} Wb
12-4 2100 A, 4200 A
12-5 0.9 N, 15 T
12-6 12.5 N, 0.49 N
12-7 9.9×10^{-6} Nm
12-8 1.04 mA
12-9 2 T
12-10 18×10^{-4} Wb

CHAPTER 13

13-1 7.95×10^{4} A, 1×10^{-2} T, 2.25×10^{-6} Wb
13-2 44.5 mA
13-3 28.4 A
13-4 45.07 mA, 1.47×10^{-3} T
13-5 (a) 199, (b) 1074
13-6 3.8×10^{-4} Wb
13-7 1.5×10^{3} A, 0.4 T, 3.6×10^{-2} Wb
13-8 202 mA
13-9 2.07 mA
13-10 20.1 A

13-11 1761
13-12 212 mA
13-13 0.634 A
13-14 4×10^5
13-15 8.8 kg
13-16 8 kg

CHAPTER 14

14-1 0.27 Ω
14-2 (a) 0.027 Ω; (b) 0.0027 Ω
14-3 13.3 A
14-4 2.5 A, 3.3 A, 5 A, 10 A
14-5 2.5 MΩ, 1 MΩ, 500 kΩ, 10 kΩ/V
14-6 95 V, 142.5 V
14-7 1.22 MΩ, 2.4 MΩ
14-8 40 kΩ, 20 kΩ, 10 kΩ
14-9 60 kΩ, 20 kΩ, 6.7 kΩ
14-10 79.5 Ω
14-11 101.86 cm, 10 mV/cm
14-12 100 μA
14-13 24.95 V

CHAPTER 15

15-1 40 V
15-2 6.25 V
15-3 1.05 A
15-4 24
15-5 10 V, 300 mA
15-6 712
15-7 1.76×10^{-8} Wb, 0.25 mV
15-8 4.74 mH
15-9 2.7 mH, 1.5 mH
15-10 17.35 mH, 17.35 mH
15-11 0.047
15-12 14.2 J
15-13 3.8 H
15-14 (a) 1.15 H; (b) 733 μH
15-15 1.33 mH
15-16 (a) 1.2 mH; (b) 800 μH
15-17 0.204
15-18 150 μH, 275 μH

CHAPTER 16

16-1 40 000 V/m, 2×10^{-6} C/m^2
16-2 50 V
16-3 (a) 88.5 pF; (b) 6.64 nF; (c) 37 V, 0.5 V
16-4 37.7 m^2
16-5 0.27×10^{-6} m
16-6 (a) 5.9 μF, 1.475 V, 2.95 V, 5.9 V, 14.75 V, 1.475 μC; (b) 185 μF, 25 V, 2.5 mC, 1.25 mC, 0.625 mC, 0.25 mC
16-7 (a) 21.4 μF; (b) 21.4 V, 3.6 V
16-8 648 μC, 324 μC, 5.7 mC

CHAPTER 17

17-1 63.2 mA, 86.5 mA, 95 mA, 98.2 mA, 99.3 mA
17-2 14.3 V, 4.1 V, 1.2 V, 0.34 V
17-3 92.9 H
17-4 1.6 ms, 0.36 ms
17-5 1.23 kΩ
17-6 24.7 V, 37.2 V, 43.5 V, 46.7 V, 48.3 V, 49.2 V, 49.6 V
17-7 9.16 mA, 3.69 mA, 1.49 mA, 0.6 mA, 0.24 mA
17-8 492 Ω
17-9 128 ms, 113 ms
17-10 5 kΩ, 1.175 s

CHAPTER 18

18-1 (a) 88 mV; (b) 880 mV
18-2 0.62 V, 0.62 V, −0.62 V
18-3 64.4 mV, 128.6 mV, 315.4 mV, 374.7 mV
18-4 7.7 V, −9 V, 7.28 V
18-5 1 kHz, 2.5 km, 25 km
18-6 277 W, 277 W, 277 W, 277 W
18-7 2.62 A, 2.36 A, 185 W

CHAPTER 19

19-1 (a) $77.8 \sin(\phi + 8.3°)$; (b) $19.59 \sin(\phi - 35°)$
19-2 $30 \sin(\phi - 15.3°)$
19-3 $90.1\angle -33.7°$, $25.5\angle 11.3°$, $2298\angle 22.4°$, $3.5\angle -45°$
19-4 $139.1 + j56.2$, $0 + j85$, $19.8 + j60.9$
19-5 $43.6\angle 16.1°$, $11.2\angle 220.7°$
19-6 $40.5\angle 9°$
19-7 $20.1\angle 15.4°$
19-8 $43.6\angle 16.1°$, $11.2\angle 22.7°$
19-9 $143.8\angle 33.9°$
19-10 $1.08\angle -122.5°$

19-11 $81.07\underline{/-22.9^\circ}$
19-12 $9.8 \sin(\phi + 11.8^\circ)$

CHAPTER 20

20-1 471.2 Ω, 212 mA
20-2 1 MHz
20-3 21.2 Ω, 4.7 A
20-4 99.5 Hz
20-5 356 mA, 24.2 V, 22.4 V, 42.7°
20-6 16.1 mA, 10.7 V
20-7 92.8 mA, 11.1 V, 4.47 V, 21.9°
20-8 0.5 μF
20-9 72.1 mA, 14.4 V, 5.44 V, 9.56 V, −16°
20-11 30.3 mA, 72.5 mA, 50.3 mA, 37.6 mA$\underline{/-36.2^\circ}$
20-13 (a) 0.85; (b) 6.35×10^{-2}
20-14 0.39 V, 1.22 mV, 319
20-15 0.46 μF
20-16 7.7×10^6

CHAPTER 21

21-1 389 mA$\underline{/19.4^\circ}$
21-2 103 V$\underline{/-70.6^\circ}$, 58.7 V$\underline{/67.9^\circ}$
21-3 47.3 V$\underline{/-29.9^\circ}$, 27.4 V$\underline{/59.4^\circ}$
21-4 18.9 mA$\underline{/-29.9^\circ}$, 10.96 mA$\underline{/59.4^\circ}$
21-5 877 Ω$\underline{/-6.9^\circ}$, 21.7 mA$\underline{/6.9^\circ}$
21-6 1.22 kΩ$\underline{/-5^\circ}$
21-7 8.1 V$\underline{/50.6^\circ}$, 2.9 V$\underline{/-87.3^\circ}$
21-8 3.9 kΩ$\underline{/29.3^\circ}$
21-9 9.94 mA$\underline{/-44.2^\circ}$
21-10 14 kΩ, 703 mH

CHAPTER 22

22-1 (a) 24.2 Ω; (b) 9.09 A; (c) 4 kW
22-2 (a) 457 vars; (b) 576 W; (c) 3.6 kvars
22-3 75 VA, 37.5 vars, 65 W
22-4 160.5 VA, 119 W, 107 vars
22-5 880 VA, 607.2 W, 637.3 vars, 15.9 μF
22-6 2.62 kVA, 1.67 kW, 2.02 kvars
22-7 256.4 A, 196 A, 1530 μF
22-8 1.8 W, 0.16 A, 0.047 A

CHAPTER 23

23-1 25.8 μA $\angle -79.1°$
23-2 3.2 V $\angle 0.1°$
23-3 1.3 mA $\angle -16.5°$
23-4 25.8 μA $\angle -79.1°$
23-5 1.3 mA $\angle -16.5°$
23-6 0.63 mA $\angle -80°$
23-7 1.48 V $\angle 23°$
23-8 3.2 V $\angle 0.1°$
23-9 1.3 mA $\angle -16.5°$
23-10 1.48 V $\angle 23°$
23-11 264 μA $\angle 81.8°$
23-12 5.7 mA $\angle 26.7°$
23-13 1.48 V $\angle 23°$
23-14 260 μA $\angle 18.5°$
23-15 4.2 A $\angle -10.5°$
23-16 11.5 mA $\angle 41°$
23-17 500.5 Ω, 2.5 mH

CHAPTER 24

24-1 500 kHz, 156 mA

24-2, 24-3

	0.25 *fr*	0.5 *fr*	0.8 *fr*	*fr*	1.25 *fr*	2 *fr*	4 *fr*
I	0.84	2.1	6.9	156	6.9	2.15	0.84 (mA)
V_L	0.33	1.67	8.76	248	13.7	6.71	5.34 (V)
V_C	5.35	6.68	13.7	248	8.76	1.7	0.33 (V)
V_R	27	67	221	4.99 (V)	221	69	27 (mV)

24-4 40 pF to 3400 pF

24-5

	0.25 *fr*	0.5 *fr*	0.8 *fr*	*fr*	1.25 *fr*	2 *fr*	4 *fr*
I	0.93	2.3	7.8	350	7.8	2.3	0.93 (mA)
V_L	0.47	2.3	12.5	700	19.5	9.2	7.44 (V)
V_C	7.44	9.2	19.5	700	12.5	2.3	0.47 (V)
V_R	18.6	46	156	7 (V)	156	46	18.6 (mV)

24-6 471 kHz
24-7 49.7, 100, 88.9
24-8 505 kHz, 495 kHz, 10 kHz

24-9 527 kHz, 533 kHz, 6 kHz, 468 kHz, 474 kHz
24-10 530.5 kHz, 200, 2.65 kHz
24-11 505.8 kHz, 191, 504 kHz, 507 kHz
24-12 15 kΩ
24-13 0.02, 110 mA, 130 mA
24-14 141.8 kHz to 142.2 kHz

CHAPTER 25

25-1 29.9 V. 36.8 V, 66.7 V, 6.9 V
25-2 6.24 V, 25 V, 60 V, 500 V, 143.5 mA
25-3 2.55×10^{-3} Wb
25-4 121 V, 1.27 A, 47°
25-5 1.24 A, 36.3°
25-6 7.06%
25-7 91.4%, 1.31 A
25-8 2.88 kΩ, 1.02 kΩ, 3.43, 4.69 Ω, 1.73 Ω
25-9 92%
25-10 3.6%

CHAPTER 26

26-1 (a) 0; (b) 0 to 100 V peak
26-2 2.7 MΩ
26-3 1.3 MΩ
26-4 254.5 Ω
26-5 180, 2.6 kΩ
26-6 2.59 A
26-7 720 Ω to 900 Ω
26-8 0.02 μF to 50 μF, 10 mH to 25 H

INDEX

J

K

L

U

V

W

X

Y

Z